S0-APC-706

MONSANTO COMPANY
INFORMATION CENTER
CHESTERFIELD

Chromatographic Analysis of Alkaloids

CHROMATOGRAPHIC SCIENCE

A Series of Monographs

Editor: JACK CAZES
Sanki Laboratories, Inc.
Sharon Hill, Pennsylvania

Volume 1: Dynamics of Chromatography
J. Calvin Giddings

Volume 2: Gas Chromatographic Analysis of Drugs and Pesticides
Benjamin J. Gudzinowicz

Volume 3: Principles of Adsorption Chromatography: The Separation of Nonionic Organic Compounds (out of print)
Lloyd R. Snyder

Volume 4: Multicomponent Chromatography: Theory of Interference (out of print)
Friedrich Helfferich and Gerhard Klein

Volume 5: Quantitative Analysis by Gas Chromatography
Joseph Novák

Volume 6: High-Speed Liquid Chromatography
Peter M. Rajcsanyi and Elisabeth Rajcsanyi

Volume 7: Fundamentals of Integrated GC-MS (in three parts)
Benjamin J. Gudzinowicz, Michael J. Gudzinowicz, and Horace F. Martin

Volume 8: Liquid Chromatography of Polymers and Related Materials
Jack Cazes

Volume 9: GLC and HPLC Determination of Therapeutic Agents (in three parts)
Part 1 edited by Kiyoshi Tsuji and Walter Morozowich
Parts 2 and 3 edited by Kiyoshi Tsuji

Volume 10: Biological/Biomedical Applications of Liquid Chromatography
Edited by Gerald L. Hawk

Volume 11: Chromatography in Petroleum Analysis
Edited by Klaus H. Altgelt and T. H. Gouw

Volume 12: Biological/Biomedical Applications of Liquid Chromatography II
Edited by Gerald L. Hawk

Volume 13: Liquid Chromatography of Polymers and Related Materials II
Edited by Jack Cazes and Xavier Delamare

Volume 14: Introduction to Analytical Gas Chromatography: History, Principles, and Practice
John A. Perry

Volume 15: Applications of Glass Capillary Gas Chromatography
Edited by Walter G. Jennings

Volume 16: Steroid Analysis by HPLC: Recent Applications
Edited by Marie P. Kautsky

Volume 17: Thin-Layer Chromatography: Techniques and Applications
Bernard Fried and Joseph Sherma

Volume 18: Biological/Biomedical Applications of Liquid Chromatography III
Edited by Gerald L. Hawk

Volume 19: Liquid Chromatography of Polymers and Related Materials III
Edited by Jack Cazes

Volume 20: Biological/Biomedical Applications of Liquid Chromatography IV
Edited by Gerald L. Hawk

Volume 21: Chromatographic Separation and Extraction with Foamed Plastics and Rubbers
G. J. Moody and J. D. R. Thomas

Volume 22: Analytical Pyrolysis: A Comprehensive Guide
William J. Irwin

Volume 23: Liquid Chromatography Detectors
Edited by Thomas M. Vickrey

Volume 24: High-Performance Liquid Chromatography in Forensic Chemistry
Edited by Ira S. Lurie and John D. Wittwer, Jr.

Volume 25: Steric Exclusion Liquid Chromatography of Polymers
Edited by Josef Janca

Volume 26: HPLC Analysis of Biological Compounds: A Laboratory Guide
William S. Hancock and James T. Sparrow

Volume 27: Affinity Chromatography: Template Chromatography of Nucleic Acids and Proteins
Herbert Schott

Volume 28: HPLC in Nucleic Acid Research: Methods and Applications
Edited by Phyllis R. Brown

Volume 29: Pyrolysis and GC in Polymer Analysis
Edited by S. A. Liebman and E. J. Levy

Volume 30: Modern Chromatographic Analysis of the Vitamins
Edited by Andre P. De Leenheer, Willy E. Lambert, and Marcel G. M. De Ruyter

Volume 31: Ion-Pair Chromatography
Edited by Milton T. W. Hearn

Volume 32: Therapeutic Drug Monitoring and Toxicology by Liquid Chromatography
Edited by Steven H. Y. Wong

Volume 33: Affinity Chromatography: Practical and Theoretical Aspects
Peter Mohr and Klaus Pommerening

Volume 34: Reaction Detection in Liquid Chromatography
Edited by Ira S. Krull

Volume 35: Thin-Layer Chromatography: Techniques and Applications, Second Edition, Revised and Expanded
Bernard Fried and Joseph Sherma

Volume 36: Quantitative Thin-Layer Chromatography and Its Industrial Applications
Edited by Laszlo R. Treiber

Volume 37: Ion Chromatography
Edited by James G. Tarter

Volume 38: Chromatographic Theory and Basic Principles
Edited by Jan Åke Jönsson

Volume 39: Field-Flow Fractionation: Analysis of Macromolecules and Particles
Josef Janča

Volume 40: Chromatographic Chiral Separations
Edited by Morris Zief and Laura J. Crane

Volume 41: Quantitative Analysis by Gas Chromatography, Second Edition, Revised and Expanded
Josef Novak

Volume 42: Flow Perturbation Gas Chromatography
N. A. Katsanos

Volume 43: Ion-Exchange Chromatography of Proteins
Shuichi Yamamoto, Kazuhiro Nakanishi, and Ryuichi Matsuno

Volume 44: Countercurrent Chromatography: Theory and Practice
Edited by N. Bhushan Mandava and Yoichiro Ito

Volume 45: Microbore Column Chromatography: A Unified Approach to Chromatography
Edited by Frank J. Yang

Volume 46: Preparative-Scale Chromatography
Edited by Eli Grushka

Volume 47: Packings and Stationary Phases in Chromatographic Techniques
Edited by Klaus K. Unger

Volume 48: Detection-Oriented Derivatization Techniques in Liquid Chromatography
Edited by Henk Lingeman and Willy J. M. Underberg

Volume 49: Chromatographic Analysis of Pharmaceuticals
Edited by John A. Adamovics

Volume 50: Multidimensional Chromatography: Techniques and Applications
Edited by Hernan Cortes

Volume 51: HPLC of Biological Macromolecules: Methods and Applications
Edited by Karen M. Gooding and Fred E. Regnier

Volume 52: Modern Thin-Layer Chromatography
Edited by Nelu Grinberg

Volume 53: Chromatographic Analysis of Alkaloids
Milan Popl, Jan Fähnrich, and Vlastimil Tatar

Volume 54: HPLC in Clinical Chemistry
I. N. Papadoyannis

Additional Volumes in Preparation

Handbook of Thin-Layer Chromatography
edited by Joseph Sherma and Bernard Fried

Chromatographic Analysis of Alkaloids

Milan Popl
Jan Fähnrich
Vlastimil Tatar

Prague Institute of Chemical Technology
Prague, Czechoslovakia

Marcel Dekker, Inc. New York and Basel

Library of Congress Cataloging-in-Publication Data

Popl, Milan.
  Chromatographic analysis of alkaloids / Milan Popl, Jan
Fähnrich, Vlastimil Tatar.
      p.   cm. -- (Chromatographic science; v. 53)
  Includes bibliographical references.
  ISBN 0-8247-8140-6 (alk. paper)
  1. Alkaloids--Analysis. 2. Chromatographic analysis.
I. Fähnrich, Jan. II. Tatar, Vlastimil.
III. Title. IV. Series.
QD421.P79 1990
547.7'2046--dc20   90-3104
                                                 CIP

This book is printed on acid-free paper.

Copyright © 1990 by MARCEL DEKKER, INC. All Rights Reserved

Neither this book nor any part may be reproduced or transmitted in any form or by any means, electronic or mechanical, including photocopying, microfilming, and recording, or by any information storage and retrieval system, without permission in writing from the publisher.

MARCEL DEKKER, INC.
270 Madison Avenue, New York, New York 10016

Current printing (last digit):
10 9 8 7 6 5 4 3 2 1

PRINTED IN THE UNITED STATES OF AMERICA

Preface

Alkaloids, as naturally occurring compounds, have attracted attention of chemists, biochemists, and pharmacologists for more than 100 years. The rapid development of chromatographic methods during the last 20 years has established chromatography as an extremely useful tool in alkaloid research. Raw plant extracts have been very successfully and quickly separated into individual compounds using these methods. This facilitates structure elucidations, especially for organic chemists. Chromatographic analysis of plants, illicit drugs, pharmaceuticals, tissues, and body fluids yields tentative identification of alkaloids in the fields of forensic chemistry, medicine, and pharmacology. Various quantitative chromatographic techniques are applied today to pharmacokinetic studies, stability tests, and production quality control of alkaloids used as pharmaceuticals. In respect to the extreme complexity of some samples, their separation and analysis present challenging problems to analytical chemists. Compared to other analytical methods such as titration, electrochemical, or spectral methods, the separation ability of chromatography becomes indispensable for samples of complex matrix.

This book is intended to aid in solving analytical tasks in the field of alkaloids by chromatographic methods. The volume is organized in such a manner that the reader becomes acquainted with analytical procedures as well as basic theory. Chapter 1 provides basic information on alkaloid classification, nomenclature, and structures for readers not familiar with the chemistry of alkaloids. Chapter 2 suggests how physicochemical properties of alkaloids to be analyzed may influence the chromatographic process. Dissociation of alkaloids is treated in detail as it strongly influences their separation. Chromatographic principles are summarized in Chapter 3. It is assumed that the reader has had some introduction to elementary thermodynamics and kinetics of chromatography, so these subjects are treated very briefly. The main body of the text consists of Chapters 4–7, dealing practically with all chromatographic techniques commonly utilized for the separation of alkaloids. When solving a given separation problem, the chromatographer may start from the published results. For common alkaloids a suitable solution often may be found in the literature. Excellent reviews covering literature on thin-layer, gas, and liquid chromatography of alkaloids up to 1982–1983 have been completed by R. Verpoorte and A. Baerheim-Svendsen ("Chromatography of Alkaloids. Thin Layer Chromatography," Journal of Chromatography Library, Vol. 23A, Gas and Liquid Chromatography, Vol. 23B, Elsevier, Amsterdam, 1983, 1984). Unfortunately, very often the attempts to transfer separations described in literature to similar or nominally the same conditions fail and modifications of the chromatographic conditions are necessary. Moreover, many of the less common alkaloids have not yet been separated in most chromatographic systems and suitable separation conditions have to be discovered experimentally. Hence in Chapters 4–6, we have attempted to summarize basic theories and factors that affect the separation of alkaloids in the respective type of chromatographic system in order to facilitate the modification of separation in the required direction and the development of new separations. It is hoped that the great variety of chromatographic systems and conditions described in these chapters will be useful in choosing the system type suitable for a given sample, available laboratory equipment, and required reliability, accuracy, and price of the analyses. All chapters contain illustrative examples which may serve as starting points in the development of the method. Emphasis has been placed on detailed description and experimental details, at least in those cases that in our opinion provide the best and most reliable separation results. This enables us to reproduce the experiment without having to consult the primary literature. A large part of Chapter 7 is devoted to the various methods applied to sample preparation and purification, which are extraordinarily important, especially for the chromatography of complex biological samples. The use of automated sample preparation for a large series of samples is also discussed.

The book contains over 1200 references, most of them to the primary literature, and if a subject is not discussed fully in the text a reference is given to provide further details. Of course, the bibliography is far from being complete. From papers applying chromatography to the analysis of alkaloids usually only those with conclusions or observations of more general validity have been included. On some topics, only recent references are given; these still give the reader access to older references.

Preface

This book will be of value to organic chemists experienced in the field of alkaloid chemistry and encountering the task of chromatographic separation and for practitioners in clinical chemistry and pharmacology, as well as for analytical chemists mastering the chromatographic techniques and applying them to the separation of alkaloids. The newcomer should be able to orient himself in the field. For specialists in the chromatography of alkaloids the book offers a survey of recent literature of the field and their classification and evaluation. We hope that this book will help the reader to analyze complex and as yet unresolved mixtures of alkaloids.

Milan Popl
Jan Fähnrich
Vlastimil Tatar

Contents

Preface	iii
Chapter 1. Classification of Alkaloids	1
I. Phenylethylamine Derivatives	2
II. Pyridine and Piperidine Alkaloids	3
III. Quinoline Alkaloids	7
IV. Isoquinoline Alkaloids	8
V. Alkaloids Derived from Miscellaneous Heterocyclic Systems	12
VI. Indole Alkaloids	15
VII. Imidazole Alkaloids	21
VIII. Diterpene and Steroid Alkaloids	22
References	24
Chapter 2. Properties of Alkaloids Relevant to Chromatography	25
I. Dissociation Properties	25
II. Physicochemical Properties	41
III. UV-vis Absorption Spectral Data	49
IV. Electrochemical Properties	57
References	64
Chapter 3. Chromatography	66
I. Principles, Classification, and Nomenclature	66
II. Basic Theory	71
III. Instrumentation in Column Chromatography	81
IV. Symbols and Relations	87
References	87

Chapter 4. Gas Chromatography	88
I. Columns and Their Preparation	89
II. Detectors	97
III. Derivatization Techniques	100
IV. GC Analysis of Alkaloids	103
V. Qualitative and Quantitative Analysis	125
References	129
Chapter 5. Liquid Chromatography	135
I. Adsorption Chromatography	136
II. Liquid Partition Chromatography	159
III. Ion Exchange Chromatography	194
IV. Ion Pair Chromatography	206
V. Detection in Liquid Chromatography	229
References	247
Chapter 6. Thin-Layer Chromatography	266
I. Technique of TLC	267
II. TLC Analysis of Alkaloids	287
References	300
Chapter 7. Applications	309
I. Sample Preparation	309
II. Sample Separation	341
References	618
Chapter 8. Conclusion	638
Index	641

Chromatographic Analysis of Alkaloids

1
Classification of Alkaloids

The alkaloids, comprising over 5000 compounds of all structural types, belong to a large class of natural products. In this class the use of trivial names entirely dominates and systematic nomenclature is missing. The trivial names of alkaloids characteristically end with -ine. According to the definition (1), almost all alkaloids occur exclusively in plants and contain at least one nitrogen atom per molecule, which is basic in character and usually constitutes part of a heterocyclic system. However, some other natural products not belonging to the alkaloids meet this specification and, conversely, some well-known alkaloids do not fully comply. A few dozen alkaloids have been isolated from animals and in certain cases the same compound has been found in both plants and animals. Some neutral or weakly basic compounds are classified as alkaloid, e.g., colchicine (pK_b 12.35) in which the nitrogen is involved in an amide group. A newer definition (2) of alkaloids is much simpler: an alkaloid is a cyclic organic compound containing nitrogen in a negative oxidation state which is of limited distribution among living organisms. This definition can be supplemented by the statement that almost all alkaloids are toxic and the large majority of them have shown pharmacological activity.

The classification of alkaloids is a difficult task even for specialists in the field; therefore, in this book the alkaloids will be classified roughly according to the type of heterocyclic system which contains the nitrogen atom. These large groups are further divided into subgroups in accordance with botanical classification of plant species in which a certain type of alkaloid occurs, e.g., tobacco alkaloids, Amaryllidaceae alkaloids, _Senecio_ alkaloids, and so on. Sometimes the name of the subgroup is derived from that of a well-known alkaloid, e.g., quinine, yohimbine, etc.

The brief review of the major alkaloid groups given in the following sections is drawn from the literature (3-5). Due to the large number and structural diversity of alkaloids, one cannot expect this review to be exhaustive.

I. PHENYLETHYLAMINE DERIVATIVES

This group of alkaloids, structurally based on 2-phenylethylamine $C_8H_{11}N$ [1], is occasionally associated with the group of isoquinoline alkaloids by virtue of being biosynthetically related to tetrahydroisoquinoline $C_9H_{11}N$ [2]. Structure [1] has been isolated from Accacia

2-Phenylethylamine [1]   Tetrahydroquinoline [2]

prominens as well as its N-methyl derivative. Among this group mescaline $C_{11}H_{17}O_3N$ [3] as a drug of abuse is of greatest importance.

Mescaline [3]

Mescaline has been found in extracts from the cactus Anhalonium lewinii in addition to N-methylmescaline $C_{12}H_{19}O_3N$ and N-acetylmescaline $C_{13}H_{19}O_4N$.

Some derivatives of phenylethylamine occur in plants as quaternary ammonium compounds, e.g., candicine $C_{11}H_{18}ON^+$ [4], which has been found in the cactus Trichocereus candicans. A special subgroup is formed by ephedrine $C_{10}H_{15}ON$ [5] and its derivatives. In the molecule of this base are two asymmetric carbon atoms, which means that ephedrine can create four optical antipodes and two racemates. In plants, e.g., Ephedra vulgaris, levorotatory ephedrine and dextrorotatory ψ-ephedrine have been found, which are not optical antipodes

Candicine [4]   Ephedrine [5]

but diastereoisomers. Ephedrine is widely used in pharmaceutical preparations. In plant species ephedrine is accompanied by norephedrine $C_9H_{13}ON$ (phenylpropanolamine) and N-methylephedrine $C_{11}H_{17}ON$.

To the ephedrine subgroup can be formally added colchicine $C_{22}H_{25}O_6N$ and other alkaloids obtained from Colchicum autumnale in which nitrogen is not involved in a heterocyclic ring. Colchicine contains an amide group and is essentially a neutral compound.

II. PYRIDINE AND PIPERIDINE ALKALOIDS

These bases form a large group containing compounds with aromatic rings as well as dihydro and tetrahydro derivatives, and hexahydro derivatives (piperidines). The large majority of this group is formed by the subgroup of piperidine derivatives in which the tropane alkaloids are included.

A. Pyridine Alkaloids

The important tobacco alkaloids occur in plants of the Nicotiana genus. The well-known compound nicotine $C_{10}H_{14}N_2$ [6] contains pyridine

Nicotine [6]   Anabasine [7]

and pyrrolidine rings. The isomer of nicotine is called anabasine $C_{10}H_{14}N_2$ [7] and has in its molecule a piperidine ring. The two optical isomers of tobacco nornicotine $C_9H_{12}N_2$ and nicotyrine $C_{10}H_{10}N_2$, can be prepared by the dehydrogenation of nicotine.

B. Dihydro- and Tetrahydropyridine Derivatives

The representative compound with a dihydropyridine ring is ricinine $C_8H_8O_2N_2$ [8], which is found in Ricinus communis. Alkaloids based

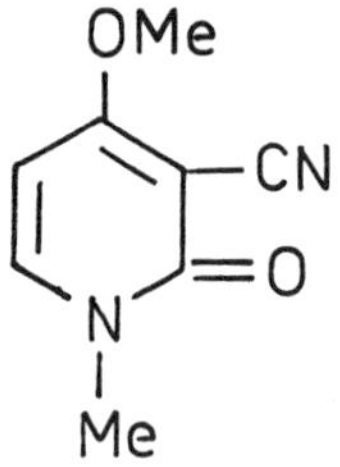

Ricinine [8] 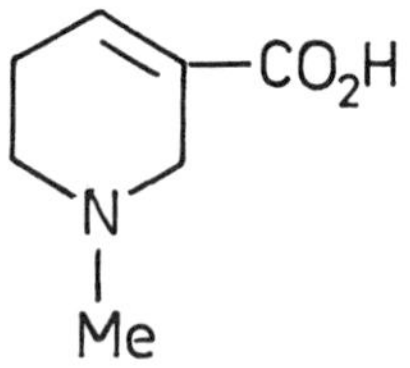   Arecaidine [9]

on tetrahydropyridine structure have been isolated from Areca catechu. They are arecaidine $C_7H_{11}O_2N$ [9], its nor derivative guvacine $C_6H_9O_2N$, and its methyl ester arecoline $C_8H_{13}O_2N$.

C. Piperidine Alkaloids

The alkaloids of this subgroup can be further divided according to the plant species from which the bases have been isolated. The bases, obtained by extraction of Piper nigrum, contain as the main alkaloid piperine $C_{17}H_{19}O_3N$ [10], which as a rule is accompanied by its stereoisomer chavicine $C_{17}H_{19}O_3N$. From the bark of Punica granatum pelletierine $C_8H_{15}ON$ [11] has been isolated. Structurally similar to

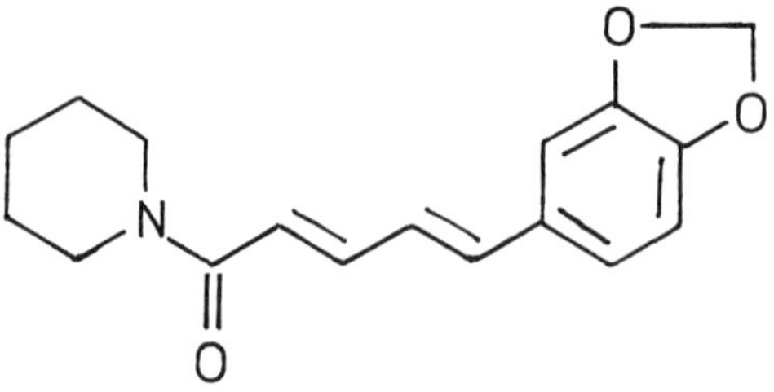

Piperine [10]

pelletierine are the bases obtained from Conium maculatum, e.g.,

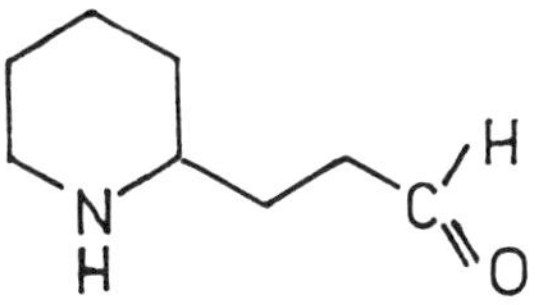

Classification of Alkaloids

Pelletierine [11]

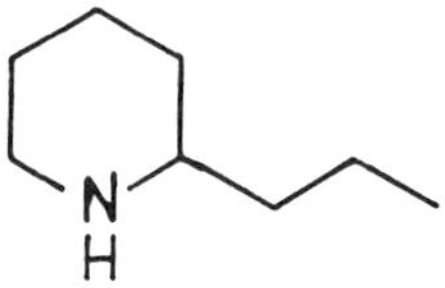

Coniine [12]

coniine $C_8H_{17}N$ [12], γ-coniceine $C_8H_{15}N$, and conhydrine $C_8H_{17}ON$. Of importance are the bases called <u>Lobelia</u> alkaloids, isolated from <u>Lobelia</u> plant species, e.g., <u>Lobelia inflata</u>. They contain levorotatory lobeline $C_{22}H_{27}O_2N$ [13], lobelanine $C_{22}H_{25}O_2N$, lobelanidine $C_{22}H_{29}O_2N$, etc.

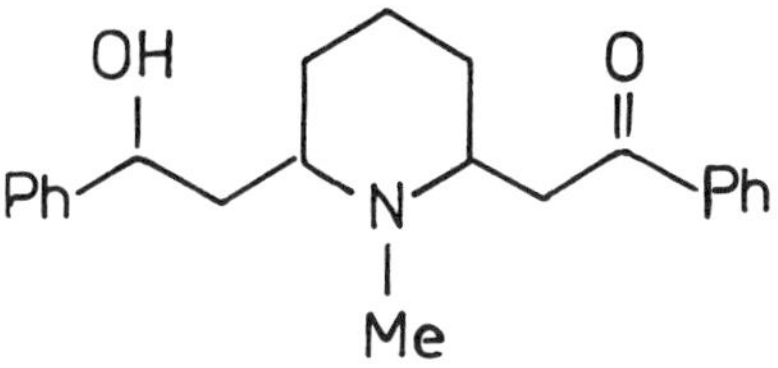

Lobeline [13]

D. Tropane Alkaloids

All bases of this subgroup possess a piperidine ring(s) as well as a pyrrolidine ring(s) in their molecular structure. The fundamental part of the structure is tropane or, more accurately, its hydroxy derivative tropine $C_8H_{15}ON$ [14], which forms esters with a large

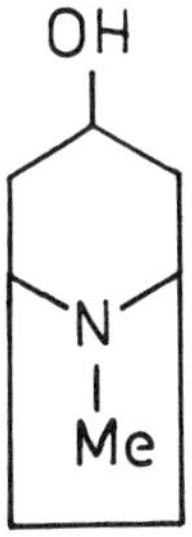

Tropine [14]

number of organic acids. These esters create the large subgroup of alkaloids called tropane alkaloids. In addition to tropine, scopine $C_8H_{13}O_2N$ [15] and methylecgonine $C_{10}H_{17}O_3N$ [16], as the fundamental parts of tropane alkaloids, have to be mentioned. As the

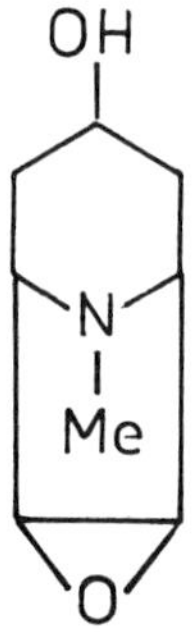

Scopine [15]

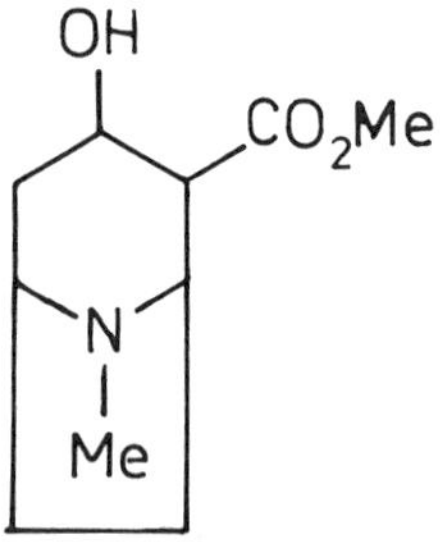

Methylecgonine [16]

free base, only tropine has been found in <u>Atropa belladonna</u>; the other two bases have been obtained by the hydrolysis of tropane alkaloids. Tropine as the most important component of tropane alkaloids itself forms two stereoisomers, tropine and ψ-tropine, differing only in the position of the hydroxyl group. For both isomers, the chair form of molecular skeleton [17] is supposed. The esters of tropine occur in many plants; however, the best known is <u>Atropa belladonna</u>. From this plant l-hyoscyamine $C_{17}H_{23}O_3N$ and its racemate atropine have been isolated. Both bases possess the same structure [18], which consists of an ester of tropine and tropic acid.

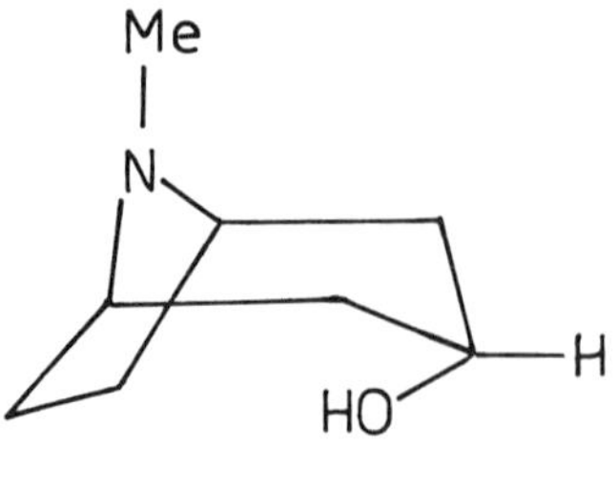

Tropine [17]

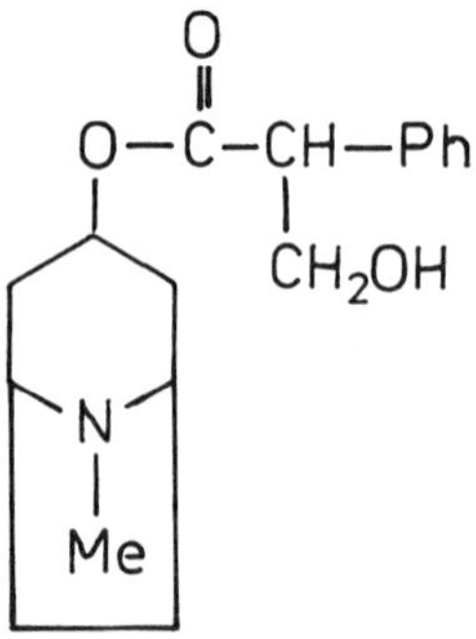

Atropine [18]

Classification of Alkaloids

The ester of tropine with atropic acid is called atropamine $C_{17}H_{21}O_2N$ and the ester of tropine with isatropic acid belladonnine $C_{34}H_{42}O_4N_2$.

Scopine esters have been isolated from <u>Datura metel</u> and other plants. The well-known base is l-scopolamine $C_{17}H_{21}O_4N$ comprising an ester of scopine and l-tropic acid. Norscopine with tropic acid gives norscopolamine $C_{16}H_{19}O_4N$.

The bases isolated from <u>Erythroxylon coca</u> are important. The drug of abuse is cocaine $C_{17}H_{21}O_4N$ [19], which is an ester of methylecgonine with benzoic acid. Other alkaloids of this plant species

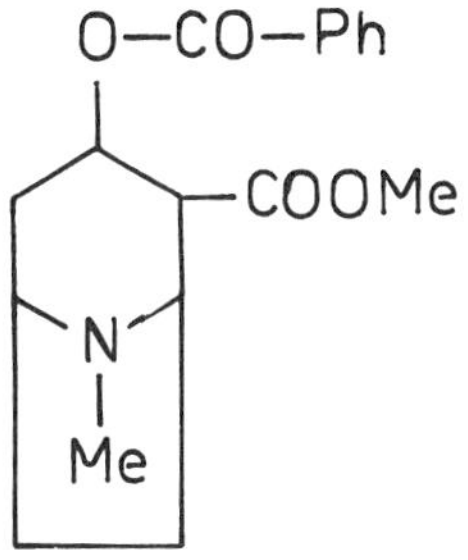

Cocaine [19]

are benzoyltropeine $C_{15}H_{19}O_2N$, an ester of tropine with benzoic acid; tropacocaine, which is a stereoisomeric derivative of benzoyltropeine; and benzoylecgonine $C_{16}H_{19}O_4N$, an ester of ecgonine with benzoic acid.

III. QUINOLINE ALKALOIDS

The alkaloids in this group are not numerous. An example would be the bases obtained from <u>Galipea officinalis</u>, e.g., cusparine $C_{19}H_{17}O_3N$ [20]. Also, some alkaloids derived from furoquinoline or pyrano-

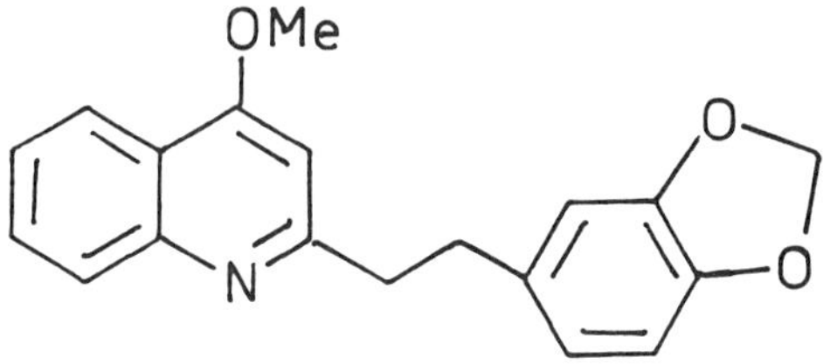

Cusparine [20]

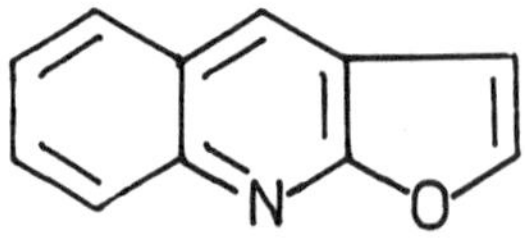

Dictamnine [21]

quinoline ring systems pertain to this group. Dictamnine $C_{12}H_9O_2N$ [21] found in <u>Dictamnus alba</u>, is a simple alkaloid with a furoquinoline ring system.

A. **Cinchona Alkaloids**

These alkaloids are an important subgroup because of their use in pharmaceuticals. Quinine $C_{20}H_{24}O_2N_2$ [22] is a levorotatory base isolated from <u>Cinchona</u> and <u>Remijia</u>. Its dextrorotatory stereoisomer is quinidine. Two other bases with the same structure are also known. Quinine and quinidine are accompanied in plants by the bases cinchonine and cinchonidine $C_{19}H_{22}ON_2$ [23]. These two alka-

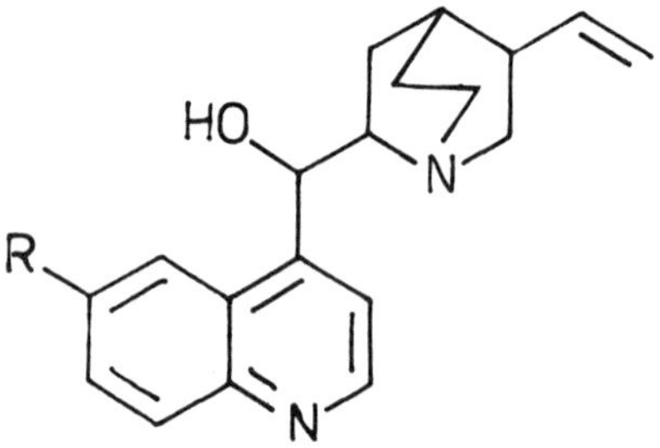

Quinine [22] R = OMe
Cinchonine [23] R = H

loids are stereoisomers, with cinchonine dextrorotatory and cinchonidine levorotatory. In plant extracts dihydro derivatives of the above-mentioned alkaloids occur which possess ethyl rather than vinyl groups in molecule.

IV. **ISOQUINOLINE ALKALOIDS**

Simple bases of this type are derived from 1,2,3,4-tetrahydroisoquinoline. As a typical example, anhalamine $C_{11}H_{15}O_3N$ [24], isolated from

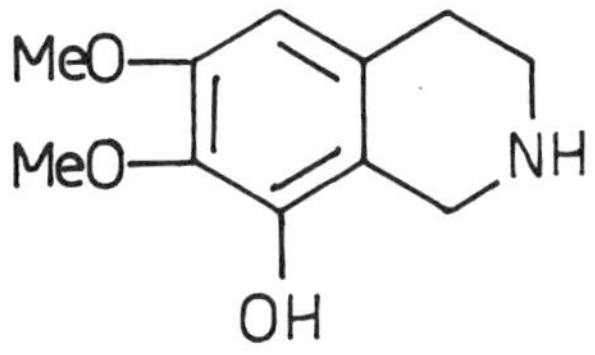

Classification of Alkaloids

Anhalamine [24]

Anhalonium lewinii, can be cited. Alkaloids of similar structure have also been found in *Salsola richteri*, e.g., salsoline $C_{11}H_{15}O_2N$.

A. Benzylisoquinoline Alkaloids

The bases in this subgroup are numerous and important. Some of them have been isolated from *Papaver somniferum*, e.g., papaverine $C_{20}H_{21}O_4N$ [25], with fully aromatic structure. The other related alkaloids are derived from tetrahydroisoquinoline: laudanosine $C_{21}H_{27}O_4N$, laudanine $C_{20}H_{25}O_4N$, laudanidine $C_{20}H_{25}O_4N$, etc. Structure [25] is closely related to the structure of another type of bases: the aporphine alkaloids. These compounds occur frequently in plants of, for example, *Corydalis* and *Magnolia*. As typical example glaucine $C_{21}H_{25}O_4N$ [26] can be mentioned.

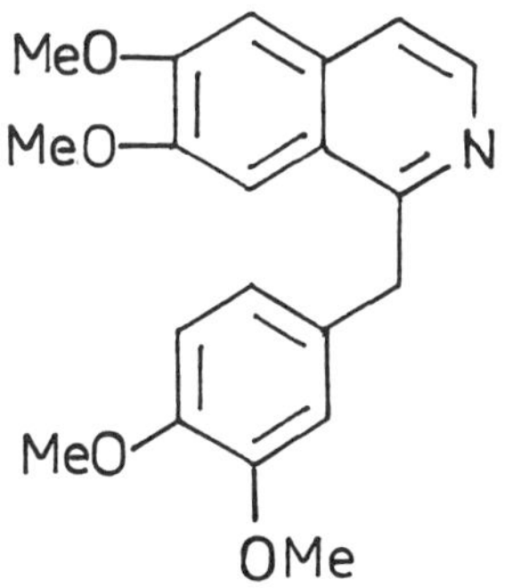

Papaverine [25] 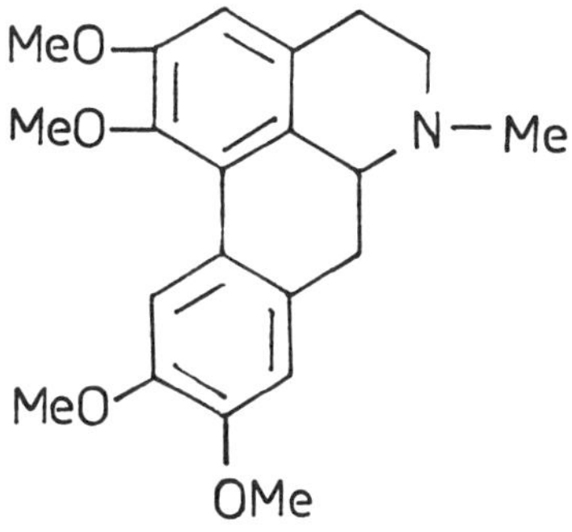   **Glaucine [26]**

From 1-benzyl-1,2,3,4-tetrahydroisoquinoline can be derived the next structure [27], which forms a fundamental part of the protoberberine alkaloids. These bases occur in *Berberis*, *Corydalis*, and other plants, and are widely distributed in nature. The name of these bases comes from the alkaloid berberine $C_{20}H_{18}O_4N^+$ [28], which is quaternary base. Palmatine $C_{21}H_{22}O_4N^+$ is also quaternary base with four methoxy groups, while canadine $C_{20}H_{21}O_4N$ is a tetrahydro derivative of berberine. Protopine bases can be classified as the alkaloids of benzylisoquinoline, although the isoquinoline ring system is absent from their structure. From the structure of protopine

Berberine [28]   Protopine [29]

$C_{20}H_{19}O_5N$ [29] its resemblance to tetrahydroberberine is evident. Protopine is widespread in many plants, e.g., Papaver somniferum. The protopine-type alkaloid cryptopine $C_{21}H_{23}O_5N$ has also been isolated from P. somniferum.

B. Morphine (Phenanthrenisoquinoline) Alkaloids

A very important subgroup of bases obtained from P. somniferum is the morphine alkaloids. Morphine $C_{17}H_{19}O_3N$ [30], a well-known component of many pharmaceutical preparations, is a levorotatory base of complex stereochemical structure with five asymmetric carbon atoms per molecule. Codeine—a methylether of morphine, $C_{18}H_{21}O_3N$ [31]—accompanies morphine in P. somniferum and can be prepared by morphine methylation. Synthetically prepared (not occurring in nature)

Morphine [30] R=H
Codeine [31] R=Me

diacetylmorphine is called heroin, and it is a known drug of abuse. Other alkaloids obtained from the opium poppy are thebaine $C_{19}H_{21}O_3N$ (structurally similar to morphine dimethylether, but possessing two conjugated double bonds) and neopine $C_{18}H_{21}O_3N$, a codeine isomer differing from codeine only in the position of its double bond. Some other alkaloids isolated from the opium poppy are mentioned in the following section.

C. Other Isoquinoline Alkaloids

Isoquinoline forms the basic part of many alkaloids with different structures. In the extract from <u>Papaver somniferum</u> a high content of the alkaloid α-narcotine $C_{22}H_{23}O_7N$ [32], which belongs to the subgroup of phthalidisoquinoline alkaloids, may be found. Genetically related to this subgroup are the bases narceine $C_{23}H_{27}O_8N$ [33],

α-Narcotine [32]   Narceine [33]

nornarceine $C_{22}H_{25}O_8N$, and others, even though isoquinoline does not appear in their molecules. These alkaloids were isolated from the opium poppy.

Some bases of isoquinoline possess more complex structures. Emetine $C_{29}H_{40}O_4N_2$ [34] is an example which occurs in Psychotria ipecacuanha. Two tetrahydroisoquinoline ring systems can be identi-

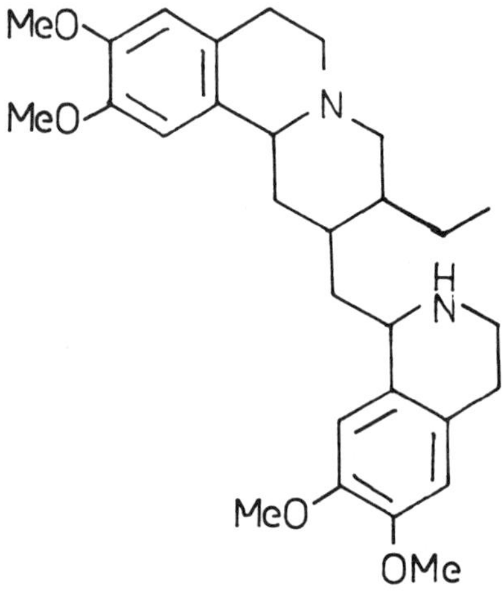

Emetine [34]

fied in the molecule of this alkaloid. Complex structures also belong to alkaloids of bisbenzylisoquinoline (two benzylisoquinolines connected by an etheric bond), of which the well-known ones are curine $C_{36}H_{38}O_6N_2$ and tubocurarine $C_{38}H_{44}O_6N_2^{2+}$ found in curare.

V. ALKALOIDS DERIVED FROM MISCELLANEOUS HETEROCYCLIC SYSTEMS

This section covers the alkaloids which cannot be formally placed in the other groups. According to the classification of heterocyclic systems, alkaloids derived from pyrrole and acridine may be included here. Bases of both types are not numerous. Among the pyrrole derivatives hygrine $C_8H_{15}ON$ [35], present in Dendrobium chrysan-

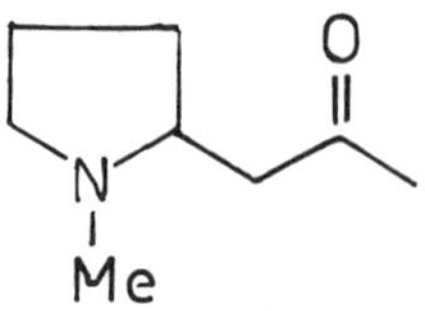

Hygrine [35]

Classification of Alkaloids

themum, can be mentioned. Bases derived from acridine occur mostly as alkoxyacridines. Among the other subgroups Senecio alkaloids, lupine alkaloids, Amaryllidaceae alkaloids, and purine bases are important.

A. Senecio Alkaloids

The numerous bases of this subgroup have been isolated from plants of Senecio and others. In nature they occur as esters of pyrrolizidine hydroxy derivatives (necines) with aliphatic acids, as is the case for the previously mentioned tropane alkaloids. Heliotridine and retronecine, which have the same formula $C_8H_{13}ON$, are the isomers of structure [36]. In Senecio alkaloids this structure is often found in

Retronecine [36]   Indicine N-oxide [37]

connection with an aliphatic acid; moreover, the major part of the alkaloid is in the form of N-oxide, e.g., indicine N-oxide $C_{15}H_{25}O_6N$ [37] obtained from Heliotropium indicum.

B. Lupine Alkaloids

These numerous bases contain, as their fundamental part, quinolizidine or, more exactly, octahydroquinolizidine. The simplest alkaloid of this subgroup is lupinine $C_{10}H_{19}ON$ [38], isolated from Lupinus

Lupinine [38]

niger and other plants. The more complex structures, derived from tetrahydroquinolizidine, can be found in cytisine, $C_{11}H_{14}ON_2$, matrine $C_{15}H_{24}ON_2$, sparteine $C_{15}H_{26}N_2$, and other related alkaloids. Sparteine has an interesting structure [39] with two condensed quinolizidine skeletal systems. For sparteine six stereoisomers are known, but not all of them have been found in nature.

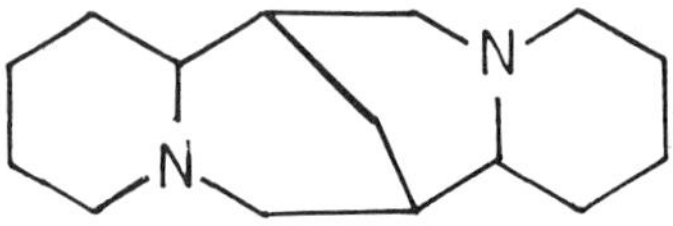

Sparteine [39]

C. Amaryllidaceae Alkaloids

The large number of bases which have been isolated from different species of Narcissus and Amaryllis is structurally based on partially hydrogenated phenanthridine ring. In Amaryllis belladonna, lycorine $C_{16}H_{17}O_4N$ occurs with the structure [40] representing one of three fundamental types of Amaryllidaceae alkaloids. As the second structural type haemanthamine $C_{17}H_{19}O_4N$ [41], obtained from Haemanthus

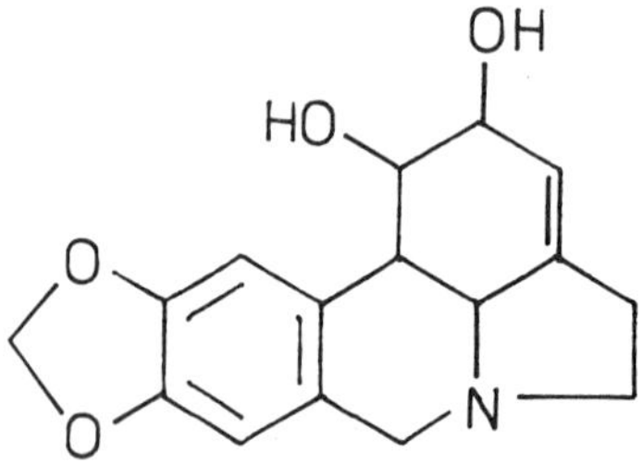

Lycorine [40] 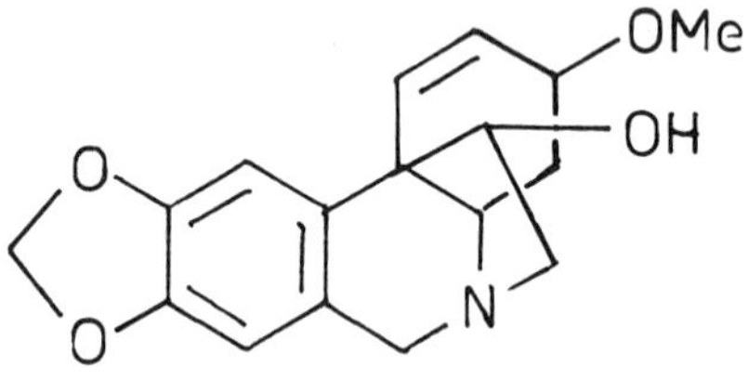   Haemanthamine [41]

puniceus, can be presented. The last structural type of Amaryllidaceae alkaloids does not contain a phenanthridine skeletal system.

D. Purine Bases

These plant bases are sometimes not regarded as alkaloids because the compounds are not related biogenetically to the amino acids.

Three compounds containing purine structure represent alkaloids of this subgroup: caffeine $C_8H_{10}O_2N_4$ [42] occurring in coffee beans, theophylline $C_7H_8O_2N_4$ [43] from tea leaves, and theobromine $C_7H_8O_2N_4$ [44] from cocoa beans.

Caffeine   [42] $R_1, R_2 = Me$

Theophilline [43] $R_1 = H, R_2 = Me$

Theobromine   [44] $R_1 = Me, R_2 = H$

VI. INDOLE ALKALOIDS

Bases with indole as the parent heterocyclic system occur frequently in nature and possess an astonishing range of structures. The assortment of this group of alkaloids into subgroups is based mostly on the plant species from which the alkaloids are obtained.

A. Simple Indole Alkaloids

The indole alkaloids are biosynthetically derived from tryptophan, as can be seen, for example, from the structure of gramine $C_{11}H_{14}N_2$ [45], isolated from barley mutants, or bufotenine $C_{12}H_{16}ON_2$ [46], found in <u>Piptadenia peregrina</u>. Harmine bases can be considered

Gramine [45]   Bufotenine [46]

the simplest indole alkaloids. All of them contain two nitrogen atoms per molecule, e.g., harman $C_{12}H_{10}N_2$ [47], tetrahydroharman $C_{12}H_{14}N_2$, harmine $C_{13}H_{12}ON_2$ [48], and its dihydro derivative harmaline

Harman [47]   Harmine [48]

$C_{13}H_{14}ON_2$. All of these alkaloids have been isolated from <u>Peganum harmala</u> as well as other plants.

B. Yohimbine Alkaloids and Related Bases

The most prevalent subgroup is the yohimbine alkaloids, which have been obtained by extraction of the bark of <u>Corynanthe yohimbe</u>. Due to the fact that one molecule of yohimbine $C_{21}H_{26}O_3N_2$ [49] contains

Yohimbine [49]

five asymmetric carbon atoms, several stereoisomers of fundamental structure exist as follows: α-yohimbine (syn. corynanthidine, rauwolscine), β-yohimbine, 3-epi-α-yohimbine (syn. yohimbene), and some others. Sometimes yohimbine and corynanthine subgroups are distinguished. However, corynanthine and corynanthidine are isomers of yohimbine and therefore only corynantheine $C_{22}H_{26}O_3N_2$ [50] and corynantheidine $C_{22}H_{28}O_3N$ (dihydro derivative of structure [50]), differing in structure from yohimbine, should be mentioned

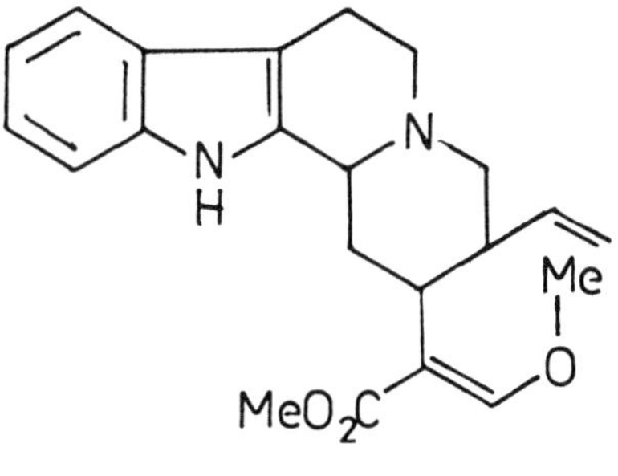

Corynantheine [50]

here. The four corynanthine alkaloids have been isolated from Pseudocinchona africana and C. yohimbe.

The bark of different species of Alstonia contains bases of which alstonine $C_{21}H_{20}O_3N$ [51] is best known. The methoxy derivative of tetrahydroalstonine has been isolated from the bark of Cinchona pelletierana and named aricine $C_{22}H_{26}O_4N$ [52].

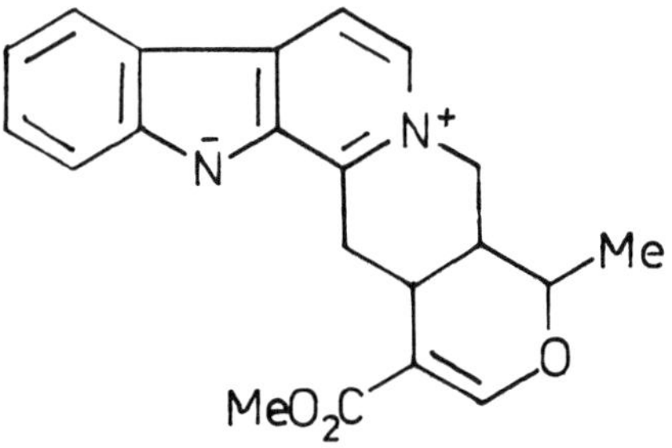

Alstonine [51]

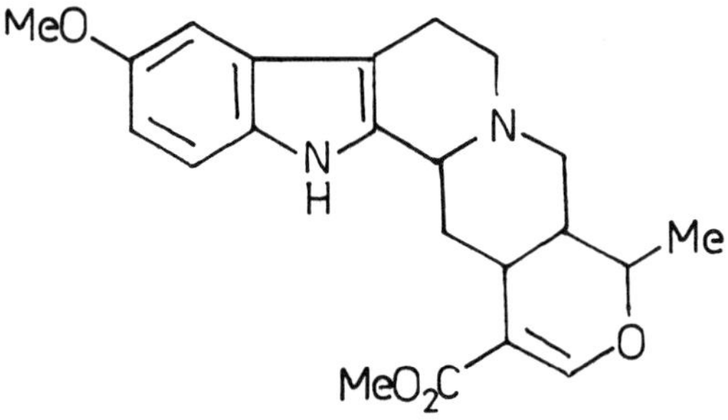

Aricine [52]

C. Rauwolfia Alkaloids

The plant species of Rauwolfia contain a large number of bases: yohimbine alkaloids, isoquinoline alkaloids, and ajmaline alkaloids. Besides yohimbine and their stereoisomers (see Sec. VI.B), ajmalicine (syn. δ-yohimbine) $C_{21}H_{24}O_3N_2$ [53] was isolated from Rauwolfia canescens and R. serpentina. Ajmalicine is isomeric with tetrahydro

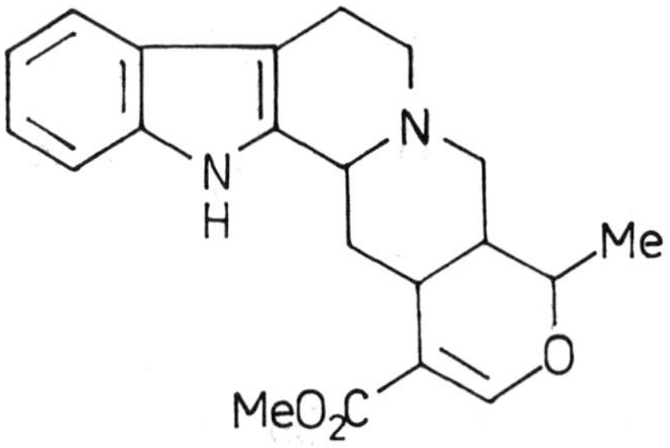

Ajmalicine [53]

derivatives of alstonine. Other yohimbine related bases found in Rauwolfia are reserpine $C_{33}H_{40}O_9N_2$ [54] and deserpidine $C_{32}H_{38}O_8N_2$ (11-demethoxyreserpine).

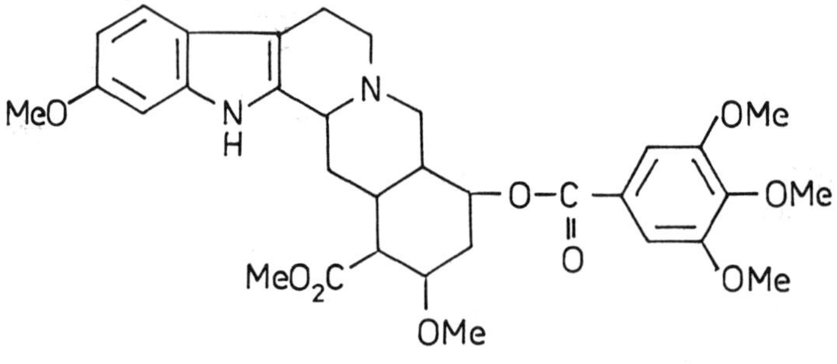

Reserpine [54]

Ajmaline alkaloids differ in structure from yohimbine and alstonine. Ajmaline $C_{20}H_{26}O_2N_2$, a dextrorotatory base occurring chiefly in R. serpentina, possesses a complex structure [55]. The same is valid for its stereoisomer isoajmaline, which behaves as a diacidic base.

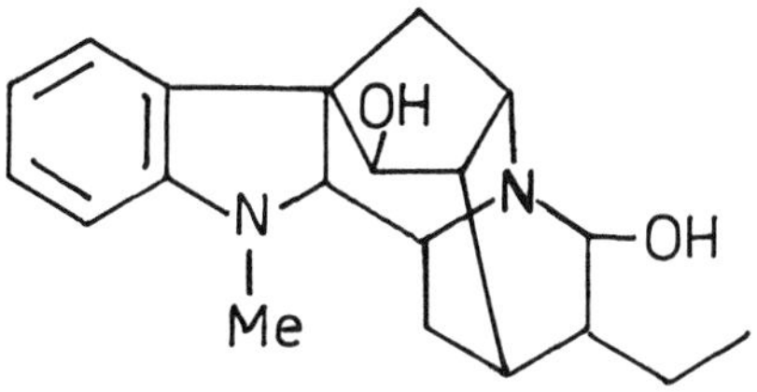

Ajmaline [55]

D. Strychnos Alkaloids

Bases of this subgroup occur in Strychnos nux-vomica. All of them are toxic; strychnine, owing to its simulating action upon the central nervous system, is used in pharmaceutical preparations. Strychnine $C_{21}H_{22}O_2N_2$ [56] illustrates the structural complexity of this alkaloid subgroup. Brucine $C_{23}H_{26}O_4N$ [57] is the dimethoxy derivative of

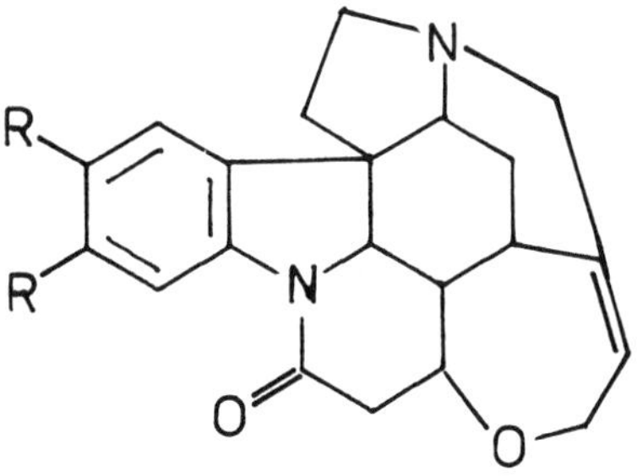

Strychnine [56] R=H

Brucine [57] R=OCH$_3$

strychnine. Both bases show the levorotatory effect. The seeds of Strychnos nux-vomica yield the other alkaloids novacine $C_{24}H_{28}O_5N_2$ and vomicine $C_{22}H_{24}O_4N_2$, which both contain a strychnine parent structure.

Bases related to strychnine have been obtained from Strychnos psilosperma. Strychnospermine $C_{22}H_{28}O_3N_2$ [58] and spermostrychnine $C_{21}H_{26}O_2N_2$ [59] are examples of such compounds.

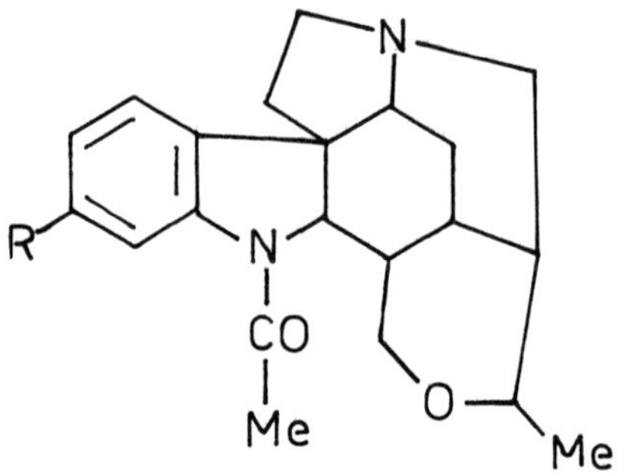

Strychnospermine [58] R=OCH₃

Spermostrychnine [59] R=H

E. Ergot Alkaloids

The alkaloids of this subgroup are derived from lysergic acid [60]. They occur in <u>Claviceps purpurea</u> as amides of this acid. Ergot alkaloids are divided into water-soluble bases, e.g., ergonovine, and water-insoluble bases, e.g., ergotamine, ergotoxine, and their isomers. Ergonovine (syn. ergometrine) $C_{19}H_{23}O_2N_3$ [61] is the simplest example of ergot alkaloids with the amidic part constituted by 2-aminopropan-1-ol. Bases of ergotamine occur in nature as amides of lysergic acid with the amidic part given by the structures [62-66].

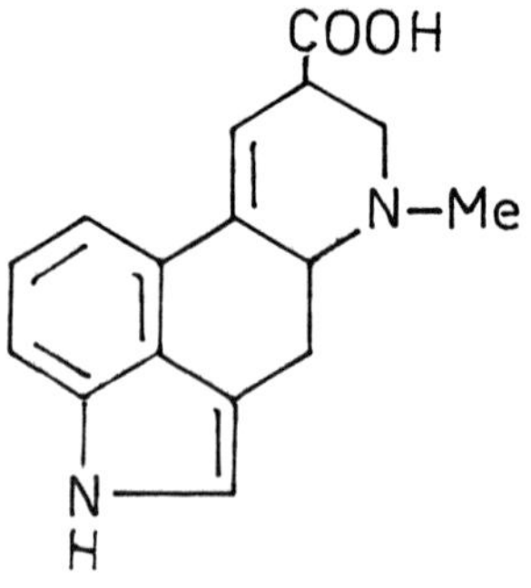

Lysergic acid [60]

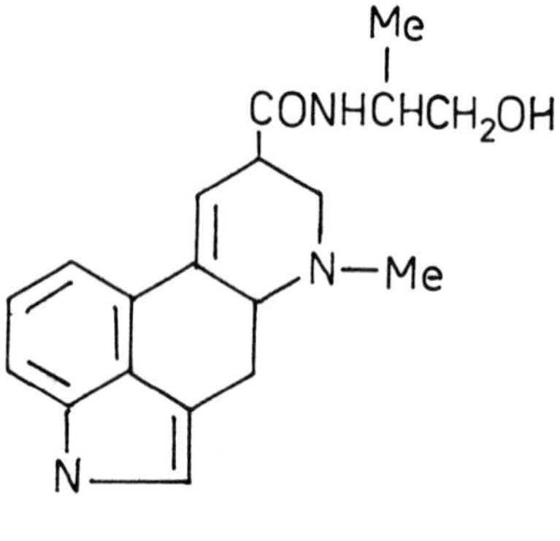

Ergonovine [61]

Classification of Alkaloids

[Structure diagram showing the amidic part with R₂, R₂ groups on CH, NH-C, CO-NH, pyrrolidine ring with N, CO, CH₂, R₁]

Ergotamine	[62]	$R_1 = CH_2\text{-}C_6H_5$ $R_2 = H$
Ergosine	[63]	$R_1 = CH_2CH(CH_3)_2$ $R_2 = H$
Ergocristine	[64]	$R_1 = CH_2\text{-}C_6H_5$ $R_2 = CH_3$
Ergocryptine	[65]	$R_1 = CH_2CH(CH_3)_2$ $R_2 = CH_3$
Ergocornine	[66]	$R_1 = CH(CH_3)_2$ $R_2 = CH_3$

Two pairs of stereoisomers, ergotamine and ergotaminine $C_{33}H_{35}O_5N_5$ [62], ergosine and ergosinine $C_{30}H_{37}O_5N_5$ [63], are classified as ergotamine alkaloids; and three pairs of stereoisomers, ergocristine and ergocristinine $C_{35}H_{39}O_5N_5$ [64], ergocryptine and ergocryptinine $C_{32}H_{41}O_5N_5$ [65], ergocornine and ergocorninine $C_{31}H_{39}O_5N_5$ [66], form the ergotoxine alkaloids. Structures [62-66] show only the amidic part of alkaloids. Ergot bases and their dihydro derivatives are frequently used in pharmaceutical preparations because they exhibit action on smooth muscle.

VII. IMIDAZOLE ALKALOIDS

A simple alkaloid of this group was isolated from Claviceps purpurea. It is called ergothioneine $C_9H_{15}O_2N_3S$ and possesses structure [67]. From the pharmaceutical point of view the bases of different plant species of Pilocarpus (e.g., Pilocarpus jaborandi) are important, because they stimulate the parasympathetic nerve endings. The chief

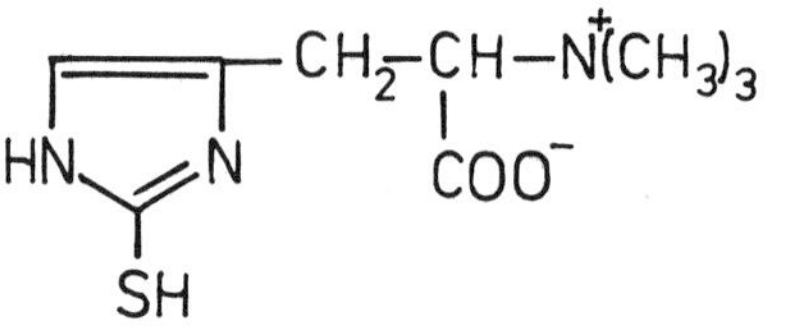

Ergothioneine [67]

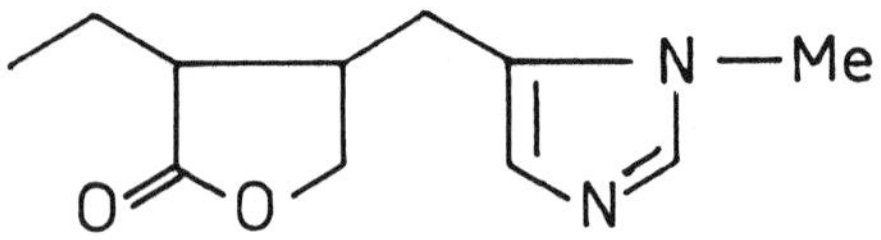

Pilocarpine [68]

alkaloid of this subgroup is pilocarpine $C_{11}H_{16}O_2N_2$ [68], usually accompanied by its stereoisomer isopilocarpine. Two pilocarpine derivatives should be mentioned here: pilocarpidine $C_{10}H_{14}O_2N_2$ (des-$\underline{N}$-methylpilocarpine) and pilosine $C_{16}H_{18}O_3N_2$. All of them have lower pharmacological activity than pilocarpine.

VIII. DITERPENE AND STEROID ALKALOIDS

The alkaloids of this subgroup are sometimes classified as pseudoalkaloids because they are not related biogenetically to the amino acids. However, their structures are the most complex of all the alkaloid groups.

A. Diterpene Alkaloids

The well-known base aconitine $C_{34}H_{47}O_{11}N$ [69], isolated from Aconitum napellus, is a representative of the Aconitum alkaloids, which is a

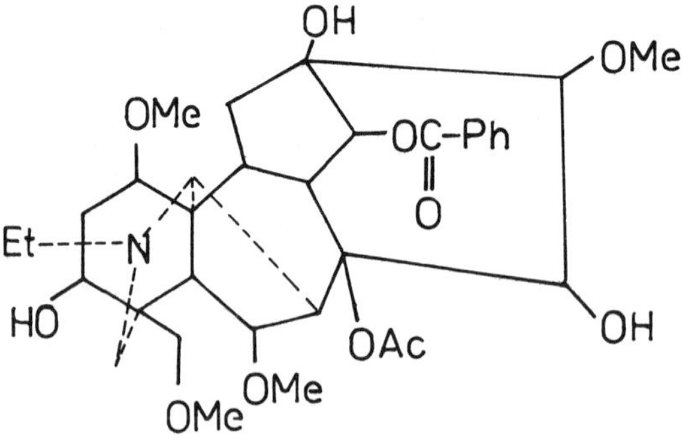

Aconitine [69]

large group of compounds, that possess diterpene skeletal structure. Jesaconitine $C_{35}H_{49}O_{12}N$, which has been found in Aconitum yesoense, is another Aconitum base. Jesakonitine contains methoxybenzoic acid, unlike aconitine, which contains benzoic acid. Bases of Delphinium staphisagria, so-called Delphinium alkaloids, e.g., delphinine $C_{33}H_{45}O_9N$, are also classified as diterpene alkaloids.

B. Steroid Alkaloids

The steroid alkaloids occur in plants in three different forms: (a) free bases, the aglycone portion of glycoalkaloids; (b) glycoalkaloids, alkaloids bonded to hexoses; and (c) alkaloids occurring as esters.

From the category of free bases, solanidine $C_{27}H_{43}ON$, [70] belonging to the Solanum (potato) alkaloids, should be mentioned.

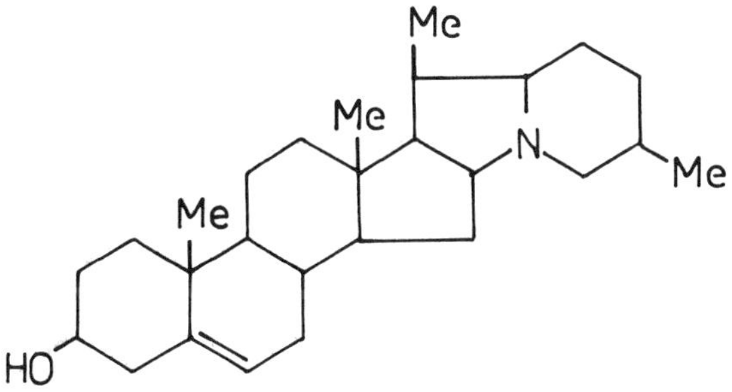

Solanidine [70]

A similar free base of tomato, tomatidine $C_{27}H_{45}O_2N_2$ [71], has been obtained as a product of tomatine hydrolysis only. Tomatine $C_{50}H_{83}O_{21}N$, a glycoalkaloid of Lycopersicum esculentum, contains the molecule tomatidine and four hexoses. Analogously, three solanines (α, β, and γ) are known with three, two, and one hexose in the molecule, respectively, in addition to solanidine.

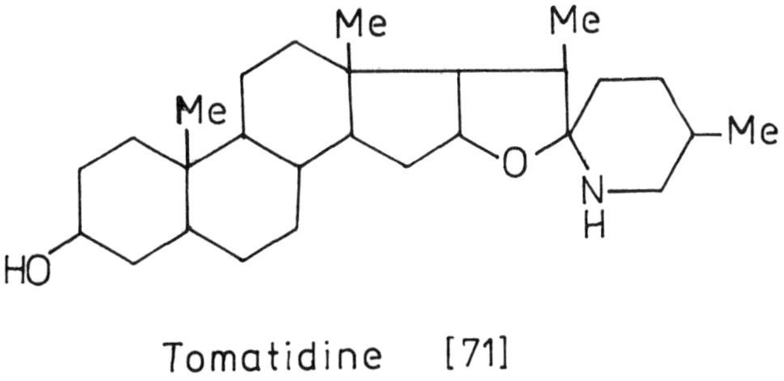

Tomatidine [71]

Veratrum alkaloids occurring in nature as esters have been extracted from Veratrum viride and other plants of the genus. The alkamines veracevine (syn. protocevine) $C_{27}H_{43}O_8N$, germine $C_{27}H_{43}O_8N$, and protoverine $C_{27}H_{43}O_9N$, esterified with different acids (e.g., acetic, methylbutyric, etc.), form the majority of Veratrum alkaloids.

REFERENCES

1. A. McKillop, An Introduction to the Chemistry of Alkaloids, Butterworths, London, 1970.
2. S. W. Pelletier, in Alkaloids: Chemical and Biological Perspectives, Vol. 1 (S. W. Pelletier, Ed.), John Wiley, New York, 1983, p. 26.
3. J. S. Glasby, Encyclopedia of the Alkaloids, Plenum Press, New York, 1975.
4. J. Staněk, Alkaloidy, Nakladatelství ČSAV, Prague, 1957.
5. R. F. Raffaut, A Handbook of Alkaloids and Alkaloid Containing Plants, John Wiley, New York, 1970.

2
Properties of Alkaloids Relevant to Chromatography

Anybody who starts with the chromatography of alkaloids faces serious problems, many of them relating to the properties of these compounds. Sometimes the difficulty begins with the choice of solvent to extract the base from the plant or to dissolve the sample. Then the analyst has to decide on either gas or liquid chromatography. When liquid chromatography is chosen, one must also select a chromatographic system, i.e., stationary or mobile phase, to be consistent with a detector type. The data needed are widely spread in the literature and it is therefore time consuming to search for them. To make the process easier we collected some data which we consider to be useful to chromatographers. Special emphasis has been placed on dissociation properties (Section I) because the majority of alkaloid analyses are carried out by liquid chromatography with the prevailing share by reversed phase or ion pair chromatography. From the standpoint of detection in liquid chromatography, the photometric and electrochemical detectors are most important. Sections III and IV review the properties of alkaloids which are important in terms of the detection method. The chapter is completed by general data given in Section II.

I. DISSOCIATION PROPERTIES

In the majority of cases basicity can be regarded as the hallmark of alkaloids. However, certain exceptions exist because some alkaloids are essentially neutral compounds. Among alkaloids of this type can be classified compounds in which nitrogen is involved in amide groups or in lactams. In addition, certain alkaloids occurring as N-oxides or quarternary nitrogen salts are weakly basic.

Alkaloids showing reasonable basicity dissociate in water and in aqueous solutions according to the equation:

$$B + H_2O \rightleftharpoons BH^+ + OH^- \qquad (2.1)$$

When water is used as a solvent, the dissociation constant K_b of monofunctional base is defined as

$$K_b = (a_{BH^+} \cdot a_{OH^-})/a_B \qquad (2.2)$$

The dissociation reaction of a base can be also described as the dissociation of conjugated acid BH^+:

$$BH^+ \rightleftharpoons B + H^+ \qquad (2.3)$$

with dissociation constant K_a. The product of both constants K_a and K_b is equal to the ionic product of water:

$$K_a \cdot K_b = K_{H_2O} \qquad (2.4)$$

Consequently, the dissociation constant of any base can be presented as K_a or in logarithmic scale as pK_a:

$$pK_a = pK_{H_2O} - pK_b = 14 - pK_b \qquad (2.5)$$

In Table 2.1 the pK_a values of some alkaloids are taken from literature (1-3). Along with these data the temperatures of measurement are presented as are remarks when a salt was measured instead of free base. Concentration data have been omitted.

A. Dissociation in Aqueous Buffers

The dissociation of alkaloids in aqueous mobile phases is extraordinarily important in reversed phase liquid chromatography and ion pair reversed phase liquid chromatography. In these separation systems the solutes occur as ions and nondissociated species. The ions are slightly retarded while the nondissociated molecules or ion pairs are retained in correspondence to their molecular structure. In a chromatographic system the molar ratio of nondissociated molecules [B] to the total concentration of an alkaloid [B] + [BH$^+$] is changed depending on the pH of the mobile phase. This dependence forms an s curve (Fig. 2.1).

Properties of Alkaloids

For an aqueous solution used as a mobile phase, the Henderson-Hasselbach equation can be applied:

$$pH = pK_a + \log([B]/[BH^+]) + \log(\gamma_B/\gamma_{BH^+}) \qquad (2.6)$$

where γ_B and γ_{BH^+} are the corresponding activity coefficients. For diluted solutions where $\gamma \doteq 1.0$, the activity coefficient can be neglected; Eq. 2.6 expresses the relation between the concentrations of both forms. The dissociated form prevails in the low-pH region while the undissociated species dominates at high pH.

Equation 2.6 states that the concentrations of both forms are equal at the inflex point and the pK_a value of an alkaloid is identical to the pH value of a mobile phase. Consequently, if we know the pK_a value for an alkaloid, we simultaneously know the relatively narrow pH region where the capacity factor of a base changes considerably. The data from Table 2.1 can be directly applied.

The drawback of this simple approach is that we usually do not use aqueous buffer solution as a mobile phase. In reversed phase liquid chromatography buffered methanol-water mixtures are used in the majority of cases, although some other organic modifiers (e.g., acetonitrile, tetrahydrofuran) are also applied. With these so-called mixed solvents the meaning and determination of pH is the key question.

For many reasons it is necessary to quantitatively express the acidity (or basicity) of the solutions containing organic modifier. The solution acidity for aqueous methanol and other polar amphiprotic solvents can be measured as the activity of solvated protons. For solvents with lower values of dielectric constant and low ionizing ability, the practical measurement of proton activity is more difficult and less reliable. In aprotic nonpolar solvents the ionization of acids and bases is negligible and solvent molecules do not participate in acid-base interactions of dissolved species. For this sort of solvent data on the strength of acids and bases are provided by comparison with a series of suitable reference bases. In this case the determination is carried out photometrically.

B. Acidity in Polar Amphiprotic Solvents

The definition-of-acidity scale is relatively simple for polar amphiprotic solvents. The most important requirement for pH definition is a measurable value of equilibrium constant K_{SH}:

$$2SH \xrightleftharpoons{K_{SH}} SH_2^+ + S^- \qquad (2.7)$$

TABLE 2.1 Values of pK_a for the Dissociation of Alkaloids in Water

Alkaloid	pK_a 1	pK_a 2	Temperature (°C)	Remark
Aconitine	8.35		15	
Arecaidine	9.07		25	As hydrochloride
Arecoline	7.41		17.5	As hydrobromide
Atropine	9.85		18	
Benzoylecgonine	11.8		25	
Berberine	11.73		18	
Brucine	8.16	2.50	15	
Caffeine	1.0		25	
Cinchonidine	8.40	4.17	15	
Cinchonine	8.35	4.28	15	
Cocaine	8.39		24	
Codeine	8.21		25	
Colchicine	1.85		15	
d-Coniine	10.9		15	
Cytisine	8.12	1.20	15	
Emetine	8.43	7.56	15	
Ergometrine	6.73		24	
Harmine	7.61		20	
Heliotridine	10.55		25	
Heroine	7.6		23	As hydrochlorride
l-Hyoscyamine	9.65		21	As sulfate
Indicine N-oxide				Essentially neutral
Isopilocarpine	7.18		15	As hydrochloride

TABLE 2.1 (Continued)

Alkaloid	pK$_a$ 1	pK$_a$ 2	Temperature (°C)	Remark
Methylecgonine	9.16		25	As hydrochloride
Morphine	8.21		25	
Narceine	3.3		25	
α-Narcotine	6.37		15	
Nicotine	8.02	3.12	25	As dihydrochloride
Nicotyrine	4.76		24	
Papaverine	6.40		25	
d,l-Pelletierine	9.40		15	
Pilocarpine	6.87		30	
Piperine	1.98		15	
Protopine	5.99			
Quinidine	8.77	4.2	15	
Quinine	8.34	4.3	20	
Retronecine	8.88		26	
Ricinine				Essentially neutral
l-Scopolamine	7.55		23	As hydrochloride
Solanine	7.54		15	
Sparteine	11.96	4.8	15	
Strychnine	8.26	2.5	25	
Thebaine	8.15		15	
Theobromine	1.0		25	
Theophylline	1.0		25	
Tropacocaine	9.88		15	
Tropine	10.33		25	
Yohimbine	7.45	3.0	23	As hydrochloride

Source: Data from Refs. 1–3.

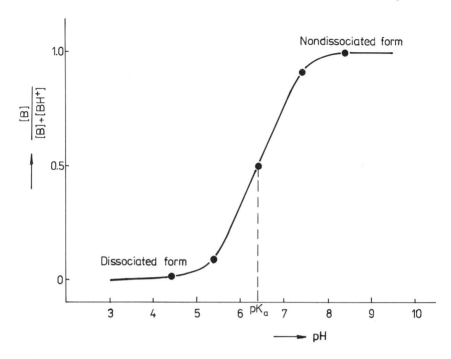

FIGURE 2.1 Dissociation curve of papaverine in water.

where SH means polar amphiprotic solvent, e.g., methanol, and K_{SH} is the autoprotolytic constant.

The acidity scale for amphiprotic solvents has been defined in a similar way as that for water, as this reference state involves an infinitely diluted solution in the given solvent, with an ionic strength approaching zero. Because the activity of individual ions is not accessible by thermodynamical measurement, some convention must be adopted regarding the activity coefficient values. A conventional paH^* scale is thus defined where an asterisk indicates that the quantity paH^* refers to the standard state in a solvent other than water. The values of paH^* are measured by means of a cell without a liquid junction but with hydrogen-silver/silver chloride electrodes. In order to obtain the standard potential of the reference electrode, conventions of chloride ion activity are applied and a diluted hydrochloric acid in a given solvent is used as the reference solution. The standard potential of the reference electrode is then used to measure the paH^* values of suitable buffer solutions.

Operational pH^* scale is based on measurements using glass and calomel electrodes. The cell must first be calibrated by buffers with known values of paH^*. The pH^* of unknown solution (X) is calculated as

Properties of Alkaloids

$$pH^*_{(X)} = pH^*_{(S)} + (E^*_{(X)} - E^*_{(S)})F/2.3RT \qquad (2.8)$$

where $pH^*_{(S)}$ is the reference value for a suitable buffer, and $E^*_{(S)}$ and $E^*_{(X)}$ are values of the emf of pH cell with electrodes immersed in either standard (S) or unknown solution (X). The more the composition and ionic strength of unknown solution differ from those of buffer, the more the determined pH^* may differ from its paH^*.

Another possibility for the determination of $pH^*_{(X)}$ of unknown nonaqueous solution is given by the equation:

$$pH^*_{(X)} = pH_{(S)} + (E^*_{(X)} - E_{(S)})F/2.3RT + \delta \qquad (2.9)$$

where $pH_{(S)}$ is the assigned pH value for aqueous reference solution, $E^*_{(X)}$ and $E_{(S)}$ are the emf measured in the mixed solvent and the aqueous reference solution, respectively, and δ is the correction factor for liquid junction potential. For methanol-water mixtures the values of δ can be found in the literature (4). This method is less reliable because organic modifier can affect the asymmetric potential of the glass electrode.

In the literature we often find so-called apparent pH values (with symbol $pH^{app}_{(X)}$) of solutions in mixed solvents. These values can be obtained as the sum of the first two terms of Eq. 2.9 with the correction factor δ being neglected.

C. Dissociation of Buffers in Polar Amphiprotic Solvents

Methods for pH^* determination in polar amphiprotic solvents described in the previous paragraph are seldom used to control the composition of mobile phase in liquid chromatography. In reversed phase liquid chromatography the pH of aqueous buffer solution is generally adjusted before mixing with organic modifier. This procedure is quite satisfactory for the reproducible preparation of mobile phase, but it should be kept in mind that pH^* and solute dissociation in mixed solvent are different from those in pure aqueous buffer. As an illustration, the logarithm of relative dissociation constant of papaverine (dissociation constant of papaverine divided by dissociation constant of phosphoric acid, both expressed in concentration units) is plotted against methanol concentration in Figure 2.2. Evidently at a given composition of phosphate buffer the addition of methanol suppresses the protonation of papaverine.

As an example let us consider the effect of methanol on pH^* of phosphate buffer—the system widely applied in liquid chromatography. As for aqueous buffer solutions, the Henderson–Hasselbach equation (2.6) can be used in methanol-water mixtures. The dissociation constant of the buffer component K^*_a determined in the given solvent and related to pH^* scale in that solvent must be used. Dissociation con-

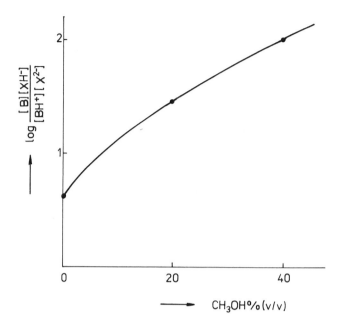

FIGURE 2.2 Dependence of relative dissociation constant (in concentration scale) of papaverine (B) and phosphoric acid ($X = HPO_4$) on methanol concentration as determined by spectrophotometry ($[X^{2-}] + [XH^-] = 2.10^{-2}$ mol liter^{-1}, $[B] + [BH^+] = 2.10^{-5}$ mol liter^{-1}).

stants K_a^* for phosphoric acid in methanol–water mixtures are given in Table 2.2. The column with heading pK_1^* states the equilibrium values:

$$H_3PO_4 \underset{\longleftarrow}{\overset{K_1^*}{\longrightarrow}} H^+ + H_2PO_4^- \qquad (2.10)$$

The pK_1^* values are equal to the pH^* values for an equimolar ratio of dihydrogen phosphate to phosphoric acid, and the same applies to the column headed pK_2^*, which states the values for an equimolar ratio of hydrogen to dihydrogen phosphate.

$$H_2PO_4^- \underset{\longleftarrow}{\overset{K_2^*}{\longrightarrow}} H^+ + HPO_4^{2-} \qquad (2.11)$$

It is evident from Table 2.2 that the pH^* of buffer in methanol–water differs fundamentally from the pH determined in aqueous solu-

TABLE 2.2 Values of pK^* for Phosphate Buffer in Methanol-Water

CH_3OH %(v/v)	pK_1^* $H_2PO_4^-$	pK_2^* HPO_4^{2-}
0	2.20	6.90
20	2.30	7.05
40	2.45	7.30
50	2.65	7.60
60	2.90	7.95
70	3.30	8.40
80	3.85	9.00

Source: Data from Ref. 5.

tion. Moreover, the differences rise with increasing methanol content in mixed solvent. A comparison of pK_1^* and pK_2^* shows that the increment of basicity is higher in the region of hydrogen phosphate than dihydrogen phosphate. The use of Table 2.2 data for the calculation of $\Delta pH_m = pH_m^* - pH_m$ (subscript m means the mobile phase, pH_m is a value for buffer in aqueous solution, pH_m^* is a value for the same buffer in a methanol-water) can be illustrated by the following example.

Example: Potassium phosphate buffer 0.03 M with pH 3.0 was mixed with methanol in the ratio 3:7 (v/v). What is pH_m^* of buffer in mixed solvent and what is ΔpH_m?

Answer: Without including activity coefficients, the following equation is valid for an aqueous solution of buffer:

$$pH_m = pK_1 + \log([H_2PO_4^-]/[H_3PO_4])$$

and similarly for buffer in 70% (v/v) Methanol:

$$pH_m^* = pK_1^* + \log([H_2PO_4^-]/[H_3PO_4])$$

The values $pH_m = 3.0$, $pK_1 = 2.20$, and $pK_1^* = 3.30$ are known (see Table 2.2) and the constant ratio of $[H_2PO_4^-]/$ can be supposed. Then pH_m^* is readily calculated as

$$pH_m^* = pH_m - pK_1 + pK_1^* = 4.10$$

and

$$\Delta pH_m = 1.10$$

The data of pK_1^* (Table 2.2) can be used for the ΔpH_m calculation until pH 4.55 in aqueous buffer solution is reached. Above pH 4.55 the pK_2^* data must be applied. However, it is necessary to emphasize the weak buffering ability of phosphate near pH 4.55, which may cause a discrepancy between calculated and experimental results.

For other buffers of the same type (e.g., acetate, succinate) the changes of ΔpH_m are similar, i.e., the pH_m^* values are always higher than the corresponding pH_m. For aqueous 0.01 M sodium acetate the value of pH = 4.7 was determined while $pH^* = 5.6$ was found in 57% (v/v) methanol solution (6).

The buffers of different charge type, dissociating in accordance with Eq. 2.3, show the shift of pH^* in the reverse direction. Table 2.3 gives the data for ammonium as well as two values for tris(hydroxymethyl)aminomethane. The effect of methanol on cationic acids is an initial increase of their strength (decrease in pH of buffer solution), and this increase continues up to methanol content 60-80% (w/w). When the content of methanol reaches 60-80%, the strength of cationic acid falls and buffer pH^* increases. From the standpoint of liquid chromatography, one can expect the change in retention behavior of cationic acid if the methanol content in mobile phase rises above 80% (v/v). For other amphiprotic solvents the same conclusion can be drawn, although the quantitative data are different.

TABLE 2.3 pK_a^* Values for Dissociation of Ammonium and Tris(hydroxymethyl)-aminomethane in Methanol-Water at 25°C

CH_3OH, % (w/w)	Ammonium	Tris
0	9.25	8.07
20	9.04	—
50	8.69	7.82
60	8.59	—
70	8.57	—

Source: Data from Ref. 6.

D. Other Factors Influencing pH* of Buffer Solution

As given by Eq. 2.6, the pH of a buffer solution is affected by the activity coefficients of buffer components. In dilute solutions the values of ion activity coefficients are determined essentially by the ionic strength of the solution. Ionic strength I is defined as

$$I = (1/2)(\sum_i c_i z_i^2) \quad (2.12)$$

where c_i is the molar concentration of ion i and z_i is its charge. It is apparent from Table 2.4 that the effect of ionic strength is distinct at low values (up to I = 0.1 mol liter^{-1}). Further changes of I in the range 0.1–0.5 mol liter^{-1} are not very expressive. Very often the buffer concentration applied in liquid chromatography is low enough for this factor to be operative.

To approximate the activity of single ions, the Debye-Hückel theory can be used in the form

$$-\log \gamma_i = A z_i^2 \sqrt{I}/(1 + Ba\sqrt{I}) \quad (2.13)$$

where $A = 1.825 \cdot 10^6/(\varepsilon T)^{3/2}$, ε is a dielectric constant, T is the thermodynamic temperature, $B = 502.9/(\varepsilon T)^{1/2}$, and a ion size parameter. The values of a ranging from 0.25 to 1.1 nm were assigned for a series of ions (7). At low ionic strength the quantity $\log \gamma_i$ is not very affected by the precise value of a. Therefore a = 0.3 nm is often used corresponding to Ba = 1 liter$^{1/2}$ mol$^{-1/2}$ for aqueous solutions at room temperature.

TABLE 2.4 pH* Values in 50% Methanol at 25°C

Ionic strength	HPO_4^{2-}	Tris
0.01	8.23	7.90
0.02	8.16	7.93
0.04	8.06	7.97
0.06	7.99	8.00
0.08	7.93	8.02
0.10	7.88	8.04

Source: Data from Ref. 6.

If the same value of parameter a is accepted for both forms (BH^{z+1} and B^z) of buffer components, the activity term of Eq. 2.6 can be written as

$$\log \gamma B^z/\gamma BH^{z+1} = [A\sqrt{I}/(1 + Ba\sqrt{I})][(z + 1)^2 - z^2]$$
$$= A(2z + 1)\sqrt{I}/(1 + Ba\sqrt{I}) \qquad (2.14)$$

For neutral and anionic acids ($z + 1 \leq 0$) the rising ionic strength shifts the pH slightly to the higher values (see Table 2.4). For compounds behaving as cationic acids (e.g., tris and nitrogen bases including alkaloids), the reverse effect has been observed.

Example: Potassium phosphate buffer containing K_2HPO_4 and KH_2PO_4, 0.01 M each, was mixed with methanol in the ratio 1:1 (w/w) at 25°C. What are the ionic strength and pH (pH^*) for both aqueous and mixed solutions? For an aqueous solution of phosphoric acid $pK_2 = 7.199$, density @ = 0.997 g cm^{-3}, and dielectric constant ρ = 73.8. For methanol-water 1:1 (w/w) mixture $pK_2^* = 8.483$, @ = 0.9125, ρ = 54.9. The value of a = 0.4 nm is used for hydrogen and dihydrogen phosphate ions (data from Refs. 8–10).

Answer: The phosphates dissociate in water as follows:

$$K_2HPO_4 \rightleftharpoons 2K^+ + HPO_4^{2-}$$

$$KH_2PO_4 \rightleftharpoons K^+ + H_2PO_4^-$$

$$I = (1/2)(2c.1^2 + c.2^2 + c.1^2 + c.1^2) = 4c$$

For aqueous solution c = 0.01 and I = 0.04. The constants of Eq. 2.14 for water at 25°C are A = 0.5115, B = 3.2914. Considering z = -2 for hydrogen phosphate ion in Eq. 2.14 and substituting into Eq. 2.6, we obtain:

pH = 7.199 − 0.243 = 6.956

After mixing with methanol, the concentration changes:

c = (0.01 · 0.5/0.997) · 0.9125 = 0.004576

and

I = 4c = 0.0183

Values A = 0.8713 and B = 3.9308 were calculated for methanol-water at 25°C by procedure given above and the pH^* can be found:

Properties of Alkaloids

$$pH^* = 8.483 - 0.292 = 8.191$$

The last factor, which is sometimes underestimated, is temperature. The changes in pH caused by temperature variation may be qualitatively assessed from Table 2.5, where the values of pK_a and pK_a^* for tris at several different temperatures are given. A variation of pH or pK_a values of about ±0.1 can be expected if the temperature variation $\Delta t = \pm 5°C$ is admitted. If the pH of mobile phase lies in the range where an alkaloid is completely dissociated or completely nondissociated, this variation is generally not important. Conversely, if we operate the mobile phase at a pH close to the inflex point of the alkaloid dissociation curve (see Fig. 2.1), this variation may bring about a large error. This is especially true when the capacity ratios of dissociated and nondissociated forms differ considerably.

It has been observed (11) that the difference $\Delta pH = \pm 0.1$ can cause the change in capacity ratio value $\Delta k = \pm 0.5$. In any event, for the operation with mobile phase at pH close to pK_a of the separated alkaloid, column thermostatting is recommended.

E. Dissociation Constants of Alkaloids in Amphiprotic Solvents

As was shown previously, the pH^* of buffer solution is substantially affected by the content of organic modifier. Results that are valid for buffers also apply to the dissociation constants of solutes. Protonated alkaloids are week cationic acids similar to ammonium and tris(hydroxymethyl)aminomethane. In this case the shift of pK_a^* in methanolic solution is toward the lower values (Table 2.3), except probably for low water content where a sharp increase in the pK_a^* of acids of all charge types is produced.

TABLE 2.5 pK_a Values for the Dissociation of Protonated Tris(hydroxymethyl)aminomethane

Temperature (°C)	pK_a in water	pK_a^* in 50% (w/w) methanol
15	8.36	8.11
20	8.21	7.96
25	8.07	7.82
30	7.93	7.68
35	7.80	7.55

Source: Data from Ref. 6.

Unfortunately, suitable data concerning the alkaloid's dissociation in the typical mobile phases used in reversed phase liquid chromatography are essentially lacking. However, a relatively comprehensive collection of pK_a^* data for alkaloids in 2-methoxyethanol (methylcellosolve) has been published (12). The pK_a^* values were determined in 80% aqueous 2-methoxyethanol by microtitration using glass and calomel electrodes. Prior to each determination the cell was calibrated with 0.05 M potassium hydrogen phthalate buffer (pH = 4.00) and each series of measurements was checked against the estimated pK_a^* of atropine. The alkaloids with $pK_a < 5.5$ were titrated directly with aqueous HCl and those with pK_a above this value were determined as hydrochlorides by titration with tetramethylammonium hydroxide. The results were calculated by a procedure similar to that represented by Eq. 2.9, omitting only the correction factor δ. To obtain the so-called apparent dissociation constants K_a^*, the titration values were worked out graphically. Some data from this set of a few hundred pK_a^* values in 80% methylcellosolve have been plotted against the corresponding pK_a values for aqueous solutions (Fig. 2.3). This plot shows the linear dependence of pK_a^* on pK_a with constant difference of 1.7 units. It is important that the ΔpK_a observed in Fig. 2.3 is constant for the whole range of pK_a values of alkaloids. Naturally, the exact value of this difference will depend on the composition of solution used. Its absolute value will increase together with an increasing content of organic modifier in the solution and will not be the same for different amphiprotic solvents. However, it seems probable that the pK_a shifts of an alkaloid caused by 2-methoxyethanol and by methanol will not be very different.

Changes of pK_a^* with the composition of the solvent are to a large extent affected by the electrostatic interactions of the ionized species participating in the dissociation reaction (13,14). A decrease in the dielectric constant of an aqueous solvent by the addition of an organic component leads to a decrease in the electrostatic stabilization of ions by the surrounding medium. This effect is more pronounced for ions with a high charge and small dimensions. The dissociation reaction

$$BH^{z+1} = B^z + H^+ \quad (2.3)$$

is expected to shift to the left for $z \leq -1$ and to the right when $z \geq 1$, and the effect is expected to increase with increasing absolute value of z. For $z = 0$ the total charge does not change in the course of the above reaction and the effect of the dielectric constant on the dissociation constant should be minimal.

Most alkaloids belong to this group. The failure of purely electrostatic models is immediately apparent because the dissociation con-

Properties of Alkaloids

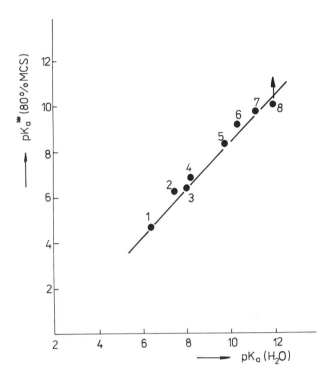

FIGURE 2.3 Values of pK_a in water vs. pK_a^* in 80% (v/v) 2-methoxyethanol for some bases. 1—narcotine; 2—yohimbine; 3—nicotine; 4—codeine; 5—atropine; 6—tropine; 7—piperidine; 8—sparteine.

stant K_a of an alkaloid increases when organic modifier is added to water. This indicates that specific solvation effects are important and applies to the solvation of H^+ by water. The aqueous solvation shell of protons is little affected by the addition of organic modifier, except at relatively low water content, and the effect of the macroscopic dielectric constant of solution on the proton stabilization is low.

On the other hand, the importance of the dielectric constant is evident from several empirical observations, some of which could be of practical value. In mixed solvents containing water (more than 50%), pK_a^* values are approximately linearly dependent on the reciprocal value of the dielectric constant if changes in water activity are taken into account (15). This dependence is common for various organic modifiers and solvent compositions. The dependence breaks down as soon as water activity drops to a value where competitive solvation of protons by other components of the solvent mixture becomes effective.

For a wider class of solvents linear dependence of the logarithm of relative dissociation constants for two acids (i.e., difference of their pK_a^* values) on the reciprocal value of the dielectric constant was ascertained (16). This is in agreement with the exclusive role of the effects of proton solvation because in the ratio of two dissociation constants the activity of H^+ is eliminated. If a similar relationship is correct, the dissociation degree of solute in buffer solution would be related to the dielectric constant in the same way (see next section).

F. Dissociation of Alkaloids in Buffered Amphiprotic Solvent

Let us now consider a problem mentioned earlier, namely, changes in the dissociation of an alkaloid in buffered aqueous solution when organic modifier is added. Applying the Henderson-Hasselbach equation (2.6) to both an alkaloid and a buffer, the dissociation of solute in aquoorganic solvent can be expressed as

$$\log ([B]/[BH^+]) = pK_{a(X)}^* - pK_{a(B)}^* + \log ([X^z]/[XH^{z+1}]) +$$
$$+ \log (\gamma X^z \gamma_{BH^+}/\gamma XH^{z+1} \gamma_B) \qquad (2.15)$$

From the preceding discussion it is evident that changes in the dissociation of alkaloid B when organic modifier is added are caused mainly by changes in the dissociation constants of both solute $pK_{a(B)}^*$ and buffer component $pK_{a(X)}^*$, and to a lesser extent by changes in the activity coefficient.

The ratio $[X^z]/[XH^{z+1}]$ can be expressed using the pH of aqueous buffer and the dissociation constant of buffer component in water. Then, when the activity coefficient term is neglected,

$$\log ([B]/[BH^+]) = pH - pK_{a(X)} + pK_{a(X)}^* - pK_{a(B)}^* =$$
$$= pH - (pK_{a(B)} + \Delta pK_{a(B)} - \Delta pK_{a(X)})$$
$$(2.16)$$

The dependence of solute dissociation on the pH of aqueous buffer is shifted with respect to that same dependence in water. The magnitude of the shift is given by the difference $\Delta pK_{a(B)} - \Delta pK_{a(X)}$, where $\Delta pK_{a(B)} = pK_{a(B)}^* - pK_{a(B)}$ and $\Delta pK_{a(X)} = pK_{a(X)}^* - pK_{a(X)}$ are differences of pK values in aquoorganic solvent and water for solute and buffer, respectively. $\Delta pK_{a(X)}$ is equal to the change of pH_m of buffer solution when organic modifier is added (see Section I.C).

Properties of Alkaloids

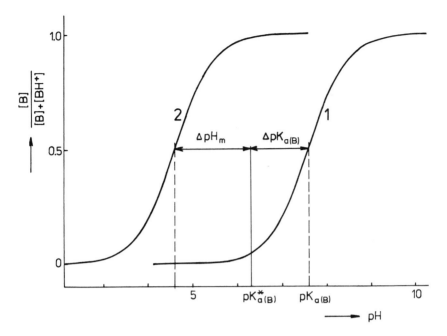

FIGURE 2.4 Dependence of alkaloid (B) dissociation degree on pH of buffer in mixed solvent ($\Delta pK_{a(B)} < 0$ and $\Delta pH_m > 0$). 1—solution in water; 2—solution in mixed solvent.

For alkaloids values of $\Delta pK_{a(B)}$ are usually negative whereas for buffer solution $\Delta pK_{a(X)}$ is typically positive (e.g., phosphate buffers). Both effects work in the same way (see Fig. 2.4). Therefore the pH of aqueous buffer must be about 2–3 units lower in order to achieve the same alkaloid dissociation in aquoorganic mobile phase as in water.

II. PHYSICOCHEMICAL PROPERTIES

As a rule, the physicochemical properties are mentioned first. In the case of chromatography of alkaloids, however, with special emphasis placed on liquid chromatography, we have considered the dissociation properties as most important. Yet, the physicochemical data remain important and therefore selected data for the common alkaloids are given in Table 2.6. The alkaloids listed represent all groups and subgroups classified in Chapter 1. Table 2.6 summarizes four items of basic information: molecular formula, melting point, boiling point, and solubility. Molecular formula has priority over

TABLE 2.6 Physicochemical Properties of Alkaloids

Alkaloid	Molecular formula	Melting point (°C)	Boiling point (°C/torr)	Solubility[a]		
				R	S	I
Aconitine	$C_{34}H_{47}O_{11}N$	202–4	—	$CHCl_3$, C_6H_6	EtOH	H_2O
Ajmaline	$C_{20}H_{26}O_2N_2$	205–7	—	$ChCl_3$		
Anabasine	$C_{10}H_{14}N_2$	—	276/760	H_2O, org. solv.		
Anhalamine	$C_{11}H_{15}O_3N$	187–8	—	H_2O, hot EtOH		
Arecaidine	$C_7H_{11}O_2N$	232	—	EtOH		
Arecoline	$C_8H_{13}O_2N$	Liquid	209/760	H_2O, org. solv.		
Aricine	$C_{22}H_{26}O_4N_2$	188–9	—	Et_2O		H_2O
Atropamine	$C_{17}H_{21}O_2N$	60–2	—	Org. solv.	H_2O	
Atropine	$C_{17}H_{23}O_3N$	118	—	$CHCl_3$	H_2O	Hexane

Properties of Alkaloids

Brucine	$C_{23}H_{26}O_4N_2$	178	—	EtOH, CHCl$_3$	Et$_2$O
Bufotenine	$C_{12}H_{16}ON_2$	146–7	320/0.1	EtOH	Me$_2$CO H$_2$O
Caffeine	$C_8H_{10}O_2N_4$	178	subl.	EtOH	
Cinchonidine	$C_{19}H_{22}ON_2$	210.5	—	EtOH	
Cinchonine	$C_{19}H_{22}ON_2$	255	—	EtOH	
Cocaine	$C_{17}H_{21}O_4N$	98	187/0.1	Org. solv.	
Codeine	$C_{18}H_{21}O_3N$	155	—	EtOH, CHCl$_3$	Et$_2$O
Colchicine	$C_{22}H_{25}O_6N$	155–7	—	CHCl$_3$, C$_6$H$_6$	
Conhydrine	$C_8H_{17}ON$	121	226/760	EtOH, CHCl$_3$	Et$_2$O
γ-Coniceine	$C_8H_{15}N$	Liquid	171/746	EtOH	
d-Coniine	$C_8H_{17}N$	Liquid	166/760	EtOH, Et$_2$O	
l-Corynantheidine	$C_{22}H_{28}O_3N_2$	117	—	Me$_2$CO	

TABLE 2.6 (Continued)

Alkaloid	Molecular formula	Melting point (°C)	Boiling point (°C/torr)	Solubility[a] R	S	I
Cytisine	$C_{11}H_{14}ON_2$	153	218/2	$CHCl_3$		
Deserpidine	$C_{32}H_{38}O_8N_2$	232	—	Et_2O		
Emetine	$C_{29}H_{40}O_4N_2$	74	—	$CHCl_3$, EtOH	H_2O	
l-Ephedrine	$C_{10}H_{15}ON$	38.1	225/760	EtOH, H_2O		
Ergocornine	$C_{31}H_{39}O_5N_5$	184 (dec.)	—			
Ergocristine	$C_{35}H_{39}O_5N_5$	170 (dec.)	—	$CHCl_3$		
α-Ergocryptine	$C_{32}H_{41}O_5N_5$	214	—			
Ergometrine	$C_{19}H_{23}O_2N_3$	163 (dec.)	—	Me_2CO, C_6H_6		
Ergosine	$C_{30}H_{37}O_5N_5$	228 (dec.)	—	EtAc		
Ergotamine	$C_{33}H_{35}O_5N_5$	214 (dec.)	—	Me_2CO		

Properties of Alkaloids

Name	Formula	MP	BP	Solvent
Germine	$C_{27}H_{43}O_8N$	220	—	MeOH
d-Glaucine	$C_{21}H_{25}O_4N$	120	—	EtOH, CHCl$_3$, C$_6$H$_6$
Guvacine	$C_6H_9O_2N$	272 (dec.)	—	H$_2$O
Harman	$C_{12}H_{10}N_2$	238	—	MeOH, CHCl$_3$, Hot H$_2$O
l-Hygrine	$C_8H_{15}ON$	Liquid	195/760	Et$_2$O
l-Hyoscyamine	$C_{17}H_{23}O_3N$	108.5	—	EtOH, C$_6$H$_6$, H$_2$O
Isopilocarpine	$C_{11}H_{16}O_2N_2$	Liquid	261/10	H$_2$O
l-Laudanidine	$C_{20}H_{25}O_4N$	185	—	C$_6$H$_6$, CHCl$_3$
l-Lobeline	$C_{22}H_{27}O_2N$	131	—	C$_6$H$_6$, H$_2$O
Lupinine	$C_{10}H_{19}ON$	69	257/760	H$_2$O, CHCl$_3$
γ-Matrine	$C_{15}H_{24}ON_2$	77	223/6	H$_2$O
Mescaline	$C_{11}H_{17}O_3N$	36	180/12	
N-Methylephedrine	$C_{11}H_{17}ON$	88	139/14	MeOH

TABLE 2.6 (Continued)

Alkaloid	Molecular formula	Melting point (°C)	Boiling point (°C/torr)	Solubility[a] R	S	I
Morphine	$C_{17}H_{19}O_3N$	256.4	—	EtOH	$CHCl_3$	
Narceine	$C_{23}H_{27}O_8N_2 \cdot 2H_2O$	155	—	H_2O, MeOH		
α-Narcotine	$C_{22}H_{23}O_7N$	176	—	$CHCl_3$, C_6H_6		
l-Nicotine	$C_{10}H_{14}N_2$	Liquid	246/730	H_2O		
Nicotyrine	$C_{10}H_{10}N_2$	Liquid	281/760			
l-Norephedrine	$C_9H_{13}ON$	51	168/2	EtOH		
l-Nornicotine	$C_9H_{12}N_2$	Liquid	130/11	$CHCl_3$		
Novacine	$C_{24}H_{28}O_5N_2$	233	—	$CHCl_3$		
Papaverine	$C_{20}H_{21}O_4N$	147	—	$CHCl_3$	EtOH	H_2O
d,l-Pelletierine	$C_8H_{15}ON$	Liquid	106/21			
Pilocarpine	$C_{11}H_{16}O_2N_2$	34	260/5	H_2O, $CHCl_3$		Et_2O

Properties of Alkaloids

Pilosine	$C_{16}H_{18}O_3N_2$	187	—	$CHCl_3$, EtOH	
Piperine	$C_{17}H_{19}O_3N$	129.5	—	EtOH, $CHCl_3$	H_2O
Protopine	$C_{20}H_{19}O_5N$	208	—	$CHCl_3$	EtOH
Pseudoephedrine	$C_{10}H_{15}ON$	119	—		H_2O
Quinidine	$C_{20}H_{24}O_2N_2$	175	—	H_2O, $CHCl_3$	
Quinine	$C_{20}H_{24}O_2N_2$	177	—	H_2O, $CHCl_3$	
Reserpine	$C_{33}H_{40}O_9N_2$	263	—	$CHCl_3$	
Ricinine	$C_8H_8O_2N_2$	201.5	Subl.	$CHCl_3$	EtOH	Hexane
α-Solanidine	$C_{27}H_{43}ON$	286	—	EtOH	
l-Sparteine	$C_{15}H_{26}N_2$	Liquid	325/154	$CHCl_3$, EtOH	C_6H_6, H_2O
Strychnine	$C_{21}H_{22}O_2N_2$	286	270/5	$CHCl_3$	H_2O, Et_2O
Thebaine	$C_{19}H_{21}O_3N$	193	—	$CHCl_3$, EtOH	H_2O
Theobromine	$C_7H_8O_2N_4$	350	Subl.		H_2O	Et_2O

TABLE 2.6 (Continued)

Alkaloid	Molecular formula	Melting point (°C)	Boiling point (°C/torr)	Solubility[a] R	S	I
Theophylline	$C_7H_8O_2N_4$	264	Subl.	Hot H_2O	EtOH	
Tropine	$C_8H_{15}ON$	63	233/760	H_2O, org. solv.		
Veracevine	$C_{27}H_{43}O_8N$	182	—	EtOH		
Vomicine	$C_{22}H_{24}O_4N_2$	282	—	EtOH, Me_2CO		
Yohimbine	$C_{21}H_{26}O_3N_2$	241	—	$CHCl_3$, EtOH	Et_2O	

[a]R, readily soluble; S, sparingly soluble; I, insoluble.
Source: Selected data from Refs. 12, 17, 18.

Properties of Alkaloids

molecular weight because it imparts more information. Melting point is considered as a basic aspect of an alkaloid. The melting temperatures are given for free bases; the bases melting below 20°C are recorded as liquids. Moreover, the compounds whose melting is accompanied by decomposition are marked. Certain alkaloids boiling at atmospheric or reduced pressure are tabulated in the subsequent column. These data are important from the point of view of an alkaloid's determination by gas chromatography. However, as can be seen from the literature survey, certain alkaloids for which the boiling points are not known (e.g., codeine, quinine, etc.) can be easily analyzed by gas chromatography. The last three columns list information concerning solubility. Owing to the lack of exact values, only qualitatively characterized solubilities are reviewed. This information can be useful for sample preparation, for selection of a suitable solvent for alkaloid extraction, and the like.

III. UV-VIS ABSORPTION SPECTRAL DATA

Photometric or spectrophotometric detectors are the most widely used in liquid chromatography as well as HPLC applications for the analysis of alkaloids. Besides detectors and chromatographic performance, the solute absorptivity is the main factor influencing the limit of detection. The quantity that is actually measured with a spectrophotometer is the absorbance A, defined as

$$A = \log (I_0/I) \qquad (2.17)$$

where I_0 and I are the intensities of incident and transmitted light, respectively. The proportionality of A to solute concentration c is given by the Lambert–Beer law:

$$A = abc \qquad (2.18)$$

which states that absorbance is directly proportional to the concentration of absorbing solution. The quantity a is called absorption coefficient (or absorptivity) for concentration c expressed in grams per liter and b is the cell width in centimeters. When concentration c is expressed in moles per liter, the absorptivity is called molar absorption coefficient (or molar absorptivity) and denoted by ε. The dimension of ε is liters mol^{-1} cm^{-1}, but as a rule it is given without units. The molar absorption coefficients range from 10^1 to 10^5, and in the literature and collections of spectra, they are usually plotted as log ε vs. wavelength.

A. UV-Vis Absorption Spectra of Alkaloids

Spectra of the most common alkaloids are arranged alphabetically in Table 2.7. From the standpoint of suitable wavelength selection there are presented the bands of maximum intensity and the bands of the longest wavelength together with corresponding molar absorption coefficients. Besides the concrete data for individual alkaloids, the absorption bands of other bases can be estimated on the basis of alkaloid structure.

The aliphatic amine group does not show any useful band at the range currently used for UV measurement. At or just below 200 nm absorption, bands with log $\varepsilon \doteq 3.5$ can be observed. Alkaloids without any other chromophore are therefore difficult to detect photometrically. The same is valid for alkaloids containing an isolated double bond which absorbs at a slightly longer wavelength. Carbonyl in acid, ester, and amide absorbs at about 215 nm with log $\varepsilon \doteq 1.7$. The increasing number of conjugated double bonds shifts the absorption bands to the longer wavelength, which is more appropriate for the operation of photometric detectors and enables a more versatile choice of chromatographic conditions. The most common chromophore in the molecules of alkaloids is a substituted phenyl group.

If only the saturated alkyl is attached to the phenyl group, the spectrum of alkaloid is very similar to those of monocyclic aromatic hydrocarbons. Besides having a more intense band at about 215 nm, a weaker absorption (typically log $\varepsilon = 2.3$) at 250–270 nm with vibrational structure is characteristic.

Polar substituents can cause the bathochromic shift up to 245 and 295 nm, and in some cases they are able to increase the intensity of the long wavelength band nearly two orders of magnitude. More effective are the substituents possessing the electrons enlarging the system of π electrons or participating in electron transitions through lone-pair electrons. As an example of favorable circumstances for this effect, α-narcotine may be presented with long wavelength absorption reaching 309 nm.

Some chromophores are more or less typical for a given group of alkaloids. This is especially true for ergot alkaloids with spectral characteristics nearly identical to those of lysergic acid. Quinoline and isoquinoline chromophores are characteristic for some members of the respective alkaloid groups. Their spectra resemble those of substituted naphthalene. Similarly, the spectra of other aza heterocyclic chromophores resemble the spectra of parent aromatic hydrocarbons with a polar substituent. Of course, quinoline and isoquinoline alkaloids with partially hydrogenated rings possess different spectra corresponding to the reduced size of the π-electron system. In a given alkaloid group the compounds with quite different spectral characteristics may be included. Table 2.8 illustrates

Properties of Alkaloids

TABLE 2.7 Ultraviolet Absorption Bands of Some Alkaloids[a]

Alkaloid	Most intense band[b] λ_{max} (nm)	log ε_{max}	Band of longest wavelength λ_{max} (nm)	log ε_{max}
Aconitine	273	3.03	280	2.95
Ajmaline	292	3.46	dtto	
Alstonine	308	4.53	368	3.65
Anabasine	260	3.41	dtto	
Anhalamine	272	2.86	dtto	
Arecaidine	214	3.92	dtto	
Arecoline	214	3.99	dtto	
Aricine	280	3.99	dtto	
Atropine	258	2.29	262	2.21
Benzoylecgonine	273	2.86	281	2.77
Berberine	265	4.39	345	4.38
Brucine	263	4.09	301	3.92
Bufotenine	277	3.79	300	3.66
Caffeine	271	3.99	dtto	
l-Canadine	288	3.83	dtto	
Candicine	278	3.25	dtto	
Cinchonidine	286	3.71	315	3.43
Cinchonine	284	3.73	315	3.46
Cocaine	273	3.02	280	2.94
Codeine	285	3.23	dtto	
Conhydrine	No chromophore			
d-Coniine	No chromophore			
l-Corynantheidine	280	3.88	dtto	
Corynantheine	280	3.89	dtto	
Cryptopine	288	3.92	dtto	
l-Curine	282	3.96	dtto	

TABLE 2.7 (Continued)

Alkaloid	Most intense band[b]		Band of longest wavelength	
	λ_{max} (nm)	log ε_{max}	λ_{max} (nm)	log ε_{max}
Cusparine	285	4.03	300	3.30
Cytisine	310	3.91	dtto	
Delphinine	273	2.96	dtto	
Deserpidine	271	4.30	dtto	
Dictamnine	309	3.93	328	3.84
Emetine	282	3.85	dtto	
l-Ephedrine	257	2.25	263	2.13
Ergot alkaloids[c]	240	4.22	310	3.95
Germine	No chromophore			
d-Glaucine	280	4.20	301	4.19
Haemanthamine	296	3.69	dtto	
Harman	288	4.22	348	3.67
l-Hygrine	No chromophore			
Isopilocarpine	216	3.75	dtto	
Laudanosine	282	3.78	dtto	
Lobelanidine	258	2.63	263	2.55
l-Lobeline	245	4.12	280	3.15
Lupinine	No chromophore			
Lycorine	293	3.68	dtto	
Mescaline	270	2.83	dtto	
Morphine	284	3.35	dtto	
Narceine	272	4.09	dtto	
α-Narcotine	309	3.69	dtto	
l-Nicotine	262	3.45	dtto	
Novacine	264	3.99	301	3.81
Palmatine	347	4.47	dtto	

TABLE 2.7 (Continued)

Alkaloid	Most intense band[b]		Band of longest wavelength	
	λ_{max} (nm)	log ε_{max}	λ_{max} (nm)	log ε_{max}
Papaverine	279	3.88	326	3.67
Pilocarpine	216	3.75	dtto	
Protopine	289	3.93	dtto	
Quinidine	332	3.69	dtto	
Quinine	332	3.68	dtto	
Reserpine	267	4.23	295	4.10
Ricinine	313	3.89	dtto	
Salsoline	285	3.55	dtto	
l-Scopolamine	258	2.29	264	2.17
α-Solanidine	No chromophore			
l-Sparteine	No chromophore			
Strychnine	254	4.10	289	3.51
Thebaine	285	3.88	dtto	
Theobromine	271	4.00	dtto	
Theophylline				
Tomatidine	No chromophore			
Tropine	No chromophore			
Veracevine	No chromophore			
Vomicine	262	3.81	297	3.57
Yohimbine				
α-Yohimbine	279–	3.86–	dtto	
β-Yohimbine	282	3.97		
δ-Yohimbine				
Pseudoyohimbine				

[a]All spectra were measured in methanol (Ref. 12).
[b]With certain exceptions of bands at short wavelength, e.g., 220 nm.
[c]Data valid for the majority of ergot alkaloids.

TABLE 2.8 Dominating Chromophores of Individual Alkaloid Groups

Group of alkaloids[a]	Dominating chromophore	Examples[b]
I. Phenylethylamine derivatives	Benzene	Mescaline [3], candicine [4], ephedrine [5]
II.A Pyridine derivatives	Pyridine	Nicotine [6], anabasine [7]
II.B Dihydro- and tetrahydropyridine derivatives	—CH=CH—C=O	Arecoline, arecaidine [9]
II.C Piperidine derivatives	1/Nonabsorbing	Pelletierine [11], coniine [12]
	2/Benzene (from substitution)	Lobeline [13], lobelanine, lobelanidine
II.D Tropane alkaloids	1/Nonabsorbing	Tropine [14], scopine [15], methylecgonine [16]
	2/Benzene (from substitution)	Atropine [18], l-hyoscyamine, cocaine [19]
III. Quinoline alkaloids	Quinoline	Cusparine [20]
III.A Cinchona alkaloids	Quinoline	Quinine [22], quinidine, cinchonine [23], cinchonidine
IV. Isoquinoline alkaloids	Benzene	Anhalamine [24], salsoline
IV.A Benzylisoquinoline alkaloids	(1) Isoquinoline	Papaverine [25]
	(2) Benzene	Laudanosine, protoberberine alkaloids [27], protopine [29], cryptopine
IV.B Morphine alkaloids	Benzene	Morphine [30], codeine [31], thebaine
IV.C Other isoquinoline alkaloids	Benzene	α-Narcotine [32], emetine [34], l-curine
V. Alkaloids derived from misc. heterocyclic systems	Carbonyl	Hygrine [35]

Properties of Alkaloids

TABLE 2.8 (Continued)

Group of alkaloids[a]	Dominating chromophore	Examples[b]
V.A Senecio alkaloids	Isolated double bond	Heliotridine, retronecine [36]
V.B Lupine alkaloids	Nonabsorbing	Lupinine [38], sparteine [39]
V.C Amaryllidaceae alkaloids	Benzene	Lycorine [40], haemanthamine [41]
V.D Purine bases	Purine	Caffeine [42], theophylline [43], theobromine [44]
VI.A Indole alkaloids	(1) Indole	Gramine [45], bufotenine [46]
	(2) Azacarbazole	Harman [47], harmine [48]
VI.B Yohimbine alkaloids	Indole	Yohimbines [49], corynantheine [50], aricine [52]
VI.C Rauwolfia alkaloids	(1) Indole	Ajmalicine [53], reserpine [54]
	(2) Benzene	Ajmaline
VI.D Strychnos alkaloids	Benzene	Strychnine [56], brucine [57], novacine, vomicine
VI.E Ergot alkaloids	Lysergic acid	Ergot alkaloids [61–65]
VIII.A Diterpene alkaloids	Benzene (from substitution)	Aconitine [69], delphinine
VIII.B Steroid alkaloids	(1) Nonabsorbing	Tomatidine [71], veracevine, germine
	(2) Isolated double bond	Solanidine [70]

[a]Classification according to Chapter 1.
[b]The number in brackets means the number of structure in Chapter 1. The alkaloids with specific chromophores were omitted from the list (e.g., ricinine [8], piperine [10], dictamnine [21], glaucine [26], berberine [28], etc.).

this fact and enables the assessment of spectral features of compounds in a given alkaloid group. The table summarizes the typical chromophoric systems in individual groups (see Chapter 1) and refers to some examples of compounds in which the chromophores are present.

Generally, absorption spectra of ionizable species can be affected by their ionization state. For most alkaloids, the basic nitrogen atom is sufficiently separated from the chromophoric group, and therefore their spectra are expected to be little affected by their protonation.

On the other hand, protonation of nitrogen that is a part of chromophore or directly attached to a chromophore markedly affects absorption spectra. To illustrate, the spectra of quinidine in phosphate buffers are given in Figure 2.5. For Na_3PO_4 solution the absorption spectrum of free base is very similar to that in Na_2HPO_4–NaH_2PO_4 buffer in which the tertiary nitrogen of quinidine is protonated. A different spectrum is observed in H_3PO_4 solution in which the protonation of nitrogen in quinoline ring takes place.

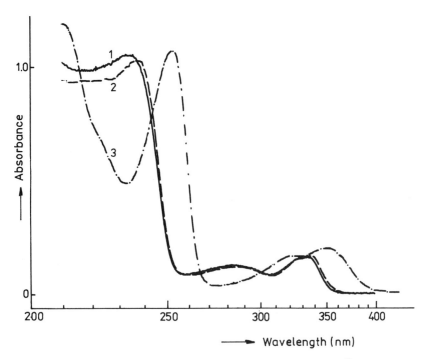

FIGURE 2.5 Absorption spectra of quinidine. $1-10^{-3}$ M Na_3PO_4 solution; $2-10^{-3}$ M Na_2HPO_4 and 10^{-3} M NaH_2PO_4 solution; $3-10^{-3}$ M H_3PO_4 solution.

Properties of Alkaloids

Amine-type nitrogen enlarges the π-electron system. Its protonation leads to the withdrawal of electrons from conjugation and therefore the blue shift is observed. As a result, the spectrum of protonated molecule then resembles the spectrum of parent chromophore with aliphatic substituent.

Another functional group occurring in alkaloids which can participate in acid—base interactions is the phenolic one. In alkaline solution this group dissociates forming phenolate ion. Its spectrum shows a red shift with respect to undissociated phenol. As an example, the spectra of morphine in acidic and alkaline solutions are given in Fig. 2.6.

IV. ELECTROCHEMICAL PROPERTIES

These properties are important from the standpoint of alkaloid detection in the column effluent and subsequent quantitative evaluation of the chromatogram. The main condition for the determination of a solute by an electrochemical detector is its ability to undergo elec-

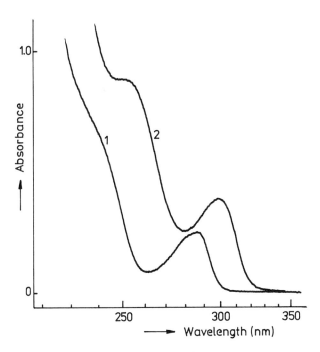

FIGURE 2.6 Absorption spectra of morphine. $1-10^{-3}$M H_3PO_4 solution; $2-10^{-3}$ M Na_3PO_4 solution.

trochemical reduction or oxidation. The response of a compound undergoing such a reaction depends on Faraday's law:

$$m = Q \cdot Mw/(z \cdot F) \qquad (2.19)$$

where m is the mass of solute converted by electrochemical reaction, Q is the number of coulombs, Mw is solute molecular weight, z is the number of electrons involved, and $F = 96,500$ coulombs mol^{-1} is the Faraday constant.

In principle two modes of operation of electrochemical detectors can be recognized. The first, not frequently applied, is based on coulometric principle. In this detector the total quantity of solute is reduced or oxidized to a product. The solute reaction is supposed to be the only occurring, and with the yield of 100%. If the molecular weight and number of involved electrons are known, then the detector needs no calibration.

Alternatively, an electrochemical detector can be operated in the voltammetric mode. When the dropping mercury electrode (DME) is used, we speak about polarographic mode. In voltammetry the flux of solute to the working electrode is limited by diffusion and only a small portion of solute (less than 1%) is converted to product.

A. Voltammetry and Polarography

Let us speak about the electrochemical determination of a solute in a voltammetric cell. There are placed two electrodes (working and auxiliary), supporting electrolyte, and a solute. Often three electrodes are used with a reference electrode added. The technique consists of applying a gradually increasing potential difference across the electrodes. The supporting electrolyte serves to conduct current without undergoing any electrode reaction and therefore the transport of solute ions by means of migration is suppressed. The current which is measured is produced by the reduction of solute ions at the cathode (or oxidation at the anode). This current depends on the concentration of the reducible (oxidizable) substance in the solution because the flux of ions to working electrode is only due to diffusion. Starting from zero applied voltage, a very small current (so-called residual current) can be measured. When the potential reaches the voltage of the corresponding electrochemical reaction, the current increases appreciably. This situation is depicted in Figure 2.7 as a typical voltammetric current-voltage curve. The high increase of current can be observed for the small increase in the applied potential. The current reaches its maximum at voltage E_p. The symbol E_p means peak potential and this value is independent of the concentration of reacting ions. For given supporting electrolyte and electrodes, the peak potentials are characteristic of ions in solution and can be used to solve any qualitative analy-

Properties of Alkaloids

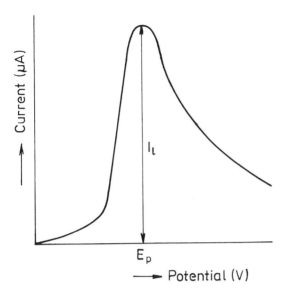

FIGURE 2.7 Voltammogram of an electrochemically active compound at a stationary electrode.

sis problem. The current corresponding E_p is referred to as the limiting current I_l. The current is directly proportional to the concentration of solute ions in solution and therefore it can be applied in quantitative analysis. After potential E_p the current slowly decreases as the concentration of reacting ions in the immediate neighborhood of working electrode is reduced. The description of voltammetry applies to a solid working electrode, e.g., glassy carbon, platinum, graphite, and the like.

When the dropping mercury electrode is used, the technique is called polarography and the current-voltage curve is slightly different (Fig. 2.8). This electrode possesses a continuously renewing surface, thus preventing the accumulation of electrolysis products. Under these conditions, when the maximum of current is reached, a steady rate of diffusion is set up. The steady current is referred to as diffusion current or limiting diffusion current I_d. Instead of peak potential E_p we measure half-wave potential $E_{1/2}$ as the characteristic of a certain ion in the solution. $E_{1/2}$ and E_p are related by the equation:

$$E_p = E_{1/2} \pm 0.029/z \; [V] \qquad (2.20)$$

where z is the number of electrons involved.

In electrochemical detectors the potential of the working electrode must correspond to the potential of the diffusion current of the respective solute, i.e., to a potential higher than $E_{1/2}$.

B. Voltammetric Determination of Alkaloids

Numerous compounds which are able to undergo reduction or oxidation at electrode can be directly determined by means of voltammetry. Besides these substances are many others with no active group in their molecule. However, some solutes of this type can be determined by the methods of indirect voltammetry. In principle two approaches are applied:

1. An active group is introduced to an inactive molecule, e.g., by oxidation, nitration, etc.
2. Catalytic waves are used for voltammetric determination.

The majority of alkaloids cannot be considered very suitable for direct voltammetric determination. Certain groups commonly occurring in these bases, e.g., tertiary or secondary amines, are not directly accessible to voltammetry; also, some other structural forms (pyridine and piperidine) cannot be included among substances that are easily determined by reduction or oxidation. Special precautions have to be made for supporting electrolyte and electrodes. For the

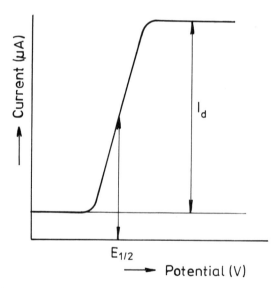

FIGURE 2.8 Polarogram of an electrochemically active compound at a dropping mercury electrode (DME).

Properties of Alkaloids

compounds undergoing cathodic reduction at fairly negative potential (e.g., ester of benzoic acid, pyridine, etc.), solvents other than water must be used.

Table 2.9 gives the values of applicable potentials for some solvents and for three different working electrodes. It can be seen that acetonitrile will be a suitable solvent for the cathodic reduction and, conversely, dimethylsulfoxide for anodic oxidation. Acetonitrile with the addition of tetraalkylammonium perchlorate or iodide may be used for the direct polarographic determination of alkaloids with benzoic or tropic acid in their molecule (Table 2.10). A value for $E_{1/2}$ of -2.14 V for atropine, reserpine, and deserpidine is in good agreement with the half-wave potential for the reduction step of benzoic acid (21). Another alkaloid introduced in Table 2.10, narceine, possesses in its molecule the carbonyl group conjugated to a benzene ring. Similar electrochemical reactions and $E_{1/2}$ values as for narceine may be expected for lobeline and lobelanine. The base colchicine contains the carbonyl group conjugated to the system of double bonds and is structurally similar to arecoline and cytisine. All the mentioned alkaloids give a reduction wave with $E_{1/2}$ valued near the half-wave potential of narceine. Some other alkaloids containing different active groups should be mentioned here, e.g., cotarnine and hydrastinine, possessing a reducible aldehyde group; papaverine, whose reduction step is attributed to the formation of dihydroquinoline; and berberine, with its complex system of conjugated double bonds producing a reduction wave.

Anodic oxidation of alkaloids has been used much more frequently than cathodic reduction. The reductive reactions occurring at potentials higher than -0.5 V are usually best performed at a mercury electrode due to its hydrogen overpotential, whereas the solid electrodes lack this advantage. The existing trend in electrochemistry toward the use of solid electrodes applies to the oxidative mode

TABLE 2.9 Applicable Potentials [V] for Different Solvents

Solvent	Electrodes		
	Hg	C	Pt
H_2O-acidic	+0.5 to −1.2	+1.4 to −1.0	+1.2 to −0.3
H_2O-alkaline	−0.2 to −2.8	+0.9 to −1.5	+0.3 to −1.1
Acetonitrile	+0.7 to −3.2	+0.4 to −1.6	+1.8 to −2.4
Dimethylsulfoxide	+0.8 to −2.6	+1.9 to −3.0	+1.4 to −1.8

Source: Data from Ref. 19.

TABLE 2.10 Direct Voltammetric (Polarographic) Determination of Alkaloids

Alkaloid	Method[a]	Working electrode	Reference electrode	Half-wave potential $E_{1/2}$ Peak potential E_p (V)	Supporting electrolyte
Atropine					
Reserpine	DCP	DME	Ag	$E_{1/2} = -2.14$	Acetonitrile +0.1 M (TBA)ClO$_4$[b]
Deserpidine					
Narceine	DCP	DME	SCE	$E_{1/2} = -1.20$	pH = 3.0
Colchicine	DCP	DME	SCE	$E_{1/2} = -1.40$	pH = 10.0
	ACP	DME	Ag/AgCl	$E_{1/2} = -1.13$	Acetonitrile + 0.01 M TEA perfluoroborate
Morphine	DCV	Pt	SCE	$E_{1/2} = +0.20$	1 M KOH
Caffeine	DCV	graphite	Hg/Hg$_2$SO$_4$	$E_p = +1.39$	acetate buffer, pH = 4.7
Cotarnine	DCP	DME	SCE	$E_{1/2} = -1.11$	pH = 4.0
Papaverine	DCP	DME	SCE	$E_{1/2} = -1.95$	50% EtOH+TMAOH[b]
Berberine	DCP	DME	SCE	$E_{1/2} = -1.1$	pH = 3.0

[a]DCP - direct current polarography, ACP - alternating current polarography, DCV - direct current voltammetry.
[b]TBA - tetrabutylammonium, TEA - tetraethylammonium, TMAOH - tetramethylammonium hydroxide
Source: Data from Refs. 19-21.

Properties of Alkaloids 63

too. The oxidative electroactivity of opium alkaloids may be discussed as one of several other examples. Morphine, with its phenolic group (3-hydroxyl group as the electroactive center), exhibits rather exceptional behavior among these compounds. It can be oxidized by means of low positive potential at a glassy carbon working electrode while most opium alkaloids are not oxidizable except at high potentials (22). Moreover, a high pH value of analyte solution, generally above 10, is required for the oxidation of codeine, thebaine, narcotine, narceine, and heroin. The difference between morphine and the others lies in the hydroxyl functionality at the 3 position of the former and an alkoxy group at the same position of the latter. Thus, most opium alkaloids may be oxidized electrochemically via oxidation of aliphatic tertiary nitrogen. Papaverine, whose nitrogen atom is situated in an aromatic ring, is not amenable to electrochemical oxidation.

However, the majority of alkaloids do not possess any group suitable for direct voltammetric determination, but many are often detected or determined by indirect voltammetry. Examples of indirect voltammetric determination of alkaloids are given in Table 2.11.

The first method of indirect voltammetry mentioned above has been used for polarographic determination of alkaloids in pharmaceutical preparations (23). As can be seen from Table 2.11, the nitration provides the possibility of detecting small quantities of morphine and quantitative nitrosation enables the polarographic determination of ephedrine. Codeine can be determined after oxidation of its tertiary amine group to the corresponding amine oxide, and purine alkaloids oxidized by bromine are also amenable to polarographic determination. However, this method is not the best from the standpoint of chromatographic separation. In the chromatographic system the behavior of the molecule with an introduced active group will probably be different from the behavior of the parent molecule. This problem can be avoided by subjecting an alkaloid to postcolumn derivatization followed by electrochemical detection. As an example, postcolumn photolysis of cocaine with an attached electrochemical detector can be introduced (24).

The catalytic waves have their origin in the effecting of adsorption or desorption. A compound not undergoing electrochemical reaction can give a wave by change in surface activity. However, it must be mentioned here that reduction and catalytic steps for an alkaloid have often been observed. In some cases the height of the catalytic hydrogen wave is not a linear function of catalyst concentration, but in other cases the catalytic step can be used for quantitative work.

Examples of alkaloids producing catalytic waves are given in Table 2.11. Quinine and other <u>Cinchona</u> alkaloids are known to produce catalytic hydrogen waves beside a reduction step. Strychnine and brucine were separated by chromatography and the appearance

TABLE 2.11 Indirect Polarographic Determination of Alkaloids[a]

Alkaloid	Half-wave potential $E_{1/2}$ (V)	Supporting electrolyte	Remark
Morphine			After nitration to nitromorphine
Ephedrine			After nitrosation to N-nitroso derivative
Codeine	$E_{1/2} = -0.84$	Acetate buffer pH = 4.7	After oxidation to amineoxide
Caffeine[b]	$E_{1/2} = -0.72$	1 M CH_3COOH	After oxidation by bromine
Quinine[c]	$E_{1/2} = -1.26$	pH = 8.0	Catalytic hydrogen wave
Strychnine		pH = 8.0	Catalytic hydrogen wave
Pilocarpine		0.001 M KCl	Catalytic wave via suppression of the oxygen maximum

[a]Direct current polarography with dropping mercury electrode (DME); half-wave potentials measured against saturated calomel electrode (SCE).
[b]Also for other purine alkaloids.
[c]Also for other Cinchona alkaloids.
Source: Data from Refs. 19, 20, and 22.

of the alkaloids in the effluent was indicated by a catalytic hydrogen wave (25). Pilocarpine can be determined as a catalyst causing the suppression of the oxygen maximum because the potential of maximum termination is related to pilocarpine concentration. Besides these examples, many other alkaloids can be detected by use of catalytic waves, e.g., reserpine, yohimbine, Haplophyllum alkaloids, Solanum glycoalkaloids, sparteine and some other quinolizidine alkaloids, etc.

REFERENCES

1. D. D. Perrin, Dissociation Constants of Organic Bases in Aqueous Solution, Butterworths, London, 1965, p. 341–352.
2. D. Dobos, Electrochemical Data, Akadémia Kiadó, Budapest, 1975.

3. J. Cíhalík, Potenciometrie, Nakladatelství CSAV, Prague, 1961.
4. R. G. Bates, M. Paabo, and R. A. Robinson, J. Phys. Chem., 67:1833 (1963).
5. A. Leitold and Gy. Vigh, J. Chromatogr., 257:384 (1983).
6. R. G. Bates in NBS Technical Note 271 Electrochemical Analysis. Studies of Acid, Bases and Salts by EMF Conductance. Optical and Kinetic Methods, R. G. Bates, ed., U.S. Department of Commerce NBS, Washington, D.C., 1965.
7. J. Kielland, J. Am. Chem. Soc., 59:1675 (1937).
8. M. Paabo, R. A. Robinson, and R. G. Bates, J. Chem. Eng. Data, 9:374 (1964).
9. M. Paabo, R. A. Robinson, and R. G. Bates, J. Am. Chem. Soc., 87:415 (1965).
10. S. Kotrlý and L. Šucha, Handbook of Chemical Equilibria in Analytical Chemistry, Ellis Horwood Ltd., Chichester, 1985, p. 90.
11. M. Popl, Le Duy Ky, and J. Strnadová, J. Chromatogr. Sci., 23:95 (1985).
12. J. Holubek and O. Štrouf, Spectral Data and Physical Constants of Alkaloids, Academia, Prague, 1965 (Vol. I)-1973 (Vol. VIII).
13. R. G. Bates, Determination of pH, Theory and Practice, John Wiley & Sons, New York, 1973.
14. L. Šafařík and Z. Stránský, Odměrná analýza v organických rozpouštědlech (Volumetric Analysis in Organic Solvents), SNTL, Prague, 1982.
15. M. Yasuda, Bull. Chem. Soc. Japan, 32:429 (1959).
16. W. F. K. Wynne-Jones, Proc. Roy. Soc. (London), Ser. A140: 440 (1933).
17. J. S. Glasby, Encyclopedia of the Alkaloids, Plenum Press, New York, 1975.
18. J. Štanek, Alkaloidy, Nakladatelství CSAV, Prague, 1957.
19. G. Henze and R. Neeb, Elektrochemische Analytik, Springer-Verlag, Berlin, 1986.
20. G. W. C. Milner, The Principles and Applications of Polarography and Other Electroanalytical Processes, Longmans, Green, London, 1958.
21. L. Meites and P. Zuman, Electrochemical Data, Vol. A, John Wiley, New York, 1974.
22. C. M. Selavka and I. S. Krull, J. Liq. Chromatogr., 10:345 (1987).
23. H. Hoffmann and J. Volke, in Electroanalytical Chemistry (H. W. Nurnberg, ed.), John Wiley, London, 1974.
24. C. M. Selavka, I. S. Krull, and I. S. Lurie, J. Chromatogr. Sci., 23:499 (1985).
25. W. Kemula, Rozc. Chem., 29:653 (1955).

3
Chromatography

People working with chromatography have yet to devise an appropriate and comprehensive definition of this method. Unfortunately, chromatography involves so many specific forms and procedures that no definition comes close. Therefore let us not belabor definition and emphasize instead the ability of chromatography to separate complex mixtures into discrete bands. Chromatography is, before anything else, a separation method, although for the absolute definition of separated components, it must be linked to other analytical methods, mostly spectroscopic.

I. PRINCIPLES, CLASSIFICATION, AND NOMENCLATURE

A. Principles

Chromatography in principle is based on the equilibrium distribution of a sample between two phases. The phases are composed of compounds other than the sample components. The volumes of both phases are much larger than that of sample and therefore the distribution coefficients of individual substances remain essentially constant. The separation of a sample by chromatography is achieved by the distribution of substances between mobile and stationary phases. The substances which are held strongly in the mobile phase pass through the system more rapidly than those held strongly in the stationary phase. Thus, the velocity of a component's movement through the system will depend on intermolecular forces holding the substance in the stationary phase or, more exactly, on the difference of forces between the solute molecules and the molecules of each phase. Because these forces for individual components are different, various substances will travel with different velocity, and in this way, they will be separated.

B. Classification

Chromatographic methods may be classified according to (a) type of mobile phase with the subdivision due to stationary phase used, (b) method of sample movement through the bed, and (c) technique. The classification which would be based on the nature of intermolecular forces is not fully accepted because generally two or more kinds of forces are operating simultaneously.

Table 3.1 summarizes the classification according to point a. Primary division is based on the type of mobile phase used: gas, supercritical fluid, and liquid chromatography. For the separation of alkaloids the group of methods belonging to liquid chromatography can be considered as most important. The methods of gas chromatography have certain limitations due to the high boiling points of many alkaloids and the low thermal stability of some of them.

TABLE 3.1 Classification of Chromatographic Methods

Mobile phase	Stationary phase	Method
Gas	Liquid	Gas-liquid chromotography (GLC)
	Solid	Gas-solid chromatography (GSC)[a]
Supercritical fluid	Liquid	Supercritical fluid–liquid chromatography (SFLC)
	Solid	Supercritical fluid–solid chromatography (SFSC)
Liquid	Liquid	Liquid-liquid chromatography (LLC)
		Liquid permeation chromatography (GPC)[a]
	Solid	Liquid-solid chromatography (LSC)
		Ion exchange chromatography (IEC)

[a]Methods usually not applied to alkaloid separation.

Supercritical fluid chromatography is less frequently used, although it has been successfully applied for the separation of alkaloids (1). From the standpoint of instrumentation, this method is an adapted gas chromatograph which uses liquefied gases (e.g., CO_2, SF_6) as the mobile phase. A fluid becomes supercritical when both its pressure and temperature exceed the critical point. A fluid in supercritical state exhibits low viscosity and is suitable for transportation of compounds whose molecular weight is too high or thermal stability too low for conventional gas chromatography. In the future one can expect supercritical fluid chromatography to prove its amplitude for alkaloid separation.

The further division of chromatographic methods according to the stationary phase used gives rise to eight methods, but two of them are generally not applicable to the chromatography of alkaloids (see Table 3.1). If we omit supercritical fluid chromatography, there remain four methods which are important. These will be dealt with in detail in the following chapters. Moreover, the method called ion pair chromatography, owing to its special importance in the separation of alkaloids, is discussed separately. On the other hand, affinity chromatography has been omitted from the list.

Considering the mode of sample movement through the bed, point b, chromatography can be divided into three types: frontal development, displacement chromatography, and elution chromatography. Briefly stated, frontal development and displacement chromatography were used in the early stages of chromatography but today find very limited application. Elution chromatography is the only type that is still of value. In elution chromatography the sample is placed at the beginning of a stationary phase (column or flat bed) and the flow of mobile phase is started. A certain part of the solute molecules is first absorbed (or adsorbed or chemisorbed, depending on the chromatographic system) and the other part moves through the bed. Stationary phase containing absorbed molecules comes into the contact with fresh, solute-free mobile phase and some molecules are reextracted into the mobile phase. Thus, elution partition chromatography may be characterized as a series of absorption-extraction steps continuously occurring during the passage of solute through the column or flat bed. Owing to differences in the intermolecular forces, sample components migrate through the bed at different velocities. Unretained components will travel at the same velocity as the mobile phase whereas the others will pass more slowly. During the solute passage spreading occurs, which is caused by diffusion and the resistance to mass transfer. If a detector is placed on the end of the column, we obtain the curve relating solute concentration to time (chromatogram), where the individual components are recorded as Gaussian curves. This is the case when the sample concentration in a column is low enough and then it is called linear elution chromatography. For higher sample concentration, peak asymmetry, mostly peak tailing, can be observed (see Fig. 3.1). In this case we speak about nonlinear elution chromatography.

Point c states classification of chromatography based on the technique used, i.e., separation either in a column or in a flat bed. Two types of columns are generally recognized: (a) packed columns, and (b) capillary columns, also called open tubular columns. Packed columns are tubes filled with adsorbent or support coated with stationary phase. Capillary columns are used mainly in gas chromatography (for details, see Chapter 4, Section I). The other columns are mentioned in Section III of this chapter.

Flat bed (or open bed) chromatography is usually divided into paper chromatography (PC) and thin-layer chromatography (TLC). Both techniques use a thin, essentially two-dimensional, bed of stationary phase where the separation takes place. The flow of liquid mobile phase results from capillary forces. Paper chromatography is not often used today and thin-layer chromatography is dealt with in detail in Chapter 6.

The relatively new technique of countercurrent chromatography (CCC) may be mentioned here, even though some authors do not consider it to be chromatography. This technique is essentially

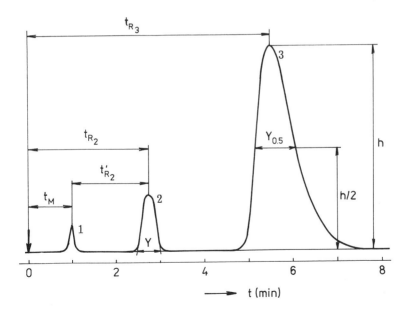

FIGURE 3.1 Hypothetical chromatogram. 0, injection point; 1, peak of unretained solute; 2, Gaussian shape peak; 3, tailing peak.

continuous liquid-liquid partition without use of any solid support. The fractionation is carried out in PTFE tubing, creating a spool which rotates in a planetary motion around a central axis. The motion causes the heavier solvent to move toward one end of the coil and the lighter phase to pass to the other end with simultaneous mixing of both phases. One end of the coil is connected to a pump delivering the solvent used as the mobile phase. The outlet from the other end of the coil is led to a detector and fraction collector. Even though countercurrent chromatography is not frequently applied, it has proved its capability for alkaloid separation (2). Samples as large as a few hundred milligrams can be separated in a single operation and no purification of raw sample is necessary.

C. Nomenclature

The basic nomenclature of column chromatography is shown by a hypothetical chromatogram (Fig. 3.1). The point of sample injection is usually marked on the chromatogram. The time between the injection point and the peak maximum is called retention time, denoted as t_R. This symbol is frequently completed by the addition of a subscript representing the corresponding peak, e.g., t_{R_2}, t_{R_3}. For an unretained substance the time between the peak maximum and the injection point is called dead time, symbolized by t_M. The time difference between the peak maximum and dead time is called adjusted retention time t_R'. All the above-mentioned quantities are measured in seconds or minutes.

The rate at which mobile phase passes through the column is expressed as volummetric flow rate or, simply, flow rate (ml/min or ml/hr) with the symbol F_m. Alternatively, the linear velocity of mobile phase denoted as u is applied. This quantity is equal to the ratio L/t_M, where L is the column length. The linear velocity is usually given in cm/sec or mm/sec. If retention time is multiplied by flow rate, we obtain retention volume V_R and, similarly, multiplication of dead time and adjusted retention time by flow rate gives dead volume V_M and adjusted retention volume V_R', respectively. All the volume quantities are measured in milliliters.

In gas chromatography, the calculation of retention volume is more complex due to the compressibility of the gaseous mobile phase. The volummetric flow rate is usually measured at the outlet of the column under atmospheric pressure, while the inlet of the column is always operated at a pressure higher than atmospheric. To enable comparison of retention volumes at different column inlet pressures, correction to mean column pressure is applied multiplying retention volumes by compressibility factor j. This factor is defined as

Chromatography

$$j = (3/2) \frac{(p_i/p_o)^2 - 1}{(p_i/p_o)^3 - 1} \tag{3.1}$$

where p_i and p_o are corresponding pressures at the inlet and outlet of the column, respectively. The product of compressibility factor j and adjusted retention volume V_R' is called net retention volume V_N.

For the purposes of column efficiency determination and quantitative analysis, the additional data may be taken from Figure 3.1. They are peak height h, peak width Y, and peak width at half-height $Y_{0.5}$. The measurement of h and $Y_{0.5}$ is obvious. On the other hand, peak width measurement requires the construction of tangents to the points of inflection of the Gaussian curve and measurement of the length of their intersection with the baseline. Peak area, necessary for quantitative analysis, can be approximated from peak height multiplied by peak width at half-height. Calculation of column efficiency is discussed in the following section.

II. BASIC THEORY

Chromatography is a separation method whose aim is the resolution of mixture components. The resolution of two adjacent peaks can be more or less complete, and it is useful to express this fact quantitatively. Therefore, the resolution $R_{i,j}$ is introduced and defined as

$$R_{i,j} = (t_{R_j} - t_{R_i})/0.5(Y_j + Y_i) \tag{3.2}$$

where t_{R_j} and t_{R_i} are the retention times of components j and i, while Y_j and Y_i are the corresponding peak widths. Because resolution is a dimensionless quantity, peak widths have to be given in the same time units as retention times. If a chromatogram plotted by the recorder is evaluated, retention distances rather than retention times are measured, and both peak widths and retention distances are employed in equal units of length.

The improvement of resolution may be achieved either by increasing column efficiency or by changing the relative retention of two unresolved components. The situation is depicted in Figure 3.2. Figure 3.2A shows two partially resolved components. Column efficiency can be improved by varying the kinetics (Fig. 3.2B), whereas relative retention can be affected by changing the thermodynamics (Fig. 3.2C). The separation can be improved, of course, by combining the two mentioned procedures.

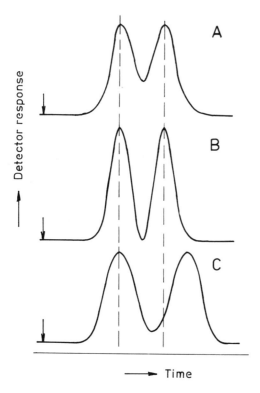

FIGURE 3.2 Resolution of two components. A, partially resolved peaks; B, resolved peaks by reducing peak width; C, resolved peaks by changing retention time.

A. Thermodynamic Aspects

As stated previously, chromatography is based on equilibrium distribution of a substance between stationary phase and mobile phase. Supposing that the equilibrium is achieved, the solute distribution may be described by means of distribution coefficient K_D defined as follows:

$$K_D = (c_A)_s / (c_A)_m \tag{3.3}$$

where $(c_A)_s$ and $(c_A)_m$ are the average concentrations of solute A in stationary and mobile phases. The concentrations are expressed in moles per unit of volume that is reasonable in gas-liquid and liquid-liquid chromatography. In liquid-solid chromatography other units have to be used for $(c_A)_s$. Therefore, in LSC the adsorption coefficient K_{ad} is applied, which is not a dimensionless quantity (for details, see Chapter 5, Section I).

More practical than the distribution coefficient, one frequently encounters in the literature the term capacity ratio or capacity factor k, which is related to K_D as follows:

$$k = (n_A)_s/(n_A)_m = (c_A)_s V_s/(c_A)_m V_m = K_D V_s/V_m \tag{3.4}$$

where $(n_A)_s$ and $(n_A)_m$ are the numbers of moles of solute A in stationary and mobile phases. V_s and V_m are the corresponding volumes of both phases in the column. The relationship between capacity ratio and retention time (or retention volume) of solute t_R and unretained component t_M is given as follows:

$$k = (t_R - t_M)/t_M = (V_R - V_M)/V_M \tag{3.5}$$

A combination of Eqs. 3.4 and 3.5 yields the important relation between retention volume V_R and distribution coefficient K_D. Supposing $V_m \equiv V_M$, we obtain

$$V_R = V_M + K_D V_s \tag{3.6}$$

This is the basic equation; with minor modifications, it is used for all methods of elution chromatography.

Today it is common practice, especially in liquid chromatography, to present the retention data as capacity ratios because their values should not depend on column dimensions. For a given solute, stationary and mobile phases, and temperature capacity ratio should be constant. The only obstacle in k determination is the problem of definition and measurement of dead time or dead volume. Until now, no universal procedure has been devised to measure it. Four main methods have been applied to dead volume estimation: measurement of retention of inorganic salts, injection of modified mobile phase, measurement of retention of unretained organic components, and mathematical determination by linearization of the retention data for homologous series of organic compounds. Relatively reliable in LC is the measurement of retention volume of the labeled component of mobile phase, e.g., deuterated water in reversed phase liquid chromatography. For the precise calculation of capacity ratio, the selection of a compound suitable for dead volume determination is of primary importance. The problems of V_M estimation in liquid chromatography have been reviewed (3). In this book, the reader may find the retention data presented as V_R/V_t, where V_t is the total volume of empty column. This approach was adopted in cases where reliable data for the estimation of V_M were not available.

If two components are to be separated, their capacity ratios must be different. The ability of a chromatographic system to separate two substances from the standpoint of thermodynamics is given by the quantity α, called selectivity ratio (also selectivity factor or retention ratio). This is defined as

$$\alpha = k_2/k_1 = V'_{R_2}/V'_{R_1} \tag{3.7}$$

The selectivity ratio compares the capacity factors of two components; or, simply, it is the ratio of adjusted retention volumes. Owing to the convention that $\alpha \geq 1.0$, the capacity factor (or adjusted retention volume) of the component eluting later is always put in the numerator. The two components can be separated if α differs from 1 and they cannot be separated when the selectivity ratio equals or nears 1. Thus, considering the improvement in the thermodynamics of the chromatographic process, we are generally trying to increase the selectivity ratios. The solution to this problem depends on the chromatographic method. With regard to the separation of alkaloids the improvement of thermodynamics will be briefly discussed for four previously mentioned chromatographic methods. The problem is considered in detail in following chapters.

Among these four methods, gas-liquid chromatography holds a rather exceptional position owing to the behavior of the gaseous mobile phase. This phase behaves as an inert, and at low pressures ordinarily encountered in GLC it also behaves as an ideal gas. Therefore, it is useless to exchange one carrier gas for another, and the only way to improve the thermodynamics of separation is to select a suitable stationary phase and temperature.

The operating temperature is of first rate importance in GLC. According to the Clausius-Clapeyrone equation:

$$\log K_D = - \Delta H/2.3RT + \text{constant} \tag{3.8}$$

The logarithm of distribution coefficient K_D is proportional to enthalpy change ΔH divided by RT, where T is the operating temperature. The enthalpy ΔH is assembled chiefly from the heat of vaporization and, due to the nonideal behavior of the system, the heat of mixing. The temperature dependence may be appreciated from the plot $\log V_N$ vs. $1/T$, since K_D is proportional to the net retention volume V_N. For a constant value ΔH a straight line will arise, and if ΔH of different components are the same, the parallel lines result. The lines of two substances with different ΔH can cross within practical limits of chromatograph operation and then the order of elution is reversed as the temperature is changed. In gas chromatography an increase in temperature always brings about a decrease in retention volume. It can more or less be stated that a temperature increase in 30 K will cut the retention volume in half.

For liquid partition chromatography the mobile phase plays an active role in solute retention. A change of mobile phase composition can bring about a change in component retention and in some cases a change in order of eluted components. The same is obviously valid for the stationary phase. Thus, the improvement of thermodynamics in LLC may be achieved either by the adjustment of mobile phase composition or by selection of stationary phase. On the other hand, temperature is not a particularly important parameter.

What has been stated for LLC should also apply to liquid adsorption chromatography. However, the number of stationary phases is today restricted to silica gel and alumina, which means that there is in practice no chance for changing the stationary phase. Therefore, the selection of mobile phase is the only possibility for improving the thermodynamics in LSC. The temperature variation is usually not applied and, if it is, then it is more by kinetic than thermodynamic considerations.

The last method, ion exchange chromatography, offers more ways to influence thermodynamics. In this method a change of stationary phase is possible, although the choice is limited. The selectivity of mobile phase may be adjusted in two ways: either by a change of mobile phase pH or by altering its ionic strength. As in other methods of liquid chromatography, the temperature is less important.

B. Kinetic Aspects

The influence of kinetics may be judged by a comparison of parts A and B of Figure 3.2. Optimization of efficiency is generally the first task for improving the separation ability of a chromatographic system.

As solute bands move through the chromatographic bed, broadening occurs which brings about the decrease of resolution and simultaneously an increased dilution of the solute by the mobile phase. The band spreading is a direct measure of column efficiency. This may be expressed as the number of theoretical plates N defined as

$$N = (t_R/\sigma_t)^2 = 16(t_R/Y)^2 = 5.54(t_R/Y_{0.5})^2 \quad (3.9)$$

where t_R is retention time, σ_t^2 is band variance, Y is peak width, and $Y_{0.5}$ is peak width in half-height. Since N is a dimensionless quantity and t_R is given in time units, σ_t, Y, and $Y_{0.5}$ must be measured in terms of the same unit of time. It is convenient to measure the retention distance in column chromatography or along the bed in thin-layer chromatography instead of t_R, and then the values of σ, Y, and $Y_{0.5}$ must be introduced in units of length. Today the chromatographs are equipped with computing integrators. This data-handling device measures the peak areas and heights, retention times, and so on. In this case calculation of the number of theoretical plates is performed by making use of the following equation:

$$N = 2\pi(d_R h/A)^2 \tag{3.10}$$

where d_R is the retention distance, A the peak area, and h its height.

The separation efficiency expressed as N is not a constant value but rather depends on the retention time or capacity ratio of the components used for calculation. By combination of a number of theoretical plates and the capacity ratio we obtain a single parameter, the number of effective plates N_{eff}:

$$N_{eff} = N(k/k + 1)^2 = 16(t_R'/Y) \tag{3.11}$$

in which t_R' is the adjusted retention time and k the capacity ratio of solute. N_{eff} better indicates the efficiency of a column, but it is not used as often as N.

For the calculation of N by means of Eqs. 3.9 and 3.10 a Guassian peak shape is supposed. However, in the chromatography of strongly polar or ionizable compounds such as alkaloids, a distortion of peak shape is frequently observed. To quantify peak tailing several procedures have been proposed. A simple, useful parameter called the peak asymmetry factor (PAF) (4) may be evaluated from the distance of the points at 10% peak height (see Fig. 3.3);

$$PAF = CB/AB \tag{3.12}$$

For peaks showing an asymmetry, the calculation of number of theoretical plates is rather problematic.

The number of theoretical plates obviously depends on column length L. To enable a direct comparison between columns of different length, another measure of column efficiency has been defined. This is the plate height H, also called the height of an equivalent theoretical plate:

$$H = L/N \tag{3.13}$$

where H is usually given in millimeters. The plate height can also be expressed as a dimensionless quantity that allows for differences in particle size. The quantity is called the reduced plate height <u>h</u>

Chromatography

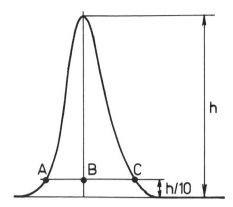

FIGURE 3.3 Peak asymmetry evaluation.

$$\underline{h} = H/d_p \tag{3.14}$$

where d_p is particle diameter for packed columns or inner-tube diameter for open tubular columns. The reduced plate height is often used in connection with reduced velocity ν, the second dimensionless parameter, taking into account the particle size

$$\nu = ud_p/D_m \tag{3.15}$$

where u is linear velocity and D_m the solute diffusivity in the mobile phase.

After introducing all measures for quantifying column efficiency, let us begin to examine the parameters influencing efficiency.

During the band movement through a bed the spreading occurs due to four different processes. These phenomena contribute to H such that the overall plate height can be expressed as the sum of incremental plate heights arising from those band-spreading processes.

1. Longitudinal molecular diffusion, yielding the contribution H_L to the total band broadening, occurs in the direction of flow. In liquid chromatography, owing to the low values of solute diffusivities, H_L is assumed to be zero.
2. Eddy diffusion with increment H_F arises from turbulence due to the different streams of mobile phase through the bed, which consists of channels of various diameters. According to this term, the band broadening is caused by large particles and heterogeneous packing. H_F is zero for open tubular columns.
3. Resistance to mobile phase mass transfer H_M occurs within a small flow stream, e.g., between two particles. The molecules in the

center of a stream travel faster than those near the edge. High values of diffusivity in gaseous phase make this term less important in gas chromatography.
4. Resistance to stationary phase mass transfer H_S is controlled by diffusion for liquid stationary phases and by adsorption-desorption kinetics in adsorption chromatography. The thickness of stationary phase causes the different solute molecules to travel different distances within the film.

The list of the processes contributing to plate height is not comprehensive. For details the reader is referred to the literature (5). Among the other phenomena, stagnant mobile phase contribution should be mentioned, although in terms of the methods used in alkaloid separation, it is less important. To facilitate efficiency estimating, the terms contributing to the overall plate height are listed in Table 3.2. There three different cases are recognized from the standpoint of rate equation. The calculation of H for the respective methods is carried out by simple addition of individual terms. The data needed for calculation are either known (d_p, d_f, $\bar{u}$, r, k) or can be estimated on the basis of literature values (D_m, D_s, k_L, k_F, k_M, k_S).

Nevertheless the calculation of individual terms is not often done. More frequently, the overall H vs. linear velocity plot is constructed using the data obtained from chromatograms. Typical curves for GC capillary and packed columns are shown in Figure 3.4 and for LC using packed columns in Figure 3.5. The plots in Figure 3.4 are remarkably different in both efficiency and shape of curve. The higher efficiencies of capillary columns are well known. One may expect capillary columns to give H two or three times smaller than that of packed columns. The shape of the curve is dependent on both the column performance and the carrier gas chosen. The curve for nitrogen generally rises more steeply for high values of $\bar{u}$ that for helium. In practice a column is usually operated at a linear velocity greater than the optimum in order to decrease retention time. This is especially true for helium, since its curve is often fairly flat and the increase in mobile phase velocity brings about only a small decrease in efficiency. Thus, for high-speed operations, helium as carrier gas is preferred. The plot for LC using microparticulate packing (Fig. 3.5) is essentially linear in the range of practical applications. At very low velocities the molecular diffusion exerts its influence and efficiency decreases. The chosen linear velocity is the compromise between efficiency, required analysis time, and acceptable operating pressure.

Chromatography

TABLE 3.2 Terms Contributing to H for Different Methods and Column Types

Method and column type	H_L	H_F	H_M	H_S
Gas chromatography packed columns	$k_L D_m / \bar{u}$	$k_F d_p$	0	$(k_S d_f^2 \bar{u}/D_s) k/(1+k)^2$
Gas chromatography open tubular columns	$2 D_m / \bar{u}$	0	$\dfrac{(1+6k+11k^2)}{24(1+k)^2} \dfrac{r^2 \bar{u}}{D_m}$	$(k_S d_f^2 \bar{u}/D_s) k/(k+1)^2$
Liquid chromatography packed columns	0	$(1/k_F d_p + D_m/k_M d_p^2 u)^{-1}$		$(k_S d_f^2 u/D_s) k/(1+k)^2$

Note: k_L, k_F, k_M, k_S—factors determined by the configuration (or nature) of the stationary phase (or packing). D_m, D_s—diffusivity of solute in the mobile phase and the stationary phase, respectively. d_f—average thickness of the stationary phase. d_p—particle diameter. $\bar{u}$—average linear velocity obtained multiplying linear velocity u by compressibility factor j. r—inner radius of the tube.

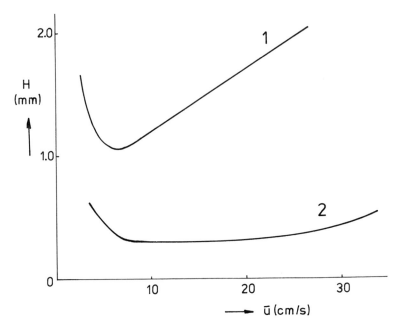

FIGURE 3.4 Plots of H vs. linear velocity in gas chromatography. 1, packed column, carrier gas nitrogen; 2, capillary column WCOT, carrier gas helium.

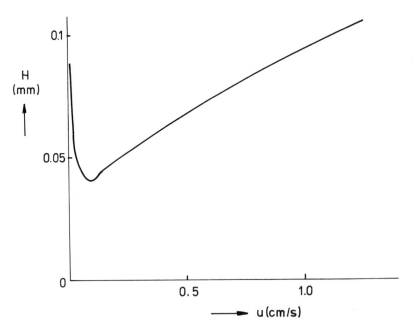

FIGURE 3.5 Plot of H vs. linear velocity in liquid chromatography. Spherical particles, $d_p = 7$ μm.

Chromatography

III. INSTRUMENTATION IN COLUMN CHROMATOGRAPHY

The apparatus for column chromatography is composed of several parts common to both GC and LC. These parts are shown schematically in Figure 3.6 and are briefly discussed in the next paragraph.

A cylinder with compressed helium or nitrogen serves in GC as the source of pressurized mobile phase. Other gases are applied rather exceptionally. In LC the mobile phase is pressurized by means of a reciprocating pump or motor-driven syringe. In some equipment two pumps are joined in a gradient device that enables one to change the composition of the mobile phase during a run. The components of a sample are then eluted by an eluent with continuously changing polarity (pH or ionic strength) at a constant flow rate.

Flow control in GC is realized by means of a combination of pressure regulator and metering valve. In many chromatographs this parameter is controlled by a microprocessor. Flow in LC is simply adjusted by pump programming.

For packed columns in both GC and LC on-column injectors can be used. Liquid samples are normally introduced by microliter syringe. Often in LC a stop-flow device must be used due to high pressure at the inlet. To overcome this problem the use of a sample

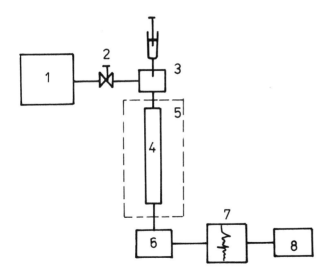

FIGURE 3.6 Scheme of a chromatograph. 1, source of pressurized mobile phase; 2, flow rate control; 3, injector; 4, column; 5, oven; 6, detector; 7, recorder; 8, data station.

TABLE 3.3 Typical Parameters of Analytical Columns Used in GC and LC

Type of column	Length L (m)	Inner diameter (mm)	Particle size d_p (μm)	Plate height H (mm)	Material
GC packed columns	0.9–3.6	2.0–4.0	90–250	0.5–2.0	Glass, stainless steel
GC capillary columns	10–60	0.25–0.75	0.2–2.0[a]	0.1–0.3	Fused silica, glass
LC packed columns	0.05–0.25	2.0–6.0	3.0–10.0	0.02–0.08	Stainless steel, glass

[a]Film thickness.

Chromatography

TABLE 3.4 Chromatographic Parameters, Symbols, and Relations[a]

(a) Parameters straightly measured

Symbol	Parameter	Dimension
d_p [dp]	Particle diameter	μm
F_m	Flow rate	ml/min
d_R	Retention distance	mm
h	Peak height	mm
L	Column length	mm, m
p_i	Inlet pressure	bar [MPa]
p_o	Outlet pressure	bar [MPa]
r	Column inner radius	mm
T [T_c]	Column temperature	K
t_M [t_o]	Dead time	sec
t_R	Retention time	sec
Y [w, W]	Peak width	mm
$Y_{0.5}$ [$w_{0.5}$, $W_{0.5}$]	Peak width at half-height	mm

(b) Parameter calculated

Symbol	Parameter	Dimension	Relation
A	Peak area	mm^2, cm^2	$A = h \cdot Y_{0.5}$
H [HETP]	Plate height	mm	$H = L/N$
j	Compressibility factor [pressure correction factor]	—	see Eq. 3.1 (GC only)
k [k', K]	Capacity ratio [capacity factor, partition factor]	—	$k = t'_R / t_M$
N [n]	Number of theoretical plates	—	$N = 5.54\, d_R^2 / Y_{0.5}^2$

TABLE 3.4 (b) (Continued)

Symbol	Parameter	Dimension	Relation
t'_R	Adjusted retention time [corrected retention time]	sec	$t'_R = t_R - t_M$
u [v]	Mobile phase velocity	cm/sec	$u = L/t_M$
$\bar{u}$ [$\bar{v}$]	Average mobile phase velocity	cm/sec	$u = L/t_M$ (GC only)
V_G	Volume of gaseous phase	ml	$V_G = j \cdot F_m \cdot t_M$ (GC only)
V_M	Dead volume [hold-up volume]	ml	$V_M = F_m \cdot t_M$
V_m	Volume of mobile phase	ml	Mostly $V_m = V_M$
V_N	Net retention volume	ml	$V_N = V'_R \cdot j$ (GC only)
V'_R	Retention volume	ml	$V_R = F_m \cdot t_R$
V_R	Adjusted retention volume [corrected retention volume]	ml	$V'_R = V_R - V_M$
V_t [V_T]	Total column volume	ml	$V_t = \pi r^2 L$

(c) Parameters not often calculated

Symbol	Parameter	Dimension	Relation
$(c_A)_m$	Average concentration of solute A in mobile phase	mol/liter	$(c_A)_m = (n_A)_m / V_m$
$(c_A)_s$	Average concentration of solute A in stationary phase	mol/liter	$(c_A)_s = (n_A)_s / V_s$
d_f	Film thickness	μm	
ΔH	Change of enthalpy	kJ/mol	see Eq. 3.8

Chromatography

TABLE 3.4 (c) (Continued)

Symbol	Parameter	Dimension	Relation
H_L, H_F, H_M, H_S	Increments to overall H	mm	see Table 3.2
$\underline{h}$	Reduced plate height	—	$\underline{h} = H/d_p$
K_D	Distribution coefficient [distribution constant, partition coefficient]	—	$K_D = (c_A)_s/(c_A)_m$
N_{eff} [n_{eff}]	Number of effective plates	—	$N_{eff} = N[k/(k+1)]^2$
$(n_A)_m$	Number of moles of solute A in mobile phase	moles	
$(n_A)_s$	Number of moles of solute A in stationary phase	moles	
$R_{i,j}$ [R_s]	Resolution	—	see Eq. 3.2
V_g	Specific retention volume	ml/g	$V_g = V_N/w_L$ corrected to 273 K
V_L	Volume of liquid phase	ml	(GLC only)
V_s	Volume of stationary phase	ml	
w_L	Weight of liquid stationary phase	g	(GLC only)
α	Selectivity ratio [selectivity factor]		$\alpha = V'_{R_2}/V'_{R_1}$
σ	Standard deviation of zone	(e.g., sec)	$\sigma = t_R/\sqrt{N}$
ν	Reduced linear velocity	—	$\nu = d_p \cdot u/D_m$

TABLE 3.4 (Continued)

(d) Other parameters (usually taken from literature)

Symbol	Parameter	Dimension
D_m [D_M]	Diffusivity in mobile phase [diffusion coefficient in mobile phase]	cm^2/sec
D_s [D_S]	Diffusivity in stationary phase [diffusion coefficient in stationary phase]	cm^2/sec
k_L, k_F, k_M, k_S	Factors of rate equation	—

[a]Alternative symbols and names are given in square brackets.

loop, also called an injection valve, is preferred in LC. The instruments operating with capillary columns in GC are equipped either with an inlet splitter injector or, for quantitative work, with a splitless injector. More details on this topic are given in Chapter 4.

Columns, their performance, materials, and packings are discussed in Chapter 4 (GC) and Chapter 5 (LC). Comparison of different analytical columns can be made based on their parameters as given in Table 3.3. The data are relevant to common commercial products only; one could turn to the literature for a much wider variety of parameters. The column may be placed in an oven, mostly used for an air bath. Column thermostatting is inevitable in GC. For many analyses a device for temperature programming is required. Such a device makes it possible to analyse a sample containing components with a wide range of boiling temperatures in a single run. The effect of this technique is frequently compared with that of gradient elution in liquid chromatography. In LC, where temperature is not critical, column thermostatting is often omitted. However, if it is required, then a column may be provided with a water-jacket.

After the column, a detector is the most important part of a chromatograph. Owing to the significance of detectors for the selective detection of alkaloids, they are discussed in detail in Chapters 4 and 5.

IV. SYMBOLS AND RELATIONS

The parameters, symbols, and relations which have been presented in this chapter are summarized in Table 3.4. Some additional parameters are also introduced here. Owing to the existence of various symbols and names with the same meaning, alternative symbols and names are given in square brackets. Note is made when the symbol applies to GC or GLC only.

Table 3.4 is divided into four parts. The first part gives the parameters which may be obtained directly from a chromatogram, by measurement of column, packing, etc. The second part presents the quantities, which are frequently calculated on the basis of data from the first part. The parameters which are not often calculated, are given in the third part. This distinction is based on the experiences of the authors only, and anybody may have another opinion. The fourth part gives few parameters which are usually taken from the literature. The assembly of parameters and their symbols is certainly not comprehensive, but to attempt an extension would be beyond the scope of this book.

REFERENCES

1. J. B. Crowther and J. D. Henion, Anal. Chem., 57:2711 (1985).
2. T. Zhang, J. Chromatogr., 315:287 (1984).
3. R. J. Smith, C. S. Nieass, and M. S. Wainwright, J. Liq. Chromatogr., 9:1387 (1986).
4. J. D. Schieke and V. Pretorius, J. Chromatogr., 132:217 (1977).
5. B. L. Karger, L. R. Snyder, and C. Horvath, An Introduction to Separation Science, John Wiley, New York, 1973.

4
Gas Chromatography

Gas chromatography (GC) is one of the best techniques for the analysis of organic substances. High sensitivity, excellent separation efficiency, and speed are the chief advantages of this method. The analyses take a few minutes or seconds, and may be carried out in nanogram amounts of sample, and sometimes less.

As its name implies, GC uses gas as the mobile phase and the components are separated in the gaseous state. Therefore any component to be determined by GC must be vaporized in a defined way. In practice this means that compounds boiling below approximately 400°C can be analyzed by GC because this temperature corresponds to the upper operation temperature of commercial instruments. It is not feasible to increase this temperature limit due to the restricted thermal stability of the stationary phases. There are some phases only capable of operating above 300°C. From this standpoint, GC analysis of high-boiling organic compounds, which includes alkaloids, is a difficult task which from the beginning had considerably limited applications. During recent years, due to the necessity of solving complex problems in biochemistry, pharmacy, and forensic analysis, the problems have been partially overcome. Nowadays several derivatization techniques are commonly used, making possible the transformation of high-boiling alkaloids into stable, more volatile derivatives suited for GC determination. Progress has also been made in the development of capillary columns, injection systems, and detectors. The problems are discussed in detail in the following sections.

Based on the stationary phase used, gas chromatography is further divided into gas-solid chromatography (GSC), where the adsorption of vaporized solute on solid adsorbent (e.g., silica gel, alumina, active carbon, etc.) is the controlling process, and gas-liquid chromatography (GLC), where the partition of vaporized sample between gaseous mobile phase and liquid stationary phase takes place. Nevertheless, both cases may be considered as marginal and the mentioned processes as dominating. Practical measurements have shown that in GLC simultaneously with partition the adsorption also occurs on supporting material or on the column wall. Gas-solid chromatographic applications are limited to the separation of permanent gases and liquids of low molecular weight. As far as the separation of alkaloids is concerned, only GLC can be used, and therefore in the next text under name of gas chromatography the meaning of GLC will be supposed.

I. COLUMNS AND THEIR PREPARATION

The columns play a crucial role in GC analysis. Two types of analytical columns—packed and capillary—were introduced in Chapter 3 and their parameters in Table 3.3. Capillary columns have the stationary phase imposed either on the inner wall or on the inert support fixed on the inner wall. The remaining space of the column is occupied by the mobile phase so that the ratio V_G/V_L ranges from 50 to 1500, whereas that for packed columns ranges from 5 to 35 (V_G, volume of gaseous mobile phase; V_L, volume of liquid stationary phase). The low value of V_G/V_L appears disadvantageous due to the resistance to mobile phase flow; hence, the possibility of reaching a high efficiency of packed columns by increasing their lengths is considerably limited. The capacity of packed columns is higher than that of capillary columns, and a sample volume of a few microliters can be directly injected onto a column. This is not the case with capillary columns, where special injectors are used to decrease the amount of injected sample. On the other hand, capillary columns 100 m long and longer are common thus making it possible to achieve the efficiency of 10^6 theoretical plates. This is more than two orders of magnitude higher than that of packed columns. Nowadays capillary columns represent the most efficient technique for separating complex organic mixtures and therefore a significant increase in their use is in effect. Owing to their small diameters and thin walls, these columns show a higher reproducibility when operated in conjunction with temperature programming.

A. Packed Columns

Packed columns are constructed mainly of glass and stainless steel. Other materials have been used (e.g., gold, titanium, etc.) but no improvement was observed. For alkaloid separations glass columns should be used. Such columns, made of hard borosilicate glass, by virtue of their transparency enable one to check the quality of the packing and to find dark carbon particles resulting from the decomposition processes of sample or stationary phase. The most important property of glass is its inertness to analyzed components. Only in a limited number of cases does the glass show the adsorption properties as a result of bonding of chromatographed substances on silanol groups occurring on the glass surface. This reaction may be of importance in the determination of components present in trace concentrations and it is possible to eliminate this phenomenon by column silanization.

In this procedure a column is filled with a solution of 5% dimethyldichlorosilane in toluene. After a few minutes the solution is poured out and the column flushed with methanol. During the reaction hydrogen of the silanol group is replaced by an organosilane group, which is less active.

$$-\overset{|}{\underset{|}{Si}}-OH + Cl-\overset{\overset{CH_3}{|}}{\underset{\underset{CH_3}{|}}{Si}}-Cl \longrightarrow -\overset{|}{\underset{|}{Si}}-O-\overset{\overset{CH_3}{|}}{\underset{\underset{CH_3}{|}}{Si}}-Cl \qquad (4.1)$$

The general preference for stainless steel columns lies in their elasticity, but at elevated temperature the catalytic decomposition of thermally unstable compounds occurs on the metal surface.

1. Supports

The support surface is loaded with a thin film of stationary phase. The nature and properties of the support may decisively influence the efficiency of the chromatographic column.

The specific area of GC supports varies from 0.1 to 20 $m^2 \ g^{-1}$ and no micropores are acceptable. Aluminosilicates are the most frequently used materials, and they will be covered in detail. Supports of another nature, e.g., halocarbons (Chromosorb T, Fluoropak, Fluorolube), glass beads (Corning), silica gel and active carbon (Carbopack) have not been applied to the GC separation of alkaloids, to the best of the author's knowledge.

Two types of aluminosilicate can be distinguished (Table 4.1). In the first type so-called pink supports are included, which are prepared by the grinding of firebrick material. These supports have very good mechanical strength, have high specific area (pores about 1 μm), and are capable of carrying up to 30% of stationary phase. They are preferentially used for the separation of nonpolar substances. When applied to the analysis of polar compounds, they show rather unsymmetric peaks. The second type includes so-called white supports. They are softer, with low specific area (pores about 8 μm), showing low adsorptivity, and hence they are recom-

TABLE 4.1 Some Commercial Supports for GLC

Support	Untreated	Acid-washed	Acid-washed silanized[a]	Special supports[b]	Maximum loading (%)	Manufacturer
Pink	Chromosorb P NAW	Chromosorb P AW	Chromosorb P AW DMCS		30	John Manville
	Chromosorb A NAW				25	
	Gas Chrom R	Gas Chrom RA	Gas Chrom RZ			Applied Science
White	Chromosorb W NAW	Chromosorb W AW	Chromosorb W AW DMCS	Chromosorb W HP	15	John Manville
	Chromosorb G NAW	Chromosorb G AW	Chromosorb G AW DMCS	Chromosorb G HP	5	
	Gas Chrom S	Gas Chrom A		Gas Chrom Q Gas Chrom P[c] Gas Chrom Z[d]		Applied Science
	Anakrom U	Anakrom A	Anakrom AS	Anakrom Q		Analabs
		Supelcon AW		Supelcoport		Supelco

[a] Treated with dimethyldichlorosilane (DMCS).
[b] Specially purified supports of high performance, if not specified otherwise.
[c] Base-washed.
[d] Base-washed, DMCS-treated.

mended for the separation of polar compounds. Further treatment may affect the support properties to a greater extent.

2. Support Deactivation

Metal oxides, commonly present in aluminosilicate supports, can act as catalysts on separated substances. These metal traces can be removed by acid washing using hot hydrochloric acid. The acid is then removed by washing with distilled water until the filtrate shows neutral reaction. If removal of the last traces of acid is required, the support should be further washed with a methanolic solution of potassium hydroxide followed by washing with distilled water. It is necessary to distinguish between "base-washed support" and "base-containing packing," the latter term referring to the stationary phase containing alkaline hydroxide. This kind of stationary phase is often applied to the separation of basic compounds such as amines and other nitrogen bases including alkaloids.

To achieve the complete deactivation of silanol groups, the support has to be carefully silanized. The acid-washed support is allowed to come in contact either with concentrated silane, with silane dissolved in toluene, or with organosilane vapors. For GC of alkaloids such inert supports as silanized aluminosilicates (e.g., Gas Chrom Q, Supelcoport, Chromosorb W HP, etc.) are preferred. Nevertheless, in some cases of alkaloid separation pink supports have also been successfully applied.

3. Packing Preparation

Essentially there are two ways to coat the support. In the first procedure, the support is added to the solution of stationary phase dissolved in a suitable solvent. Then the mixture is well mixed by gentle stirring and the solvent evaporated under reduced pressure. In the second procedure, the mixture of support and dissolved stationary phase is filtered and the wet-coated support dried.

As was already mentioned, stationary phases containing alkaline hydroxide are often prepared for alkaloid separation. In such cases the support is added to the solution containing both stationary phase and potassium hydroxide. The subsequent solvent evaporation has to be performed in the absence of air, mostly in the atmosphere of nitrogen. Carbon dioxide from air can react with potassium hydroxide by forming carbonate, which is less active as a tail reducer than hydroxide. The selected stationary phase has to be compatible with potassium hydroxide present at the support surface. An example would be the partial decomposition of silicone phases in the presence of alkaline hydroxide. On the other hand, Carbowax and Apiezon L are often used in combination with KOH without any problem.

Gas Chromatography

If the support is to be coated with two stationary phases, they can be introduced simultaneously, provided that a suitable solvent for both phases exists. In the adverse case the phases must be introduced successively using an evaporation procedure.

Empty columns can be filled either under vacuum or under pressure. If phases containing potassium hydroxide are involved, the filling under pressure of nitrogen is preferred. Before the first use, the packed columns have to be conditioned by programmed heating in the flow of carrier gas. The columns are heated to a temperature higher than that of operation, but the temperature limit of the stationary phase must be considered. The time required for column conditioning is usually at least 24 hr.

B. Capillary Columns

The introduction of capillary columns into GC has brought a qualitative jump in efficiency. In comparison with packed columns having 10^3-10^4 theoretical plates, a capillary column 50 m long easily accomodates about 250,000 plates. Use of these columns brings also about changes in methodology and instrumentation (from sample injection up to detection). The small column diameter made it possible to decrease the sample size while simultaneously increasing efficiency. In the following sections the changes in instrumentation arising from the application of capillary columns will be briefly discussed.

1. Sample Injection Systems

Due to the small void volume of capillary columns, the injection systems commonly used for packed columns cannot be employed. Even sample volumes of an order of 10^{-1} μl are too high for capillary columns, and this fact slowed down the development of capillary chromatography at the beginning. At that time the inlet splitter injection was used, which was simple from the standpoint of construction but inconvenient for trace analysis. Only since the first splitless injection system was developed (1,2) has the full application of capillary GC in trace analyses become possible. Nowadays the following injection systems are used: (a) inlet splitter injection, (b) splitless injector, (c) direct injector, (d) on-column injector, and (e) solid injector.

The inlet splitter injector is based on a simple splitter (capillary or needle valve) inserted between the injector chamber and the capillary column. The major part of a vaporized sample is vented in atmosphere and a smaller part is led onto column. The described system is inconvenient for quantitative analysis of mixtures boiling in a wide range of temperatures because of discrimination of the low-boiling components. Some manufacturers have improved the splitter injector by thorough mixing in the splitter, which guarantees reproducible and nondiscriminatory sample introduction. The split ratio can be varied by keeping the column pressure constant and adjusting the mass flow to obtain the desired split ratio. Splitter injectors are best for concentrated samples if dilution with a solvent is impossible, e.g., peaks close the solvent, and for sample resolution and peak shape.

The splitless injection technique is based on the separation of low-boiling solvent from the components to be analyzed. The injection is carried out on a cold column with the splitter closed, so that the solvent vapors pass through the column at the same time the analyzed components are retained at the column inlet. After solvent removal, the splitter is opened and the column temperature increased (usually the temperature program is started), which causes elution of the analytes. This injector can also be operated with a solvent that has the highest boiling point of the components present. The splitless technique is advantageous for diluted solution analysis because the components are preconcentrated on the cold column inlet. The important feature of this injector is septum flushing, which removes the remaining solvent vapor and high-boiling components released from the septum.

In the direct capillary injector the sample is injected into a small-volume insert which connects directly to the capillary column. Since there is no splitting involved and no loss of sample, this system is convenient for quantitation. The sample is limited by the capacity of the column and hence this injector is preferably used with wide-bore capillary columns.

The on-column injector is in principle similar to the previous type, but instead of an insert the on-column syringe is employed with a long, flexible, fused silica needle, making it possible to deposit the sample 1.3 in. beyond the column entrance. For analysis of concentrated samples the injection volume can be as small as $0.2 \mu l$. Naturally there is no sample splitting and hence the sample is completely transferred to the column. This injector is convenient for very dilute samples and for thermolabile compounds because the injector does not act as a vaporizer.

The solid injector has been developed to deliver high-boiling components. The sample solution is loaded onto the tip of a glass needle, the solvent is evaporated, and its vapors are vented off through a restrictor just above the inlet of the capillary column. The needle is then dropped down into a heated chamber where solute evaporation takes place. This injector can be used for very high-boiling solutes as well as for very dilute solutions because preconcentration of the solute on the needle is possible. The capillary column in combination with the solid injector always shows lower efficiency due to the certain dead volume of this injection system.

The choice of injection system depends mainly on the sample type. As far as the alkaloids are concerned, any of five injection systems mentioned above is suitable. Some limitations have already been stated with respect to the sample boiling point and sample concentration for solid injector and splitter, respectively. Owing to the high boiling point of the majority of alkaloids and their good solubility in low-boiling solvents, the splitless injection system may be considered very convenient for alkaloid GC determination, especially in cases where quantitative analysis of diluted solution is to be carried out.

2. Capillary Columns

As capillary columns are generally denoted, the tubes have an inner diameter of less than 1.0 mm. The stationary phase is fixed on the inner wall of the tube or in close proximity to it on a porous support. The nomenclature of capillary column types is rather complex and more than 10 different names can be found in the literature. However, essentially two types of columns are distinguished: wall-coated open tubular (WCOT) and porous layer open tubular (PLOT). The WCOT columns have an inner wall coated by a thin coherent film of stationary phase with a layer thickness from 0.1 to 1.0 μm. This type of column exhibits high efficiency and low resistance to the mobile phase flow, but its capacity and thermal stability are low. These ready-to-use columns are offered by a number of manufacturers. The PLOT columns have an inner wall coated with a porous layer and only this layer is coated with stationary phase. The porous layer on the inner wall can be created either chemically by etching of the wall or by depositing pulverized material. This type of column exhibits a higher capacity and thermal stability, but its efficiency is usually lower than the WCOT type. PLOT columns can rarely be found on the market.

In the past the materials of capillary columns varied from stainless steel to plastics. Nowadays two materials are used: fused silica and glass. The material influences the column quality in two ways. First, the inner-wall roughness and its chemical properties affect the adhesion of the stationary phase. Second, the active centers on the inner wall (silanol groups, metal ions) influence the quality of the separation process (peak tailing, irreversible sorption, etc.). Many scientists have grappled with this problem. Inner-wall preparation for capillary columns is a tedious procedure requiring a great deal of experience, and therefore commercially produced columns are used in most cases.

The present state of capillary column development is associated with the rising use of immobilized stationary phases. Increased thermal stability and resistance to organic solvents are the main features of these phases. These columns may be freed of impurities by flushing with suitable solvents, which is particularly important for chromatographers employing splitless or on-column injection systems. Immobilization in situ for polysiloxane phases is usually performed either by radical reaction (3) or by γ-radiation (4). Along with polysiloxane phases, the immobilization of Carbowax 20M has also been described (5).

C. Operation Conditions

Column selection is the first issue to resolve before the analysis begins. It depends on the sample to be analyzed and relies on the chromatographer's experience. In order to analyze alkaloids by GC with the use of packed columns, either nonpolar (OV-101, SE-30, OV-1, Apiezon L) or a middle-polarity (Carbowax 20M, OV-225) stationary phase can be recommended. Very often used are certain phases (e.g., Apiezon L, Carbowax 20M) in combination with potassium hydroxide, which is especially suitable for alkaloids possessing high pK_a values. White supports that have been carefully deactivated by silanization should be used, and for analytical separations the content of stationary phase should not exceed 5%.

With the use of highly efficient capillary columns, selection of the stationary phase is not so important. It seems that nonpolar silicone phases are increasing in popularity for the separation of alkaloids and the results obtained with them are very good. Commercial WCOT columns give a fast analysis due to much higher permeability and lower resistance to mass transfer than packed columns. The GC separation can be affected by the nature of carrier gas and its linear velocity. In the case of WCOT columns, hydrogen as carrier gas allows the highest linear velocity at low inlet pressures and gives a lower slope of the Van Deemter curve than helium and nitrogen. The only drawback is the risk of explosion in the event of column leak. The advantage of hydrogen is obvious, and for operation with WCOT columns hydrogen is recommended.

If components differing considerably in volatility are to be chromatographed, column temperature programming should be used. Only in this way can good resolution and an acceptable time of analysis be achieved. However, two problems are associated with temperature programming for a packed column. First, the back-pressure inside the column increases as the temperature increases, and so a mass flow controller has to be used to provide a constant flow of carrier gas against a changing back-pressure. Second, the bleed from the column increases as the temperature rises, causing a simultaneous increase in the baseline. To compensate for this, the signal from the analytical column has to be combined with one from the identical reference column connected to a detector polarized in the reverse direction. The resulting differential signal is then recorded as a chromatogram.

Gas Chromatography

II. DETECTORS

For alkaloid determination by gas chromatography, the selective nitrogen-phosphorus detector (NPD) and the electron capture detector (ECD) can be applied, as can the universally utilized flame ionization detector (FID) and the more sophisticated combination GC with mass spectrometer (GC-MS).

A. Flame Ionization Detector

The principle of this well-known detector can be briefly stated: when organic compounds are burned in a hydrogen-air flame, ions are produced. A number of mechanisms have been suggested to describe that, but the net result may be summarized as follows:

$$\text{Organic sample} + H_2 + O_2 \longrightarrow CO_2 + H_2O + (\text{ions})^+ + (\text{ions})^- + e^-$$

$$\text{ions}^- + e^- = \text{current flow}$$

The ions are collected at polarized electrodes and the produced current is amplified and recorded. Often the detector jet, made of metal, serves as one of the electrodes. The FID exhibits a wide range of linearity (up to 10^7), a low limit of detection (approximately 1 ng of component can be determined), and the ability to respond to all organic compounds. However, its response depends on the molecular formula and the structure of the substance being detected. As a rule the response of a compound decreases with the increasing content of hereroatoms in the organic molecule. As far as alkaloids are concerned, their responses are not much different, at least for the purpose of rough estimation. Exceptions may be expected for low molecular weight compounds with a high content of nitrogen, such as purine bases. In any case, for accurate quantitative analysis the calibration and determination of response factors are necessary.

B. Nitrogen-Phosphorus Detector

There are two types of specific detectors for organic nitrogen and phosphorus compounds: the alkali flame ionization detector (AFID) and the thermoionic-specific or bead detector (TSD).

The AFID is similar in construction to the FID. The column effluent is mixed with a supply of hydrogen immediately below the jet provided with the salt tip. Independently supplied air supports combustion. The flame fulfills the dual purposes of burning the sample and volatilizing the alkali metal salt. The AFID operates on the principle that enhanced response to nitrogen or phosphorus compounds is obtained if alkali metals such as potassium, cesium, and especially rubidium are introduced to the flame. The precise mechanism of the reactions occurring within the flame is not yet fully understood. Compared with FID, this detector uses a hydrogen-rich flame and the sensitivity and selectivity are considerably influenced by changes in the flow rates of various gases. The response also depends on the position of the salt tip relative to the flame.

The TSD is the alternative option. Instead of a salt tip, there is an electrically heated ceramic-alkali bead situated above the jet flame. A mixture of gases surrounding the bead provides an environment in which the decomposition of organic nitrogen and phosphorus compounds produces ions. The ions are collected and the current flow is measured using a standard FID electrometer.

The linearity of the N-P detector is typically 10^3-10^5, and this combined with its high selectivity toward organonitrogen and organophosphorus compounds means that it is very convenient for the analysis of alkaloids. However, this detector is less easy to operate due to the need for careful optimization. Moreover, one has to consider the possible alteration to the detector response caused by other heteroatoms present in the sample molecule. Of the two types of N-P detectors, the thermoionic-specific type is easier to operate and consequently it appears to be gaining in popularity.

C. Electron Capture Detector

In principle the electron capture detector (ECD) is an ionization chamber where ionization of gases by a radioactive β-emitting source (^{63}Ni or tritium) takes place:

$$N_2 + \beta \longrightarrow N_2^+ + e^-$$

carrier gas   low energy
          high energy

The electrons produced are collected by the application of a potential between the radioactive source and the anode situated in the upper part of the chamber. The established current is known as the "standing current." When an electrophilic compound (usually one containing fluorine, chlorine, bromine, or iodine) is eluted into the chamber, a reaction occurs between the compound and the free electrons:

$AB + e^- \longrightarrow AB^-$ nondissociative reaction

$AB + e^- \longrightarrow A + B^-$ dissociative reaction

The result of these reactions is a replacement of fast-moving electrons by slow-moving negative ions. These ions have a higher probability of recombining with positive ions and in this way the standing current of the detector is decreased.

There are three existing modes of ECD operation: (a) DC mode, (b) pulsed mode, and (c) pulse-modulated mode. The ECD is extremely sensitive to halogenated compounds, being capable of measuring as little as picogram amounts of solute, but its linearity is poor (usually about 10^3). The pulse-modulated mode has increased the linearity of ECD to 10^4. This detector is very convenient for determining the trace amounts of alkaloid derivatized with a reagent containing three or more atoms of fluorine or chlorine. It should be remembered that sample amounts greater than 10^{-8} g should not be used since an overloaded ECD may take a very long time to recover.

D. GC-MS Coupling

The GC-MS coupling is the most effective technique for analyzing mixtures of organic compounds. Mass spectrometry provides a very high sensitivity, as it is capable of determining a substance present in an amount of approximately 10^{-10} g. The theoretical bases and details of instrumentation can be found elsewhere (6).

In a GC-MS system the operating conditions of both instruments have to conform to the mutual coupling. A molecular separator is inserted between the instruments because the GC operates approximately at atmospheric pressure and the MS at 10^{-5} Pa. The molecular separator separates away the molecules of carrier gas and lets the sample molecules proceed to the ion source. Narrow capillary columns with a volumetric flow rate of carrier gas up to 1 ml min^{-1} can be coupled directly to MS without any separating device. Nevertheless, low flow rates are preferred and, if necessary, the analysis can be accelerated by the use of a higher temperature. A higher flow rate of carrier gas always means a higher dilution of the sample and a loss of sensitivity. The capillary column operating with an inlet pressure 0.2 MPa and an outlet pressure of 10^{-4} Pa loses about 12% of its sensitivity, but the analysis time is several times shorter. For operation with GC-MS coupling, columns containing thermally stable stationary phases are used. Due to the necessity of keeping the column bleed as small as possible, silicone phases are preferentially employed.

III. DERIVATIZATION TECHNIQUES

Gas chromatographic analysis of organic compounds is limited by the boiling point of the substance determined and its thermal stability because by no means should the substance be allowed to decompose. However, many alkaloids exhibit low volatility and low thermal stability, and hence their GC determination is complicated. In some instances it is advantageous to make use of alkaloid derivatization by a suitable reagent. The aim of this operation is not only to improve the physical properties (volatility and thermal stability) of the compounds to be chromatographed, but to increase their sensitivity to detection by a selective detector, mainly ECD. This is done by inserting certain atoms (fluorine, chlorine) in the molecule of the derivatized substance. The sample is usually derivatized in a special reaction tube stoppered with sealing cap or, in some cases, directly in the injection chamber of gas chromatograph, by so-called on-column derivatization. The derivatization techniques which have been used in the GC determination of alkaloids are discussed below.

A. Trimethylsilylation

The trimethylsilylation reaction enables substitution of an active hydrogen in a molecule of solute for the $-Si(CH_3)_3$ group. In this way the polarity of the solute molecule and the capacity for hydrogen bond formation is suppressed, and the derivatized product has a lower boiling point. Silylation reagents generally decompose with water and hence tubes, syringes, and solvents used for reaction must be completely water-free. The water content in the sample and solvents used can be assessed from the peak height of hexamethyldisiloxane (bp 101°C) resulting as a product of the hydrolysis reaction. Silylation takes from several seconds to a few hours. The determination of the time necessary for the reaction to occur is performed experimentally. The reaction mixtures are analyzed by GC 5, 15, and 30 min and 1, 4, and 8 hr from the time the reaction was started. The constant peak height of the reaction product is evidence that the reaction is complete. When trimethylsilyl derivatives are evaporized on a metal surface, their decomposition sometimes occurs. Therefore, an all-glass inlet is recommended and the carrier gas must be carefully dried. The solvent, which must also be carefully dried, can be pyridine or dimethylformamide, and in special cases dimethyl sulfoxide, tetrahydrofuran, or acetonitrile are also employed.

The following reagents were used to prepare trimethylsilyl derivatives of alkaloids:

N,O-Bis(trimethylsilyl)trifluoroacetamide (BSTFA)
N,O-Bis(trimethylsilyl)acetamide (BSA)

1,1,1,3,3,3-Hexamethyldisilazane (HMDS)
N-Trimethylsilyl-N-methyltrifluoroacetamide (MSTFA)
Trimethylchlorosilane (TMCS)
Trimethylsilylimidazole (TMSI)

The BSTFA and MSTFA are supplied either as pure compounds or with the addition of 1% TMCS, which acts as a catalyst. These reagents are advantageous because they form byproducts with low-boiling points. The HMDS reacts slowly but in this case TMCS can also be used as catalyst. The TMCS reacts very slowly, but, as mentioned above, in combination with other reagents it is a strong catalyst.

The sample to be silylated has to be dried. Frequently, after evaporation of an extract, the residue is used as the starting material for the reaction. The sample in the mass from 10^{-1} to 10 mg is placed in a stoppered tube and silylating reagent in surplus from 10 to 50 times is added. The use of a tube whose inner wall is covered by Teflon is advantageous. Then the tube is shaken vigorously for 30 sec and the mixture allowed to stand for 5 min. The mixture may be injected directly into the gas chromatograph. If necessary, the reaction kinetics may be established as mentioned previously. In some cases so-called on-column silylation can be performed. This means that a syringe is filled successively with sample solution and silylating reagent, e.g., TMSI, and the mixture is immediately injected into the gas chromatograph.

Trimethylsilylation is applied mainly to the analysis of tropane alkaloids (Section II.D of Chapter 1), Cinchona alkaloids (Section III.A of Chapter 1), isoquinoline alkaloids (Section IV of Chapter 1), and some indole alkaloids (Section VI of Chapter 1).

B. Alkylation

The alkylation reaction proceeds via substitution of active hydrogen on nitrogen or oxygen atom by an alkyl. As far as the alkaloids are concerned, alkyl derivatives with up to five carbon atoms in the alkyl group have been described. Because of their high response in ECD, pentafluorobenzyl derivatives are also prepared.

Methylation may be carried out directly in the injection chamber, where the sample and the reagent are simultaneously charged. Trimethylanilinium hydroxide and a temperature of about 200°C are commonly used for this purpose. In the same way, trimethylammonium hydroxide can be applied. Another means for alkaloid methylation is thermal decomposition of the respective tetramethylammonium salts in the injection chamber (flash heater) of the gas chromatograph at about 350°C. Similarly, by use of tetrapropylammonium hydroxide and tetrabutylammonium hydroxide, propyl and butyl derivatives may also be prepared.

If the alkylation is to be performed prior to the injection, then the same procedure used in the case of silylation is applied. The sample is either evaporated to dryness or a small volume of the sample is placed in a tube and the surplus of reagent added. After mixing and allowing the mixture to stand for a time at room or slightly elevated temperature, the mixture may be directly injected onto the GC column. The time necessary for the reaction to occur is usually determined experimentally. The following compounds are used as alkylation reagents: dimethylformamide + dimethylacetal, dimethyl sulfate + sodium hydride, and diazomethane for methylation; ethyl iodide for ethylation; butyl iodide, dimethylformamide + di-n-butylacetal for butylation, etc.

Pentafluorobenzyl derivatives are prepared by means of pentafluorobenzyl bromide or chloride using an extractive alkylation technique.

Alkylation is utilized as the derivatization technique first of all for purine bases (Section V.D of Chapter 1) as well as for tropane alkaloids (Section II.D of Chapter 1) and Cinchona alkaloids (Section III.A of Chapter 1). However, this concerns only the alkaloids with hydrogen situated on a nitrogen atom.

C. Acylation

As in the alkylation procedure, during acylation a hydrogen situated on a nitrogen atom reacts, but in this case it is substituted for an acyl group (-COR). For the operation with ECD it is advantageous to work with a reagent that possesses a few fluorine atoms in its molecule. From this point of view two types of reagents are distinguished: acylating and perfluoroacylating.

Acetic and propionic anhydrides are convenient reagents for acylating a wide variety of compounds. Usually they are used with pyridine. Trifluoroacetic anhydride (TFAA), bp 40°C, and trifluoroacetylimidazole (TFAI), bp 140°C, are employed for the preparation of trifluoroacetyl derivatives. For pentafluoroacylation can be used pentafluoropropionic anhydride (PFPA) or pentafluoropropionylimidazole (PFPI), while heptafluorobutyric anhydride (HFBA), bp 110°C, or heptafluorobutyrylimidazole (HFBIM) is used for the preparation of heptafluoroacyl derivatives.

Even the acylation can be performed as a precolumn derivatization or an on-column reaction. For the first procedure, about 1-2 mg of sample should be dissolved with 200 mg of reagent in 2 ml of toluene and allowed to react 15-30 min at 60°C in a closed vial. Cold solution may be washed with water, dried over $MgSO_4$, and injected; or the reaction mixture may be evaporated in order to remove the surplus of reagent, and the residue reconstituted in a convenient solvent and injected.

On-column acylation can be carried out in either one or two steps. For the one-step mode PFPI or HFBIM is usually employed, and the sample and reagent are injected simultaneously. In the two-step mode the sample is charged as the first and after about 5 sec the reagent (usually acetic or propionic anhydride) is injected.

Acylation and perfluoroacylation are widely employed for the derivatization of isoquinoline bases, mainly opium alkaloids, but some other alkaloids have also been derivatized in this way.

IV. GC ANALYSIS OF ALKALOIDS

This section deals with some practical aspects of the GC of alkaloids with regard to the specificity of individual groups and subgroups. For this reason the classifications of Chapter 1 will be followed. One point to be noted at the beginning. During the last 20 years the majority of alkaloids were separated by GC, or at least attempts were made to do so. On the other hand, relatively few alkaloids are now analyzed by this technique, assuming that the present literature reflects the situation reliably. The author believes that the advantages of GC are undeniable: higher column stability and reproducibility of retention data, availability of universal detectors with a wide range of linearity, higher absolute efficiency of capillary columns, etc. Conversely, high boiling points and thermal instabilities are the limiting factors of GC. That is why alkaloids are analyzed by GC whenever the use of severe conditions or special devices is not required. Naturally, there are questions as to what are severe conditions and what is a special device. Again, referring to the current literature, a column temperature of 250–280°C should be considered the upper limit. This temperature is convenient for the analysis of, say, opium alkaloids including heroin and papaverine, but is not very suitable for the analysis of Cinchona alkaloids. The author admits that this point of view is dependent on the personal experience of each chromatographer and the equipment available.

Whenever possible, nonderivatized alkaloids should be analyzed. Derivatization is often a time-consuming operation, which carries the risk of losses from an incomplete reaction. However, in the case of trace amounts of the alkaloid to be determined, the preparation of a fluorinated derivative is fully justified. The author should like to emphasize another point regarding the gas chromatography of alkaloids. If GC is to be used, then it should be capillary GC. Particularly when the analyses of plant extracts, biological fluids, and samples of similar character are to be carried out, the high column efficiency is necessary owing to the number of potentially interfering compounds or to the number of alkaloid isomers.

As far as the choice of detector is concerned, FID is very suitable for most analyses. For alkaloids with a higher relative abundance of nitrogen in the molecule, e.g., purine bases, the use of a N-P detector causes an increase in sensitivity and the suppression of components that do not contain nitrogen in their molecules. The use of perfluorinated reagents requires employment of ECD with its preference for trace analysis.

Readers can construct their own opinions based on the following examples and the current literature.

A. Phenylethylamine Derivatives

The most important bases of the phenylethylamine derivatives are Ephedra alkaloids, which can be separated by GC either underivatized or after derivatization. The stationary phases used are methylsilicones (SE-30, OV-1, OV-17) or, more often, the polar Carbowax phases (4000, 6000, 20M) with the addition of alkaline hydroxide.

The GC analysis of ephedrine and related bases as underivatized compounds is possible, but ephedrine and ψ-ephedrine are usually not separated. A simple determination of ephedrine without differentiation of diastereoisomers in pharmaceutical formulations may be carried out by a recommended method (7) using a packed colum 1 m × 4 mm containing 2% Carbowax 6000 on Chromosorb G AW DMCS impregnated with 5% KOH. A similar method was applied by Pickup and Paterson (8) who used a column containing 8% Carbowax 20M on Chromosrob W impregnated with KOH. The column was operated at 180°C. Alm et al. (9) analyzed 27 drugs, among them ephedrine and norephedrine, on two capillary columns of different polarities. The chromatograms obtained from both columns are shown in Figure 4.1A and B. The increased response of caffeine in Figure 4.1B should be noticed because both chromatograms were obtained from one injection by installing two capillary columns in a common split-splitless injector.

If ephedrine and ψ-ephedrine are to be separated, either sample derivatization or the use of a chiral stationary phase should be employed. Gilbert and Brooks (10) separated two couples of diastereomeric ephedrines (ephedrine, ψ-ephedrine, norephedrine, and nor-ψ-ephedrine) as N-acetyl-O-trimethylsilyl derivatives on a 5-m-long column packed with 1% OV-17 on GasChrom Q. At 170°C the four derivatized components were resolved and their indices varied from 1865 to 1945 units. For other examples of GC determination of ephedrine, the reader is referred to Section II.A of Chapter 7.

As far as the other phenylethylamine alkaloids are concerned, Peyote alkaloids (mescaline, N-methylmescaline, hordenine, anhalamine, etc.) were separated using packed columns with different stationary phases (11). To determine mescaline in the plasma of rabbits after intravenous injection, GC-MS was applied (12). Deuterated mescaline as internal standard was added to plasma and the sample extracted with benzene. The benzene solution was evaporated and the residue acylated by the addition of 0.2 ml trifluoroacetic anhydride and 0.2 ml ethyl acetate, and heated to 60°C for 30 min. Then the solution was evaporated and the residue reconstituted in 20 μl of methanol. The analysis was carried out on a column packed with 2.5% QF-1 on Varaport at 195°C. The fragmentographic detection was capable of attaining the detection limit of 5 ng ml^{-1}.

B. Pyridine and Piperidine Alkaloids

Gas chromatography has often been applied to the analysis of Tobacco alkaloids (Section II.A of Chapter 1), and tropane alkaloids (Section II.D of Chapter 1), less frequently to piperidine bases (Section II.C of Chapter 1) and practically not at all to determine dihydro- and tetrahydropyridine derivatives (Section II.B of Chapter 1), although, for example, the boiling point of arecoline is 209°C.

Pyridine alkaloids are bases that are often analyzed by GC and the determination of Tobacco alkaloids by this technique was performed as early as 1958 (13). Since that time many papers on the GC analysis of nicotine and related alkaloids have been published and on the basis of these the following generalization can be drawn.

Tobacco alkaloids are very volatile compared to other alkaloids and can be easily determined by direct GC analysis. Their derivatization, if any, is occasionally carried by means of perfluorinated reagents in order to increase the solute response for ECD. Nicotine and related bases may be determined using packed or capillary columns. As stationary phase for packed columns various Carbowax types have been used on white supports impregnated with KOH, although silicone phases like SE-30 and SE-52 have also been applied. The packed columns are usually 1.0-2.0 m long and operated at a temperature of about 200°C. There are no strict limitations on stationary phase selection in capillary GC. Non-polar silicone phases have become popular and are frequently used today. The present trend toward immobilized stationary phases of the siloxane type can be noted. Along with nicotine, the following alkaloids have been determined: nornicotine, metanicotine, cotinine, myosmine, anabasine, anatabine, and nicotyrine.

The determination of nicotine in tobacco samples and fresh tobacco leaves was described by Severson et al. (14), who compared Soxhlet extraction and sonification for sample preparation with better results for the latter process. The methanol extracts were separated on capillary columns with Carbowax 20M or Superox $_{TM}$-4. For experimental details, see Figure 4.2A and B, that show the chromatograms obtained with FID and NPD. Using an autosampler,

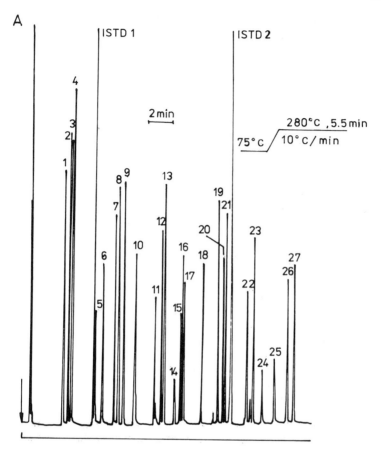

FIGURE 4.1 GC separation of illicit drugs and similar compounds. A, Column 11 m × 0.2 mm coated with SE-54, FID. B, Column 10 m × 0.2 mm coated with OV-215, NPD. Split-injection 1:40, helium 40 cm/sec. 1, amphetamine; 2, phentermine; 3, propylhexedrine; 4, methylamphetamine; 5, norephedrine; 6, ephedrine; 7, phenmetrazine; 8, phendimetrazine; 9, amfepramone; 10, benzocaine; 11, phenacetine; 12, methylphenidate; 13, pethidine; 14, caffeine; 15, phenazone; 16, lidocaine; 17, phencyclidine; 18, procaine; 19, methadone; 20, dextropropoxyphene; 21, cocaine; 22, codeine; 23, diazepam; 24, acetylmorphine; 25, heroin; 26, fluorazepam; 27, papaverine. (Reproduced from J. Chromatogr., Ref. 9, by permission of Elsevier Science Publishers B.V.)

Gas Chromatography

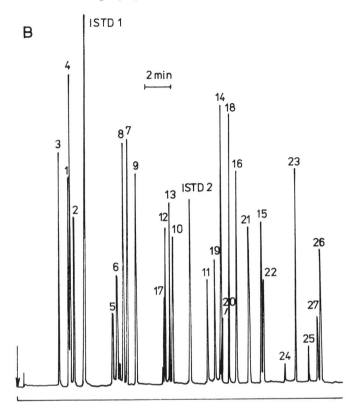

FIGURE 4.1 (Continued)

it was possible to analyze up to 45 samples per day. The analysis of trace alkaloids in tobacco leaf lamina was carried out by Matshushima et al. (15), whereby 17 alkaloids were identified in dichloromethane extracts by means of IR, nuclear magnetic resonance (NMR), and GC-MS techniques. The alkaloids were separated on a 30 m × 0.27 mm glass column coated with OV-101 and operated at a programmed temperature from 100 to 290°C at 3 K/min with flame thermoionic detector. Raisi et al. (16) published the quantitative analysis of nicotine in several cigarette brands using a packed column. For details, the reader is referred to Section II.B in Chapter 7.

Nicotine in tobacco smoke can be determined after its adsorption on cigarette filter or on glass fiber filter. Then nicotine is usually washed out by a suitable solvent and determined in the same way as given above.

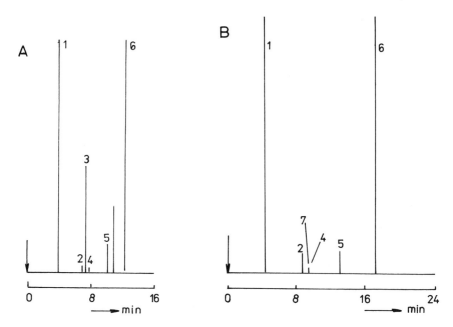

FIGURE 4.2 GC analysis of methanol extract of NC 2326 tobacco. A, FID; B, NPD. Column 15 m × 0.25 mm coated with Superox$_{TM}$-4, temperature program 130–200°C at 4 K/min, split injection 60:1, helium 25 cm/sec. 1, nicotine; 2, nornicotine; 3, neophytadiene; 4, anabasine; 5, anatabine; 6, 2,4'-dipyridyl (internal standard); 7, myosmine. [After Severson et al. (14).]

The determination of Tobacco alkaloids in biological fluids requires sample preparation similar to that for other alkaloids. However, nicotine and related bases may be isolated by steam distillation. This is in principle the method of Burrows et al. (17), further improved by Falkman et al. (18). The sample of blood was made alkaline and distilled with steam. The acidified distillate was purified by extraction with dichloromethane, after which the distillate was made alkaline and nicotine was extracted with dichloromethane. Falkman et al. (18) separated nicotine on a column packed with 8% Carbowax 20 M + 2% KOH on Chromosorb W AW DMCS at 150°C. Thompson et al. (19) used a 12-m-long capillary column deactivated with Carbowax 20M and coated with dimethylsilicone for the determination of nicotine and cotinine in nanogram quantities in tissue homogenates. Two solvents, ethyl acetate and toluene, were used for the extraction of nicotine and cotinine, respectively. Nicotine was determined at a temperature program of 80–150°C at 4

K/min while cotinine required 120–200°C at 4 K min. A N-P detector and MS detection were employed. A capillary column coupled with MS was used by Daenens et al. (20) for the analysis of cotinine in biological fluids. Quantitation of cotinine was effected after liquid-liquid extraction of 0.25–1.0 ml of biological sample with a trideuterated cotinine as internal standard. A column coated with Carbowax 20, a temperature program of 40–224°C, and MS detection were employed. A Carbowax capillary column was also used by Davies (21). For details, see Section II.B of Chapter 7.

The derivatization of nicotine is rarely carried out. Hartvig et al. (22) reported the reaction of nicotine with trichloroethyl chloroformate in the presence of pyridine at 90°C. The resulting δ-trichloroethylcarbamate derivative of nicotine is very sensitive in ECD and picogram quantities may be determined. On the other hand, the product is partially decomposed by heat in an injection chamber.

Piperidine alkaloids rank among the components preferentially analyzed by LC techniques. The only recommendation for GC analysis of piperine can be found in the paper of Verzele et al. (23) who separated black pepper extract by capillary GC. The chromatogram is given in Figure 4.3. Even though they (23) claim the superiority of GC for this purpose, there are no other papers to confirm it.

Tropane alkaloids, as has already been stated, are suitable for GC determination. Atropine and scopolamine can be analyzed directly, but their analysis is accompanied by some problems because they partially dehydrate by high temperature in the injection chamber yielding apoatropine and aposcopolamine, respectively. Besides being dependent on temperature, this process is influenced by the quantity of glass wool in the inlet of column packing. In a column without glass wool the process is substantially suppressed, even at 350°C. For the successful analysis of atropine and scopolamine the use of silanized columns and supports is recommended. The choice of solvent may influence the formation of apo derivatives. Better results were attained with chloroform than with ethanol. For the separation of atropine and related alkaloids, silicone stationary phases, such as OV-17 or SE-30, were preferred by Moffat et al. (24). The determination of atropine and scopolamine in the extract of Belladonna was reported by Wyatt et al. (25), who separated the purified sample on a column packed with 3% OV-17 on Gas Chrom Q. Nevertheless, to avoid thermal decomposition these alkaloids have often been derivatized prior to GC determination. Grabowski et al. (26) applied trimethylsilylation by BSA prior to the analysis of homoatropine methyl bromide in tablets and elixirs. The same kind of derivatization, but by means of MSTFA, was carried out by Liebisch et al. (27) to determine tropine, ψ-tropine, nortropine, scopoline, atropine, and scopolamine.

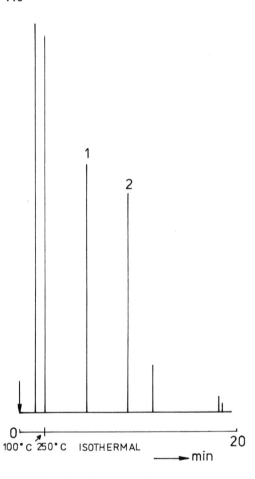

FIGURE 4.3 Glass capillary GC analysis of black pepper extract. Column 25 m × 0.5 mm coated with HTS-OV-1, 250°C isothermal, on-column injection at 100°C, hydrogen 4.6 ml/min. 1, tetrahydropiperine (internal standard); 2, piperine. [After Verzele et al. (23).]

Cocaine represents the type of alkaloid which can be directly separated by GC without the necessity of using severe conditions or derivatization. For the determination of cocaine mostly silicone stationary phases have been used; either nonpolar (SE-30, OV-1, OV-101, SE-52, SE-54) or of middle polarity (OV-25 and OV-225). The column operating temperatures range from 175 to 240°C. Unfortunately, of the two chief cocaine metabolites—benzoylecgonine and ecgonine—the latter one cannot be analyzed directly by GC and its derivatization is inevitable. Ecgonine can be acylated, e.g., by heptafluorobutyric anhydride and benzoylecgonine is often derivatized in the same way. The determination of cocaine in plant material was described by Turner et al. (28). The sample was extracted by refluxing with ethanol and purified by reextraction. The extract was separated on a column containing 6% OV-1 on Chromosorb W. Cocaine was often analyzed on capillary columns, e.g., on a 20 m × 0.35 mm column coated with SE-30 (29), using flash-heater trimethylsilylation in order to determine narcotic drugs. Owing to its structure, cocaine was not derivatized. The column operating temperature was programmed from 50 to 240°C at 5 K/min and the last component, heroin, was eluted at 40 min. Some other papers deal with capillary GC of cocaine (30,31). The paper of Plotczyk (30) discusses the possibility of analyzing cocaine in street drugs using on-column injection on a column 25 m × 0.32 mm with SE-54 and temperature program 80°C for 0.5 min, 80–280°C at 20 K/min. An interesting procedure for determining cocaine in urine, plasma, and red blood cells was described by Javaid et al. (32). They extracted the drug into cyclohexane from a slightly alkaline sample. Cocaine was then reduced with lithium aluminum hydride in order to make possible its acylation by pentafluoropropionic anhydride. Derivatized cocaine was chromatographed on a 10-ft-long column packed with 5% OV-225 on Gas Chrom Q at 110°C. Sensitive detection was achieved with the use of ECD. For other examples of GC determination of cocaine and related bases, the reader is referred to Section II.B of Chapter 7.

C. Quinoline Alkaloids

Among the quinolines the greatest interest has been devoted to GC determination of Cinchona alkaloids. The GC analysis of these bases can be briefly characterized as possible but not advantageous for the following reasons. (a) By using packed columns with nonpolar stationary phases (e.g., SE-30) the components of diastereomeric pairs—quinine-quinidine and cinchonine-cinchonidine—are not separated. (b) To separate them a column with a polar stationary phase (e.g., neopentyl glycol succinate, XE-60) has to be applied, but the operating temperature (usually in the range 220–270°C) will be close to the upper temperature limit of the stationary phase used. Moreover, the application of short columns (1–2 m) packed with a silanized support with low-content (1–3%) stationary phase is necessary.

Kazyak and Knoblock (33) separated Cinchona alkaloids and other drugs on a 6 ft × 4 mm column packed with 1% SE-30 on Anakrom ABS (acid- and base-washed, silanized) at 225°C. Later Brochmann-Hansen and Fontan (34) tested a 3-ft-long column containing 1%

SE-30 on Gas Chrom P at the same temperature. In both cases the chromatographic system did not separate diastereomers. Separation was achieved when the polar stationary phase (1% of XE-60, cyanoethyl silicone) was used at 230°C (34). Brochmann-Hansen and Fontan (35) also showed the possibility of separating Cinchona diastereomers on a column containing Gas Chrom Q successively coated with 1% polyvinylpyrrolidone and 1% neopentyl glycol succinate at 230°C. The differentiation of quinine and quinidine can be attained without separation, as shown by Furner et al. (36), who using GC-MS coupling were able to distinguish these compounds in the selected ion monitoring mode.

Better results in the separation of Cinchona alkaloids have been achieved by the use of capillary GC. Figure 4.4 is a chromatogram

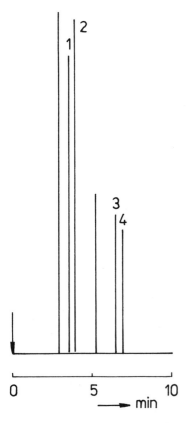

FIGURE 4.4 Glass capillary GC analysis of Cinchona alkaloids. Column 30 m × 0.3 mm coated with RSL-903 (highly polar polyaromatic sulfone), 280°C isothermal, solid injector, hydrogen 5 ml/min. 1, cinchonine; 2, cinchonidine; 3, quinidine; 4, quinine. [After Verzele et al. (23).]

of four fully separated bases which was obtained using a capillary column made of soft glass and etched by hydrochloric acid. The important feature of the procedure, according to the source (23), was the use of a solid injector (falling needle). The applications of capillary columns can also be found in other papers (30,37).

Derivatization of Cinchona alkaloids has also been employed. Thus, Smith et al. (38) prepared trimethylsilyl derivatives of the bases and their dihydro analogs by means of MSTFA and separated them on a column containing 3% OV-225. Methylation by trimethylanilinium hydroxide was also investigated by Midha and Charette (39).

It can probably be concluded that GC is useful when a single Cinchona alkaloid is to be determined, e.g., the content of quinine in heroin, but it is not convenient, say, for the analysis of plant extracts.

D. Isoquinoline Alkaloids

Due to the presence of morphine and other opium bases, the isoquinoline alkaloids have registered the highest number of GC applications. Following the classification of Chapter 1, of the subgroup IV.A, papaverine, some protoberberine and aporphine alkaloids have been analyzed by GC. Of subgroup IV.B, morphine and opium alkaloids have been separated by this technique and more than 100 pertinent references can be found in the literature. Of the last subgroup IV.C, only α-narcotine has been subjected to GC analysis, but it is commonly separated together with other morphine alkaloids.

Papaverine due to its fully aromatic system may be directly analyzed by GC, even though a high operating temperature (250–300°C) is required for its elution. However, there are some losses by adsorption and thus the presaturation of column packing by several injections of papaverine solution is recommended (40). Due to the high temperatures silicone stationary phases (OV-1, OV-17, SE-30, OV-101, etc.) are preferred. In order to determine papaverine in blood and plasma (41) the alkaline sample was extracted with toluene, purified by reextraction, and analyzed on a column containing 2% OV-101 on Chromosorb W HP at 275°C. Before analysis the adsorption centers of the column were deactivated as mentioned above, and papaverine was detected by the use of both FID and NPD.

Aporphine and tetrahydroprotoberberine alkaloids were separated as trimethylsilyl derivatives (42) on column packed either with 3% OV-1 or with 3% OV-17, both on Gas Chrom Q support and operated at 260°C. The GC determination of berberine in urine was reported by Miyazaki et al. (43). Lyophilized sample was extracted with ethanol and the extract purified on polymeric resin column. Berberine was separated as a free base on column with 1% Dexsil 300 on

Gas Chrom Q at 230°C. Mass spectrometric detection with chemical ionization made it possible to quantify even 1 ng berberine/ml of urine.

Morphine and opium bases have been analyzed by means of GC, first in the crude drug itself (opium), in heroin, and then in biological materials and pharmaceutical formulations. For the separation of morphine alkaloids mostly silicone stationary phases (SE-30, SE-52, OV-1, OV-17, QF-1), nonpolar or middle-polar have been used. Retention times on polar stationary phases have been too long even when low amounts of stationary phase were employed. Most of these alkaloids can be analyzed directly by GC, but from the standpoint of quantitation some compounds, particularly morphine, have to be derivatized prior to the separation. Derivatization, mainly trimethylsilylation and acylation, makes it possible to suppress the losses caused by adsorption and to improve the limit of detection. If morphine is to be analyzed directly by GC, a specially prepared column should be employed (44), but the preparation is quite tedious. Trimethylsilylation can be performed either as a precolumn reaction or as an on-column derivatization. As far as acylating reagents are concerned, the perfluorinated kinds are generally employed.

In order to determine morphine and codeine in opium, Nakamura and Noguchi (45) invented a rapid method whereby trimethylsilyl derivatives were prepared by means of BSA before analysis. The sample was then separated on a column with 3% OV-17 on Gas Chrom Q or with 3.8% UCW-98 on Gas Chrom Q. Analysis times were about 7 min at 240°C on the first column and about 5 min at 230°C on the second one. In another paper (46) the problem of morphine determination in opium extract was solved by on-column silylation using TMSI. Before the GC analysis of opium, Sperling (47) separated the sample solution by LC on Celite in two portions. The first fraction was analyzed directly and the second one after trimethylsilylation with BSA. Gas chromatographic determination of the first fraction was performed on a column 6 ft × 4 mm packed with 3% OV-1 on Gas Chrom Q at 240°C using FID. The retention times were: codeine 3.41, thebaine 4.66, papaverine 9.08, and narcotine 21.33 min. The second fraction containing trimethylsilylated morphine was analyzed on the same column at 233°C and the retention time of morphine-TMS was 5.38 min. Some other examples of opium alkaloid analysis using acylated samples and capillary columns can be found in Chapter 7 (see Section II.D, Refs. 209 and 237).

Analyses of heroin and the impurities present in heroin have often been carried out with GC. Heroin itself, giving a sharp, symmetric peak, can be directly separated by this technique. Law et al. (48) analyzed the impurities in heroin by means of packed columns containing 2.5% SE-30. Trimethylsilyl derivatives were separated at 250°C and detected by FID. Codeine, morphine, diamorphine, and papaverine were determined within 10 min. Impurities in heroin were determined by Neumann and Gloger with capillary GC (49). The impurities were extracted into toluene, derivatized by BSTFA or BSA, and separated on a 12 m × 0.2 mm column coated with SE-54. Eluted derivatized alkaloids were determined either with FID or NPD. Another example of heroin analysis is given in Chapter 7 (see Section II.D, Ref. 253).

Morphine and codeine has frequently been determined in blood, urine, and other biological samples. Cimbura and Koves (50) isolated morphine from blood on polymeric sorbent (XAD-2), acylated with acetic anhydride and derivatized alkaloid separated on a packed column containing 3% OV-17 on Chromosorb W HP. The analysis was performed at 240°C using a N-P detector. Edlund (51) determined morphine and codeine in blood by the use of capillary GC. A prepared sample was acylated by PFPA and separated on a 25 m × 0.36 mm column with OV-1 stationary phase at 220°C. A solid injector (falling needle) and ECD were used. The determination of codeine, morphine, and 10 potential urinary metabolites by GC-MS is described by Cone et al. (52). After hydrolysis the alkaloids were isolated by extraction, trimethylsilylated, and separated on columns containing 3% Silar-5CP, 3% Silar-10CP, 3% OV-225, or 3% OV-17 on Gas Chrom Q. The best results were obtained with Silar-5CP at 250°C. The detection method was MS with chemical ionization. Alkaloid concentrations as low as 10 ng ml^{-1} could be determined. GC-MS coupling has frequently been applied to the determination of morphine in serum, cerebrospinal fluids, etc. (53,54). According to Ref. 53, morphine was derivatized with BSTFA and after separation on CP Sil 8 capillary column at 300°C, the masses m/e 340 and m/e 342 were monitored. Reference 54 describes the derivatization of morphine with PFPA and separation on OV-17 capillary column. Other examples of the determination of morphine and codeine in biological fluids can be found in Chapter 7 (see Section II.D, Refs. 58, 65, and 208, Figs. 7.12 and 7.13). The quantitation of some compounds by capillary GC is also discussed in a monograph (55).

E. Alkaloids Derived from Miscellaneous Heterocyclic Systems

Four subgroups (Senecio, lupine, Amaryllidaceae, and purine alkaloids) are presented in this section along with carbazole alkaloids, which were not mentioned in Chapter 1.

The only reference found in the literature (56) gives the conditions of carbazole alkaloid separation: nonpolar silicone stationary phases (OV-17 and SE-30) at low concentration (3%), FID and programmed operating temperature either 220–295°C at 4 K/min, or 150–245°C at 4 K/min. The names of the carbazole alkaloids are given in Chapter 7, Section II.E, Table 7.16.

Senecio alkaloids (also called pyrrolizidine alkaloids) are not among the compounds frequently analyzed by GC. They may be chromatographed either directly or after their derivatization. The first systematic study was performed by Chalmers et al. (57), who distinguished three different types of Senecio alkaloids: nonester bases, esters with monocarboxylic acids, and macrocyclic diester alkaloids. They separated all these alkaloids on 6-ft-long columns packed with 4% SE-30 on Gas Chrom P. For free bases they used the column temperature 140°C, and for both types of esters 205°C. Column silanization and the use of an all-glass injector were emphasized. Pyrrolizidine alkaloids were determined mostly in plant materials. Lüthy et al. (58) determined them in Petasites hybridus and P. albus; Stengl et al. (59) in methanolic extract of Symphytum officinalis. Reference 59 describes the separation of trimethylsilylated alkaloids on a column 1.8 m × 2 mm with 4% SE-30. The stability of pyrrolizidine alkaloids in hay and silage was investigated by Candrian et al. (60), who extracted the material with methanol and determined seneciphylline, platyphylline, jakobine, O-acetylseneciphylline, and jacozine by GC and GC-MS. The components were separated on a 20-m-long capillary column coated with SE-54 which was operated at the following temperatures: 200°C for 1 min, from 200 to 250°C at 5 K/min. Capillary GC was also used by Bicchi et al. (61), who determined nonderivatized pyrrolizidine alkaloids in Senecio inaequidens. Five alkaloids were determined: senecivernine, senecionine, integerrimine, retrorsine, and retrorsine analog. The 20 m × 0.32 mm glass column coated with OV-1 was operated at a temperature program of 120°C for 1 min, 120–230°C at 5 K/min, 230°C hold for 20 min. Deinzer et al. (62) described the analysis of pyrrolizidine alkaloids in goat's milk. In this case the sample was hydrolyzed with $Ba(OH)_2$ and pyrrolizidine alkaloids were converted to retronecine. This was then separated by ion exchanger, derivatized with HFBA, and analyzed with either 1.8 m × 2 mm packed column containing 7% OV-101 on Chromosorb W HP at 130°C or 10 m × 0.2 mm capillary column with SP-2100 at 115°C. ECD and GC-MS coupling were used for detection and identification, respectively.

The lupine (quinolizidine) alkaloid situation is in some ways similar to that of Senecio alkaloids. Gas chromatographic analyses of these bases are infrequently done and when they are done they deal with plant materials. On the other hand, lupine alkaloid derivatization has not beeen applied and the boiling points of some of these substances are relatively low (e.g., sparteine). For their separation silicone stationary phases have generally been used in both packed and capillary columns. Frequently GC-MS coupling has been applied for positive identification. This technique was used by Cho and Martin, who separated and identified 20 lupine alkaloids (63) on a 1.3 m × 4 mm packed column with 10% QF-1 at 230°C. In their paper (63) the five most abundant ions are given for 20 analyzed alkaloids. GC-MS was also used by Keller and Zelenski (64) to determine the alkaloids of Lupinus argenteus. They employed a 2 m × 2 mm packed column with 3% OV-17 on Gas Chrom Q at programmed temperatures from 125 to 265°C at 4 K/min. The retention times of some lupine alkaloids relative to sparteine are given in Table 4.2. As can be seen, the elution order was the same in both cases, even though the polarity of QF-1 is higher than that of OV-17. Capillary GC coupled with MS was used by Priddis (65) for the analysis of extract of lupinseed Lupinus angustifolius. The separation of 12 alkaloids is shown in Figure 4.5. Also, in this case sparteine, isolupanine, and lupanine have been eluted in the same order as that presented in Table 4.2. Toxic alkaloids from the Calcaratus subspecies of Lupinus arbustus were analyzed by Hatfield et al. (66), who used a capillary column 10 m × 0.25 mm coated with OV-17 with temperature programming from 130 to 230°C at 4 K/min.

Amaryllidaceae alkaloids due to their low volatility cannot be determined directly by GC and therefore the methods of LC are preferred for their analysis. Millington et al. (67) separated trimethylsilyl derivatives of lycorine, ambelline, criglaucine, and other Amaryllidaceae alkaloids by GC-MS coupling. In order to analyze the extract of Crinum glaucum, they used a packed column containing 3% OV-1 on Gas Chrom Q with temperature programming from 215 to 250°C.

TABLE 4.2 Relative Retention Times of Some Lupine Alkaloids

Alkaloid	Stationary phase	
	QF-1 (63)	OV-17 (64)
Sparteine	1.00	1.00
Δ^5-Dehydrolupanine	4.57	3.20
α-Isolupanine	5.71	3.36
Lupanine	5.92	3.76
Thermopsine	7.64	5.06
Anagyrine	8.50	5.61

Source: Data based on Refs. 63 and 64.

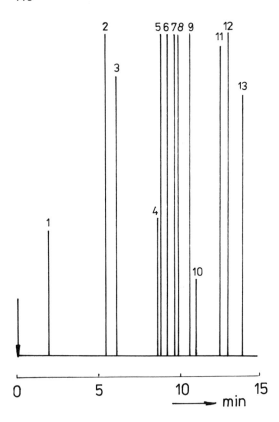

FIGURE 4.5 GC analysis of 12 lupine alkaloids. Column 12 m × 0.2 mm coated with OV-101, temperature program 60°C, hold 1 min, 60–120°C at 30 K/min, hold 0.5 min, 120–260°C at 10 K/min, hold 2 min; split-injection 60:1, helium 2 ml/min, FID. 1, epilupinine; 2, sparteine; 3, caffeine (internal standard); 4, 13-hydroxysparteine; 5, cytisine; 6, tetrahydrorhombifoline; 7, angustifoline; 8, α-isolupanine; 9, lupanine; 10, 17-hydroxylupanine; 11, multiflorine; 12, 17-oxolupanine; 13, 13-hydroxylupanine. [After Priddis (65).]

In contradistinction to Amaryllidaceae alkaloids, purine bases can be directly determined by GC, which has often been used for this purpose. Nevertheless, the derivatization of theobromine and theophylline has also been performed, mainly to improve the limit of detection and to suppress the adsorption losses. From the standpoint of material analyzed, caffeine has been determined in coffee, plant materials, drinks, pharmaceuticals, and plasma. Theophylline, owing to its wide application in medicine, has been assessed mostly in biological fluids and pharmaceuticals. Theobromine has been determined in cocoa and in marketed cocoa products.

The molecules of theobromine and theophylline differ from the caffeine molecule in one methyl group; therefore it is possible by methylation to convert these two alkaloids to caffeine. This method was described by Khayyal and Ayad (68) who used it to determine the sum of three bases as caffeine. For methylation 0.2 M trimethylanilinium hydroxide in methanol was employed and caffeine was determined on a column with 3% OV-17 on Gas Chrom Q at 180°C.

A simple procedure, usually extraction, has been used as a first step for the determination of caffeine in pharmaceuticals, coffee, cola drinks, tea, and so on. Alkalinized sample is extracted by dichloromethane or chloroform and an aliquot portion then charged onto a column. A similar method may be used in order to determine caffeine in plant materials, but the extract has to be purified. Stewart (69) analyzed the extracts from flower buds of several citrus cultivars and from leaves of the Valencia orange. Caffeine was isolated and identified using HPLC, GC, UV spectrometry, and MS. Gas chromatographic analysis was performed on a packed column containing 1% OV-17 on Supelcoport with temperature programming from 150 to 250°C at 8 K/min. Delbeke and Debackere described a simple, rapid method for GC determination of caffeine in urine (70). A urine sample (2 ml) was made alkaline and transferred to Extrelut-1 column for solid phase extraction. Caffeine was eluted with 6 ml of a dichloromethane−methanol 9:1 mixture, and the eluate was evaporated and reconstituted in 0.2 ml of ethyl acetate. The analysis was effected on a 2 m × 2.5 mm column with 3% OV-7 on Chromosorb W HP at 195°C with the use of NPD. In a later paper (71) the automated sample preparation was described as making use of centrifugally based extractor/concentrator with extraction cartridges, but the chromatographic method remained unchanged. Gas chromatographic determination of caffeine in breastmilk and blood plasma was reported by Stavchansky et al. (see Chapter 7, Section II.E, Ref. 48).

Although direct GC analysis of theophylline is possible, e.g., GC assay of theophylline in plasma (72), derivatization is still often applied. Methylation by means of trimethylanilinium hydroxide is generally employed because this reagent makes on-column derivatization possible. This method was used by Dusci et al. (73) for the determination of theophylline in plasma. The sample was then separated on a packed column with 3% OV-225 at 235°C. Floberg et al. (74) used theophylline ethylation for the assay of caffeine and theophylline in plasma. Alkaloids in the presence of tetramethylammonium ions were extracted into dichloromethane and alkylated with ethyl iodide. The prepared sample was separated on a 25-m-long capillary column coated with OV-225 at programmed temperatures from 170 to 210°C. Propylation of theophylline and its metabolites was applied by Tserng (75) in order to determine them in biological fluids. The components were converted to their respective N-propyl derivatives using propyl iodide in dimethylformamide with K_2CO_3 catalysis. The derivatized sample was separated on a packed column with 3% of methylphenyl silicone. The detection was effected by MS using electron impact ionization (70 eV) and selected ion monitoring. The following ions were monitored: caffeine (m/e 194), theophylline (m/e 222), 3-methylxanthine (m/e 250), 1,3-dimethyluric acid (m/e 280), and 1-methyluric acid (m/e 308). The detection limits for individual components ranged from 0.5 to 10 ng.

F. Indole Alkaloids

The alkaloids of this group are for the most part not amenable to analysis by GC and therefore GC techniques has not often been applied to them. Usually the bases must first be derivatized and in many cases this has been effected by trimethylsilyl reagents.

Some simple indole alkaloids were determined in the ethanol extracts of Phalaris species by Audette et al. (76). Prepared samples were chromatographed on a 6 ft × 1/8 in. column packed with 1% DEGS on silanized Chromosorb W AW at 180°C. Among other components, gramine (t_R = 0.52 min), hordenine (1.08), tryptamine (6.58), and bufotenine (74.00) were identified.

Vinca alkaloids, frequently used in pharmaceuticals, were separated after their derivatization with BSA by Gazdag et al. (77). The analysis was carried out on a 1 m × 3.2 mm packed column containing 3% OV-101 on Gas Chrom Q with temperature programming from 200 to 330°C at 5 K/min. Capillary GC for the determination of apovincaminic acid in plasma was applied by Polgár and Vereczkey (78). The sample of plasma with the addition of tetrabutylammonium hydroxide was extracted with chloroform, and the acid was converted to methyl ester with diazomethane and analyzed. A 10 m × 0.25 mm column coated with SP-2100 operated at 220°C and a N-P detector were used for the analysis. The determination of vinpocetine (apovincaminic acid ethyl ester) in plasma by capillary GC-MS was carried out by Hammes and Weyhenmeyer (for details, see Chapter 7, Section II.F, Ref. 415).

Yohimbine alkaloids generally have not been analyzed by GC; the only reference to yohimbine retention data can be found in the paper of Ardrey and Moffat (79) (see Table 4.3).

Gas chromatography is not very convenient for the analysis of Rauwolfia alkaloids. Some of the methods which have been applied for this purpose are neither easy nor reliable. Settimj et al. (80) hydrolyzed reserpine to 3,4,5-trimethyloxybenzoic acid and after reaction with diazomethane the respective methyl ester was determined. Phillipson and Hemingway (81) analyzed the alkaloids of plants of the genus Uncaria by means of TLC, GC, MS, and UV spectrometry. They used a 2-ft-long column containing 5% SE-52 which was operated at 240°C. Among the other components the retention times of the following bases were determined: ajmalicine (t_R = 20.6 min), isoajmalicine (10.7), and tetrahydroalstonine (17.5). Under identical conditions the retention times of harman alkaloids (simple indole alkaloids convenient for GC analysis) were as follows: harman (t_R = 0.6 min), harmine (1.5), and harmaline (1.4). The high retention times and low resolution of short columns show that for the separation of Rauwolfia alkaloids the LC methods are preferable. In order to determine sandwichine, isoajmaline, ajmaline, and tetraphyllicine in Rauwolfia vomitoria, Forni (82) derivatized alkaloids by trimethylsilylation and separated them on a 2-m-long column containing 3% OV-17, but the column temperature had to be increased to 270°C.

Strychnos alkaloids, although possessing relatively large molecules, are suitable for GC analysis. For their separation nonpolar silicone stationary phases have generally been applied. Bisset and Fouché (83) examined the alkaloids of Strychnos nux vomica and S. icaja by the use of a 2 ft × 1/8 in. column containing 5% SE-52 on Aeropak 30 at 250°C. The following retention times were found: spermostrychnine t_R = 2.57 min, strychnospermine 4.39, strychnine 5.85, brucine 15.2, vomicine 21.2, and novacine 32.1. Bisset and Fouché (83) discuss the relationship between the retention times and chemical structure of Strychnos alkaloids. Strychnine and brucine were determined in pharmaceutical preparations by Sondack and Koch (84), who applied a 0.9-m-long column with 3% OV-1 on Gas Chrom Q at 280°C. Miller et al. (85) analyzed alfalfa for strychnine residues using a 1.2 m × 2.0 mm column with 1.5% OV-17 on Gas Chrom Q at 275°C. Clinical and toxicological observations in a nonfatal case are represented by the assay of data of Edmunds et al. (86). A column containing 3% OV-1 was operated at 265°C and, using FID, a detection limit of 0.05 mg liter^{-1} was attained. It can be concluded that Strychnos alkaloids can be separated by GC technique by the use of a short column containing 1-3% of nonpolar silicone phase at a temperature of about 270°C.

Gas chromatography can be denoted as quite inconvenient for the separation of ergot alkaloids with the exception of the clavines. These bases can be directly analyzed by GC, but for those containing a hydroxyl group derivatization is recommended. Agurell and Ohlsson (87) investigated three stationary phases (JXR, SE-30, XE-60) for the separation of clavine alkaloids. Satisfactory results were obtained with a 6 ft × 3 mm column containing 5% XE-60 on Gas Chrom P at 225°C. Gas chromatographic determination of ergometrine, the simplest ergot alkaloid, is described by Sondack (88), who converted it to a trimethylsilyl derivative by means of TMSI in the presence of N-trimethylsilyldiethylamine and dry pyridine. The

TABLE 4.3 Retention Indices of Some Alkaloids

Alkaloid	SE-30 or OV-1 (79)	Methylsilicone 250°C (37)
Acetylcodeine	2510	2530
Ajmaline	2705	—
Apoatropine	2050	—
Apomorphine	2530	—
Atropine	2199	2269
Benzoylecgonine	2570	2582
Berberine	2070	—
Brucine	3280	—
Caffeine	1810	1840
Cocaine	2187	2233
Codeine	2376	2422
Cotarnine	1808	—
Dihydroergotamine	2315	—
Ephedrine	1363	—
Ethylmorphine	2411	2504
Harman	1952	—
Harmine	2291	—
Hyoscine	2303	—
Hyoscyamine	2192	—
Laudanine	2695	—
Mescaline	1688	—
N-Methylephedrine	1400	—
6-Monoacetylmorphine	2537	2545
Morphine (BSTFA derivative)	2454	2513
Nicotine	1348	—
Norephedrine	—	2107
Papaverine	2825	2818

TABLE 4.3 (Continued)

Alkaloid	SE-30 or OV-1 (79)	Methylsilicone 250°C (37)
Physostigmine	1804	—
Pilocarpine	2014	—
Psilocin	1980	—
Psilocybin	2059	—
Quinidine	2784	2807
Quinine	2803	2815
Sparteine	1801	1830
Strychnine	3119	3063
Thebaine	2517	—
Theobromine	1999	—
Theophylline	1999	1962
Tropine	1193	—
Tubocurarine	2495	—
Yohimbine	3269	—

sample was then separated on a 1.2-m-long column containing 1% OV-1 on Gas Chrom Q at 260°C. Derivatized ergometrinine could also be distinguished. Gas chromatographic determination of ergotoxine and other ergot alkaloids including their dihydro derivatives is practically impossible because of their low volatility and thermal instability. Some procedures developed for GC analysis of ergot alkaloids are based on the separation and determination of the products of alkaloid decomposition, which take place in the injection chamber. Van Mansfelt et al. (89) conducted a detailed investigation of the decomposition of six ergot alkaloids at an injection port temperature of 300°C and separated the products on a column with 3% SE-30 at 225°C. This method can be useful, but it is certainly not as convenient and reliable as other direct chromatographic methods might be. Dihydroergot alkaloids were also separated as decomposition products of their methanesulfonate derivatives (see Chapter 7, Section II.F, Ref. 440, Fig. 7.26).

G. Imidazole Alkaloids

Due to its volatility, pilocarpine can be directly separated by GC using polar or nonpolar stationary phases at a temperature of about 200°C. Brochmann-Hansen and Fontan (35) also investigated the GC separation of pilocarpine along with that of other alkaloids. Unfortunately, pilocarpine tends to peak tailing and epimerization giving isopilocarpine. The tailing may be overcome by acylation, e.g., with HFBA, as reported by Bayne et al. (90). Derivatized pilocarpine was separated at 190°C on a 1.8 m × 2 mm column made of silanized glass and packed with 3% OV-17 on Chromosorb W. Using ECD quantities as low as 50 pg of pilocarpine could be determined.

H. Diterpene and Steroidal Alkaloids

Of the two subgroups, the diterpene alkaloids have not been analyzed by GC. As far as steroidal alkaloids are concerned, a few papers deal with the GC determination of free bases (aglycone part of glycoalkaloids). Bushway et al. (91) determined solanidine in milk. Their method is based on alkaline saponification of milk for the purpose of releasing steroidal alkaloid as a free base. After saponification the base was isolated by liquid-liquid extraction and separated on a 1.2 m × 2 mm column packed with 3% OV-17 on Chromosorb W HP at 265°C. Similarly, Van Gelder (92) determined total C_{27} steroidal alkaloid composition of Solanum species. In this case both packed and capillary columns were used. A 1 m × 2 mm column containing 10% SE-30 Chromosorb W HP was operated with temperature programming from 260 to 300°C at 5 K/min. A 50 m × 0.2 mm capillary column coated with CP Sil 5 was operated isothermally at 290°C. Solasodine in Solanum extracts was determined by Marini et al. (93), who used a 3 m × 4 mm column containing 3% SE-30 on Chromosorb W (acid- and base-washed) at 220°C. Tomatidine was used as internal standard. Sometimes the derivatization of steroidal alkaloids is applied primarily by means of fluorinated trimethylsilyl agents, which enable the use of ECD and in this way improve the detection limit.

Gas chromatographic determination of steroidal alkaloids as free bases does not pose a big problem. Conessine can be mentioned as another example in this class (see Chapter 7, Section II.H, Ref. 475, Fig. 7.29). On the other hand, the GC separation of glycoalkaloids with molecular weight of about 1000 seems to be very difficult. The only good method, which was elaborated by Herb et al. (94), involves the permethylation of glycoalkaloids with methyl sulfate, sodium hydride, and methyl iodide with the subsequent separation carried out on a 90 cm × 2 mm column packed with 3% Dexsil 300 on Supelcoport at programmed temperature up to 350°C. It is probable that glycoalkaloids will be determined by a method of LC

Gas Chromatography 125

rather than one of GC, even though LC detection of these compounds is not highly sensitive.

V. QUALITATIVE AND QUANTITATIVE ANALYSIS

A. Qualitative Analysis

Qualitative analysis is based on the assumption that the retention time of any compound remains constant under fixed chromatographic conditions. These conditions are column dimensions, packing, temperature, and carrier gas flow rate. The direct comparison of the retention time of an unknown compound with that of a standard, performed under identical conditions, may be considered the first step toward identification. Often this is done by the simple measurement of sample spiked with the given standard. However, the identical retention time measured on a single column only does not prove the identity of a compound, since there is the possibility that several substances have the same retention time. To enhance the likelihood of identification it is necessary to repeat the measurement with the use of a column packed with stationary phase differing in polarity, since two compounds rarely have the same retention time on different phases. For positive identification some other methods, i.e., spectroscopic ones, have to be combined with GC. Frequently, GC-MS and GC-IR are employed to make a positive identification, especially when the sample is completely unknown.

In order to extend the applicability of the chromatographic data it is possible to express them either as relative retention times or as retention (Kovats) indices.

Relative retention times do not require that chromatographic conditions be set too precisely. The relative retention time of a component is given by the ratio of retention time of this component to that of a compound selected as a reference. Assuming that the variations in operating parameters affect both the sample and the reference retention times, then the variations will be less significant. As the reference may be chosen either one of the components occurring in the sample or one which is added and which emerges at the point when no other components are eluted. Relative retention data are usually calculated using the corresponding adjusted (corrected) retention times t_R'.

$$r_{is} = \frac{t_{R_i} - t_M}{t_{R_s} - t_M} = \frac{t_{R_i}'}{t_{R_s}'} \tag{4.2}$$

The dead time, which must be known, is determined to be the retention time of methane or air. Though relative retention times are more useful and reproducible than simple retention times, their drawback resulting from the arbitrary choice of reference compound is obvious. At certain times, e.g., when a sample containing components that boil at a wide range of temperature is analyzed, two reference compounds can be used for the low- and high-temperature region.

To overcome this shortcoming, a system of retention indices was introduced (95) which uses n-alkanes as a set of reference compounds. The alkanes are well-suited to be standards as a result of their stability and availability. When the logarithms of adjusted retention times of n-alkanes are plotted against the respective carbon numbers, a straight line results. In order to determine the retention index of an unknown compound, chromatography with at least two n-alkanes, eluted before and after the compound is necessary. Then it is possible to construct the calibration graph and to interpolate an apparent carbon number for the unknown component. Conventionally this number is multiplied by 100 and the product is known as the retention index.

The retention index I of an unknown compound can be calculated as follows:

$$I_x = 100n \frac{\log(t'_{R_x}/t'_{R_z})}{\log(t'_{R_{z+n}}/t'_{R_r})} + 100z \qquad (4.3)$$

where t'_R is the respective adjusted retention time, x is an unknown compound, z and z+n are n-alkanes with z and z+n carbon numbers, respectively, and n is the difference in carbon number.

Some authors have attempted to collect the retention indices for certain groups of compounds. Such a data bank for alkaloids, drugs, and some other substances, mostly of toxicological interest, was presented by Ardrey and Moffat (79). There have been 1318 retention indices collected, all of them measured on packed columns containing either SE-30 or OV-1 as stationary phase. The drawback of this collection is the lack of temperature data. Owing to the fact that the retention indices are temperature-dependent with the dependencies being different for various compounds, the above mentioned data (79) can be considered as approximate. The retention indices of some alkaloids discussed in that paper are given in the first column of Table 4.3.

If properly specified, the retention indices should be reproducible with better than 1.0-unit accuracy. From this point of view, the data of Lora-Tamayo et al. (37) are exact. They were obtained using a siloxane-deactivated fused silica capillary column coated with cross-linked methyl silicone at 250°C and a helium flow rate 1 ml

Gas Chromatography

min^{-1}. The data have been extracted from clinical and forensic specimens after 2 years of daily routine analysis. The isothermal mode was chosen in order to reduce the analysis time, since it has been shown that retention indices obtained in isothermal runs coincide closely with those determined under temperature programming. The second column of Table 4.3 gives the data of Lora-Tamayo et al. (37) concerning the alkaloids. As compared with the data of Ardrey and Moffat (79), these figures are generally higher with an average difference of about 30 units.

B. Quantitative Analysis

Quantitative analysis in gas and liquid chromatography is based on measurement of the peak area. The area of any peak displayed on a chromatogram is a measure of the amount of the particular component present in the sample. The first step in quantitative analysis is therefore to estimate the area under the peak, which is proportional to the quantity of the component present. In some cases peak height is measured instead of area. This method is simple and rapid but should be used for the quantitation of symmetric (Gaussian) peaks only. Moreover, the height of a peak depends also on the operating conditions (column temperature, carrier gas flow rate) and the technique of sample injection along with component quantity. If two peaks partially overlap or the baseline shows a drift, the change of baseline course should be taken into account and the use of peak height for quantitation becomes less acceptable.

In earlier times several methods were elaborated in order to estimate peak area, e.g., triangulation, planimetering, cut and weight method, etc. These procedures have since been abandoned, and digital integrators or more sophisticated data station systems make possible the easy and precise measurement of peak area. While the formerly used tedious methods were capable of attaining repeatability from 1.5 to 4.0%, the digital integrators achieve a relative deviation of ±0.5% or better.

Quantitation in GC is based on the assumption that detector response changes linearly with the concentration of sample component at the constant sample volume. Many detectors give a linear response over a wide range of concentrations. The extent of this region varies with the type of detector and with the compound to be determined. Flame ionization detectors exhibit the linearity over seven orders of solute concentration, while electron capture detectors usually attain about three orders. Kucharczyk et al. (96) determined the relative response factors of several compounds for nitrogen-phosphorus detector. Although a value of 6.44 has been found for caffeine, a value 0.72 has been obtained with the same amount of codeine phosphate (compounds relative to carisoprodol with response factor 1.0). Clearly, for N-P detector response the relative abundance of nitrogen is of primary importance. As can be seen from the previous example, for the majority of cases the detector should be calibrated for each component to be analyzed. Moreover, the flow rate stability of carrier gas and the high reproducibility of injected amounts must be attained.

To determine the alkaloids quantitatively, the external standard method (also called the method of absolute calibration) and the internal standard method are preferentially employed. Sometimes the method of standard addition is also used.

The external standard method is based on the successive injections of known amounts of sample and those of pure components to be determined (standard) under identical conditions. The evaluation from peak areas may be performed either by means of calibration curve or by direct comparison. In the first case the calibration curve is obtained by several injections of the same volume of solutions containing various concentrations of standard. The peak areas are then plotted against the absolute amounts of standard. If two or more components of the sample are to be determined, then two or more calibration curves have to be constructed. The second procedure requires determination of only one point of the calibration curve, assuming that the curve intersects the origin of coordinates. The absolute amounts of solute are calculated from peak areas found in the sample analysis using the calibration curves or simply the corresponding response factors for linear dependencies. The external standard method depends on the high precision of injection repeatability. Manual injection using a microsyringe is generally not reproducible enough; therefore an injection valve or autosampler should be used. This method is often used in the quality control of pharmaceutical products, raw materials, plant extracts, etc., where mostly the main components are analyzed. The relative standard deviations vary from ±0.5 to ±2.0%.

The internal standard method is based on the comparison of solute peak area with that of internal standard. This method is especially important in pharmacokinetic and metabolism studies, and it is therefore discussed in detail in Section E of Chapter 5 as the main method of quantitation in liquid chromatography. The chief advantage of this method is its capability of minimizing the errors arising from sample preparation. Repeatability of quantitative results in the order of ±2.0% to ±5.0% can be attained in such a case.

The method of standard addition is based on the addition of a known amount (known volume and concentration) of pure compound to be determined to the known volume of sample. First a given volume of sample is analyzed and after that the same volume of a mixture of sample and the standard solution is injected. The concentration of analyzed component is then calculated as

$$c_i = \frac{V_s c_s / V_i}{[A_{is}/A_i(1 + V_s/V_i) - 1]^{-1}} \tag{4.4}$$

where V_s and V_i are the volumes of standard and sample solutions mixed together before injection, A_i and A_{is} are the peak areas of component i corresponding to the sample solution and to the mixture of sample and standard solutions, respectively, and c_s is the concentration of component i in standard solution. Also in this case the injection repeatability is of primary importance.

There are several sources of error in quantitative analysis. One area of difficulty is the precise determination of peak area, which at present depends mostly on having reliable integration algorithms. Also, injection technique is frequently a decisive factor. The error of injection repeatability should be less than ±5%, even when the internal standard method is employed. When using capillary columns the possibility of component discrimination should be kept in mind. From this point of view, either direct-injection or on-column injection systems should be preferred, but a properly operated splitless injector gives reliable results too.

REFERENCES

1. K. Grob and G. Grob, J. Chromatogr. Sci., 7:584 (1969).
2. K. Grob and G. Grob, J. Chromatogr. Sci., 7:587 (1969).
3. P. Sandra, G. Redant, E. Schlucht, and M. Verzele, J. HRC & CC, 4:411 (1981).
4. G. Schomburg, H. Husmann, S. Ruthe, and M. Herraiz, Chromatographia, 15:599 (1982).
5. M. Horka, V. Kahle, K. Janák, and K. Tesařík, J. HRC & CC, 8:259 (1985).
6. B. J. Gudzinowicz, M. J. Gudzinowicz, and H. F. Martin, Fundamental of Integrated GC-MS, Marcel Dekker, New York, 1976.
7. Joint Committee, Pharm. Soc. and Soc. Anal. Chem., Analyst, 100:136 (1975).
8. M. E. Pickup and J. W. Paterson, J. Pharm. Pharmacol., 26:561 (1974).
9. S. Alm, S. Jonson, H. Karlsson, and E. G. Sundholm, J. Chromatogr., 254:179 (1983).
10. M. T. Gilbert and C. I. W. Brooks, Biomed. Mass Spectr., 4:226 (1977).
11. J. Lundström and S. Agurell, J. Chromatogr., 36:105 (1968).
12. G. van Peteghem, A. Heyndrickx, and W. van Zele, J. Pharm. Sci., 69:118 (1980).

13. L. D. Quin, Nature, 182:865 (1958).
14. R. F. Severson, K. L. McDuffie, R. F. Arrendale, G. R. Gwynn, J. F. Chaplin, and A. W. Johnson, J. Chromatogr., 211:111 (1981).
15. S. Matsushima, T. Ohsumi, and S. Sugawara, Agric. Biol. Chem., 47:507 (1983).
16. A. Raisi, E. Alipour, and S. Manouchelvri, Chromatographia, 21:711 (1986).
17. I. E. Burrows, P. J. Corp, G. C. Jackson, and B. F. J. Page, Analyst, 96:81 (1971).
18. S. E. Falkman, I. E. Burrows, R. A. Lundgren, B. F. J. Page, Analyst, 100:99 (1975).
19. J. A. Thompson, M.-S. Ho, and D. R. Petersen, J. Chromatogr., 231:53 (1982).
20. P. Daenens, L. Laruelle, K. Callewaert, P. De Schepper, R. Galleazzi, J. van Rossum, J. Chromatogr., 342:79 (1985).
21. R. A. Davis, J. Chromatogr. Sci., 24:134 (1986).
22. P. Hartvig, N. O. Ahnfelt, M. Hammarlund, and J. Vessman, J. Chromatogr., 173:127 (1979).
23. M. Verzele, G. Redant, S. Qureshi, P. Sandra, J. Chromatogr., 199:105 (1980).
24. A. C. Moffat, A. H. Stead, and K. W. Smalldon, J. Chromatogr., 90:19 (1974).
25. D. K. Wyatt, W. G. Richardson, B. McEwan, J. M. Woodside, and L. T. Grady, J. Pharm. Sci., 65:680 (1976).
26. B. F. Grabowski, B. J. Softly, B. L. Chang, and W. G. Haney, Jr., J. Pharm. Sci., 62:806 (1973).
27. H. W. Liebisch, H. Bernasch, and H. R. Schütte, Z. Chem., 13:496 (1973).
28. C. B. Turner, C. Y. Ma, and M. A. Elsohly, Bull. Narc., 31:71 (1979).
29. A. S. Christophersen and K. E. Rasmussen, J. Chromatogr., 174:454 (1979).
30. L. L. Plotczyk, J. Chromatogr., 240:349 (1982).
31. I. Barni Comparini, F. Centiny, and A. Pariali, J. Chromatogr., 279:609 (1983).
32. J. I. Javaid, H. Dekirmenjian, J. M. Davis, and C. R. Schuster, J. Chromatogr., 152:105 (1978).
33. L. Kazyak and E. C. Knoblock, Anal. Chem., 35:1448 (1963).
34. E. Brochmann-Hansen and C. R. Fontan, J. Chromatogr., 19:296 (1965).
35. E. Brochmann-Hansen and C. R. Fontan, J. Chromatogr., 20:394 (1965).
36. R. L. Furner, G. B. Brown, and J. W. Scott, J. Anal. Toxicol., 5:275 (1981).
37. C. Lora-Tamayo, M. A. Rams, and J. M. R. Chacon, J. Chromatogr., 374:73 (1986).

38. E. Smith, S. Barkan, B. Ross, M. Maienthal, and J. Levine, J. Pharm. Sci., 62:1151 (1973).
39. K. K. Midha and C. Charette, J. Pharm. Sci., 63:1244 (1974).
40. D. E. Guttmann, H. B. Kostenbauer, G. R. Wilkinson, and P. H. Dube, J. Pharm. Sci., 63:1625 (1974).
41. V. Bellia, J. Jacob, and H. T. Smith, J. Chromatogr., 161:231 (1978).
42. J. L. Cashaw, K. D. McMurtey, L. R. Meyerson, and V. E. Davis, Anal. Biochem., 74:343 (1976).
43. H. Miyazaki, E. Shirai, M. Ishibashi, and K. Niizima, J. Chromatogr., 152:79 (1978).
44. H. V. Street, W. Vycudilik, and G. Machata, J. Chromatogr., 168:117 (1979).
45. G. R. Nakamura, T. T. Noguchi, J. Forensic Sci., 14:347 (1974).
46. K. E. Rasmussen, J. Chromatogr., 120:491 (1976).
47. A. F. Sperling, J. Chromatogr., 294:297 (1984).
48. B. Law, J. R. Joyce, T. S. Bal, C. P. Goddard, M. Japp, and J. Humphreys, Anal. Proc., 20:611 (1983).
49. H. Neumann and M. Gloger, Chromatographia, 16:261 (1982).
50. B. Cimbura and E. Koves, J. Anal. Toxicol., 5:296 (1981).
51. P. O. Edlund, J. Chromatogr., 206:117 (1981).
52. E. J. Cone, W. D. Darwin, and W. F. Buchwald, J. Chromatogr., 275:307 (1983).
53. R. H. Drost, R. D. van Ooijen, T. Ionescu, and R. A. A. Maes, J. Chromatogr., 310:193 (1984).
54. A. W. Jones, Y. Blom, U. Bondesson, and E. Änggard, J. Chromatogr., 309:73 (1984).
55. Glas Capillary Chromatography in Clinical Medicine and Pharmacology (H. Jaeger, ed.), Marcel Dekker, New York, 1985.
56. B. K. Chowdhury, A. Mustapha, and P. Bhattacharrya, J. Chromatogr., 329:178 (1985).
57. A. H. Chalmers, C. C. J. Culvenor, and L. W. Smith, J. Chromatogr., 20:270 (1965).
58. J. Lüthy, U. Zweifel, P. Schmid, and C. Schlatter, Pharm. Acta Helv., 58(4):98 (1983).
59. P. Stengl, W. Wiedenfeld, and E. Röder, Dtsch. Apoth.-Ztg., 122:851 (1982).
60. U. Candrian, J. Lüthy, P. Schmid, C. Schlatter, and E. Gallasz, J. Agric. Food Chem., 32:935 (1984).
61. C. Bicchi, A. D'Amato, and E. Cappeletti, J. Chromatogr., 349:23 (1985).
62. M. L. Deinzer, B. L. Arbogast, D. H. Buhler, and P. R. Cheeke, Anal. Chem., 54:1811 (1982).
63. Y. D. Cho and R. O. Martin, Anal. Biochem., 44:49 (1971).
64. W. J. Keller and S. G. Zelenski, J. Pharm. Sci., 67:430 (1978).

65. C. R. Priddis, J. Chromatogr., 261:95 (1983).
66. G. M. Hatfield, D. J. Yang, P. W. Ferguson, and W. J. Keller, J. Agric. Food Chem., 33:909 (1985).
67. D. S. Millington, D. E. Games, and A. H. Jackson, Proc. Int. Symp. Gas Chrom. Mass Spectrom., 1972, 275.
68. S. E. Khayyal and M. M. Ayad, Anal. Lett., 16:1525 (1983).
69. I. Stewart, J. Agric. Food Chem., 33:1163 (1985).
70. F. T. Delbeke and M. Debackere, J. Chromatogr., 278:418 (1983).
71. F. T. Delbeke and M. Debackere, J. Chromatogr., 325:304 (1985).
72. R. E. Chambers, J. Chromatogr., 171:473 (1979).
73. L. J. Dusci, P. Hackett, and I. A. McDonald, J. Chromatogr., 104:147 (1975).
74. S. Floberg, S. Lindström, and G. Lönnerholm, J. Chromatogr., 221:166 (1980).
75. K. Y. Tserng, J. Pharm. Sci., 72:526 (1983).
76. R. C. S. Audette, J. Bolan, H. M. Vijayanagar, R. Bilous, and K. Clark, J. Chromatogr., 43:295 (1969).
77. M. Gazdag, K. Mihályfi, and G. Szepesi, Fresenius Z. Anal. Chem., 309:105 (1981).
78. M. Polgár and L. Vereczkey, J. Chromatogr., 241:29 (1982).
79. R. E. Ardrey and A. C. Moffat, J. Chromatogr., 220:195 (1981).
80. G. Settimj, L. Di Simone, and M. R. Del Guidice, J. Chromatogr., 116:263 (1976).
81. J. D. Phillipson and S. R. Hemingway, J. Chromatogr., 105:163 (1975).
82. G. P. Forni, J. Chromatogr., 176:129 (1979).
83. N. G. Bisset and P. Fouché, J. Chromatogr., 37:172 (1968).
84. D. L. Sondack and W. Koch, J. Pharm. Sci., 62:101 (1973).
85. G. Miller, J. Warren, K. Gohre, and L. Hanks, J. Assoc. Off. Anal. Chem., 65:901 (1982).
86. M. Edmunds, T. M. T. Sheehan, and W. V. Hoff, Clin. Toxicol., 24:245 (1986).
87. S. Agurell and A. Ohlsson, J. Chromatogr., 61:339 (1971).
88. D. L. Sondack, J. Pharm. Sci., 63:584 (1974).
89. F. J. W. Van Mansfelt, J. F. Greving, and R. A. De Zeeuw, J. Chromatogr., 151:113 (1978).
90. W. F. Bayne, L.-C. Chu, and F. T. Tao, J. Pharm. Sci., 65:1724 (1976).
91. R. J. Bushway, D. F. McGann, and A. A. Bushway, J. Agric. Food Chem., 32:548 (1984).
92. W. M. J. Van Gelder, J. Chromatogr., 331:285 (1985).
93. D. Marini and F. Balestrieri, Boll. Chim. Farm., 123:83 (1984).
94. S. F. Herb, T. J. Fitzpatrick, and S. F. Osman, J. Agric. Food Chem., 23:520 (1975).

95. E. Kovats, Helv. Chim. Acta, 41:1915 (1958).
96. N. Kucharczyk, F. H. Segelman, E. Kelton, J. Summers, and R. D. Sofia, J. Chromatogr., 377:384 (1986).

5
Liquid Chromatography

Various forms of liquid chromatography are commonly distinguished, differing in the nature of the intermolecular forces acting on the molecules of solute in stationary and mobile phases. The usual classification system distinguishes the following forms: liquid adsorption chromatography (LSC), liquid partition chromatography (LLC) with two different modes—normal phase (NP-LLC) and reversed phase (RP-LLC), ion exchange chromatography (IEC), size exclusion chromatography or gel permeation chromatography (GPC), and affinity chromatography. Of these forms, size exclusion and affinity chromatography can be omitted for the separation of alkaloids. On the other hand, ion pair chromatography can be classified as a new, independent form, even though its nature allows its placement among the other forms mentioned above. In the real chromatographic systems it is sometimes the case that two or three separation mechanisms are operating simultaneously. The separation of alkaloids is a typical example of this. Therefore the classification into particular sections that follows could be considered formal. To achieve any classification, the stationary phase used was the deciding criterion. Thus, in the section on adsorption chromatography (Section I) is presented the separations of alkaloids where silica gel or alumina served as stationary phase. The real separation mechanism is also discussed. In the section of liquid partition chromatography (Section II) are discussed applications whereby chemically bonded stationary phases are used, and in the ion exchange chromatography section (Section III) are described separations using ion exchangers. The section on ion pair chromatography (Section IV) reviews separations in which a suitable pairing ion was added to the mobile phase.

I. ADSORPTION CHROMATOGRAPHY

In liquid adsorption chromatography the sorption of solute is the result of various forces that exist between molecules of solute and adsorption sites and between molecules of eluent and adsorption sites. It is supposed that these centers are from the very beginning occupied by molecules of eluent. Any solute molecule X_m must displace a certain number n of adsorbed eluent molecules S_{ad} from the adsorbent sites, giving an adsorbed solute molecule X_{ad}:

$$X_m + nS_{ad} \rightleftarrows X_{ad} + nS_m \tag{5.1}$$

Equation 5.1 reduces the net energy of adsorption on the difference of solute adsorption energy and adsorption energy of n eluent molecules. Since the adsorbent surface is completely covered by adsorbed eluent, the area occupied by adsorbed solute must be equal to that covered by n molecules of eluent. This theory, treated in detail by Snyder (1), is expressed in a well-known equation valid for liquid adsorption chromatography:

$$\log K_{ad} = \log V_a + \alpha(S^o - A_s \varepsilon^o) \tag{5.2}$$

where K_{ad} is the adsorption coefficient, V_a and α describe the properties of adsorbent, S^o and A_s are the parameters of solute, and ε^o is the parameter of eluent. In adsorption chromatography it is useful to substitute the volume of stationary phase for the weight of adsorbent, and therefore K_{ad} is a ratio of adsorbed moles of solute per gram of adsorbent to moles of solute in 1 ml of eluent. V_a is the volume of adsorbed eluent per gram of adsorbent supposing monolayer adsorption. Parameter α expresses the activity of adsorbent and by definition $\alpha = 1$ for fully thermally activated adsorbent. Adsorbents, partially deactivated by addition of water, have α values in the range 0.5-0.9. S^o is the solute adsorption energy, which actually means a dimensionless free energy of solute adsorption on a given adsorbent. This energy is defined for a standard adsorption system, pentane eluent, adsorbent type specified, $\alpha = 1.00$. Parameter A_s is the area of adsorbed solute molecule in units of 0.085 nm^2. Solvent strength parameter ε^o is a dimensionless adsorption energy of eluent covering the adsorbent surface of 0.085 nm^2. By definition ε^o is zero for pentane and in the range 0-1 for other solvents. Table 5.1 presents ε^o and A_s data for various solvents. The relative order of ε^o is generally the same for different adsorbents, but these ε^o values vary somewhat from one adsorbent to another. Table 5.1 gives the ε^o values for silica gel and alumina. Product $A_s \varepsilon^o$ represents the dimensionless adsorption energy of n

Liquid Chromatography

eluent molecules covering the same area as the solute molecule. Equation 5.2 can be used for adsorbent standardization and for the description of the chromatographic system. Using the relation for capacity factor:

$$k = K_{ad} \frac{w}{V_m} \tag{5.3}$$

and known values (w, weight of adsorbent; V_m, eluent volume in column), we can calculate the adsorption coefficient. The S^o and A_s values for many compounds can be found in the literature (1), and ε^o values for various solvents are given in Table 5.1. From the retention data of two or three compounds, the parameters α and V_a can be calculated.

A. Separation on Silica Gel

Silica gel has for a long time been the most widely used adsorbent in adsorption chromatography. Chromatographic and other properties of silica gel have been extensively studied. A summary of contemporary knowledge is provided by Unger's book (2). Some workers use the essentially synonymous terms silicic acid and/or silica. All these materials have the empirical formula $SiO_2 \cdot xH_2O$, which is common to other silica species of different crystal structure, surface composition, and porosity. In chromatography amorphous silica particles are used that have a spherical or irregular shape and a highly porous structure. The main characteristics of silica as chromatographic material are particle size and specific surface area. Specific surface area (typically in the range 200–400 m^2 g^{-1}) is related to the pore structure of the silica. Materials with large specific surface areas have small average pore sizes and vice versa. This is illustrated in Table 5.2, in which typical properties of some commercial silicas are summarized.

Surface structure is characterized by the presence of hydroxyl groups bonded to silicon atoms. These silanol groups are active sites for the solute adsorption. For crystalline silica the mean distance between surface hydroxyl groups is about 0.5 nm. In amorphous silica some hydroxyls are sufficiently close together (less than 0.3 nm) to interact via formation of a hydrogen bond (Fig. 5.1). The number of hydrogen-bonded hydroxyl groups is greater in narrow-pore silicas whereas free hydroxyls prevail in wide-pore silicas.

Under normal conditions surface hydroxyls are hydrated with water. To be activated before chromatographic use, silica is usually heated at 150–200°C for 2–6 hr. This procedure is sufficient to remove most of the physically adsorbed surface water. Heating

TABLE 5.1 Properties of Some Solvents

Solvent	Boiling point (°C)	Viscosity at 20°C 10^{-3} Pa s (cP)	Refractive index n_D^{20}	UV cutoff^a (nm)	Molecular area 0.085 nm^2	Elution strength ε^o	
						Silica	Alumina
Pentane	36.1	0.24	1.358	210	5.9	0.00	0.00
Isooctane	99.2	0.50	1.404	210	7.6	0.01	0.01
Cyclohexane	81.0	1.00	1.427	210	6.0	0.03	0.04
Tetrachloro-methane	76.7	0.97	1.466	265	8.6	0.14	0.18
Benzene	80.1	0.65	1.501	280	6.0	0.25	0.32
Diethylether	34.6	0.23	1.353	220	4.5	0.29	0.38
Chloroform	61.5	0.57	1.443	245	5.0	0.31	0.40

Liquid Chromatography

Solvent							
Methylene chloride	40.2	0.44	1.424	245	4.1	0.32	0.42
Tetrahydrofuran	64.5	0.55	1.408	220	5.0	0.35	0.45
Acetone	56.2	0.32	1.359	330	4.2^b	0.43	0.56
Ethyl acetate	77.1	0.45	1.370	260	5.7^b	0.45	0.58
Acetonitrile	80.0	0.37	1.344	210	10.0^b	0.50	0.65
Isopropanol	82.4	2.3	1.380	210	8.0^b	0.63	0.82
Methanol	64.7	0.60	1.329	210	8.0^b	0.73	0.95
Water	100.0	1.00	1.333	210		1.00	1.00

[a] For given wavelength the transmittance falls to 10% in a cell of optical pathlength 1 cm.
[b] For silica gel the molecular area is 10.0.

TABLE 5.2 Some Data on Properties of Commercial Silica Gels

Name	Particle size (μm)	Shape[a]	Specific surface area (m²/g)	Pore size (nm)	Specific pore volume (ml/g)	Total porosity (relative value)	Packing density (g/ml)	pH[b]
Grace silica gel	5	I	500	8				4.5[c]
Hypersil	5; 7	S	170; 200	10; 11.5	0.7		0.53	9.0; 6.9[c]
Lichrosorb Si-60	5; 6; 7; 10; 30	I	400; 475; 482; 500; 550	6.0	0.75; 0.76; 0.85			7.8; 6.9[c]
LiChrosorb Si-100	4–7; 5; 10;30	I	278; 309; 320; 400	10; 11.1	1.0; 1.02; 1.15; 1.2	0.83	0.34	7.0
LiChrospher Si-100	5; 10; 20	S	256; 370	10; 12	1.2			5.3
LiChrospher Si-500	10	S	45	50	0.88			9.9
LiChrospher Si-1000	10	S	19	100	0.72			9.2
Nucleosil 100	5; 7.5; 10	S	300	10	1.0			5.7

Liquid Chromatography

Material	Particle size (μm)	Shape[a]	Surface area (m²/g)	Pore diameter (nm)	Pore volume (mL/g)	Density	pH[b]	pH[c]
Partisil 5, 10	5; 6; 10; 20	I	400; 412	4–5; 5; 5.5; 6.0	0.55; 0.60; 1.0	0.88		7.5; 3.5c
Porasil	8–12; 10	I	300–350; 350; 400	10	1.0; 1.1		0.37	7.2
Sil-60	5; 10; 15; 20	I	500	6	0.75		0.75	
Spherisorb S	5; 10; 20	S	180; 183; 190; 200; 220	8; 8.1; 10; 10.9	0.52; 0.6	0.61	0.6	9.5; 8.9c
Spherosil XOA 1000		S	860; 1096	1.8; 3.5	0.5; 0.78			
Spherosil XOA 600	5; 5–7; 10; 20	S	550; 600 ± 10%	3; 7.0; 8; 8.3	0.7–1; 1.06; 1.2			6.4c
Spherosil XOA 400	10	S	350–500; 465	9.4; 10; 10.8	1.05; 1.26			8.1
Spherosil XOA 200	10	S	125–250; 170	15; 16.0; 22.4	0.90; 1.04			
Zorbax-BPSil		S	300	5.6	0.5	0.71	0.60	3.9

[a]S, spherical; I, irregular.
[b]pH of the 1% (w/w) aqueous suspension in double-distilled water (7).
[c]pH of the suspension of 1 g of silica in 15 ml double-distilled water (pH 5.1) (9).
Source: Data from Refs. 2–9.

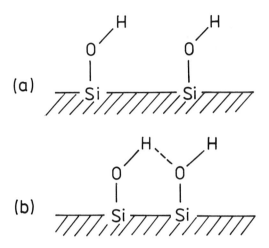

FIGURE 5.1 Free and hydrogen-bonded silanol groups on the silica gel surface.

above 200°C produces siloxane groups by dehydration of hydrogen-bonded hydroxyls. This reaction also occurs for free hydroxyls by heating above 400°C. The loss of surface hydroxyl groups by the formation of siloxane groups in the narrow pores is irreversible and is accompanied by a decrease in surface area (3). The surface concentration of hydroxyls for all fully hydroxylated silicas is approximately 8-9 µmol m^{-2}, which corresponds to the number of 5 hydroxyls per nm^2 (2).

Hydrogen-bonded hydroxyls are usually considered to be stronger adsorption sites than the free hydroxyls. Such an inhomogeneity on the surface can cause great nonlinearity of adsorption isotherm and so impair column efficiency and load capacity. Fully activated silica is therefore scarcely used in chromatography. Instead, deactivation with water or modification of silica surface with polar organic modifiers (e.g., aliphatic alcohols, acetonitrile, tetrahydrofuran, esters, or dichloromethane) (10-13) is applied. Such modifiers are preferentially adsorbed on most active centers deactivating them. When using dry-packed columns, activity of the silica can easily be controlled by addition of the desired amount of water to fully activated silica and by equilibration of the material before packing. With slurry-packed or commercial columns, control of surface activity can be achieved only be equilibrating the column with eluent of defined water content. This procedure is in most cases laborious and time consuming, especially for nonpolar mobile phases due to low solubility and/or high retention volumes of water in these solvents. Full equilibration of the column may require up to approximately 10^3 column volumes of eluent. Alkaloids, however, are separated using polar mobile phases and equilibration of column in such a case is not a serious problem.

Surface hydroxyl groups are also responsible for the acid-base and ion exchange properties of silica gel (2). Weak acidity of the silica surface is caused by deprotonization of the silanol groups in the ionizing solvent, e.g., in aqueous medium:

$$\equiv Si-OH + H_2O \rightleftharpoons \equiv SiO^-, H_3O^+$$

where the comma indicates that the hydronium ions are localized at the surface of the silica by electrostatic forces to retain overall electroneutrality. When in contact with electrolyte solutions, the exchange of hydronium ions with other cations is effective. In infrared experiments the mean value $pK_a = 7.1 \pm 0.5$ was determined (14,15). Nevertheless the acidity is not the same for all silanol groups, rather it decreases with increasing surface ionization (16,17). Cation exchange reactions take place in the pH range 2-6. Not all surface silanols are ionizable. Even at high pH values only 10-20% of the silanol groups are ionized and can participate in ion exchange. The surface concentration of exchangeable silanols is similar in most of the fully hydroxylated silica gels. This fact forms the basis of the method of approximate surface area determination elaborated by Sears (18), who observed a linear relationship of the specific surface area and amount of sodium hydroxide necessary to adjust the silica suspension in 20% sodium chloride to pH = 9. The slope of this relationship corresponds to the surface concentration of approximately 1.3 exchangeable acid centers per nm^2 (2.1 $\mu mol\ m^{-2}$, respectively).

The initial surface ionization states of silica can differ for various commercial products (7,9). The large range of pH values for silica suspensions in water (see Table 5.2) reflects the differences in production procedures and final treatments. Retention and column efficiency for polar compounds are often low for these reasons. Acid washing of the packing may improve peak shape of acidic compounds (e.g., phenols), whereas separation of compounds with basic character may require base treatment (7). Recently, Schwarzenbach reviewed the preparation, properties, and applications of buffered silica gel (9).

The surface hydroxyl groups are of prime importance in the separation of alkaloids and other compounds with basic character. In aprotic organic solvents nitrogenous base could be hydrogen-bonded to surface silanols:

$$\equiv SiOH + NR_3 \rightleftharpoons \equiv SiOH \ldots NR_3$$

or the formation of hydrogen-bonded ion pairs could be assumed:

$$SiOH + NR_3 \rightleftharpoons SiO^- \ldots HN^+R_3$$

with both reactions leading to extremely high retention.

Owing to these strong interactions the separation of alkaloids on silica can scarcely be performed using solvents of only moderate polarity. Thus dihydrochelerythrine and N-methylflindersine have been isolated from plant extracts and separated on the silica gel column using hexane—ethyl acetate 4:1 as a mobile phase (19). Aconitine could have been eluted with diethyl ether partially saturated with water (20) but, as a rule, a more polar component in the eluent mixture is required to compete with hydrogen bonding of solute on adsorbent surface. Usually methanol (21–30), other aliphatic alcohols (25,26,28,31), acetonitrile (21,30), acetone (28,33,34) or tetrahydrofuran (29, 35) have been used to suppress alkaloid retention.

Early measurements of the elution times for 24 alkaloids in chloroform-methanol and diethyl ether—methanol mixtures of different ratios have indicated a continuous decrease of retention volumes with increasing methanol concentration (22). Some deviations can possibly be ascribed to poor control of the water content in the mobile phase and/or in silica gel. Capacity factors measured for some eburnane alkaloids using chloroform with relatively low content (2, 5, and 8% v/v) of methanol (30) have yielded an approximately linear relationship (except for apovincamine and apovincaminic acid ethyl ester) when their reciprocal values have been plotted against methanol concentration (Figure 5.2). This behavior is in agreement with a simple model for adsorption from binary solvent mixtures (36). Retention of 13 indole alkaloids examined in binary mixtures hexane-tetrahydrofuran, methylenechloride-diethylamine, chloroform-methanol, and tetrahydrofuran-methanol have resulted in a more complex pattern (29). In the medium concentration range capacity factors k have been correlated with a molar concentration X_S of polar solvent of the mobile phase via an equation based on a simplified Snyder relationship (1,37–39):

$$\log k = c - n \log X_S \tag{5.4}$$

where c and n are constants. Values of n for hexane-diethylamine have been in the range 3–6.5, indicating multisite adsorption of solute on the silica surface. For other solvents lower values of n have been found (0.2–3.0). Experimental dependencies often intersect each other, which means that the elution order of compounds is changed. In some cases lower k values in comparison with Eq.

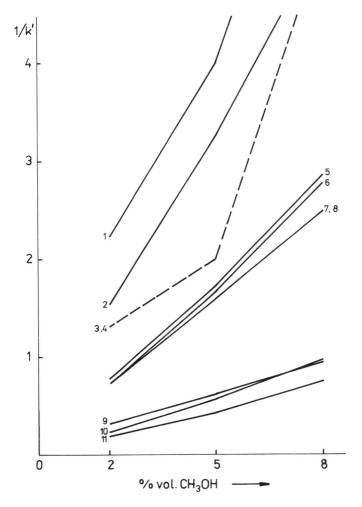

FIGURE 5.2 Dependence of reciprocal values of capacity factors 1/k of some eburnane alkaloids on methanol concentration in the mobile phase chloroform-methanol. Column 250 × 4.6 mm packed with Li-Chrosorb Si-60, 5 μm. 1, (+)-cis-Dehydrovincamine; 2, (+)-cis-vincamone; 3, 4, (+)-cis-apovincamine and (+)-cis-apovincaminic acid ethyl ester; 5, (+)-cis-dehydroepivincamine; 6, (+)-cis-vincamenine; 7, 8, (+)-cis-vincamine and (+)-cis-vincaminic acid ethyl ester; 9, (2)-cis-isovincanole; 10, (+)-cis-epivincamine; 11, (+)-cis-vincanole. (Data from Ref. 30.)

5.4 have been determined in the low-concentration region. The mixtures of chloroform-methanol have shown an anomalous upward curvature of log k vs. log X_S for several alkaloids at high methanol concentrations. This complicated retention behavior indicates complex interactions of multifunctional solute and solvent molecules in mobile phases and of solute and silica surface. Nevertheless, the key role of steric accessibility of the basic nitrogen atom (29) and of its type (28) for the effective retention of solute has been well recognized. Differences in the steric hindrance of the nitrogen atom often enable good separation of structural and stereoisomers (26,28-30). An amide group as in 2-oxyindole structure (29) and hydroxyl groups (28,29) contribute markedly to the solute retention.

Separations with a polar modifier in the mobile phase are sometimes adversely affected by peak tailing (21-24, 40). However, there are in most cases ways to eliminate peak tailing. The reasons for peak tailing are not usually discussed in detail but several possibilities exist. Besides slow equilibration of solute between mobile and stationary phase, isotherm linearity can be restricted due to such side reactions as association, when solute molecules attract each other or attract other components in the mobile or stationary phase. Water and other polar components of the mobile phase are preferentially adsorbed on the silica gel surface. Under these conditions the surface layer is sufficiently polar to allow partial ionization of alkaloids. When the system is not buffered, a degree of ionization and the distribution coefficient are dependent on solute concentration. Tailing caused by undefined ionization of solute in the stationary phase can be affected by the addition of acidic or basic component to the separation system.

The addition of an acidic component is less common. As an example mixtures of chloroform-ethyl acetate-methanol-acetic acid 60:50:20:3 have been applied to separate ergotamine and ergocristine on a "Corasil" column without problems of tailing (21). Formic acid (41,42) along with acetic acid (43,44), eventually with their ammonium salts (45,46) or diethylamine (47-49) has also been used. The acid component not only enhances the ionization of basic solute in the stationary phase and effectively eliminates the acidity of silanol groups, but also leads to the formation of ion pairs, which probably remain undissociated even in the stationary phase. Retention is therefore only slightly affected by the concentration of acid as has been observed for some basic antiarrythmic drugs eluted by means of a dichloroethane-isopropanol-aqueous perchloric acid 84.2:15:0.8 mixture (50). [Superior efficiency has been also mentioned with regard to a methanolic mobile phase compared to an isopropanolic or butanolic one (50).] A decrease in retention was caused by an increasing concentration of alcohol in the mobile phase (50,51). Retention of basic compounds also depends on the type of acid used.

Capacity factors of apomorphine on LiChrosorb Si-100 column have increased in the range 1.19–9.57, when the series perchloric, nitric, hydrochloric, trichloroacetic, methanesulfonic, and monochloroacetic acid was used in the mobile phase dichloromethane-methanol-1 M aqueous acid 91.6:8:0.4 (51). The elution of alkaloids in the form of ion pairs is sometimes advantageous for their detection. A small amount (0.1%) of 70% perchloric acid added to the mobile phase dichloromethane-hexane-methanol 60:35:5.5 induced fluorescence of quinidine (52), which does not fluoresce in the form of unionized free base.

Silica gel has also been applied as a support in liquid-liquid ion pair chromatography. The stationary phase 0.2 M $HClO_4$, 0.8 M $NaClO_4$ in situ loaded on silica gel column was used in the chromatographic assay of quinidine (50,53) and n-propylajmaline (54). Picric acid (0.03–0.06 M) in citrate buffer pH 5–6 enabled the separation and sensitive UV detection of some tropane alkaloids exhibiting otherwise low absorptivity (55–58). The comparison of chromatographic data with extraction coefficients determined by batch experiment showed only a qualitative agreement with theory, which means that some other retention mechanism is also operative (57). These separation systems have drawbacks in terms of their stringent requirements for temperature control, the necessity for saturation of the mobile phase with stationary phase, and flow rate limitations to keep down stationary phase bleeding.

The most widely applied approach to the elimination of peak tailing is the addition of a basic component to the mobile phase. This suppresses the ionization of alkaloid molecules. Acidic centers on the adsorbent surface are occupied by a basic component which eliminates their interaction with basic nitrogen of the solute. Other adsorption sites are also deactivated. Thus, the addition of alkali to the mobile phase eliminates tailing of basic solutes (23,59–61), decreases their retention (20,29,40,59–61), and increases sample recovery (24).

Ammonia as a basic modifier has often been applied (20,24,62, 63). Retention data for 21 drugs including some alkaloids have been measured on μPorasil columns with the mobile phase chloroform-concentrated ammonia 1000:2. Satisfactory column efficiency has been obtained for most of the basic drugs (64). Mobile phases containing ammonia possess poor long-term stability due to the high volatility of ammonia. A decrease in ammonia concentration leads to increased retention (65). On the other hand, for preparative purposes it is favorable that ammonia can easily be removed by solvent evaporation.

Diethylamine is the second alkaline modifier often used for the separation of alkaloids (66). Separation of Cinchona alkaloids using mobile phase hexane-dichloromethane-methanol-diethylamine 66:31:2:

0.65 is shown in Figure 5.3 (66). Decreasing capacity factors of some indole alkaloids with rising concentration of diethylamine in the mobile phase dichloromethane-diethylamine comply approximately with Eq. 5.4 above (29). Also for the mobile phase containing diethylamine in diethyl ether, partially saturated with water, a decrease capacity factors of some alkaloids was observed (20). Water content is the second parameter governing retention in a similar manner and should be thoroughly controlled (20,67).

Besides diethylamine as the alkaline modifier, for the separation of Cinchona alkaloids (Figure 5.3), methanol as the polar modifier has been used (66). The methanol content can be easily controlled and a small, variable amount of water in the components of the mobile phase has a negligible effect on separation in the presence of

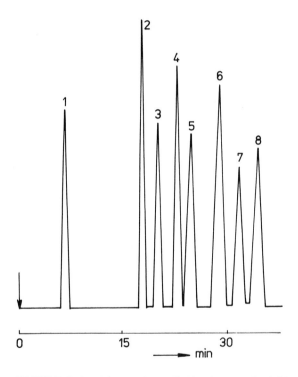

FIGURE 5.3 Separation of Cinchona alkaloids. Column 250 × 4.6 mm packed with Hypersil, 5 μm, mobile phase hexane-dichloromethane-methanol-diethylamine 66:31:2:0.65, 1 ml min^{-1}, UV detection at 312 nm. 1, Quinidinone; 2, quinidine; 3, cinchonine; 4, hydroquinidine; 5, cinchonidine; 6, quinine; 7, hydrocinchonidine; 8, hydroquinine. All 2 μg. (After Ref. 66).

methanol. Other factors which can cause irreproducible separation are temperature and sample size. An approximately linear dependence of ln k on 1/T has been established with similar slopes for cinchonine, quinine, quinidine, and cinchonidine. An increase in temperature has caused an _increase_ in capacity ratio similar to the decrease in methanol concentration in the mobile phase. A 2-3% higher retention was observed as sample size decreased from 4 μg to 20 ng. Changes in the hexane-dichloromethane ratio facilitate control of the system's selectivity (66).

Triethylamine has been used as the basic modifier in the separation of some pharmaceutical products including ephedrine (68) and caffeine (69). Optimization of the mobile phase composition based on the concept of isohydric solvents has been applied (70). Water content in the series of isohydric solvents is adjusted to equal that amount of water adsorbed on silica. Isohydric solvents can therefore be changed without requiring lengthy equilibration of the column. It should be emphasized that isohydric solvents are not saturated with water to the same degree, but the percentage of saturation decreases with increasing solvent polarity. An optimization procedure has been established that employs the linear dependence of capacity factors on the reciprocal value of the molar fraction of water in isohydric solvents (68,69).

Triethylamine is sometimes reported to be less effective in the improvement of peak width and symmetry (59,66), and a higher concentration of secondary or tertiary amine in the mobile phase can be incompatible with UV detection at shorter wavelengths (60,71). Mixtures of diethylamine with chloroform readily form colored solutions (59). Thus, other basic modifiers such as cyclohexylamine (60) or ethanolamine (59,61) have been sometimes used.

Similar to the situation with water, silica gel must be equilibrated with basic modifier before use. It may require high volumes of eluent (20,60).

If the ionization ability of the mobile phase improves, e.g., when sufficient methanol is added to a solvent of medium polarity, an increase in the retention of basic compounds is usually observed. As noted above, an upward curvature of an otherwise decreasing dependence of the capacity factor on the methanol/chloroform ratio was observed for some indole alkaloids (29). With pure methanol as the mobile phase, many basic compounds were strongly retained or could not be eluted from the silica gel column at all (72-74). Evidently, some new retention mechanism, presumably ion exchange, becomes operative.

If salt $BHNR_3$ of nitrogenous base NR_3 and acid BH is dissolved in an ionizable solvent, dissociation occurs (solvation of ions is omitted in the following equations):

$$BHNR_3 \rightleftharpoons B^- + HN^+R_3$$

When such a solution will be injected onto the silica column containing partially ionized silanol groups, ion exchange may occure:

$$SiO^-,H^+ + B^- + HN^+R_3 \rightleftharpoons SiO^-,HN^+R_3 + H^+ + B^-$$

Similarly, free base NR_3 is at least partially ionized in a protogenous solvent SH such as methanol:

$$SH + NR_3 \rightleftharpoons S^- + HN^+R_3$$

which after injection onto the column will lead to the exchange:

$$SiOH^-,H^+ + S^- + HN^+R_3 \rightleftharpoons SiO^-,HN^+R_3 + SH$$

In both cases protons produced by these exchange reactions are neutralized by the counterions present in the mobile phase and migrate through the column with mobile phase, leaving protonized solute molecules bound to the silica gel surface via strong electrostatic forces. These molecules can be freed again only by ion exchange with other cations present in the mobile phase. As common solvents have low values of autoprotolytic constants (75), the concentration of cations H^+ in pure solvents is also very low (e.g., 1.1×10^{-7}, 3.5×10^{-9}, and 7.9×10^{-17} at 25°C for water, methanol, and acetonitrile, respectively) and so slightly ionized weak bases are eluted as tailing peaks with pure ionizable solvents. In order to elute molecules retained strongly by the ion exchange mechanism, some electrolytes must be added to the mobile phase.

Separations on silica gel columns using methanol-containing ammonia, ammonium buffer, perchloric acid, or other ionic compounds have come into wide use in forensic laboratories for the screening and determination of basic drugs (64,74,76–79). Figure 5.4 shows the separation of opium extract using the chromatographic system developed by Jane (76). Additional areas of application are numerous, e.g., the separation of hallucinogenic components of Psilocybe mushrooms was achieved using a similar system (80). A mobile phase containing primarily acetonitrile was used in the determination of heroin and its metabolites in blood (81).

The most important factors affecting separation with these systems are the pH and ionic strength of the mobile phase. At high pH values the retention of basic compounds usually decreases since ionization is suppressed and ionic interactions responsible for reten-

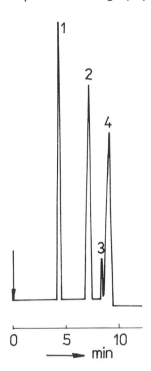

FIGURE 5.4 Separation of an opium sample. Column 250 × 4.6 mm, packed with fines obtained from Partisil, 6 μm, by sedimentation, mobile phase methanol–2 N ammonia solution–1 M ammonium nitrate solution, 27:2:1. 1 ml min^{-1}, UV detection at 278 nm. 1, Narcotine; 2, thebaine, 3, codeine; 4, morphine. (After Ref. 76.)

tion are eliminated. Hence, if some basic compound is to be retained, the pH of the mobile phase must be sufficiently low to achieve ionization of the solute. If separated compound is only partially ionized, the mobile phase has to be well buffered in order to keep the degree of ionization constant and independent of sample concentration. At lower pH values a slow decrease in retention has been observed, which is usually ascribed to the lower ionization of surface silanol groups (72,74,78,82). However, effective separations of basic compound are possible even in strongly acidic media, e.g., using 0.01–0.1% v/v perchloric acid in methanol as the mobile phase (74).

All available experimental data show that the retention of basic compounds decreases with an increasing concentration of ionic modifier in the mobile phase (72–74,78,81–83). Data on the effect of cation and anion type on retention are not as uniform. Only minor differences in retention have been observed for various anions present in the mobile phase (72,74,80), but a comparison of bromide and perchlorate ions has shown distince differences (73). Useful changes in selectivity are probably made possible by choosing suitable anions (74). Phosphoric acid has not been found suitable as a substitute for perchloric or sulfuric acid in strongly acidic mobile phases (74).

Little effect of changing inorganic cations has been reported (72,73). However, the type of amine used in a methanol-water 8:2 mobile phase has made a large difference in the retention of tricyclic antidepressants (82). The addition of dichloromethane (83) or chloroform (74) to the methanolic mobile phase has not changed capacity factors to a large extent, but selectivity can be improved. The addition of water to a primarily methanolic mobile phase can cause decreased retention (49,72) or no significant effect (74), but increased retention is often also found (82,83), as would be expected on reversed phase packings. The elution of compounds with higher polarity before those of lower polarity is also behavior associated with reversed phase materials (74). The introduction of hydroxyl groups to the solute structure has decreased retention time and the same has made dealkylation of the aminic nitrogen. The elution order of structurally similar amines has been primary < secondary < tertiary (74).

Similar dependencies and chromatographic behavior of basic compounds on silica gel with primarily methanolic mobile phases have also been ascertained for those with primarily aqueous mobile phases (49,84). The addition of methanol or acetonitrile to the mobile phase usually results in decreased basic solute retention. The similarity to reversed phase systems can also be seen in the fact that retention increases as hydrophobicity of the solute increases (84,85). Positive increments to the logarithm of capacity factor have been reported for organic carbon atoms (0.2, 0.1, and 0.4 for carbon atom in alkyl group, aryl group, and methyl on ammonium nitrogen, respectively), whereas negative values -0.3 to -0.4 have been observed for the carboxyl or hydroxyl group (84). A reversal of the elution order and minima on the curve of the capacity factor vs. volume fraction of water in the mobile phase has been found for opium alkaloids eluted with buffered methanol-water mixtures (49) and some glycoalkaloids eluted with acetonitrile-water mixtures containing ethanolamine (85). Similar minima have also been observed for crown ethers (86). A continuous change of retention behavior from "normal" to "reversed" can therefore by observed in some cases.

Since the factors affecting the separation of basic compounds on silica gel using mobile phase with ionization ability are numerous and experimental results display considerable diversity, it is not surprising that there are also differences of opinion regarding the main retention mechanism in these systems. Ion exchange with silanol protons is probably responsible for the strong retention of organic bases when pure solvent is used as mobile phase, but it does not necessarily mean that it is also the dominating retention mechanism when ionic components are present. There is common agreement that the dominating retention mechanism is based on ionic interactions. Besides ion exchange with cations bound to surface silanol groups (49,72,74,78,82), ion interaction (82), ion pair formation (74,78), and ion pair adsorption (73,84) or partition (72) have been stressed. Straight phase partition (49), dipole interactions, van der Waals forces, and hydrogen bonding (78) might also be of considerable importance. Sorption of cations and anions present in the mobile phase may substantially modify properties of an adsorbent surface (84). This is especially true for long-chain amines (82,87, 88), which form an organic layer on the silica gel surface which enables typical reversed phase separation onto "dynamically modified" silica (87,88). The sorption of anions is also possible to a large extent, as can be deduced from the difficulties in washing out perchloric acid from the column (74). The possibility of a "dynamic ion exchange" mechanism on adsorbed acid should not be overlooked. Hydrophobic properties displayed by the silica gel surface in aqueous mobile phases may be ascribed to the siloxane groups (82,86).

The practical aspects of the chromatography of alkaloids and other organic bases on silica gel using a methanolic mobile phase are quite favorable. Selective retention of basic components means that other components present, e.g., in biological or pharmaceutical samples, are not retained and the column can be used for a long time without excessive contamination. Reproducibility, efficiency, and flexibility are usually excellent. For basic compounds, the efficiency of the silica gel column is usually superior to that of a column containing a chemically bonded phase (72,74,82).

Surprisingly, the stability of silica gel columns in methanolic mobile phases has also been reported as satisfactory (73,74,76,79, 83), even at high pH values using ammonium as cation (78), and the same is valid for acidic aqueous mobile phases (84). Silica dissolves readily in aqueous solvents at a pH higher than 8–9, but even at lower pH values the concentration of dissolved silica in the mobile phase can be in the range 10–100 ppm (2,89–91). Concentration depends on particle size, flow rate, temperature, pH, and type and concentration of salt (90–92). Dissolution of the silica gel leads to damaged packing, which is indicated by a successive reduction of column efficiency and an increase in pressure drop. The addition of an organic solvent such as methanol to the aqueous mobile phase reduces these effects. Column degradation can also be considerably eliminated using silica gel precolumn (49,74,90), in which mobile phase is saturated with silica, or guard column, which also removes particles and strongly retained contaminants that occasionally arise in the samples (74,90). In every case flushing of the system with solvent containing no salt is strongly recommended, and not only with respect to the lifetime of the column. Corrosion and salt crystallization can damage the HPLC pump, sampling valve, and detector (79,83,84).

Some less common modes of silica gel application should also be mentioned. The separation of some purine and strychnos alkaloids was performed by partition chromatography (93), where Corasil I and Corasil II coated with Poly G-300 and heptane-ethanol mixtures saturated with stationary phase were used. Adsorption was partially effective because elution orders for each support differed (93).

In an attempt to eliminate baseline drift and breakthrough of the stationary phase after injection of a large volume of sample, dynamic coating was applied (94). Concentration of the stationary phase in the mobile phase corresponding to 10-60% saturation was also used and the column packed with a dry support. This mode of operation is essentially equivalent to a common adsorption chromatography, differing only in the use of one particular component of the mobile phase.

Changes in alkaloid retention caused by complex formation have been observed for silver-impregnated silica. Rapid regeneration of this system after gradient elution has been reported (95). Supercritical liquid chromatography was recently examined using caffeine, codeine, and cocaine as test compounds (96). A silica gel column (20 cm × 2.1 mm) packed with silica gel with mobile phase carbon dioxide-methanol 9:1 was coupled to a UV detector and a mass spectrometer. The dependence of capacity factors on back-pressure was evaluated.

B. Separation of Alkaloids on Alumina

Alumina, like silica gel, is a polar adsorbent, but it is less frequently used than silica. For chromatography, γ-alumina is utilized which, according to the degree of activation, has its surface covered with hydroxyl groups and chemisorbed water to a certain extent. Alumina is amphoteric in nature and its surface contains active sites with positive or negative charges. A specific surface area varies between 100 and 200 m^2 g^{-1}. The application of alumina is limited to the pH range 2-10. Below pH 2.0 and above pH 10.0 the stability of alumina is poor. Commercially produced aluminas for HPLC are supplied in the form of spherical beads with particle diameters of 10 or 5 μm. Sorbent activation is carried out by heating at 400°C for 6-16 hr.

Liquid Chromatography

Separation on alumina may be governed by three different principles: (a) adsorption, (b) steric exclusion (97), and (c) ion exchange (98,99).

(a) Adsorption chromatography on alumina enables the separation of nonpolar or mildly polar solutes using eluents of weak or medium polarity ($\varepsilon^\circ = 0$–0.40; see Table 5.1). Activated alumina with the addition of 1–4% H_2O is suitable for the separation of aromatic hydrocarbons, where the adsorption energy increases with the rising number of aromatic carbon atoms in the solute molecule. Addition of water to alumina leads to a decrease in adsorption energy. Acidic solutes ($pK_a < 5.0$) tend to sorb irreversibly on alumina.

(b) Alumina possesses regular, cylindrical pores whose diameter depends on the process of preparation. In the literature (1, 97) values of 2.7–13 nm have been mentioned. Using alumina with pores of 13-nm diameter, proteins in the range 1000–2000 have been separated (97). Pure water has served as mobile phase.

(c) The AlOH groups on the alumina surface have an amphoteric nature, which can be demonstrated by the following equilibrium states (98):

$$\equiv AlO^- H^+ \rightleftharpoons \ \equiv AlOH \rightleftharpoons \ \equiv Al^+ OH^-$$

The addition of acid or base may modify alumina to cation exchanger or anion exchanger:

$$\begin{array}{ccc}
\underset{\underset{-Al-\overline{O}-Al-\overline{O}-}{|O|}}{Na} & \xleftarrow{\text{NaOH}} \underset{\underset{-Al-\overline{O}-Al-\overline{O}-}{|O|}}{H} \xrightarrow{\text{HCL}} & \underset{-Al-\overline{O}-Al-\overline{O}-}{Cl \quad\quad H}
\end{array}$$

The transition among these states takes place at a pH where the net charge of the surface is zero. At lower pH the charge of the surface is positive and at higher pH it is negative. The pH value corresponding to zero point of charge (ZPC) is not fixed but depends on the nature of the anion buffer. For example, for citrate buffer ZPC is localized at pH 3.5, for acetate and phosphate ($H_2PO_4^-$) 6.5, for borate 8.3, and for carbonate 9.2. The pH value, where the first ionized form of the buffer is the most abundant, is the pH value of ZPC.

For the separation of alkaloids on alumina via ion exchange, the protonized form of an alkaloid has to be considered:

$$B + H_2O \rightleftharpoons BH^+ + OH^-$$

Ion BH^+ can interact with the cation exchanger, which means that a mobile phase with pH above alumina's isoelectric pH is required. However, alkaloids are weak bases and are not protonized at a pH above their pK_a values. Within these limits of pH values the following mechanism takes place:

$$\begin{array}{c} Na \\ | \\ |O| \\ | \\ -Al-\overline{O}-Al-\underline{O}- \\ | \quad | \end{array} + BH^+ \rightleftharpoons \begin{array}{c} BH \\ | \\ |O| \\ | \\ -Al-\overline{O}-Al-\underline{O}- \\ | \quad | \end{array} + Na^+$$

In this case the surface charge is negative and alumina acts as a cation exchanger. The retention of alkaloids can be controlled by the addition of a suitable base to the mobile phase. Protonized alkaloids compete with base cation for sites on cation exchanger and their retention can be adjusted by varying the base concentration.

The literature contains scarce references to the separation of alkaloids on alumina (24,60,99,100). Three papers (24,60,100) deal with adsorption chromatography, the last one (99) with ion exchange chromatography. Alumina Woelm of particle size 18–30 μm was applied in two cases (24,60) and the columns employed were long: 50 × 0.23 cm and 61 × 0.95 cm. Chan et al. (60) separated cocaine, quinidine, mescaline, and opium alkaloids on alumina using a mixture of cyclohexane and cyclohexymaline (99.8:0.2) as eluent. Kingston and Li (24) investigated the separation of indole alkaloids from Tabernaemontana holstii. A mixture of chloroform-hexane (9:1) served as mobile phase. According to authors: "An adequate but not complete separation of the components was obtained." The last example of straight phase adsorption chromatography was reported by Quercia et al. (100). The particle size, column length, and mobile phase composition correspond to the requirements of modern HPLC. A 25 × 0.2 cm column packed with Micropak Al-5 was eluted with a mixture of pentane-methanol 98:2 or 97:3. Analysis of dihydrogen derivatives of ergot alkaloids in pharmaceutical preparations was carried out. Based on a review of the experience with adsorption chromatography on alumina, it can be stated that this technique does not seem promising for the separation of alkaloids.

On the other hand, good results have been achieved using alumina as a cation exchanger. Recently, Laurent et al. (99) reported the separation of heroin and opium on this sorbent. A 250 × 4 mm column packed with Spherisorb A5Y (Chrompack) was washed with 1 M sodium chloride solution, then with 100 ml of aqueous buffer (concentration higher than 0.1 M), and finally with 20 ml of diluted buffer containing organic modifier. This procedure is always repeated when the pH of mobile phase has been changed to another value. The measurements were carried out using citrate buffer (citric acid and tetramethylammonium hydroxide-TMA) and three organic modifiers (0–60%): methanol, acetonitrile, and tetrahydrofuran. Some of the results are given in Table 5.3. The authors (99) assumed that at given conditions (pH 5.0, citrate buffer) all alkaloids are ionized except caffeine. Ions of solutes compete with TMA for active sites on the alumina surface, which acts as cation exchanger. The enhanced concentration of TMA facilitates the decreased retention time of alkaloids. The addition of organic modifier affects the separation in two ways. The increase of organic modifier concentra-

TABLE 5.3 Retention Data (V_R/V_t) of Heroin and Opium Compounds on Alumina[a]

Compound	Mobile phase[b]			
	20% MeOH	40% MeOH	20% CH$_3$CN	60% CH$_3$CN
Caffeine	0.7	0.7	0.8	0.8
Papaverine	2.1	1.4	1.6	1.0
Narcotine	2.3	2.1	1.9	1.1
Procaine	2.3	2.6	2.0	2.1
Thebaine	2.6	3.0	2.0	2.1
Acetylcodeine	2.5	3.1	2.0	2.1
Heroin	2.6	3.3	2.0	2.1
6-Monoacetyl-morphine	2.7	3.4	2.3	2.4
Codeine	2.8	3.5	2.4	3.0
Morphine	3.0	3.9	2.4	3.7
Strychnine	3.1	4.1	2.4	2.6

[a]Estimated values on the basis of retention times (total column volume V_t = 4.15 ml).
[b]Citrate buffer (citric acid and tetramethylammonium hydroxide) 0.01 M of pH 5 was used as aqueous component.
Source: Ref. 99.

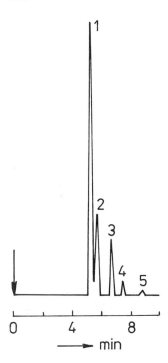

FIGURE 5.5 Separation of heroin sample on alumina. Column 250 × 4.6 mm packed with Spherisorb A5Y, mobile phase 0.01 M tetramethylammonium hydroxide, citric acid to pH 5 in water-acetonitrile-methanol 75:12.5:12.5. 1, Heroin; 2, acetylcodeine; 3, 6-monoacetylmorphine; 4, strychnine; 5, morphine. (After Ref. 99.)

tion enhances the solvation of TMA in the mobile phase while simultaneously reducing its competition for active sites. On the other hand, the addition of organic modifier increases the apparent pH of the mobile phase and decreases the effective pK_a of the solute (101). From the results given in Table 5.3 it can be concluded that the addition of organic modifier increases the retention of all solutes, except caffeine, narcotine, and papaverine. Caffeine was not ionized and its retention volume corresponded approximately to the dead volume of the column. Narcotine and papaverine have pK_a values (5.5–6.5) close to the pH 5.0 of the mobile phase used. At high concentration of organic modifier in the mobile phase the ionization of these solutes is suppressed and their retention volumes approach that of the dead volume. According to Laurent et al. (99), the choice of pH, TMA concentration, and the nature and concentration of an organic modifier provides the separation system with great flexibility. The authors arrived at optimum composition of the mobile phase for separation of illicit heroin samples: methanol-acetonitrile-water (12.5: 12.5:75), pH 6.5, and a concentration of citric acid and TMA of 0.01 M. As an example, the chromatogram of a heroin sample from the Far East is shown in Figure 5.5.

II. LIQUID PARTITION CHROMATOGRAPHY

Conventional liquid-liquid partition chromatography using a support material coated with liquid stationary phase is at present virtually omitted from chromatographic practice. The reasons are its limited resistance against solvent strength and flow rates, its incapacity to utilize gradient elution, its severe requirements for precise thermostatting of the column, and its dependence on precolumns. Nowadays chemically bonded stationary phases based primarily on silica gel matrix as well as porous polymeric packing materials are available in many types and possessing different properties. These packings are capable of substituting classical liquid-liquid systems in nearly all cases and it will be dealt with in this section.

An all liquid-liquid chromatographic system can be classified as either normal phase (NP) or reversed phase (RP). Typical RP systems are roughly characterized by the following features: Stationary phase is formed by hydrocarbonaceous material of aliphatic or aromatic nature. Water containing a varying amount of organic modifier is used as mobile phase. The increase in the content of organic modifier in the mobile phase decreases retention. The more extensive the hydrocarbonaceous moiety in the solute molecules, the greater the retention observed. The presence of such polar functionalities as ether, ester, and hydroxyl groups in solute molecules decreases retention.

Typical stationary phases for normal phase chromatography carry polar functional groups like hydroxyl, amino, nitro, or nitrilo. Strong retention is observed with the use of nonpolar mobile phases such as alkanes. The increase in polarity of the mobile phase resulting from the use of more polar solvents causes the decrease in retention. Elutropic series are of similar order as for adsorption chromatography on silica or alumina (see Table 5.1). The presence of polar functional groups in the solute molecule leads to an increase in retention. Many of the normal phase packings contain a sufficient amount of hydrocarbonaceous skeleton to make possible the operation with aquoorganic mobile phases in the reversed phase mode.

The question arises as to whether separation on chemically bonded packings is really governed by the partitioning mechanism. Academically interesting but practically limited, the study of retention mechanism has gained considerable attention. Despite this, the theory of liquid chromatography on bonded phases is not yet fully understood even for neutral solutes and alkyl-bonded reversed phase systems. Since these systems are currently the most widely used ones, they will be characterized here in some detail.

The first description of separation mechanism assumed that liquid-liquid partitioning between the mobile and stationary phases behaved similarly to liquid alkane. Preferential solvation or swelling of alkyl chains by the organic component of the mobile phase leads to the formation of mixed stationary liquid (102-105). The composition and properties of this phase probably change along with its thickness and depend on the composition of the mobile phase. The changes in its volume have been observed and they create uncertainties in the precise determination of column dead volume (106-111). Besides the alkyl chains and the components of mobile phase, silica gel surface and especially unreacted surface silanol groups have to be taken into account as the components of the stationary phase.

Other descriptions emphasize the adsorption of solute on the interphase between the alkyl region and the mobile phase or directly on individual alkyl chains (112-115). Mutual interactions within the molecules of aqueous mobile phase are much stronger than their interactions with solute molecules and therefore the solute molecules are repulsed into the stationary phase. Competitive interactions in the mobile phase are therefore dominant in determining solute retention and weak interactions with nonpolar alkyl chains of stationary phase are often neglected. A retention model incorporating these features was developed by Horváth et al. (116,117) applying the solvophobic theory of Sinanoglu (118,119). The theory relates retention to the hydrophobic molecular surface area of the solute in contact with aqueous mobile phase and the decrease in surface area when the solute molecule comes in contact with alkane stationary phase.

The retention mechanism on stationary phases with short bonded alkyl groups like methyl can probably be well approximated by adsorption. It was shown by NMR (120,121) and neutron-scattering (122) experiments that bonded longer alkyl chains are partially movable, and it is therefore questionable as to what extent the retention mechanism may be similar to the adsorption on solid surfaces. On the other hand, the thin layer of chemically bonded alkanes is hardly expected to possess exactly the same properties as a bulk liquid (120,123,124). A high specific surface area of RP packings offers great potential for adsorption-like phenomena to occur. Theories that are based on the unified treatment of partition and adsorption in liquid chromatography have therefore been proposed (125-127).

Based on various theories a large number of parameters have been applied to describe retention behavior in RP liquid chromatography. In accordance with the liquid-liquid partition mechanism, solvent extraction systems have often been used to predict retention. For various groups of compounds the values of log k can be linearly related to the hydrophobicity parameter as measured by the partition between octanol and water (128-139). In this connection recent fluorimetric evidence should be mentioned in the sense that pyrene molecules in chemically alkyl-bonded phases are surrounded by an environment of polarity similar to that of octanol (140). The hydrophobicity of molecules can be approximately evaluated using the system of Rekker's hydrophobic fragmental constants (141,142) or Hansch's hydrophobic substituent constants (143,144). In many cases the hydrophobicity parameters, which are often used in studies on the quantitative structure-pharmacological activity relationship, were assessed using RP liquid chromatography (145-149). For nitrogen compounds only a limited applicability of hydrophobic parameter values for the prediction of retention was found (150).

The theory of regular solutions (151-153) has been applied to RP chromatography on alkane-bonded stationary phases (135,154-158) with qualitative agreement. Unfortunately, the published values of this solubility parameter often differ considerably, which leads to large discrepancies between calculated and experimental retention data (135,155,158). The solubility parameter concept was originally proposed for nonpolar media with dominating dispersion interactions, but it was also extended to include dipole and acid-base interactions enabling a deeper description of solvent polarity (155,159-163).

In agreement with the supposed dominant role of mobile phase interactions, the retentions of similar compounds were correlated with their solubilities in mobile phase (114) and in water (114,130,139). Correlations between various parameters describing molecular surface area and solubility in water have been reported for classes of similar compounds (164-166), and therefore molecular surface area should also correlate with retention data, as confirmed experimentally in many cases (116,130,131,139,167,168). This correlation follows from solvophobic theories as mentioned above. Molecular surface area is sometimes approximately calculated additively from group surface area increments (169).

Other physicochemical parameters such as Hammet's constants (168), dipole moments (167), molecular polarizabilities, ionization potential (170), and molar and van der Waals volumes (131,168,171) have been also correlated with retention data. Molar volumes and van der Waals volumes can be approximately calculated as the sum of structural increments (169). Topological parameters such as connectivity indices of different orders (130-132,136,137,139,168,172, 173), Wiener numbers, correlation factor (168,173-177) or shape parameter (176-179), defined as the ratio of the length to the width of the molecule, have been used in attempts to express the effect of molecular branching and shape on retention. Also, parameters based on quantum chemical calculations have been applied, such as the general index of molecular complexity (180) or submolecular polarity and dispersive interaction descriptors (181).

Some of the approaches mentioned above are consistent with the concept of a linear free-energy relationship, which expresses the change of free energy for the transfer of solute molecule from mobile to stationary phase as a sum of increments from structural groups. This concept was originally applied to partition chromatography by Martin (182). If correct, it will be invaluable for the prediction of retention behavior as it would permit one to derive log k values from structural increments. In most cases it generally holds for homologous series where log k values are essentially linearly related to the number of carbon atoms in the molecule (130,183-185) with nearly identical slopes in every series (110,186). Unfortunately, the increment for a given functional group is generally dependent on the other structural features of the molecule and this greatly reduces the predictive ability of the concept. For example, the concept failed for disubstituted benzenes in RP chromatography due to mutual interactions of functional groups. Deviations were treated with partial success using Taft's approach (187,188). This concept does not prove correct in the case of many alkaloids which are isomeric but differ considerably in their retention. In fact, the separation of isomers is only possible as a result of deviations from the simple additivity relationship.

The retention volume and plate number for a given solute are dependent on the properties of the packing, the mobile phase composition, and temperature. The properties of alkyl-bonded packings are determined by many factors requiring thorough control in the manufacturing process in order to obtain reproducible results. Different properties have been observed not only for packings with similar specifications from different suppliers but often for different batches of the same product (189-192).

The relation between retention characteristics and structure of alkyl-bonded sorbent is complex and depends on both the properties of the supporting silica gel and the conditions of the derivatization reaction. Specific surface area, pore volume, and concentration of the surface silanol groups of silica gel are important. The derivatization can be performed by a great variety of procedures (2,124). The reaction of alkylchlorosilanes with surface silanols is in widespread use. Tri- and dichloroalkylsilanes are capable of mutual condensation unless the reaction is performed under strictly anhydrous conditions. Stationary phase produced in this way may acquire a partially polymeric character. However, arguments have been presented that this polymerization reaction is often negligible and the most "polymeric" phases are in fact "monomeric" (193).

The following hydrolysis of untreated chlorines leads to the formation of secondary silanols, which eventually modifies the polarity and selectivity of "polymeric"-type stationary phases.

Monochlorotrialkylsilanes, e.g., chlorodimethyloctadecylsilane, react uniquely with surface silanols forming a bonded stationary phase of the monomeric or brush type. Unreacted silanol groups are undesirable in many applications. In order to eliminate them, additional end-capping by reaction with a small reactant such as trimethylchlorosilane is often applied. For steric reasons the reaction with surface silanol groups is by no means complete. Maximal surface coverage is about 4 μmol m^{-2}, which means that approximately 50% of surface silanol groups remains unreacted even for end-capped packings (2), but their accessibility may be reduced. Some data regarding the chemistry and properties of alkyl-bonded and other RP packings are given in Table 5.4.

Capacity factors of solutes retained by RP mechanism increase with the length of the bonded alkyl chain under otherwise constant surface coverage and other conditions (115,167,194,195). A linear relationship between capacity factors and carbon content was reported in some cases (195,196), but more complex dependencies were also described (194,197). Retention also depends on the number of reactive groups of the reagent used for modification, type of reaction product (monomeric or polymeric), and porosity of the base silica. Selectivity is also affected by the same factors.

There are some indications that mass transfer is slower for highly loaded stationary phases and phases with long alkyl chains, and their efficiencies are slightly reduced (194). The same is valid for polymeric phases in comparison to the monomeric phases.

Some packing materials are supplied in several modifications differing in the average pore diameter of the base silica. Wide-pore packings are mainly devoted to the separation of high-molecular-weight substances which are excluded from the pores of common packings. However, their surface areas and carbon contents are lower than those of conventional narrow-pore packings.

The separation of alkaloids and other basic compounds is strongly affected by the presence of residual silanol groups, whose acidity pertains after alkylation of the surface even though their concentration decreases preferentially during the derivatization reaction (198). The interaction with basic compounds by ion exchange is possible and should be taken into account (197,199). Tailing peaks and changes in retention volume caused by injection of large amounts of solute are indications of this effect. In the dependence on alkyl layer structure the accessibility of surface silanols is restricted, thus slowing down the rate of mass transfer. Unless the accessibility of silanols is completely eliminated, the separation of basic compounds is usually adversely affected. Some manufacturers supply stationary phases

TABLE 5.4 Some Packings for Reversed Phase Liquid Chromatography

A. Octadecylsilane Phases

Name	Base material	Surface area[a] (m²/g)	Pore size[a] (nm)	Pore volume[a] (ml/g)	% C	End-capping	Particle size (μm)	Shape[b]
μBondapak C18	Porasil	see Table 5.2			10.0	No	10	I
LiChrosorb RP-18	LiChrosorb Si-60	see Table 5.2			22.0	No	5; 7; 10	I
LiChrospher 100-RP-18	LiChrospher 100	see Table 5.2			—	—	10	S
MicroPak CH	LiChrosorb Si-60	see Table 5.2			22.0	Yes	10	I
MicroPak MCH					12.0	Yes	5; 10	I
Nucleosil C18	Nucleosil 100	350	10	1.0	14	Yes	3; 5; 7; 10	S
Nucleosil 120 C18	Nucleosil 120	200	12	0.65	11	Yes	3; 5; 7; 10	S
Nucleosil 300 C18	Nucleosil 300	100	30	0.8	6.0	Yes	5; 7; 10	S
Octadecyl =Si-60	Si-60	550	6	0.9	—	No	3; 5	I
Octadecyl =Si-300	Si-300	280	30	1.5	—	Yes	3; 5	I
Octadecyl =Si-500	Si-500	70	50	1.5	—	No	10	I
Octadecyl=Daltosil 100	Daltosil 100	300	12	1.0	—	Yes	4	S
ODS Hypersil, C18	Hypersil	see Table 5.2			10	Yes	3; 5; 10	S

Liquid Chromatography

Name	Base material	% C				End-capping	Particle size (μm)	Shape
Partisil-10-ODS-3	Partisil 10	see Table 5.2			10	Yes	10	I
Perisorb RP-18	Perisorb A	14	0.05		10	No	30–40	Sc
Radial-Pak C18	Radial-Pak Si	200	—		7	No	10	S
Spherisorb S ODS	Spherisorb S	see Table 5.2				No	5; 10	S
Spherisorb S5 ODS2	Spherisorb S	see Table 5.2				Yes	5	S
Spherosil C18	Spherosil XOA 600	see Table 5.2			22	No	5–7	S
Supelcosil LC18	Supelcosil LC-Si	171	—		11.2	Yes	3; 5	S
Ultrasphere ODS	Ultrasphere Si	—	—		12.5	Yes	3; 5	S
Vydac-201 TPB	Vydac 101 TPB	80	30	0.6	—	No	10	SI
Vydac-218 TPB	Vydac 101 TPB	80	30	0.6	—	Yes	10	SI
Zorbax ODS	Zorbax Sil	see Table 5.2			15.0	No	8	S

B. Octylsilane Phases

Name	Base material	% C	End-capping	Particle size (μm)	Shape
LiChrosorb RP-8	LiChrosorb Si60	13	No	5; 7; 10	I
MOS-Hypersil C8	Hypersil	7.0	No	3; 5; 10	S
Nucleosil C8	Nucleosil 100	9.0	No	5; 7; 10	S

TABLE 5.4 (B) (Continued)

Name	Base material	% C	End-capping	Particle size (μm)	Shape
Nucleosil 120-C8	Nucleosil 120	7.0	No	3; 5; 7; 10	S
Nucleosil 300-C8	Nucleosil 300	2.0	No	5; 7; 10	S
Octyl=Si60	Si60	—	No	3	I
Octyl=Si100	Si100	—	No	3; 5	I
Octyl=Daltosil 100	Daltosil 100	—	Yes	4	S
Perisorb RP-8	Perisorb A	—	No	30–40	S[c]
Supelcosil LC-8	Supelcosil LC-Si	6.6	Yes	3; 5	S
Ultrasphere Octyl	Ultrasphere Si	6.5	Yes	5	S
Zorbax C8	Zorbax Sil	15.0	No	6	S

C. Other Hydrocarbon Phases

Name	Base material	% C	End-capping	Particle size (μm)	Shape[b]	Remark
Diphenyl=Si-100	Si-100	—	No	3; 5; 10	I	
Hamilton PRP-1	styrenedivinylbenzene	—	—	10–15	S	
Hitachi Gel 3011	styrenedivinylbenzene	—	—	10–15	S	—CH$_2$OH groups
LiChrosorb RP-2	LiChrosorb Si-60	—	No	5; 7; 10	I	

Name	Base material	% C		Particle size (μm)	Shape	
Nucleosil C_6H_5	Nucleosil 100	8.0	No	7	S	
Nucleosil 120 C_6H_5	Nucleosil 120	6.0	No	7	S	
Perisorb RP-2	Perisorb A	—	No	30–40	S[c]	
Phenyl = Si-100	Si-100	—	No	3; 5; 10	I	
Phenyl-Hypersil	Hypersil	5.0	No	3; 5; 10	S	
SAS-Hypersil, C1	Hypersil	2.6	No	3; 5; 10	S	
Vydac-214 TPB	Vydac 101 TPB	—	Yes	10	SI	C_4 groups
Vydac-219 TPB	Vydac 101 TPB	—	Yes	10	SI	C_6H_5 groups

D. Other Nonhydrocarbon Phases

Name	Base material	% C	Particle size (μm)	Shape[b]
APS-Hypersil, NH_2	Hypersil	2.2	3; 5; 10	S
APS-Hypersil 2, NH_2	Hypersil	2.0	3; 5; 10	S
μBondapak CN	Porasil	—	10	I
μBondapak NH_2	Porasil	2.4	10	I
CPS-Hypersil, CN	Hypersil	4.0	3; 5; 10	S
LiChrosorb CN	LiChrosorb Si-60	—	5; 7; 10	I
LiChrosorb NH_2	LiChrosorb Si-60	—	5; 7; 10	I

TABLE 5.4 (D) (Continued)

Name	Base material	% C	Particle size (μm)	Shape[b]
Nucleosil CN	Nucleosil 100	—	5; 10	S
Nucleosil 120-CN	Nucleosil 120	—	7	S
Nucleosil NH$_2$	Nucleosil 100	—	5; 10	S
Nucleosil 120-NH$_2$	Nucleosil 120	—	7	S
Ultrasphere Cyano	Ultrasphere Si	—	5	S
Ultrasil NH$_2$	Ultrasil	—	5	S

[a]Data related to the base material.
[b]I, irregular; S, spherical; SI, spheroidal.
[c]Pellicular packing.

Liquid Chromatography

with strongly deactivated silanol groups especially designed for the separation of basic compounds (200) (see Table 5.4), but even in this case controlling the effect of the silanol groups by means of mobile phase composition is recommended.

Besides the wide range of methods for determining surface silanol groups, there are simple chromatographic tests to evaluate their accessibility. In a system of alkyl-bonded packing with a nonpolar mobile phase, the retention of a polar solute can be used as a measure of this effect. Most often nitrobenzene as the test probe is used with heptane or hexane as the mobile phase (196,201,202). However, the nitrobenzene test does not fully account for the behavior of the packing toward basic compounds as determined by capacity factors and plate numbers of <u>Cinchona</u> alkaloids as the test probes under RP conditions (203).

The presence of trace metals is another fact which can adversely affect the separation of specific compounds (204,205). The complexing of these solutes with traces of metals degrades their separation. Acetylacetone was proposed as the test probe in the chromatographic method indicating the presence of metals forming acetylacetonates (206). Naphthalene and 1-nitronaphthalene present in the testing mixture served to simultaneously detect changes in the silanol group interactions and in the phase ratio. Nevertheless, it was shown that the trace metals present in packing are insignificant with respect to amine separation, in contrast to the metallophilic interactions with stainless steel frits. Screens have been recommended as a substitute for frits in column endings for the separation of amines (207).

The structure and type of alkyl-bonded layer also affect the stability of the packing material. Since silica gel readily dissolves in alkaline aqeuous media (89,90), the use of silica-based packings is restricted to the approximate pH range 2-8. Primary, secondary, or tertiary amines are less detrimental than strong bases such as sodium hydroxide and quaternary ammonium hydroxides (89). Because of the protection of the underlying silica surface, packings with longer chain length, having high surface coverage and endcapping, tend to be more stable. The dissolution of silica is less important in mobile phases rich in organic modifier. The lifetime of a column can be prolonged by the use of a silica-packed precolumn which presaturates the mobile phase with silicic acid in the same way as for silica gel separation columns (90).

The dissolution of silica support is not the only process that occurs during the aging of RP columns. According to chromatographic testing and exhaustive spectral analyses, the increase in surface silanol groups caused by the loss of silane groups and the hydrolysis of surface siloxane groups was evidenced (208). Acidic conditions (pH < 3) are known to promote the cleavage of silane from silica, especially for monomeric stationary phase, and considerable phase stripping is observed (209,210). The effect is more serious when methanol rather than acetonitrile is the organic modifier. Polymeric packings based, say, on polystyrene-divinylbenzene copolymer are more suitable than those based on silica. They make possible the performance of separations with a much wider pH range (approximately 1.0–14.0). Considerable improvements have been achieved in the stability and efficiency of polymeric packings so that these parameters are only slightly worse than those of silica-based packings. Moreover, under acidic conditions on silica-based packings the changes in retention and separation factor values were observed even for neutral compounds, indicating some changes in the structure of the alkyl layer (197).

Retention in RP chromatography can be primarily controlled by the composition of the mobile phase, mainly the concentration of organic modifier in that phase. In the restricted concentration range an approximately linear decrease of the logarithm of capacity factor with organic modifier concentration is generally observed as follows:

$$\log k = a - b\phi \tag{5.5}$$

where a and b are constants, and ϕ is the volume fraction of organic modifier in aquoorganic mobile phase. The value of b is dependent on the solute, organic modifier, and type of packing material (211). For a given solute b can be used as a measure of solvent strength of the organic modifier. Average values as proposed by Snyder (211) are given in Table 5.5 and more or less follow the eluotropic series in RP chromatography. The value of a represents the log k

TABLE 5.5 Average Values of b (Eq. 5.5), Polarities P', and Selectivities x_e, x_d, and x_n ($x_e + x_d + x_n = 1$) According to Snyder (211, 230) for Various Organic Modifiers

Solvent	b	P'	x_e	x_d	x_n
Methanol	3.0	5.1	0.48	0.22	0.31
Acetonitrile	3.1	5.8	0.31	0.27	0.42
Acetone	3.4	5.1	0.35	0.23	0.42
Dioxane	3.5	4.8	0.36	0.24	0.40
Ethanol	3.6	4.3	0.52	0.19	0.29
Isopropanol	4.2	3.9	0.55	0.19	0.27
Tetrahydrofuran	4.4	4.0	0.38	0.20	0.42

value extrapolated to pure water and often differs from the experimental value and from the values obtained with other organic modifiers. This probably reflects the differences in the character of stationary phase when a different organic modifier participating in its solvation is used.

In the wider concentration range the quadratic function

$$\log k = a - b\phi + c\phi^2 \tag{5.6}$$

often shows better agreement with experimental data. The curvature is usually more expressive for less polar modifiers and more polar solutes (155,172). Quadratic function (5.6) has been derived from regular solutions theory (155) and from the interaction indices model (212). Other dependencies, e.g., linear in $\log \phi$ or logarithm of molar concentration, were also proposed (213,214) as were more complex semiempirical dependencies based on solvophobic theory (116, 117).

In the low range of organic modifier content the deviations from supposed simple dependencies are frequently observed (155,215-217). Significant changes in the composition of the stationary phase (106, 218-220) when a low amount of organic modifier is added to water are probably responsible for these deviations. The majority of RP packings are not wetted with these mobile phases and this presumably leads to slow mass transfer between stationary and mobile phases (196). After a sudden change in the mobile phase composition from pure organic solvent to pure water, a long time is necessary for the retention volumes of solutes to achieve stationary values (221-223). The complete removal of adsorbed organic modifier from the surface layer is difficult using pure water. Lengthy equilibration was also observed at low methanol concentrations (221). The difficulties encountered with the use of pure water in ion pair RP chromatography have been described (224). Moreover, the retention times in pure water are dependent on the column's thermal history (222). When a column was heated to 70-80°C and then returned to the original temperature, decreased retention was observed. At the same time, entrapping of solute molecules in the alkyl layer during column cooling was reported (222). Therefore it seems advisable to avoid the use of purely aqueous mobile phases whenever possible and to keep the organic modifier concentration at least at 10%.

The large deviations from the common RP dependencies of retention on mobile phase composition are encountered when another retention mechanism becomes operative. Minima on the log k vs. ϕ curves were ascertained for crown ethers and some peptides (86, 225). Similar behavior was also established for other amino compounds (226-228) and polar solutes (155,229). These compounds are capable of interacting with surface silanol groups and the resulting mixed retention mechanism is believed to be responsible for such anomalies. The effect is usually more pronounced for less polar organic modifiers.

Selectivity of a given column can be altered by changing the type of organic modifier. There is general agreement that the differences in solvent selectivity toward polar compounds are caused by differences in the ability of solvent to participate in dipole and acid-base interactions. To express these differences, an extended solubility parameter treatment was proposed (154,155,159-163). Another approach based on Rohrschneider parametrization of stationary phases in gas chromatography was developed by Snyder (230). His data on solvent polarity parameter P' and selectivity parameters x_e, x_d, and x_n expressing solvent affinity toward ethanol, dioxane, and nitromethane are given in Table 5.5 for several RP modifiers.

When considering the effect of organic modifier on selectivity, the properties of the stationary phase and its interaction with organic modifier have to be taken into account. In the solvation of stationary phase not only the interaction with alkyl chains but also the ability of organic modifier to interact with residual silanol groups is important. The importance of stationary phase solvation is apparent from the anomalous behavior of ternary mobile phases (156,231, 232), where, e.g., a low concentration of tetrahydrofuran in a mobile phase containing methanol and water effectively eliminates methanol from the stationary phase (218). Selectivity of the system can be substantially affected by the addition of a small amount of the third component to the mobile phase (154,218). Retention data for ternary mixtures generally cannot be linearly interpolated between the systems with individual organic modifiers (128,156,231). Nevertheless, anomalies in the properties (density, interfacial or surface tension) of ternary mixtures have also been reported (232). One must usually adopt an empirical approach when choosing a suitable separation system for a given sample. Although not fully transferable, published data on the effect of organic modifiers on the functional group selectivities might be of value (128,154,156,233,234).

Solvent strength and selectivity are not the only properties to be considered in the choice of the organic modifier. In some cases the possibility of operating a photometric detector in the short-wavelength range may be decisive. Properties of acetonitrile are superior in this respect, althought the elimination of dissolved oxygen improves the transmission of other solvents (235). The lower viscosity of acetonitrile in comparison to alcoholic solutions is favorable, but its higher toxicity and price are detrimental.

The temperature dependence of the capacity ratio in partition chromatography is given by the van't Hoff equation:

$$\ln k = -\frac{\Delta H}{RT} + \frac{\Delta S}{R} + \ln \psi \tag{5.7}$$

where ΔH and ΔS is retention enthalpy and entropy, respectively, and ψ is the phase ratio. If these values are independent of temperature the plot of log k vs. 1/T should be linear and from its slope retention enthalpy ΔH could be evaluated. Such plots have been observed for many compounds in various chromatographic systems with the application of chemically bonded stationary phases (102,103,183,236-240). Since the phase ratio ψ is difficult to assess for these packings, the retention entropies are uncertain and only their differences for various compounds are meaningful.

For a class of similar compounds, van't Hoff plots often intersect at one point corresponding to the so-called compensation temperature (183,236,240). This effect is the consequence of a strong correlation between enthalpy and entropy which is observed in many physicochemical systems. Differences in enthalpies between two compounds are partially compensated by differences in entropy. Compensation temperature is dependent on the class of compared compounds and the composition of the mobile phase (239,240).

There are, however, numerous cases in which the linear van't Hoff plots were not observed. If, in a mixed retention mechanism, partial processes are characterized by different retention enthalpies, then the nonlinear van't Hoff plot results (225). In systems where buffering of the mobile phase strongly affects separation, the temperature dependence of the dissociation constants of buffering components can be another reason for plot curvature (241). Conformational changes in the solute molecule can also lead to the nonlinearity of the plot (242). Recently, considerable deviations from linearity were found for neutral nonpolar compounds (alkanes, alkylbenzenes) when van't Hoff plots were constructed on a wider temperature range (222,243). Together with calorimetric (244), gas chromatographic (245), and NMR spectral measurements (246), changes in the structure of the alkyl-bonded stationary phase resembling phase transitions were identified as the source of these deviations (223).

These observations show the limitations of the simplified theoretical descriptions of the separation mechanism. A theory corresponding to experimental data at one temperature does not necessarily hold at another temperature for the same packing and mobile phase. Enthalpy-entropy compensation reveals the substantial role of entropy in the separation. As the difference in enthalpy and entropy terms determines log k value, the errors in both terms will add and yield a great error. The correlation of retention data with theories yielding both enthalpic and entropic terms would therefore be difficult. On the other hand, theories predicting only an enthalpic (or entropic) term can be compared with retention data in a semiquantitative manner due to the correlation of both terms in a class of similar compounds.

In practice, temperature is rarely used to control separation in RP liquid chromatography. Column thermostatting is required only to maintain the reproducibility of separations. Low temperatures can be useful for the separation of thermally labile compounds. Elevated temperatures are sometimes used to increase separation efficiency as the rate of mass transfer is accelerated or to decrease mobile phase viscosity (237). Limited solubility of sample components may be another factor necessitating elevated temperature. Simultaneously, however, processes leading to column deterioration, such as dissolution of silica or cleavage of the bonded phase, are also more efficient.

Besides alkyl-bonded chains, many other functional groups are being bonded to silica gel to produce a chromatographic packing material. The most common of these are phenyl, nitrilo, amino, nitro, and diol. Chiral molecules are bonded in packings assigned for the separation of optical isomers. The tradename of the packing does not fully express the surface chemistry. In many cases the functional group is not directly bonded to the silica. A spacer group, e.g., alkyl chain, provides a sufficiently large distance from the silica gel surface and prevents changes in group properties due to the proximity of the surface. Better coverage can also be achieved by these means for bulkier functional groups as the spacer chain offers them some mobility and eliminates excessive strain. In fact, nitro derivatives contain nitrophenyl groups in most cases.

Phenyl-bonded packings are typically used in RP mode in a way similar to that of alkyl-bonded phases. The carbon content is usually lower than that for the corresponding alkyl-bonded packing. Therefore, capacity factors are often lower for phenyl phase (234, 247). The presence of aromatic nucleus in the stationary phase will increase the affinity of the packing for the groups preferentially interacting through dispersion forces, e.g., to the aromatic part of the molecules. Although some results indicate similar behavior (248), selectivity is probably governed in a more complex way (247). The role of the organic modifier is substantial as can be clearly seen from the exceptional differences for tetrahydrofuran (247). Selectivity to other functional groups is also generally different from that of alkyl-bonded and nitrilo phases (234).

Similar conclusions can be drawn for the polar-bonded packings when used in the RP mode. Compared to the alkyl-bonded phases, the composition of stationary phase is complex. The presence of polar groups substantially affects the transfer of mobile phase components into the stationary phase. Due to the presence of accessible polar groups the effects connected with the mixed retention mechanism are expected to appear more frequently than for alkyl-bonded phases. The amount of bonded phase and its tendency to interact through dispersive interactions determines to the first approximation the degree of retention (248). As expected, the selectivity of the chromatographic system differs substantially for various bonded groups. A considerable effect of the type of organic modifier on functional group selectivity was also demonstrated (234). Packings with a bonded amino group can be used as a weak anion exchanger. Amino groups also partially deactivate the acidic properties of surface silanol groups, which is important when basic compounds are separated. All these effects cannot be fully predicted by current theories. Therefore an experimental approach based on analogy with published data (if available) must be adopted when an actual separation task is presented.

In the straight phase mode when eluted with nonpolar solvent, silica-based chemically bonded packings essentially behave as partially deactivated silica (249,250). Contrary to bare silica, the addition of water to the mobile phase is usually not necessary and equilibration of the column with mobile phase is accomplished more readily. The adsorption-displacement model developed for adsorption chromatography by Snyder was successfully applied to octadecyl (251), nitrilo (252,253), amino (250,254,255), and nitro (256) stationary phases. This theory describes quantitatively the decrease in retention that occurs when a polar modifier is added to the mobile phase. For an amino phase the effect of the polar modifier was shown to be closely related to the sorption of the modifier on the stationary phase (254).

Amino and nitro columns exhibit enhanced affinity toward polycyclic aromatic systems. For more polar compounds, which are eluted with the addition of more polar solvent, the interactions with bonded polar groups becomes efficient. Under these conditions significant differences between individual packings result in a specific selectivity pattern for each solvent–polar modifier combination (253).

As for the stability of polar-bonded packings, the restrictions are the same as for alkyl-bonded silica-based packings. Moreover, the specific sensitivity of bonded groups to chemical attack requires additional precautions. A bonded primary amine group is able to condense with reactive aldehydes and ketones. Acetone, for example, must not be used in the mobile phase for this column. If impurities of this type are present in the solvents used, e.g., formaldehyde in methanol, column deterioration can be expected. For the same reason, the sample injected should be free of carbonyl-containing solutes such as ketosteroids or reducing sugars. Primary amines can furthermore undergo oxidation. Degassing of the mobile phase is recommended and peroxides present in some solvents (ethers, tetrahydrofuran) must be eliminated (6). For the nitrile group acidic eluent must be avoided to prevent hydrolysis of CN groups.

A. Separation of Alkaloids in Reversed Phase Systems

Initial studies on RP chromatography using silica-based alkyl-bonded phases were unsuccessful in separations of basic compounds including alkaloids and this chromatographic technique was deemed unsuitable for this purpose (77,226). With the increasing number of applications of RP chromatography in various area of analytical chemistry, the conditions that enable the successful separation of most alkaloids were brought into existence. At present, most chromatographic analyses of alkaloids are performed using RP chromatographic systems, which implies the widespread use of this technique (6). In many respects RP separations of alkaloids on silica-based packings are carried out similarly to the separations on bare silica with similar mobile phases. Alkyl coverage seems to only partially modify the retention and selectivity of the system (74,82). Nevertheless, a changed selectivity might be just what is required. Also, the specific properties of RP packings with respect to the requirements of sample preparation may contribute to the popularity of RP chromatographic analyses of alkaloids.

In mobile phases used in RP chromatographic systems alkaloid B can be partially present in the ionized form BH^+. The ratio of both forms at a given pH^* is determined by the dissociation constant K_a^* of conjugate acid BH^+:

$$BH^+ \xrightleftharpoons{K_a^*} B + H^+$$

When describing the retention of ionizable solutes, the presence of all forms in both mobile and stationary phases has to be taken into account (257–259). For simplicity's sake only dissociation to the first degree will be considered here. The capacity factor k can then be expressed as:

$$k = \frac{(n_B)_s + (n_{BH^+})_s}{(n_B)_m + (n_{BH^+})_m} = \frac{x_B(n_B)_s}{(n_B)_m} + \frac{x_{BH^+}(n_{BH^+})_s}{(n_{BH^+})_m}$$

$$= k_B x_B + (k_{BH^+}) x_{BH^+} \qquad (5.8)$$

where n denotes the amount of B or BH^+ in the stationary (s) or mobile (m) phase, $x_B = (n_B)_m/[(n_B)_m + (n_{BH^+})_m]$, and $x_{BH^+} = (n_{BH^+})_m/[(n_B)_m + (n_{BH^+})_m]$ fraction of the respective form of base in the mobile phase. The values of k_B and k_{BH^+} represent (hypothetical) capacity factors of the free and protonized bases, respectively. By substituting for x_B and x_{BH^+} from the Henderson-Hasselbach equation (2.6) and neglecting the activity coefficient, we obtain

$$k = \frac{k_{BH^+} + k_B \cdot 10^{(pH^* - pK_a^*)}}{1 + 10^{(pH^* - pK_a^*)}} \qquad (5.9)$$

This S-shaped dependence of k on mobile phase pH^* has actually been observed in many cases (257-261), as shown in Figure 5.6. Here the capacity factor of quinine is plotted against the pH of aqueous buffer used in the mobile phase preparation. The curve fitted to the experimental points according to Eq. 5.9 is drawn in the Figure 5.6 for the best values $k_{BH^+} = 1.14$, $k_B = 9.81$, and half-value $pH_{1/2} = 6.15$. Half-value $pH_{1/2}$ corresponds to the pK_a^* in Eq. 5.9 but is related to the aqueous buffer instead of the mobile phase. Therefore $pH_{1/2}$ is shifted with respect to the pK_a^* value

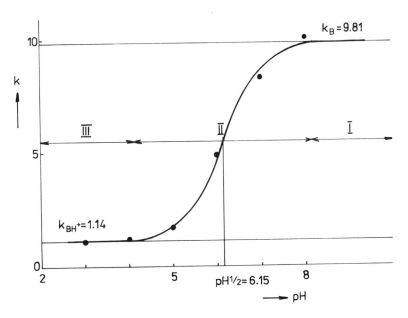

FIGURE 5.6 Capacity factor of quinine as a function of the pH of phosphate buffer. Stationary phase Separon SIX C18 5 μm (Laboratory Instruments Works, Prague), mobile phase methanol-0.025 M NaH_2PO_4 adjusted to pH required by 10% wt NaOH or 10% wt H_3PO_4, 70:30 v/v. (Based on data from Ref. 261.)

or to the value $pK_a = 8.34$ tabulated for water (see Table 2.1) due to the effect of organic modifier as discussed in Chapter 2, Section I.F.

The use of Eq. 5.9 assumes that k_B and k_{BH^+} are independent of pH. In fact, changes in the stationary phase, such as the dissociation of surface silanol groups, can affect k_B and k_{BH^+}, and considerable changes in both values can be hidden by total dependence. Nevertheless, it can be stated that RP separations of alkaloids can be performed in three different pH regions. In the first region (I in Figure 5.6) pH^* of the mobile phase is sufficiently high that alkaloid is present solely as a free base and will behave essentially as any other neutral solute. In agreement with this behavior, k_B decreases when the concentration of organic modifier in the mobile phase increases (260-263) (Figure 5.7A). The presence of a polar nitrogen group will decrease the retention of the analyte in a pure RP mechanism. However, undissociated residual silanol groups can interact with basic nitrogen of neutral alkaloid molecules through hydrogen bonding or acid-base interactions in a similar way as for bare silica (see Chapter 5, Section I.A). The physical state of surface silanol groups depends on the composition and concentration of ionic components in the mobile phase, which can substantially affect equilibrium for silanophilic interactions. The k_B value is therefore expected to depend on the concentration of buffer and other salts in the mobile phase. In fact, the decrease in k values with the increase of concentration of phosphate buffer was observed (261) (Figure 5.8). In a 70% methanolic mobile phase, a pH of 7 for the aqueous buffer renders the alkaloids nearly undissociated and k values thus approximately represent k_B values (although the change in ionization due to ionic strength difference also partially contributes to decreased retention). An ordinary salting-out effect would lead to increased retention (116,258). There are probably other reasons besides the presence of residual silanol groups for the dependence of retention on the concentration of electrolytes in the mobile phase, since this effect was even observed for the RP separation of alkaloids on styrene-divinylbenzene copolymer (262).

In the transient region II (Figure 5.6) both ionized and neutral forms of alkaloid are present in the mobile phase. Unless k_B and k_{BH^+} have similar (e.g., low) values, the alkaloid capacity factor is strongly affected by the pH^* of the mobile phase. In order to achieve sufficient loadability of the column and reproducibility of the separation, effective buffering of the mobile phase is necessary. On the other hand, the separation of alkaloids from compounds differing in pK_a^* values can be optimized by adjustment of mobile phase acidity.

In the acidic region III (Figure 5.6) the protonized form of alkaloid BH^+ dominates the separation. At first sight it seems improb-

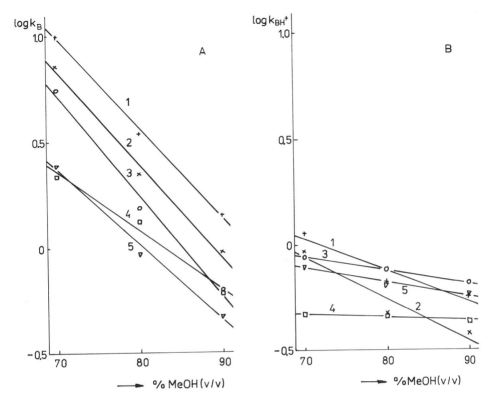

FIGURE 5.7 Effect of methanol content on the capacity factor of (A) free base B, (B) protonized form BH^+ of some alkaloids. 1, Quinine (+); 2, emetine (x); 3, dihydroergocornine (o); 4, codeine (□); 5, yohimbine (Δ). Values of k_B and k_{BH^+} were evaluated by fitting Eq. 5.9 to experimental points taken from Ref. 261. Conditions as in Figure 5.6 except for varying methanol content.

able that ionized species would be retained by nonpolar RP packing of alkyl or phenyl type. In fact, there are in principle two mechanisms enabling ionized species to be retained. Besides the protonized nitrogen atom with localized charge, alkaloid molecules contain an organic moiety which can interact with an organic RP layer. This interaction is essentially a surface phenomenon since the ionized nitrogen atom together with the solvation shell cannot penetrate the alkyl layer. Whereas the organic part of the ionic solute comes in contact with the nonpolar surface of the sorbent, the ionized nitrogen atom accompanied by an oppositely charged ion and all the ionic atmosphere probably remains in the adhering polar medium. This double-layer-type adsorption is well established in RP ion pair chro-

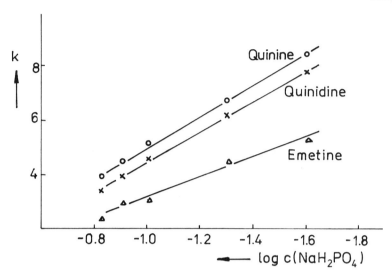

FIGURE 5.8 Dependence of alkaloid capacity factors on the concentration of NaH_2PO_4 in the buffer component (pH = 7) of the mobile phase. Other conditions as in Figure 5.6. (Reproduced from the Journal of Chromatographic Science, Ref. 261, by permission of Preston Publications, A Division of Preston Industries, Inc.)

matography for cationic and anionic surfactants (see Chapter 5, Section IV). Ionized organic acids have essentially been separated by this mechanism (264). Because the interaction of the organic part of the solute with the nonpolar surface of the sorbent is substantial, water-rich mobile phases promote retention. For alkaloid ion BH^+ the process can be formally described as

$$BH_m^+ + A_m^- \rightleftharpoons BH_s^+ + A_s^-$$

where indices m and s denote the species in mobile and stationary phases. Cation A^- is retained together with solute BH^+ because the system's electroneutrality must not be violated. Although the stationary phase formed by the surface layer of the sorbent and the adhering eluent is not well defined, the above equation represents that the retention of ionized solute BH^+ is affected by the concentration of other ionic components in the mobile and stationary phases.

The presence of an electrolyte of sufficient concentration is of great importance. Otherwise the surface charge produced by initial sorption of ionic solute eliminates the access of other ionized molecules of the same charge to the surface and to the pores of the sorbent. It is a frequent phenomenon that diluted ionic compounds are eluted with retention volumes lower than column dead volume due to ion exclusion (107,264). The addition of electrolyte to the mobile phase eliminates this effect. The presence of electrolyte C^+A^- in the mobile and surface stationary phases makes it possible to formulate the retention as

$$BH_m^+ + C_s^+ \rightleftharpoons BH_s^+ + C_m^+$$

hence, essentially as a dynamic ion exchange mechanism.

The second retention mechanism is ion exchange, which can take place if ionized silanol groups (82,265,266) or, eventually, other acidic centers (267) are present on the sorbent. Then the cations C^+, which neutralize the $\equiv SiO^-$ groups in the neighborhood of sorbent surface, can be exchanged by ions of alkaloid BH^+:

$$BH_m^+ + \equiv SiO^-, C^+ \xrightleftharpoons{K_{ex}} C_m^+ + \equiv SiO^-, BH^+$$

Here, similarly as in Section I.A of this chapter, the comma denotes the bonding of cations to the surface due to electroneutrality requirements. Just as in common ion exchange chromatography, the capacity factor k_{BH^+} can be expressed as

$$k_{BH^+} = \psi K_{ex} \frac{[C^+]_s}{[C^+]_m} = \frac{K_{ex} \, n_{SiO^-}}{V_m [C^+]_m} \qquad (5.10)$$

if the concentrations of a solute in both mobile and stationary phases are negligible in comparison with the concentration of eluting cation C^+. In this equation K_{ex} is the equilibrium constant of the exchange reaction, $\psi = V_s/V_m$ is the phase ratio, V_m is the volume of mobile phase in the column, and n_{SiO^-} is the total number of ion exchange sites in the column. In the case of uni-univalent exchange, the capacity factor should therefore be inversely proportional to the concentration of eluting ions in the mobile phase $[C^+]_m$. If the ion exchange is operating within the column, an electrolyte must be present in the mobile phase. Otherwise $([C^+]_m \longrightarrow 0)$, the sorption of solute is irreversible and solute is not eluted from the column at all. At lower concentrations of eluting cations the retention of ionized form can be greater than the retention of neutral form as in the case of several basic compounds studied at high methanol content in the mobile phase (265). Simultaneously, a decrease in retention was observed at low pH (265,266). Such a decrease can be explained by an exchange of the eluting cation C^+ at the surface by protons H^+ followed by the suppression of ionization of the surface silanol groups in agreement with the low acidic strength of these groups:

$$SiO^-, H^+ \rightleftharpoons SiOH$$

In this way the number of exchangeable sites decreases. A large concentration of eluting cations in the mobile phase largely eliminates the ion exchange retention mechanism.

Often systems are studied in which two or more eluting cations are present in the mobile phase and only the concentration of one cation is changed or eventually one cation is substituted for another with a different elution strength. Moreover, if another retention mechanism is operating, concentration dependence is of the type:

$$k = \frac{a + b[C^+]_m}{1 + c[C^+]_m} \tag{5.11}$$

where a, b, and c are constants and $[C^+]_m$ is the concentration of the eluting cation studied. This dependence has actually been observed in several systems (86,265,268,269). An equation of this type can be derived based on various suppositions (86,268).

Even in the case of an ion exchange mechanism the interaction of organic parts of ions with the nonpolar surface of RP packing contributes to the stabilization of retained organic ions. Thus, the addition of methanol to the mobile phase leads to decreased retention (261,263,266,270) (Figure 5.7B). At the same time some amines or ammonium compounds are more effective eluting agents than the inorganic cations (266,270). In the concentration region where the interactions with silanol are not completely eliminated, the concentration of amine in the mobile phase can effect the separation of silanophilic compounds from other components of the sample. A surprising behavior in this respect was observed on a µBondapak C18 column; retention of ephedrine was low on this column even at zero triethylamine concentration in contrast to other columns (270).

The separation of alkaloids and other basic drugs is frequently characterized by broad and tailing peaks. The elution volumes and peak shapes often depend on the amount of injected solute in such cases. Curvature of the sorption isotherm is known to have such an effect. The mixed retention mechanism can operate in this unfavorable manner (86,271) if at least one participating mechanism is characterized by strong interactions and poor linearity of the isotherm. This may be caused, for example, by a limited number of sorption sites. Unless of a low concentration, the presence of ionized solute molecules in the mobile and/or stationary phase changes the concentration of other ionic components and affects the distribution coefficient. Ionic separations are therefore prone to isotherm nonlinearity. In fact, the increase in concentration of buffer components in the mobile phase usually improves the peak symmetry and sharpness to some extent (203). Another factor that decreases efficiency is slow mass transfer.

The effect of silanol on efficiency is well recognized but not terribly well understood (82,241). Bare silica offers better separation efficiency than the same silica with bonded phase (77,82). The high surface coverage exacerbates this unfavorable effect. Both the decrease in number of silanol groups and their accessibility for higher polar and ionic solutes can contribute to this behavior.

The efficiency of a column for the separation of basic compounds can be substantially improved by the addition of an organic base as amines or quaternary ammonium compounds (203,241,268,270-272). This effect may be at least partially ascribed to the masking of silanol groups (241). Amines containing hydroxyl groups and short-alkyl-chain amines have been less effective in improving efficiency (272). However, the best results for trimethylamine, triethylamine, dimethylamine, and diethylamine are reported in Ref. 271. In addition, dibutylamine, hexylamine, $\underline{N}$-methylhexylamine, and $\underline{N},\underline{N}$-dimethylhexylamine substantially reduced retention and peak tailing (272). Bulky substituted nitrogenous bases, e.g., tetrabutylammonium ion, had little effect (268). However, amines are incompatible with short-wavelength photometric detection.

Frequent anomalies in the chromatographic behavior of alkaloids require caution in quantitative analysis. The calibration curves should be investigated throughout the required concentration range and checked for linearity. In particular, the measurements based on peak height evaluation could be adversaly affected by changes in column efficiency due to the high amount of injected solute and other components of the sample. Recovery might also be a problem. For example, thebaine was reported to adsorb on column active sites containing μBondapak C18 and eluted with methanol-0.3% $(NH_4)_2CO_3$ 4:1. The column required presaturation with thebaine for quantitative work (273).

Other difficulties can result from slow column equilibration, as has been observed chromatographically (261) and by effluent pH measurement (267). It was reported that a slow equilibration rate is not a kinetically determined phenomenon (267), but this effect is still far from being fully understood. In any case, similar phenomena should be taken into account when a gradient elution method is to be developed.

Since the separation of alkaloids is to a large extent controlled by factors which are not greatly emphasized from the standpoint of the production of common chromatographic sorbents, it is practically impossible to reproduce a given separation with another column of similar type. The differences among columns of various tradenames are extreme (203,274). The transfer of the elaborated method from one column to another will probably require the adjustment of separation conditions with an unpredictable final result. Often only a slight change in the mobile phase composition or the addition of another organic modifier can result in a desired effect (275,276).

Several examples of the separation of Cinchona and opium alkaloids are given on the Figures 5.9—5.12. The advantageous use of a highly alkaline mobile phase with styrene-divinylbenzene gel without danger of column damage is demonstrated in Figure 5.9 (262). Separation was performed at an elevated temperature. Adequate stability of Bondapak Phenyl column in methanolic mobile phase of pH 10.5 was claimed and resulted in the successful separation of opium alkaloids in neutral form (Figure 5.10). In contrast with the separation systems using alkyl-bonded phases, morphine was sufficiently retained, with other alkaloids being eluted in reasonable time without the need for gradient elution (277).

Cinchona alkaloids are separated in Figures 5.11 and 5.12 (203) under acidic conditions in ionized forms and, in some cases, even as double-charged species (quinine, quinidine). The low content of methanol compared to that of the above-given examples is noteworthy. In part it outweighs the high content of inorganic buffer required to suppress excessive tailing. From the seven columns tested with sodium phosphate buffer, peak tailing was least on the columns containing µBondapak C18 and LiChrosorb RP-8 (Figure 5.11). Superior separation using hexylamine as cation is evident from Figure 5.12.

Figure 5.13 shows the possibility of separating the optical isomers of alkaloids on RP columns after derivatization of sample with chiral reagent. Diastereoisomeric derivatives of (±)-ephedrine obtained by reaction with 2,3,4,6-tetra-O-acetyl-β-D-glucopyranosyl-isothiocyanate were separated on an Ultrasphere ODS column (278). Among other reagents used for this purpose, R-α-methylbenzyliso-thiocyanate was applied to ephedrine (279). The addition of a chiral component, such as β-cyclodextrine, to the mobile phase (280) was applied to the separation of optical isomers of nicotine analogues (Chapter 7, Ref. 117). The use of chiral ion pair reagent will be mentioned in Section IV of this chapter.

B. Separation of Alkaloids on Polar Stationary Phases

Stationary phases with polar functional groups are used less frequently than reversed phases for the separation of alkaloids. Prin-

Liquid Chromatography

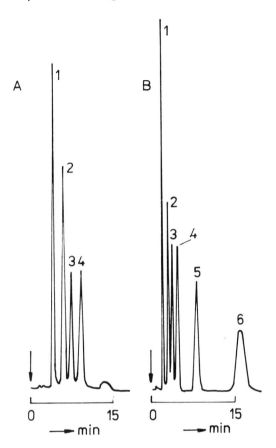

FIGURE 5.9 Separation of opium alkaloids on macroporous styrene-divinylbenzene copolymer Hitachi Gel 3010, 10 μm. Column 220 × 4.6 mm, mobile phase 60% acetonitrile, 0.020 M NH_3 at 60°C; flow rate (A) 1 ml min^{-1}; (B) 2 ml min^{-1}, UV detection at 254 nm. 1, Morphine; 2, codeine; 3, papaverine; 4, yohimbine; 5, narcotine; 6, reserpine. (Reprinted with permission from Analytical Chemistry 52 (1980) 1963, Ref. 262. Copyright 1980 American Chemical Society.)

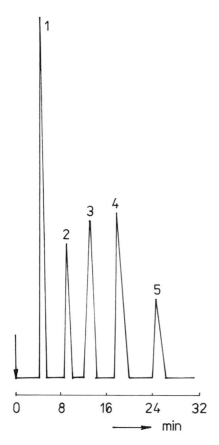

FIGURE 5.10 Separation of principal opium alkaloids on μBondapak phenyl, 10 μm. Column 300 × 3.9 mm. Mobile phase 1% sodium acetate, 7.0 mM triethylamine (pH 11) -methanol 42:58, flow rate 1.0 ml min^{-1}, UV detection at 254 nm. 1, Morphine; 2, codeine; 3, papaverine; 4, thebaine; 5, noscapine. (After Ref. 277.)

Liquid Chromatography 187

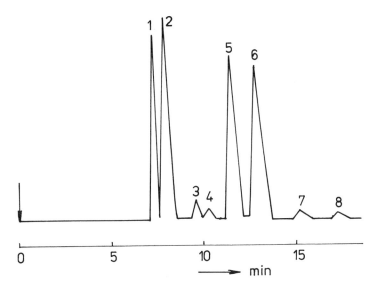

FIGURE 5.11 Separation of major Cinchona alkaloids and their dihydro derivatives on LiChrosorb RP-8 Select B, 7 μm. Column 250 × 4 mm, mobile phase 0.1 M potassium dihydrogen phosphate, pH 3.0–acetonitrile 85:15, 1.0 ml min^{-1}, UV detection at 220 nm. 1, Cinchonine; 2, cinchonidine; 3, hydrocinchonine; 4, hydrocinchonidine; 5, quinidine; 6, quinine; 7, hydroquinidine; 8, hydroquinine. (After Ref. 203.)

cipally, their use in both reversed phase (RP) and normal phase (NP) modes is possible but rapid transfer of a single column between respective modes is not generally recommended. When used in RP mode the retention mechanism and the dependence of retention on experimental conditions are similar to those for the separation on RP packings. In the NP mode the rules for separating on alumina or silica with mobile phases of low polarity are applicable. Since experimental data on the retention behavior on these packings are rare, this approach is useful although approximate. The differences in capacity factor values and selectivities from those of the usual RP and NP packings are of course encountered, and they are the main reason for utilizing these packings.

As compared to other packings discussed in this section, nitrile phase has been applied in numerous cases. The published data are in agreement with the expected behavior. The capacity factors of opium alkaloids on Nucleosil 10 CN column decreased when the concentration of acetonitrile in the mobile phase buffered with ammonium acetate to pH 5.8 increased from 20% to 40% (281). In contrast

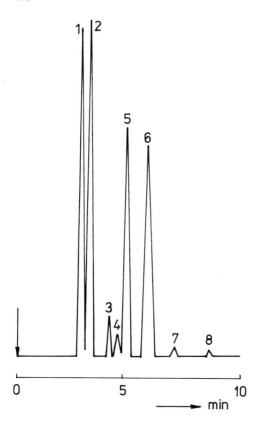

FIGURE 5.12 Separation of major <u>Cinchona</u> alkaloids and their dihydro derivatives on Novapak C18, 4 µm. Column 150 × 3.9 mm, mobile phase 7% acetonitrile in 0.05 M hexylamine adjusted to pH 3.0 with orthophosphoric acid, 1.0 ml min^{-1}, UV detection at 220 nm. Peak identities as in Fig. 5.11. (After Ref. 203.)

with the C18 column the retention of morphine was sufficient, although otherwise the separations were very similar. At 20% acetonitrile the retention times increased when the buffer pH was changed from 4 to 7, as was expected from the suppressed ionization of alkaloid. A similar pH dependence has also been found for morphine on the Radial-Pak µBondapak CN column eluted with mobile phase 0.05 M KH_2PO_4 buffer-acetonitrile 8:2. However, pseudomorphine showed maximal retention at pH 5.3 (282). The separation of opium alkaloids on nitrile column is given in Figure 5.14 (281). The method for the determination of the main alkaloids in gum opium consisting of µBondapak CN column and mobile phase 1%

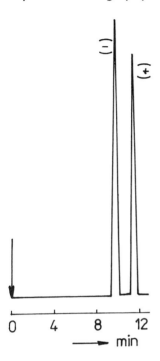

FIGURE 5.13 Separation of optical isomers of ephedrine after derivatization with 2,3,4,6-tetra-$\underline{O}$-acetyl-β-$\underline{D}$-glucopyranosyl isothiocyanate. Column 150 × 4.6 mm packed with Ultrasphere ODS, 5 μm, mobile phase 2.33 g liter^{-1} monobasic ammonium phosphate-acetonitrile 6:4, 1.0 ml min^{-1}, UV detection at 254 nm. The enantiomeric identity is given. (After Ref. 278.)

sodium acetate, pH 6.78 adjusted with glacial acetic acid-acetonitrile-1,4-dioxane 75:20:5 was recently reported (283). The changes in retention of atropine and apoatropine on Nucleosil 5 CN column with pH, sodium acetate buffer concentration, and methanol content in the mobile phase were similar to those with RP packings (284).

The separation of four Cinchona alkaloids on a Spherisorb CN column was optimized by the adjustment of methanol, acetonitrile, and tetrahydrofuran concentrations in the mobile phase (285). The fourth aqueous component was buffered to pH 7 with sodium phosphate. In the optimized separation all four alkaloids were sufficiently retained to be separated from polar components of plant extract (Figure 5.15). Capacity factors and characteristics for oxidative amperometric detection have been determined for 43 basic drugs including several alkaloids using LiChrosorb CN column and mobile

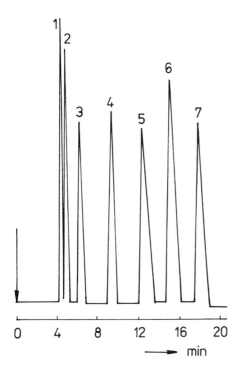

FIGURE 5.14 Separation of seven opium alkaloids on Nucleosil 10 CN. Column 300 × 4 mm, mobile phase 1% ammonium acetate, pH 6.3–acetonitrile–dioxane 79:16:5, 1.5 ml min^{-1}, UV detection at 254 nm. 1, Narceine; 2, morphine; 3, codeine; 4, cryptopine; 5, thebaine; 6, narcotine; 7, papaverine. (After Ref. 281.)

phase acetonitrile-phosphate buffer (pH 3, ionic strength = 0.05) 4:6 containing 0.001 M sodium chloride (286).

The operation of nitrile phase in the NP mode was applied to the separation of alkaloids in a few cases. The retention of eburnane and ergot peptide alkaloids on a µBondapak CN column eluted with hexane-chloroform-acetonitrile mixture was similar to that obtained on silica (30,287). The addition of diethylamine to the mobile phase in concentrations as low as 0.001 mol liter^{-1} led to a considerable decrease in alkaloid retention (287). The retention times for papaverine, caffeine, cocaine, yohimbine, heroin, strychnine, and other drugs were reported together with the method of HPLC assay of papaverine in blood using a Varian Micropak CN column and mobile phase n-hexane-dichloromethane-acetonitrile-propylamine 50:25:25:0.1 (288).

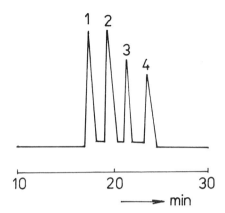

FIGURE 5.15 Separation of the four <u>Cinchona</u> alkaloids on Spherisorb CN, 5 μm. Column 250 × 4.6 mm, mobile phase 0.0068 M H_3PO_4 buffered to pH 7.0 with 1 M NaOH-methanol-acetonitrile-tetrahydrofuran 50:28.7:17:3.3, 1.5 ml min^{-1}, 50°C, UV detection at 231 nm. 1, Quinine; 2, quinidine; 3, cinchonidine; 4, cinchonine. (After Ref. 285.)

Phases with amine functional groups when used with RP eluents display substantially different selectivity from that of the usual RP packings. An approximately reversed elution order was observed for opium alkaloids with elution of morphine coming last (Figure 5.16). Papaverine was retained slightly and coeluted with α-narcotine (289). These properties turn out to be advantageous in the analysis of illicit heroin (290). On the aminopropyl-bonded silica S5NH$_2$, methanol produced much broader and more tailing peaks than acetonitrile. The addition of tertiary alkylammonium salt resulted in a shorter retention time of morphine and sharper peaks. With mobile phase acetonitrile-0.005 M tetrabutylammonium phosphate 85:15 the separation was satisfactory. Again, α-narcotine and papaverine were eluted first whereas morphine was the most retained opium alkaloid (290).

An amine stationary phase [μBondapak NH_2 (291), Radial-Pak amino column (292)] was applied to the separation of potato glycoalkaloids with mobile phase tetrahydrofuran-3.4 g $liter^{-1}$ KH_2PO_4-acetonitrile 50:30:20 (291) and tetrahydrofuran-acetonitrile-water-methanol 55:30:10:5 or 50:25:15.5:9.5 (292). Pyrrolizidine-<u>N</u>-oxide alkaloids of <u>Symphyti</u> radix were separated on μBondapak-NH_2 column with mobile phase acetonitrile-water 92:8 (293) or in reduced form on MN-Nucleosil-NH_2 column with mobile phase dichloromethane saturated with 1% ammonium carbonate-isopropanol 20:1 (294).

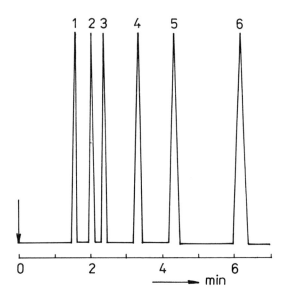

FIGURE 5.16 Separation of opium alkaloids on Zorbax NH_2 (Du Pont). Column 250 × 4.6 mm, mobile phase acetonitrile–0.025 M KH_2PO_4 75:25, 2 ml min^{-1}, UV detection at 286 nm. 1, Papaverine (k = 0.157); 2, thebaine (0.450); 3, narceine (0.728); 4, codeine (1.371); 5, morphine (2.029); 6, tyrosine (3.41, internal standard). (After Ref. 289.)

Studies on the separation of ergot alkaloids (295,296) on Micro-Pak NH_2 and LiChrosorb NH_2 columns with diethyl ether or chloroform containing ethanol or isopropanol as mobile phase showed a decrease in retention with increasing concentration of ethanol. The configurations stabilized by intramolecular hydrogen bonding to nitrogen atoms were substantially less retained. In addition, other changes in intramolecular hydrogen bonds affected the retention. Alkaloids containing benzyl groups were more retained than alkyl-substituted ones. The smaller the alkyl group substituent, the higher were the retention volumes (296). The compounds containing primary or secondary hydroxyl groups had larger retention volumes, whereas those containing tertiary hydroxyl groups had lower retention volumes compared to the parent compounds (295). Semisynthetic ergot preparations were analyzed and isolated on a Perkin-Elmer LC column NH_2 and MicroPak NH_2 columns with mobile phases hexane-ethanol 7:3 and diethyl ether-ethanol 4:1, respectively. Unlike the RP-18 column, methylation of the indole nitrogen lowered retention (297).

LiChrosorb DIOL stationary phase was applied in the study of fluorometric detection by ion pair formation with phosphate buffer as the mobile phase. Dihydroergotamine, emetine, ephedrine, and other amines were studied. Retention increased with eluent pH in the range 2.0-6.0 (298).

A nitrophenyl phase with mobile phase dichloromethane-methanol was shown to have advantageous properties in the separation of piperine and its cis-trans isomers and was applied to the analysis of pepper extract (299).

A "carbohydrate" column (Waters Assoc.) was applied for the separation of glycoalkaloids (291,300). Methanol-isopropanol and isopropanol-cyclohexane mixtures were tested as the mobile phase. Increasing retention was observed with higher concentrations of isopropanol and cyclohexane, respectively (300). The determination of α-chaconine and α-solanine in potatoes was performed using mobile phase tetrahydrofuran-water-acetonitrile 56:14:30. Increased sensitivity may be achieved by using acetonitrile-water 85:15 as an eluent since it permits UV detection at 200 nm (291).

The application of nonpolar styrene-divinylbenzene copolymer in the RP mode was mentioned in the preceding section (II.A). If a polar functional group is introduced into the copolymer structure, the packing may be operated in the NP mode in a way analogous to that of, say, silica gel. Using a Hitachi Gel 3010-O (porous polystyrene modified by hydroxymethyl) column with eluent $\underline{n}$-hexane-ethanol-triethylamine 70:30:0.5, the determination of individual ergot alkaloids was possible (34). The substitution of hexane in the mobile phase by cyclohexane yielded lower retention. With mobile phase $\underline{n}$-hexane-chloroform-triethylamine 5:95:0.5, a different elution order of alkaloids was obtained.

When used with polar mobile phases organic gels containing polar functional groups behave in the same way as nonpolar packings in RP mode. The separation of basic drugs was studied on the styrene-divinylbenzene-methyl methacrylate terpolymer substituted with hydroxymethyl groups (260). Caffeine was included in the set of tested compounds. The addition of water to the methanolic mobile phase containing 1% of ammonia solution increased retention in most cases. The retention of caffeine, however, decreased only slightly. By stepwise substitution of methanol by acetonitrile the minimum in retention was found for the mixed mobile phase. Acetonitrile turned out to be a stronger eluting agent. The typical S-shaped dependence of k values on pH^* in the pH^* range 1.2-12.8 was observed for basic drugs except for some border points. The retention of caffeine as a weak base was not affected by mobile phase pH^* in the range studied. A higher efficiency than that of polymeric packings of medium and low polarity was claimed (260).

The sorption of alkaloids on hydrophilic and hydrophobic gels of the Sephadex series was studied (301). The concentration of electrolytes in the mobile phase did not influence sorption and the neutral solutes were sorbed more strongly than their ionized forms. Retention was also a function of the matrix concentration of the dextran gel.

Cyclodextrins form inclusion complexes with aromatic compounds in aqueous media, and cyclodextrin polymers demonstrate this property when employed in chromatographic separations (302). Indole alkaloids have been separated on β-cyclodextrin polymeric packing using citrate (303,304) or phosphate (304) buffer as the mobile phase at pH 4.0–5.5. Not all carbohydrate-type packings are suitable for these separations. For example, on Sephadex G-25 column little difference in the retention of alkaloids was found. The salt concentration and temperature (5–35°C) did not significantly affect retention. Since β-cyclodextrin is a chiral compound, the separation of enantiomers on this packing can be accomplished in some cases. A partial separation of (+)-vincadifformine and (−)-vincadifformine was achieved (303), but the efficiency was poor. The separation of enantiomers was dependent on the particle size of the packing, pH, and the type of buffer. Racemic vincadifformine and quebrachamine were resolved to their enantiomers on a preparative scale in good yield and optical purity (304). Recently, the study was extended to α-cyclodextrin polymer on which enantiomers of vincamine were separated (305). The selectivity factor and elution volumes increased as the pH of the mobile phase increased from 4.5 to 7.0. The upper pH limit is given by the limited solubility of studied alkaloids in water in the form of free base.

Another specialized group of packings that enables the separation of optical isomers is the bonded chiral packings. Ephedrine was analyzed on an α-aminopropyl packing modified with (R)-N-(3,5-dinitrobenzoyl)phenylglycine (306). Both ionically and covalently bonded packings of this type are commercially available. The enantiomers of ephedrine were resolved as their cyclic oxazolidine derivatives produced by condensation with 2-naphthaldehyde. An ionically bonded column was eluted with hexane-isopropanol 99.5:0.5. The theoretical model for the resolution of ephedrine and related α,β-aminoalcohols as enantiomeric oxazolidine derivatives is discussed in Ref. 307.

III. ION EXCHANGE CHROMATOGRAPHY

Ion exchangers were used as packing materials in the early column chromatographic separations. Considerable advances in general chromatographic theory and instrumentation as well as in the development of ion exchange technique have since been achieved (308,309). The applications of ion exchangers in the field of alkaloids appeared soon after these exchangers became commercially available. The industrial scale isolation of nicotine from waste gases in tobacco plants (310) and the development of portable extraction unit for the recovery of alkaloids from Cinchona bark (311) can be mentioned as examples. In the earliest analytical applications ion exchangers were used to isolate free base of alkaloid followed by acid-base titration (312,313). Simple ion exchange columns are currently used for the isolation of alkaloids from analyzed material or in order to transform them to the desired salt, e.g., for identification purposes. However, the direct analyses of alkaloids by means of ion exchange chromatography are not so frequent by now as might be expected owing to their basic character. The ion exchange separation of alkaloids has never reached the level of spread and perfection associated with, say, the ion exchange analysis of amino acids.

Ion exchange chromatography is performed on packing containing charge-bearing functional groups and only ionized solutes are supposed to be separated. The charge of the exchanger group is opposite to that of solute ions. For the separation of alkaloid cations BH^+, a cation exchanger containing ionized acidic groups such as sulfonic group $-SO_3^-$, is required. In chromatographic applications the column is equilibrated with an electrolyte containing at least one type of cation C^{z+} and the respective anions. Cation C^{z+} squeezes out the cations originally present in the ion exchanger and the ion exchanger phase then contains fixed anions R^- which are neutralized by the equivalent amount of cation C^{z+}. If an alkaloid that is generally protonized to the nth degree BH_n^{n+} is introduced to the mobile phase, the exchange of cations between mobile phase and ion exchanger occurs according to equilibrium:

$$z(BH_n^{n+})_m + nC_s^{z+} \rightleftarrows z(BH_n^{n+})_s + nC_m^{z+}$$

characterized by equilibrium coefficient

$$K_{ex} = \frac{[BH_n^{n+}]_s^z [C^{z+}]_m^n}{[BH_n^{n+}]_m^z [C^{z+}]_s^n} \tag{5.12}$$

where indices m and s denote the mobile and stationary ion exchanger phases, respectively. If the concentration of a solute is sufficiently low that it does not appreciably affect the concentration of eluting cation C^{z+} in the stationary and mobile phases, the distribution of solute between both phases is independent of solute concentration. The capacity factor of BH_n^{n+} can then be expressed as follows:

$$k = \psi K_{ex}^{1/z} \frac{[C^{z+}]_s^{n/z}}{[C^{z+}]_m^{n/z}} \tag{5.13}$$

which is a generalized form of Eq. 5.10. As the concentration $[C^{z+}]_s$ is determined by the capacity of exchanger (if the penetration of eluting electrolyte into the exchanger phase is neglected), the capacity factor of the solute is inversely proportional to $[C^{z+}]_m^{n/z}$. This facilitates the estimation of the charge ratio n/z by evaluation of the dependence of the capacity factor on the concentration of eluting cation in the mobile phase. However, the capacity of the weak ion exchanger and hence $[C^{z+}]_s$ depends on the composition and pH of the mobile phase. Even for strong ion exchangers the dependence expressed by Eq. 5.13 is as approximate as the rough simplifications used in the model because the effect of activity coefficients, swelling of the exchanger, and activity of solvent have not been taken into account. Also the concept of homogeneous stationary phase is doubtful, especially for chemically bonded silica-based ion exchangers. Moreover, anions and corresponding additional cations of the mobile phase partially penetrate the cation exchanger phase. For electrolytes of 1:1 valence type C^+A^- this process can be expressed as

$$C_m^+ + A_m^- \rightleftharpoons C_s^+ + A_s^-$$

with the corresponding equilibrium coefficient

$$K_d = \frac{[C^+]_s [A^-]_s}{[C^+]_m [A^-]_m} \tag{5.14}$$

While $[C^+]_m = [A^-]_m$, cations in the stationary phase are primarily neutralized by the anionic groups of ion exchanger. The concentration of electrolyte C^+A^- given by $[A^-]_s$ is therefore suppressed. The invasion of electrolyte to the exchanger phase is negligible at low concentrations of eluting electrolyte when the exchanger phase contains essentially the salt of an ion exchanger only, whereas anions of the eluting electrolyte are excluded. At higher concentrations the distribution of electrolyte to the exchanger phase becomes significant. The exclusion of ions from the exchanger phase is utilized in the separation technique known as Donnan exclusion chromatography.

Besides ionic interactions, other forces contribute to the solute retention. The dispersion forces between solute ions and exchanger matrix affect the affinity of the ion exchanger toward the solutes and are responsible for the differences in solute retention. These interactions are especially important for organic ions. Due to dispersion forces even neutral solutes or oppositely charged ions (e.g., anions on cation exchanger) can be retained by the exchanger matrix and under favorable circumstances the ion exchanger can be used to separate them. As the majority of ion exchangers possess a porous structure, the sieving effect is also effective. Solutes of high molecular weight cannot fully penetrate the porous structure and are excluded from the ion exchanger phase by the same principle that operates in gel permeation chromatography.

The retention of solutes in ion exchange chromatography can be further affected by secondary equilibria in both mobile and stationary phases. The formation of complexes is often encountered in inorganic ion exchange chromatography. Many metals can be separated in this way, e.g., in the form of anionic chloride complexes or using complex-forming reagents. The most common secondary equilibria are acid–base reactions. The dissociations of weak bases or acids are pH-dependent with corresponding behavior in ion exchange separations. Even so, weak ion exchangers such as cation exchangers containing carboxyl groups are affected by the mobile phase pH because the high concentration of H^+ ions leads to the formation of undissociated carboxylic groups and to a decreased ion exchanger capacity. In other words, weak cation exchangers are characterized by a high affinity toward H^+ ions due to the formation of undissociated functional groups. Because they are too strong, H^+ ions cannot be used as eluting cations, and weak cation exchangers can exchange ions only at high pH values. Also some other ions, as soon as they are able to form complex species (ion pairs) with functional groups of ion exchanger, show high retention.

Since the interaction of solute and ion exchanger is complex, it is generally impossible to predict retention. For inorganic cations selectivity series were established which are more or less independent of the type of cation exchanger unless specific interactions occur. The ions of higher valency are preferentially retained over ions of lower valency by the exchanger at low concentration of the eluting electrolyte. It can be judged from Eq. 5.13 that for $[C^{z+}]_s/[C^{z+}]_m > 1$ the expression $[C^{z+}]_s^{n/z}/[C^{z+}]_m^{n/z}$ increases as n increases. Simultaneously it is evident that the polyvalent cations are stronger eluting agents than the univalent ones. Inorganic cations of small radii in the solvated state and high polarizability are preferentially retained. For organic cations the retention depends on their structure in a more complex way. The interaction with organic matrix of the exchanger contributes substantially to retention in a way similar to that for reversed phase packings or rather for packings with polar functional groups eluted with a mobile phase commonly used in RP-LC.

In ion exchange chromatography retention is primarily controlled by the concentration of eluting electrolyte. Theoretically any retention of solute can be adjusted by this parameter (see Eq. 5.14). A gradient in ionic strength should be applied if the sample contains ions differing widely in their affinity toward ion exchanger. The high-capacity exchangers require a higher concentration of eluting electrolyte. Various types of eluting ions generally differ in their eluting strength and selectivity. The mobile phase pH must be adjusted by a suitable buffer if weak acids or bases are to be separated. The suppression of solute ionization eliminates ion exchange retention. However, hydrophobic interations may increase and outweigh this effect. The ion exchange capacity of weak exchangers is dependent on the pH and overall composition of the mobile phase. Separations may be performed at elevated temperature in order to accelerate mass transfer and hence improve column efficiency.

Organic solutes are often separated using aquoorganic mobile phases. The more organic modifier is added to the mobile phase, the weaker is the reversed phase contribution to retention. In addition, the ionization of solute and ion exchanger, as well as exchanger swelling, is affected by the organic modifier. Very often the addition of organic modifier is necessary because of insufficient solubility of the sample in pure water.

Ion exchangers are produced in a wide variety of types and properties. The characteristics of 921 ion exchange materials have been summarized (314). Some of the exchangers that have been used for the separation of alkaloids are included in Table 5.6. Various criteria can be used to classify ion exchangers. The strength of an exchanger is determined by its functional groups. Strong acidic cation exchangers contain sulfonic acid groups, whereas weak cation exchangers contain carboxylic acid functional groups in most cases. The first chromatographic applications often involved resins based on styrene polymer crosslinked with divinylbenzene with a capacity of about 5 mmol g^{-1}. Dry polystyrene-based resins swell when placed in solvent. Swelling is extreme for high-capacity resins of low crosslinking and these packings are too soft to be used in the form of microparticles as required for HPLC. At a high degree of crosslinking the swelling is lower but the particles of resin become brittle and the rate of mass transfer is slow. The degree of swelling is also dependent on pH, concentration, and type of electrolyte and on the presence of organic component in the mobile phase. The packed columns cannot simply be transferred from one form to another because even minor changes in the mobile phase composition result in mechanical damage of the column bed. The efficiency of porous polymeric packings is usually poor due to slow mass transfer and a long diffusion path inside the particles. To

TABLE 5.6 Characteristics of Some Ion Exchangers Used for the Separation of Alkaloids

Type[a]	Name	Particle size (μm)	Particle type[b]	Functional group	Matrix[c] (% cross-linking)	Capacity (mmol/g)	pH range
SCX	Aminex A Series	13±2 (A5) 7-11 (A7)	PP	$-SO_3^-$	PS-DVB (8%)	5	
SCX	SE-Sepha-dex-C25	125-420	PP	$-(CH_2)_2SO_3^-$	Dextran polysaccharide	2.3	>2
CH	Chelex 100	37-74	PP	$-CH_2N(CH_2COOH)_2$	PS-DVB		
SCX	Bondapak CX/Corasil	37-50	S, SP, CB	$-SO_3^-$	Silica		
SCX	Zipax SCX	25-37	S, SP	$-SO_3^-$	Silica	0.0032; 0.005	
SAX	Zipax SAX	25-37	S, SP	$-NR_3^+$	Silica	0.012	
WAX	Zipax WAX	25-37	S, SP	$-NH_2$	Silica	0.012	
SCX	Partisil SCX	10	I, TP, CB	$-benzene-SO_3^-$	Silica	1[d]	1.5-7.5
SCX	Nucleosil SA	5, 10	S, TP, CB	$-SO_3^-$	Silica	1[d]	1-9
SCX	Vydac 401TP	10	S, TP, CB	$-SO_3^-$	Silica	1[d]	

[a]SCX = strong cation exchanger; CH = chelating exchanger; SAX = strong anion exchanger; WAX = weak anion exchanger.
[b]PP = porous polymer type; S = spherical; I = irregular; SP = superficially porous; TP = totally porous; CB = chemically bonded.
[c]PS-DVB = polystyrene divinylbenzene copolymer.
[d]Approximate values based on elemental analysis.
Source: Data from Refs. 2, 6, and 314.

overcome these difficulties pellicular and superficially porous packings have been developed. Pellicular ion exchange packings possess inert spherical particles covered by a layer of ion exchange material. The inert core of superficially porous packing is covered by a porous layer of supporting material which is coated with a thin film of chemically bonded or mechanically held ion exchanger. In this way the migration distance of the solute within the exchanger is diminished and equilibrium is rapidly achieved. Similar intentions led to the development of surface sulfonated polystyrene-divinylbenzene beads and packings prepared by agglomeration of small latex particles on the surface of resin beads (315). Packings of this type are used currently in ion chromatography, but the exchange capacity of surface-type exchangers is substantially lower than the bulk types.

The main advantage of totally porous silica gel microparticles with chemically bonded ionized groups is their excellent mechanical stability. Swelling is no problem in both aqueous and organic media. Fine-particle packings can be operated at flow rates common to other types of HPLC. However, the poor chemical stability of this packing is detrimental because it can be used only at pH approximately 2.0–8.0. The low capacity of silica-based packings is advantageous in analytical applications when a sensitive detector is used. In such a case a lower concentration of eluting electrolyte is used for the required retention. This is favorable for direct conductivity detection or indirect photometric detection, and diminishes the risk of side equilibria.

A. Ion Exchange Chromatography of Alkaloids

Chromatographic separations of alkaloids on classical ion exchangers (resin-based) are possible but require a long time. For example, the elution of opiates and diluents in illicit narcotic mixtures on SE-Sephadex C25 strong cation exchanger with 0.2 M NaH_2PO_4, pH 4.6, required about 22 hr (316). A higher speed of analysis can be achieved using microparticulate resinous packings. Separations are seldom based on a pure cation exchange mechanism. In the study of pilocarpine isomerization to isopilocarpine, Aminex A-7 column was eluted with 0.2 M trometamine buffer containing 5% isopropyl alcohol at pH 9 (317). Pilocarpine was practically in the form of free base only at this pH (pK_a = 6.87, see Table 2.1) and thus the ion exchange mechanism was strongly suppressed. Weakly basic xanthine derivatives were separated on Aminex 50W-X4 column (318–320) also as neutral compounds (caffeine) or weak acids (if an unsubstituted aromatic -NH group was present—xanthine, theophylline, theobromine). Typical S-shaped curves were observed for the pH dependence of k values of xanthine (319) and hypoxanthine

(320), whereas the retention of caffeine was independent of pH from 3.0 to higher values, similarly to its behavior on styrene-divinylbenzene packings without sulfonic groups (319). The low retention of anionic forms of weak acids is in agreement with an essentially reversed phase retention mechanism and Donnan exclusion from the exchanger phase. Retention decreased when ethanol or acetonitrile was added to the mobile phase (320) and was dependent on the form of ion exchanger (318-320). The retention was lowest on the exchanger in the $N(CH_3)_4^+$ form (320). The analyses of caffeine in coffee (318) and caffeine and theophylline in blood serum were performed at an elevated temperature of about 65°C.

Some studies were performed on pellicular and superficially porous ion exchangers. Reserpine and other Rauwolfia alkaloids were separated on pellicular cation exchanger ION-X-SC at 60°C (321). The mobile phase of pH 7.0 or pH 8.0 or gradient in pH within these limits was used. At this pH the alkaloids have not been fully protonized. Methanol (30%) was added to the mobile phase to enhance solubility of free bases, which otherwise tended to adsorb on the packing and tailed badly. A simultaneous increase of pH and buffer concentration shortened the retention times. The efficiency obtained would not be satisfactory for today's chromatographers.

Ion exchange separation of opium alkaloids was optimized using Bondapak CX/Corasil column (322). Unlike the dextran exchanger SE-Sephadex-C25, where interactions of alkaloids with matrix were weak (323), the addition of organic modifier was necessary for ion exchangers with aromatic matrices. The retention could be then varied according to the concentration of organic modifier in the mobile phase and decreased with increasing acetonitrile concentration and increasing salt concentration in the mobile phase. Using 30% acetonitrile in the mobile phase, the retention of morphine, codeine, and narcotine changed only slightly at pH 3-7, whereas retention of thebaine and papaverine decreased at pH 7. The best efficiency was obtained at 50°C. At the same temperature shallow minima were observed for some alkaloids on the dependence of retention volume on temperature.

Separations at a high pH of the mobile phase where the dissociation of alkaloids is suppressed are also possible. Eight components of illicit diamorphine were separated in 12 min on Zipax SCX column using the gradient in pH of 0.2 M borate buffer (pH 9.3-9.8) and in concentration of acetonitrile (0-12%) and n-propanol (0-2%) (324). The chromatographic behavior of morphine, heroin, 6-(O-acetyl)morphine, and methadone was studied on Zipax exchangers (325). Even at high mobile phase pH in the range from 9.2 to 9.8, when only a small percentage of alkaloid molecules have been in the ionized form, the ion exchange contributed substantially to retention. Under these conditions the logarithm of the capacity factor should be linearly dependent on the pH of the mobile phase with a slope equal to -1.0, if only the ion exchange mechanism is operating. The linearity was observed but the slopes were -1.5, -1.0, and -0.6 for morphine, 6-($\underline{O}$-acetyl)morphine, and heroin, respectively. The capacity factors were linearly related to the reciprocal ionic strength of the mobile phase in agreement with Eq. 5.10. Some deviations from the theory can be attributed to the contribution of retention mechanism similar to that of reversed phase systems. The retention increased by successive replacement of hydroxyls by $-OCOCH_3$ groups and decreased when acetonitrile or $\underline{n}$-propanol was added to the mobile phase. The efficiency of the column improved in the presence of organic modifier in the mobile phase. However, only a limited concentration of organic modifier in the mobile phase is permitted on Zipax exchangers. The retentions of morphine and 6-monoacetylmorphine on the strong anion exchanger Zipax SAX were weak and decreased with the acidity of the mobile phase. Donnan exclusion, the ionization of alkaloids, as well as the decreased ionization of phenolic groups might contribute to this behavior (325). Xanthine derivatives were separated on anion exchanger coated on Zipax support at pH 9.2 in 0.01 M $Na_2B_4O_7$. As expected, they were eluted in the order of increasing strength as weak acids: theobromine, xanthine, and theophylline (326).

Similar dependencies were observed on totally porous silica-based ion exchangers. Using a Partisil SCX strong cation exchanger, the combination of reversed phase and ion exchange mechanisms was ascertained for several alkaloids and other drugs (327). The linear dependence of retention volume on reciprocal ionic strength and the decrease of retention with the rise in methanol concentration were demonstrated. For morphine, tubocurarine, and some other basic drugs higher retentions were observed at pH 7 than at pH 3 and pH 5 at 40% methanol and ionic strength 0.1 M. This indicates that, under these conditions, the retention of undissociated alkaloids by partition is stronger than that of ionized alkaloids by ion exchange. Also, the plate number was dependent on pH and methanol concentration. Similar results were obtained for dopamine-derived tetrahydroisoquinoline and tetrahydroberberine alkaloids on three types of microparticulate silica-based cation exchangers (328). The linear dependence of capacity factors on reciprocal ionic strength was observed for Nucleosil SA and Vydac 401 TP columns, but not for Partisil SCX column. The column efficiency was adversely affected by pH $>$ 5.5. The organic modifiers decreased retention in the order butanol $>$ dioxane $>$ isopropanol $\simeq$ acetonitrile $>$ ethanol. At elevated temperatures the capacity factors and the plate height decreased.

As an example, the separation of hallucinogenic components of the Psilocybe mushroom (329) is given in Figure 5.17. An analyt-

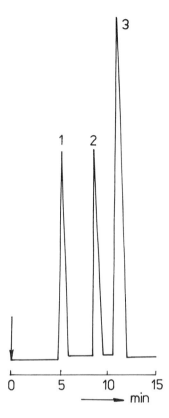

FIGURE 5.17 Ion exchange chromatography of Psilocybe alkaloids. Column 250 × 4.6 mm packed with Partisil SCX, 10 μm, mobile phase 0.2% $(NH_4)_3PO_4$ + 0.1% KCl, pH 4.5 in water-methanol 80:20, 1 ml min^{-1}, 50°C, UV detection at 267 nm. 1, Psilocybin; 2, psilocin; 3, dimethyltryptamine. (After Ref. 329.)

ical Partisil SCX column was protected by a precolumn packed with pellicular beads. A recent gradient separation of nicotine metabolites on the same type of column (330) is shown in Figure 5.18. In this separation a pellicular cation exchange guard column was used. The type of electrolyte in the mobile phase affected the efficiency of the column. The addition of triethylamine lowered retention as expected and improved the efficiency as well. Slightly longer retention times were observed when buffer pH was increased. Interestingly, the addition of methanol above 30% increased the retention of N-methylnicotinium iodide and N,N'-dimethylnicotinium iodide.

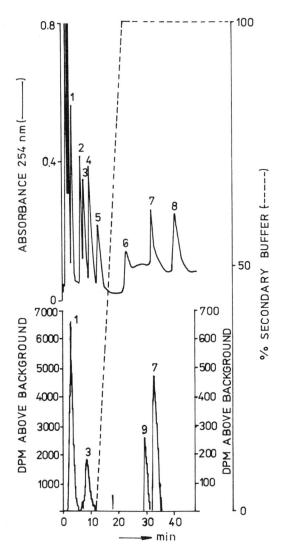

FIGURE 5.18 Ion exchange chromatography of 6-hr guinea pig urine after intraperitoneal administration of [2'-^{14}C]nicotine coinjected with authentic standards. Column 250 × 4.6 mm packed with Partisil-10 SCX with guard column, gradient elution 1 ml min^{-1}, with gradient course shown by broken line. Primary buffer methanol–0.3 M sodium acetate, pH 4.5, 30:70, secondary buffer as the primary one with 1% (v/v) triethylamine and pH adjusted to 4.5. UV detection at 254 nm and ^{14}C radioactivity detection. 1, Cotinine; 2, nornicotine; 3, nicotine-N'-oxide; 4, nicotine; 5, N-methylcotinium; 6, N'-methylnicotinium; 7, N-methylnicotinium; 8, N,N'-dimethylnicotinium; 9, unidentified metabolite. (Reproduced from Journal of Chromatography, Ref. 330, by permission of Elsevier Science Publishers B.V.)

Another technique utilizing ion exchangers should be mentioned here although it is no longer applicable to the column chromatography of alkaloids. If ions of copper, nickel, zinc, cadmium, or other metal are loaded on the ion exchange column, the coordination of amines or other suitable compounds to these ions should be possible in the exchanger phase and responsible for their retention. A mobile phase containing another component capable of coordinating to these metals, most often ammonia, is commonly used as an eluent. In this technique of ligand exchange chromatography (331,332), retention is dependent on the ability of compounds to compete with ammonia molecules in the coordination sphere of metal ions, which enables the specific separation of compounds. Retention decreases with increasing concentration of ammonia in the mobile phase.

Copper(II) ions create the most stable complexes and are most often applied. The resins containing carboxyl groups held metal ions much more strongly then strong-acid polystyrene sulfonate exchangers. If the bleeding of the metal from the column is high, the addition of the metal salt to the mobile phase is necessary. Therefore, its concentration can be used to control retention. The bleeding of the ions is negligible when chelating ion exchangers are used. These materials contain functional groups capable of tightly binding metals via chelate formation. The functional groups occupy the coordination sphere of ions to a large degree and solutes are retained to a lesser extent. Xanthine derivatives were studied in a system of this type containing Chelex 100 resin, which was applied to the determination of caffeine in coffee and tea (333). For the ability of solutes to act as ligands in metal complexes, the accessibility of basic nitrogen is decisive. Substitution on the nitrogen atom or in that neighborhood decreases retention (331). The behavior of several alkaloids in various ligand exchange systems was studied (334), with a mobile phase consisting of ammonia solution in aqueous alcohol. The alcohol was added to improve the solubility of alkaloids. Retention volumes were inversely proportional to the ammonia concentration and decreased with rising alcohol concentration in the mobile phase. Retention was also dependent on the structure of the resin matrix.

Ion exchange in the analysis of alkaloids is currently not as popular as reversed phase or ion pair chromatography. However, rapid developments in the field of ion chromatography may renew interest in its application. For example, the techniques of indirect detection (335) have not yet been applied to alkaloid separations. Indirect photometric detection might be useful in the analysis of alkaloids that absorb weakly in the UV region. When applied to ion exchange chromatography, UV-absorbing elution ion is used which causes the background absorption of the eluate. The presence of solute in the eluate decreases the concentration of the eluting ion and hence the background absorption. Solutes are then detected as negative peaks on the baseline. The decrease is stoichiometric if the invasion of eluting electrolyte to the exchanger phase is negligible, and thus universal calibration on a molar basis is possible (335,336).

IV. ION PAIR CHROMATOGRAPHY

This section is devoted to the group of methods in which an ionic surface-active component is used to modify retention of ionic solutes of opposite charge. Names such as soap chromatography (337), paired ion chromatography (338), solvent-generated ion exchange chromatography (339,340), hetaeric chromatography (341), and ion interaction chromatography (342,343) have appeared to denote this technique. Variation in terminology mainly reflects differences of opinion regarding the principle retention mechanism. Although the role of ion pair formation in the given separation system is often questionable, we use the term ion pair chromatography in accordance with a common practice in the liquid chromatography literature (344).

An ion pair is a neutral complex produced from ionizable solutes of opposite charge. Basic solute, e.g., alkaloid, dissolved in amphiprotic solvent is present in the solution as a free base B or in the protonized form BH^+. If the pairing acid is also present in the solution, either in undissociated form PH or as dissociated species P^-, the formation of ion pairs BHP can occur according to the equation

$$B + PH \rightleftharpoons BHP$$

characterized by formation constant

$$K = \frac{[BHP]}{[B][PH]} \quad (5.15)$$

or according to the equation

$$BH^+ + P^- \rightleftharpoons BHP$$

with formation constant

$$K_{+/-} = \frac{[BHP]}{[BH^+][P^-]} \quad (5.16)$$

Both constants describe the same phenomenon and are related to each other as follows:

$$K_{+/-} = K\frac{K_{a,BH^+}}{K_{a,PH}} \tag{5.17}$$

where K_{a,BH^+} and $K_{a,PH}$ are the dissociation constants of protonized base BH^+ and acid PH in the given solvent. Formation constant K will be preferred in nonpolar solvents where species dissociation is negligible and $K_{+/-}$ constant in polar solvents if both components are present in their ionic forms.

Initial theoretical models of ion pair chromatography sprung from analogy with ion pair liquid-liquid extraction, where the concentration of ionized species in organic phase can be neglected in comparison with the concentration of neutral aggregates of ions. The ionized pairing reagent P^- is present in the polar phase rich in water. This can be either stationary phase in normal phase chromatography or mobile phase in reversed phase chromatography. Some applications of normal phase ion pair chromatography were mentioned in the Section I.A of this chapter. In this section we therefore restrict our discussion to reversed phase systems. If ionized solute BH^+ is introduced to a mobile phase containing pairing ion (hetaeron) (341), ion pair formation may occur. Electrically neutral ion pairs are then retained by nonpolar stationary phase. If retention of ionized unpaired solute is negligible, the capacity factor of the solute is given as

$$k = k_{BHP} x_{BHP} = k_{BHP} \frac{K_{+/-}[P^-]_m}{1 + K_{+/-}[P^-]_m} \tag{5.18}$$

where k_{BHP} is the capacity factor of the ion pair, x_{BHP} is the fraction of the solute in the mobile phase which is bound in ion pairs, $K_{+/-}$ is the formation constant of ion pairs in the mobile phase, and $[P^-]_m$ is pairing reagent concentration in the mobile phase. A similar description was already applied to the ion pair systems in liquid-liquid partition chromatography (345,346). However, it was demonstrated that its applicability is limited by the surface phenomena (57,347,348). A high specific surface area of silica-based or polymeric reversed phase packings makes the surface phenomena even more important. It was soon demonstrated that amphiphilic ion pair reagents are retained by the packings for reversed phase liquid chromatography (337,342,349,350). The retention of pairing ion P^- must be accompanied by the retention of oppositely charged ions C^+ of mobile phase electrolyte:

$$P_m^- + C_m^+ \rightleftharpoons P_s^- + C_s^+$$

characterized by distribution coefficient

$$K_d = \frac{[P^-]_s [C^+]_s}{[P^-]_m [C^+]_m} \qquad (5.19)$$

where m and s denote mobile and stationary phases. For example, by adsorption of anionic surfactants, such as sodium salts of alkylsulfonic acids, the sodium cations are retained together with sulfonate anions. The charge of ionic components is sometimes neglected in equations describing this system (337,341,351). Then the parameters describing equilibria in which net charge is transferred between phases are generally not constant but rather are dependent on the concentrations of other ionic components in the system. Alternatively, the phenomena involving charge transfer between phases may be considered, but in such a case the difference in electrostatic potential between phases, which depends on their composition, has to be taken into account. In this section we use only the equations in which the total charge transferred between mobile and stationary phase is zero.

The adsorbed detergent together with its counterions on the surface of the reversed phase packing forms a layer which is akin to an ion exchanger (349,352). For anionic surfactant P^- accompanied by counterion C^+, the cation exchange with solute ion BH^+ is then possible with equilibrium constant K_{ex}:

$$C_s^+ + BH_m^+ \rightleftharpoons C_m^+ + BH_s^+$$

just as in ion exchangers (see Section III of this chapter) or for sorption of ionized solutes in reversed phase liquid chromatography (Section II). Similarities between reversed phase ion pair chromatography and reversed phase chromatography of ionized solutes at high inorganic electrolyte concentrations in the mobile phase have been established (351). For the capacity factor it follows that

$$k = \psi K_{ex} \frac{[C^+]_s}{[C^+]_m} \qquad (5.20)$$

If C^+ and P^- are the only ions present in the stationary phase, $[C^+]_s = [P^-]_s$ and

$$k = \psi K_{ex} \frac{[P^-]_s}{[C^+]_m} \tag{5.21}$$

Sometimes ions in the surface layer are denoted as being connected in some way with their counterions. Symbols of the type $(P^-C^+)_s$ or $(PC)_s$ are used (339,340,353–355), which may, however, lead to the misunderstanding that ions are bound together as ion pairs. The only binding force in the mechanism discussed here is the necessity for electroneutrality, namely, that the total concentration of cations be equal to the total concentration of anions as in any uni-univalent electrolyte solution. The presence of free ions in the stationary phase is considered in the ion interaction model (342,343) and in models of retention based on the Stern–Gouy–Chapman theory of the electrical double layer (356–360). As the ion exchanger phase is formed by dynamic coating of hydrophobic support by amphiphilic ions, the term "dynamic ion exchange" would be suitable to denote this mechanism. However, in our opinion this term is often misused for mechanisms in which "ion-pair formation takes place between the solute and hetaeron bound to the stationary phase" (341). In the phenomenological description used here such a mechanism would be written as

$$PC_s + BH^+_m \rightleftharpoons BHP_s + C^+_m$$

with equilibrium coefficient

$$K_{pe} = \frac{[BHP]_s [C^+]_m}{[BH^+]_m [PC]_s} \tag{5.22}$$

Here PC_s and BHP_s represent a complex of pairing ion and mobile phase counterion and solute, respectively, in the stationary phase. Such a complex is identical with the ion pair or at least indistinguishable from it. The capacity factor of solute is then given (supposing that concentration of ion pairs in the mobile phase is negligible compared to the concentration of free ions) as

$$k = \psi \frac{[BHP]_s}{[BH^+]_m} = \psi K_{pe} \frac{[PC]_s}{[C^+]_m} \tag{5.23}$$

As has been already pointed out (361), dynamic ion exchange in this sense represents only another formulation of the ion pair mechanism. If the adsorption of ion pairs PC is described as

$$P^-_m + C^+_m \rightleftharpoons PC_s$$

with equilibrium constant

$$K_{pd} = \frac{[PC]_s}{[P^-]_m [C^+]_m} \quad (5.24)$$

substitution from Eq. 5.24 to 5.23 leads to

$$k = \psi K_{pe} K_{pd} [P^-]_m \quad (5.25)$$

This is just Eq. 5.18 for $K_{+/-}[P^-]_m \ll 1$ because

$$k_{BHP} K_{+/-} = \psi K_{pe} K_{pd} = \frac{(n_{BHP})_s}{(n_{BH^+})_m [P^-]_m} \quad (5.26)$$

Equations 5.23 and 5.18 are therefore equivalent except that they express the retention of solute by means of a different set of equilibrium coefficients and parameters.

Comparison of Eqs. 5.23 and 5.21 reveals a close similarity in the theoretical dependencies of the ion interaction and ion pair mechanisms which hinders their resolution. In both cases at a given concentration of eluting cation C^+ in the mobile phase, the dependence of capacity factor on the concentration of pairing ion P^- in the mobile phase should follow the adsorption isotherm of ionized pairing ion P^- and ion pair PC, respectively. Whether the ion-pairing reagent is adsorbed as free ion or in the form of an ion pair seems to be important from the standpoint of retention mechanism. At least partial ionization of anthraquinone-2-monosulfonic acid sorbed on a macroporous styrene-divinylbenzene resin was proved by its catalytic action on the inversion of sucrose (362). By electrophoretic measurements the ionic adsorption of diphenylguanidinium on Amberlite XAD-2 macroporous styrene-divinylbenzene resin (356) and octyl sulfate on ODS-Hypersil (361) was demonstrated. Considering ionic adsorption, models based on the theories of electrical double layer are capable of describing adsorption of diphenylguanidinium and benzylammonium cations on Amberlite XAD-2 resin (356) and of other ions on various sorbents (360). The models of double-layer adsorption also reveal the principal limitations of those phenomenological treatments which apply the concept of homogeneous stationary phase and do not consider the space structure of the stationary phase.

In fact, sorption of such strong acidic pairing reagents as alkylsulfonic acids in the form of ion pairs with, say, sodium ions seems highly improbable. However, it does not mean that the formation of ion pairs does not contribute to the retention of solute at all. The high value of constant K_{pe} may outweigh the low concentration $[PC]_s$ in Eq. 5.23. Similarly, the formation of ion pairs in the mobile phase cannot be excluded (363). Both mechanisms should generally be taken into account, i.e., the presence of ionic form and ion pair of solute in both stationary and mobile phases. Then

$$k = \frac{\psi\{[BH^+]_s + [BHP]_s\}}{\{[BH^+]_m + [BHP]_m\}} = k_{BH^+} x_{BH^+} + k_{BHP} x_{BHP}$$

$$= \frac{\psi\{K_{ex}[C^+]_s + K_{pe}[PC]_s\}}{\{[C^+]_m (1+K_{+/-}[P^-]_m)\}} \qquad (5.27)$$

The retention dependencies are usually discussed using Eq. 5.21 or 5.23 eventually extended to include other phenomena. The provision is made for nonzero retention of solute in the absence of pairing reagent by insertion of an additional term in the numerator of these equations. Experimental retention data can be directly compared by means of equations similar to 5.21 or 5.23 if the amount of adsorbed pairing reagent is determined. Linear or slightly curved dependencies of k on the ratio $(n_p)_s/[C^+]_m$, where $(n_p)_s$ denotes the total amount of pairing reagent retained by the column irrespective of its form, have been observed (357,361). Data for polyvalent ions were in agreement with the more general form of this dependence (340). It was, however, demonstrated (364,365) that $(n_p)_s/[C^+]_m$ is not a universal parameter for the description of retention and that it depends on the composition of mobile phase electrolyte in a more complex way. Also the observed changes in selectivity cannot be explained simply by Eq. 5.21 or 5.23 (365).

If the retention is to be correlated with the composition of the mobile phase only, some assumptions about the adsorption isotherm of the pairing reagent have to be adopted. The adsorption of pairing reagent is fully described neither by Eq. 5.19 nor by Eq. 5.24 since the surface saturation occurs at a high pairing ion concentration. Typical experimental adsorption isotherms are given in Figure 5.19 (366). However, more complex forms of adsorption isotherm were also observed (367,368). The extent of adsorption depends on the presence of other ionic components in the mobile phase. The increasing concentration of electrolyte leads to increased adsorption of pairing ion (340,355,356,364,365,369). The effect is de-

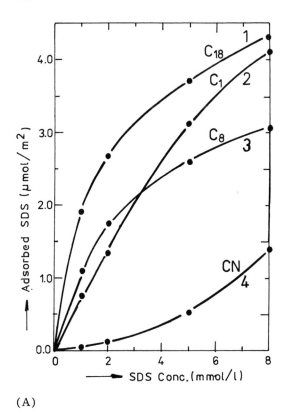

(A)

FIGURE 5.19 Adsorption isotherms of sodium dodecyl sulfate from aqueous mobile phase on various silica-based stationary phases at 30°C. (A) Concentrations under critical micellar concentration (CMC). (B) Above CMC (see page 213). 1, ODS Hypersil (specific surface area 105 m^2 g^{-1}, bonded moiety octadecyl); 2, SAS Hypersil (104 m^2 g^{-1}, trimethyl); 3, MOS Hypersil (129 m^2 g^{-1}, octyl); 4, CPS Hypersil (115 m^2 g^{-1}, cyanopropyl); 5, Hypersil (150 m^2 g^{-1}, unbonded). (Reprinted with permission from Analytical Chemistry 58 (1986) 1356, Ref. 366. Copyright 1986 American Chemical Society.)

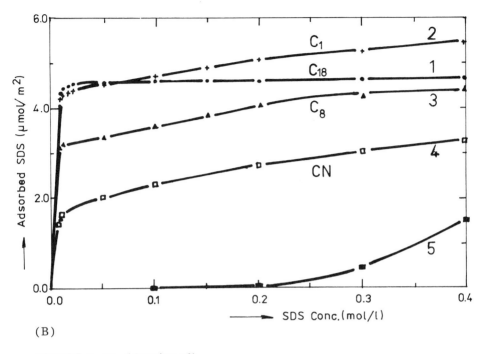

(B)

FIGURE 5.19 (Continued)

pendent on the type of electrolyte (356,369). Some results were plotted as a function of the product $[P^-]_m[C^+]_m$ (370) as proposed by Eqs. 5.19 or 5.24. The capacity factor of benzenesulfonates, which corresponds to the slope of their adsorption isotherm at zero concentration, was found to increase with the ionic strength I of the mobile phase according to Eq. 5.28:

$$\frac{1}{k} = \frac{1}{C_1} + \frac{1}{C_2\sqrt{I}} \tag{5.28}$$

as predicted by the Stern–Gouy–Chapman double-layer theory (356, 371).

The adsorption of ion-pairing reagent decreases with the addition of the organic modifier to the mobile phase (340,342,357,368, 372–375). At the same time the ion-pairing reagent partly removes organic modifier present in the stationary phase of reversed phase packing (357). In mobile phases without an organic modifier the adsorption of pairing reagent on alkyl-bonded packings is not very reproducible and equilibration times are long (224,376).

The large hydrophobic moiety in the molecular structure of the ion-pairing reagent promotes its adsorption (342,355,357,367,374). The maximal exchange capacity of the column with adsorbed surfactant is higher for pairing ions with longer alkyl chains (361,367, 368). Figure 5.20 (361) compares the adsorption isotherms for octyl-, decyl-, and lauryl sulfates on ODS-Hypersil (361). In many cases the adsorption isotherm may be approximated by the Langmuir isotherm (352,356,372,375). Assumptions about the saturation of a limited number of adsorption sites available for pairing reagent leading to Langmuir isotherm have frequently been used in theoretical models of ion pair chromatography (341,361,372,377,378). Although the Freundlich isotherm fails to predict slope at $[P^-]_m = 0$, it has been applied to the expression of experimental isotherms (337,352, 357,361,375,379) and the interpretation of retention data (377,380,

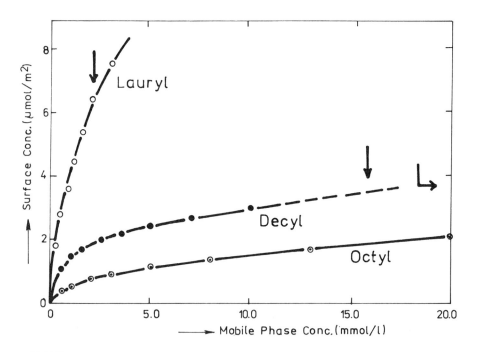

FIGURE 5.20 Adsorption isotherms of alkylsulfonates on ODS Hypersil (105 m^2 g^{-1}) from water-methanol 80:20 containing 0.02 M sodium phosphate, pH 6.0. Total concentration of Na held constant at 0.05 M by the addition of NaCl. Critical micellar concentrations are indicated by arrows (100 mM for octyl sulfate). (Reproduced from Journal of Chromatography, Ref. 361, by permission of Elsevier Science Publishers B.V.)

381). However, other experimental data on the adsorption of pairing reagent could not be approximated by Langmuir or Freundlich isotherm (364,365,373,374,382). The assumption of two types of adsorption site may improve experimental agreement (370,373). The models of an electrical double layer lead to a modified Langmuir isotherm (356,357).

In most cases the dependence of capacity factor on the concentration of pairing reagent closely resembles the adsorption isotherm for the pairing reagent. At higher concentrations a more or less significant decrease in retention is sometimes observed (337,339, 341,354,357,365,367,372,383,384) unlike that for the adsorption isotherms. There are several possible explanations for this decrease. If the pairing reagent is added to the mobile phase together with its counterion (337,339,354,364,365,367,384), the increasing value of the denominator in Eqs. 5.21 and 5.23 will decrease retention when the stationary phase approaches saturation. The ion pair formation in the mobile phase, which can be included in the denominator of these equations as the factor $(1 + K_{+/-}[P^-]_m)$, may bend the dependencies downward at higher $[P^-]_m$ values (339,372). If, however, ion interaction is the prevailing retention mechanism, then the formation of ion pairs in the mobile phase will be accompanied by their simultaneous formation in the stationary phase. This process may be more probable in the less polar stationary phase than in the mobile phase. Hence, the second term in the numerator of Eq. 5.27 will probably be more effective than the term $K_{+/-}[P^-]_m$, in the denominator and the increase in retention is expected. In other words, it seems probable that uncharged ion pairs of solute and pairing reagent will be better retained by hydrophobic interactions than ionized solutes by combined ionic, hydrophobic, and other nonionic interactions.

Another possible explanation for retention decrease emphasizes the fact that pairing reagent added to the mobile phase to a certain extent acts as organic modifier (367,368,384). A slight decrease in the retention of neutral compounds by the addition of pairing ion was observed (361,365,382–385). However, specific interactions such as hydrogen bonding with adsorbed pairing reagent may cause a selective increase in the retention of some nonionized solutes (369). The decrease in retention at high concentration of pairing ion is sometimes ascribed to micelle formation in the mobile phase above critical micelle concentration or to the formation of lower aggregates (dimers, trimers, etc.) of pairing reagent under critical micelle concentration (340,386). Solute molecules penetrating micellar phase or interacting with aggregates of pairing reagent in the mobile phase contribute to the decrease in capacity factor. At present the term micellar chromatography is used to denote chromatography with aqueous mobile phases without organic modifier but containing a detergent in concentration above critical micellar concentration (387–390). When dealing biological samples some advantages were observed (388). However, low efficiency due to slow mass transfer in these systems is detrimental (391).

The formation of higher aggregates of solute with pairing ion in the stationary phase was assumed in the so-called dynamic complex exchange model in order to explain some deviations from simple theories (386). The decrease of solute retention at high concentrations of pairing reagent cannot be explained in this way. The neglect of activity coefficients and the approximate character of models in terms of a homogeneous stationary phase should be born in mind as the possible source of deviations from theoretical behavior. The presence of residual acidic surface silanol groups on silica-based packing may further complicate the description of the system.

In the above considerations the complete dissociation of solute, pairing, and eluting reagents in both mobile and stationary phases was assumed. If any of these species is a weak acid or base, the behavior of the system will be pH-dependent (339,341,372,380,381). Incomplete dissociation of solute can be included in Eq. 5.27 as in Eqs. 5.8 and 5.9. The decreased dissociation of solute in the mobile phase will decrease retention due to ion interaction and the ion pair mechanism. Retention due to these two mechanisms may be higher or lower than the retention of undissociated solute. Both a decrease and an increase of retention might therefore be encountered when changing the mobile phase pH in the dependence on other chromatographic conditions. In the ion interaction mechanism at low pH values, H^+ ions may act as additional eluting ions for cationic samples and decrease retention. If a weak acid is used as pairing reagent for cationic samples, its lower dissociation at low pH values decreases the capacity of the adsorbed exchanger layer and the decrease in retention is expected in a way similar to that for weak cation exchangers in IEC. Even the adsorption of weak acidic pairing reagent should be pH-dependent and the interpretation of system behavior would be difficult. The formation of ion pairs with ionized solutes and retention by the ion pair mechanism will decrease as a result of suppression of ionization of weak acidic pairing reagent. If, however, both pairing reagent and solute are undissociated, the ion pairs can be formed via reaction characterized by formation constant K (Eq. 5.15) and contribute to retention.

It follows from the preceding discussion that the primary factors used to control retention in reversed phase ion pair LC are the types and the concentrations of pairing reagent and of eluting ion. The more hydrophobic the pairing reagents, the more they are effective in increasing retention (341,365,367,372,384,392). At a given concentration the long alkyl surfactants will yield a higher retention of solute. Interestingly, common dependencies for some basic solutes have been observed when their capacity factors were plotted against the concentrations of alkylsulfonates of various chain lengths in the stationary phase, if the concentration of eluting cation was held constant (361,365,382). Hence, the greater effect of more hydrophobic pairing ions comes mainly from their stronger adsorption to the reversed phase packing. Use of pairing ions of different structural types had another effect on solute retention. Changes in selectivity accompanying changes in the type or concentration of pairing reagent may be at least partially due to changes in nonionic interactions in the stationary phase caused by adsorption of the pairing reagent. Substantial differences in selectivity were observed when using chelated metals as pairing reagent (393, 394). In some situations mixed pairing reagents may be beneficial (343,395).

The increasing concentration of eluting ions decreases solute retention (341,355,364,365,369). Although the adsorption of pairing reagent is increased to a certain extent, this is counterbalanced by enhanced concentration of eluting cations in the mobile phase. In some cases the linear plots of capacity factors against reciprocal concentrations of eluting ions in the mobile phase have been obtained (339,350,361,396). Such dependencies follow from Eqs. 5.21 and 5.23 if the concentration of pairing ion is high and stationary phase is close to saturation with it. Then $[P^-]_s$ or $[PC]_s$ is approximately independent of $[C^+]_m$. The strength of the eluting cations resembles that in ion exchange chromatography (378). Organic cations are more efficient in solute elution and change system selectivity (354). A substantial increase in efficiency was not observed (354).

The addition of organic modifier to the mobile phase suppresses retention (339,343,355,365,383,384,392) due to a decreased amount of adsorbed pairing reagent to the stationary phase and diminished hydrophobic interactions, which normally contribute to the stabilization of solute ions in the stationary phase or are responsible for ion pair retention. As in reversed phase chromatography, linear or quadratic function can be used to express the dependence of the logarithm of the capacity factor on the volume fraction of organic modifier in the mobile phase (397-399). Hydrophobicity of the solutes determines to a large extent their relative retention (400). The applicability of the functional group concept was studied extensively and the correlation of group increments with Hantsch π values was demonstrated (398,399,401).

In most cases, at elevated temperature the retention times are shortened in reversed phase ion pair chromatography (339,402), as in other types of chromatography. Maxima on temperature dependencies were also observed (392). The amount of pairing reagent retained by stationary phase decreases (340) and thus contributes to the observed decrease in retention. Van't Hoff plots permitted the evaluation of overall enthalpy and entropy changes of the retention process in ion pair chromatography and revealed enthalpy-entropy compensation similar to that for neutral solutes in reversed phase chromatography (398,399,401). Just as in reversed phase chromatography, the retention and selectivities are dependent on the type of packing material (339,392,399,401). The retention capability can be related to the carbon loading of the support (399,401). Various packings also differ in efficiency and inclination to form skewed and tailing peaks for ionized solutes.

As in other separation methods for ionized solutes, the amount of injected sample is strongly limited by concentrations of ionic components in the mobile and stationary phases. These concentrations should not be significantly changed by the presence of solute (268, 370); otherwise changes in retention volume and peak shape result. More retained samples are more likely to exhibit these effects (372).

As regards the practical aspects of ion pair chromatography (403,404), the importance of column equilibration in this procedure must be emphasized (350,361,372,405,406). It may be a long time before the equilibrium of highly hydrophobic pairing ion is achieved. Moreover, if the front of the pairing ion is eluted from the column, it does not necessarily mean that other ionic components are also equilibrated. Sometimes further washing of the column with the mobile phase is required to obtain reproducible retention. Although possible (355,407,408), the usefulness of gradient elution is therefore limited. Pairing reagents of shorter alkyl chain length offer faster equilibration and better solubility in the mobile phase, but a higher concentration is necessary to achieve the required effect. An eventual precipitation after mixing pairing reagent with buffer components or organic modifier should be checked in a long-term test. Even if the precipitate is not formed immediately after mixing, the precipitation may occur slowly in the chromatograph reservoir thus damaging pump and column. The solubility of potassium salts of alkylsulfonates is limited (396). Precipitation after mixing sodium dodecyl sulfate and potassium phosphate buffer was dependent on the purity of sulfate reagent (409). The solubility of surfactant in acetonitrile is usually worse than that in methanol. Column washing by a mobile phase with a higher content of organic modifier must be controlled because precipitation or unwanted changes in pH may degrade the column (376). The presence of pairing reagent may adversely affect the stability of silica-based packings and may narrow the usable pH range (403). The mobile phase should never stand in the column for a prolonged time. The adsorption spectra of some pairing reagents have been presented (408) in order to choose a suitable reagent with respect to photometric detector compatibility.

The parameters which can be varied in reversed phase ion pair chromatography in order to control the retention are numerous. High flexibility together with high efficiency contributes to the popularity of this technique for the separation of ionic solutes. If neutral compounds are to be analyzed with ionized solutes, one should first adjust the retention of neutral solutes by controlling the concentration of organic modifier. Then, according to the selection of ion-pairing reagent, eluting ion, and their concentrations the retention of ionized solutes can be brought to the required range while retention of neutral solutes changes only slightly. If a change in solute dissociation might produce a desired change in retention or selectivity, buffer pH can be adjusted.

A. Ion Pair Chromatography of Alkaloids

Ambiguities regarding the retention mechanism in ion pair chromatography stimulated many systematic studies on this subject. Only a few of them, however, deal with alkaloids. The most exhaustive studies seem to be those examining systems with sodium alkylsulfonates as the pairing reagents in mobile phases containing acetic acid (410–415). Mobile phases of this type have frequently been prepared using PIC reagents of Waters Associates. Besides analysis of basic compounds, these systems allow the simultaneous analysis of neutral or weakly acidic compounds. With increasing organic content of the stationary phase, the solutes are more strongly retained (412). Longer retention times were observed on alkyl-bonded phases than on those containing phenyl, cyclohexyl, or nitrile (412,413). The difference was smaller for short-chain alkylsulfonates. The increased chain length of pairing ion led to the enhanced retention of basic compounds. There was no significant enhancement of retention on nitrile phase (413). The increase in log k values was similar for bases of the same charge but the steepest dependence was observed for double-charged bases. In this respect the anomalous behavior of morphine and aminopyrene could not be explained (413). Increasing the percentage of organic modifier in the mobile phase reduced retention (412,413) with an approximately linear decrease of log k values (413). The selectivity for compounds differing in aliphatic character usually decreased with methanol content in the mobile phase and for some pairs of compounds with an increasing chain length of the pairing ion (414).

The system initially used by Lurie (410), consisting of a µBondapak C18 column and mobile phase water-methanol-acetic acid 59:40:1 with 0.005 M sodium heptanesulfonate at pH 3.5, enabled the successful separation of opium alkaloids (Figure 5.21) and many other drugs. The poor separation of some compounds of interest required an improvement of the system accomplished by substituting 0.02 M

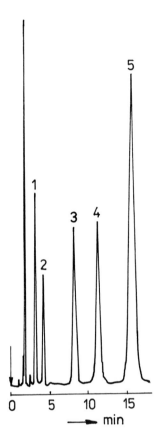

FIGURE 5.21 Ion pair chromatography of opium alkaloids. Column 300 × 4 mm packed with μBondapak C18, mobile phase water-methanol (containing 1.1 g liter^{-1} of sodium-1-heptanesulfonate)-acetic acid 59:40:1, pH 3.5, 2 ml min^{-1}, UV detection at 254 nm. 1, Morphine; 2, codeine; 3, thebaine; 4, narcotine; 5, papaverine. (Reprinted with permission from Journal of Association of Official Analytical Chemists, Ref. 410.)

methanesulfonic acid for pairing reagent in the above mobile phase and by introducing an alternative mobile phase with a water-methanol-acetic acid ratio of 79:20:1. In both mobile phases the pH was adjusted to 3.5 with 2 M sodium hydroxide. The increased concentration of methanesulfonic acid was applied in order to reduce the variation in retention times with sample concentration (415). Whether the systems employing methanesulfonic acid should be collectively referred to as ion pair chromatography is not clear because the retention is reported to be essentially independent of methanesulfonic acid concentration (414,415).

Mobile phases of similar type have been applied to the semipreparative chromatography of drugs, e.g., LSD (411). The mobile phase containing 2.5 mM octanesulfonate and 0.5% acetic acid in methanol-water 56:44 and µBondapak C18 column were used for the determination of emetine (416). By changing the methanol concentration from 30% to 70% the decreasing retention of emetine was observed for four alkylsulfonates examined as pairing reagent. The retention increased in the order pentane-, hexane-, heptane-, and octanesulfonate. If the capacity factor was adjusted to 4.2 for each alkylsulfonate by methanol content in the mobile phase, the number of theoretical plates for emetine increased in the same order of pairing reagents. A methanol concentration of 60% was used when emetine was determined as a fluorescent product obtained by oxidation (416). On the same type of column with 0.005 M heptanesulfonate (PIC B-7 reagent) in methanol-water 5:5 and 4:6 and in acetonitrile-water 25:75, the retention times for 21 morphinans were determined (417). Retention times were higher for solutes of lower polarity. Enhancement of the organic modifier resulted in decreased retention. Acetonitrile turned out to be a more efficient organic modifier than methanol (417). The same pairing reagent in water-methanol-acetic acid 71:28:1 was used for the determination of morphine sulfate and contaminants in pharmaceuticals (418). Five µBondapak C18 columns tested displayed some variation in retention time, tailing, and resolution.

By using of acetonitrile instead of methanol and phosphoric acid instead of acetic acid in the analysis of illicit heroin (419), faster flow rates could be applied and UV detection at 220 nm was possible. For a mobile phase containing 0.02 M methanesulfonic acid in water-acetonitrile-phosphoric acid 87:12:1 (pH 2.2), the retention data on µBondapak C18 and Partisil 10 ODS-3 columns and the ratio of detector responses at 220:254 nm were presented for 47 heroin addulterants and byproducts (419). Phosphoric acid was used to acidify the mobile phase for the determination of morphine, diamorphine, and their degradation products in pharmaceutical preparations (420). On the Ultrasphere ODS column morphine and pseudomorphine were best separated using 0.01 M sodium pentanesulfonate–acetonitrile–

phosphoric acid 69.95:30:0.5 as the mobile phase. Increasing the concentration of phosphoric acid reduced retention and sharpened the peaks. The presence of sodium pentanesulfonate in the mobile phase improved efficiency, eliminated peak broadening and tailing, and, surprisingly, shortened the retention times of both solutes (420). The expected retention enhancement due to the presence of pentanesulfonate was probably suppressed by the presence of sodium cations. Similar behavior was observed for strychnine in stomach content once an assay became available (421). Whereas the dependence of log k on acetonitrile content in 5 mM phosphate buffer pH 3.0 was not linear and displayed a minimum at about 70% acetonitrile, the addition of 0.01 M sodium octenesulfonate substantially decreased retention and the dependence became linear. In addition, better resolution was attained by the reduction of peak tailing. The efficiency of the system was further increased by the addition of 0.001 M tetrabutylammonium hydroxide. Simultaneously retention times and the differences between them decreased (421). A potassium phosphate buffer (pH 4.5) and methanol 35:65 mixture containing 0.005 M sodium heptanesulfonate was used in the chromatography of reserpine and rescinnamine with electrochemical detection (422). The change of pH from 3.5 to 6.0 resulted in a similar retention time of reserpine but a rapid increase was observed upon further increase of the pH to 7.5. Reserpine seems better retained in this system as free base than in protonized form even in the presence of pairing reagent. The separation of alkaloids present in opium extracts with a mobile phase that contains phosphate buffer is presented in Figure 5.22 (423).

Statistical optimization methods were applied to the separation of morphine, codeine, narcotine, and papaverine (424). From the four parameters studied, the pH of the mobile phase and the concentration of phosphate buffer turned out to be less important than the concentrations of methanol and camphorsulfonic acid as pairing reagent. As expected, at a higher concentration of methanol a lower retention was obtained. The dependence of retention on the concentration of camphorsulfonic acid was rising at lower concentrations. Above approximately 0.015 M camphorsulfonic acid (at 38–44% methanol) the decrease in retention was observed (424). In an assay of the combined formulation of ergometrine and oxytocine (425), the mobile phase was buffered at pH 5.0 with phosphoric acid–triethylamine buffer. At 40% acetonitrile and 0.05% w/v sulfonates or sulfates, increasing log k values along with an increasing alkyl chain length of pairing reagent were observed. Good separation was obtained on an ODS Hypersil column with 0.01% w/v of sodium dodecyl sulfate in 0.25 mM buffer–acetonitrile 60:40. However, with repeated injections the peak of oxytocin drifted and quantitation was violated by the presence of chlorbutol. Thermostating con-

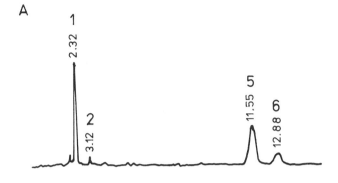

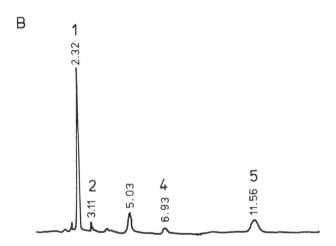

FIGURE 5.22 Ion pair chromatography of Afghan (A) and Hungarian (B) opium samples. Column 150 × 4.6 mm packed with Nucleosil 5 C18, mobile phase 0.05 M KH_2PO_4 (pH 3.0) containing 0.005 M sodium octanesulfonate-acetonitrile 70:30, 0.9 ml min^{-1}, 60°C, UV detection at 220 nm. 1, Morphine (k = 1.3); 2, codeine (2.1); 3, narceine (5.5); 4, thebaine (5.9); 5, narcotine (10.4); 6, papaverine (11.7). (Reprinted with permission from Magyar Kemiai Folyoirat, Ref. 423.)

trol of the system did not suppress this drift. Similar separation on HSCP C18 and Partisil ODS packings required a higher concentration of pairing reagent and buffer, but the drift was not observed. With other packings the separation was inadequate. When sodium tetradecyl sulfate was substituted for dodecyl sulfate, a higher concentration of phosphate buffer was required and the differences between various packings were less pronounced. After use columns were washed with five- to 10- column volumes of 83 mM phosphoric acid–acetonitrile 60:40 to remove triethylamine (425). A continuous decrease of both LiChrosorb RP-2 and Zorbax TMS columns resolution and separation was observed when hyoscyamine and hyoscine were determined in leaves, fruits, and seeds of Datura (426). A mobile phase consisting of 5 mM d-10-camphorsulfonic acid in phosphate buffer–acetonitrile–diethylamine 94:6:0.6 was used. An adjustment of the acetonitrile proportion in the mobile phase was necessary during column life in order to conserve the initial column performance (426). Some differences in the behavior of various packings were noted in the study on the determination of pyridostigmine, neostigmine, and endrophonium compounds in biological fluids (427). The retention of pyridostigmine and neostigmine on all tested columns decreased when tetramethylammonium chloride was added to the mobile phase consisting of sodium phosphate buffer–acetonitrile with 5 mM sodium heptanesulfonate. On MCH column (not endcapped) the retention also decreased for a mobile phase free of pairing reagent. The addition of tetramethylammonium reduced tailing on MCH and LiChrosorb RP-8 columns. No effect on peak shape was noted for RP-18 and Ultrasphere Octyl columns. An approximately linear decrease of log k values with acetonitrile concentration 16–22% was observed. Home-packed RP-18 and RP-8 columns deteriorated rapidly, while an Ultrasphere Octyl column proved to be very stable (427). The separation of apomorphine, apocodeine, isoapocodeine, and N-n-propylnorapomorphine was never as good on µBondapak C18 as on µBondapak Phenyl column (428). On a C18 column the peaks showed extensive tailing and a linear relationship between the amount of injected apomorphine and peak height could not be established. On the phenyl column excellent linearity was found using mobile phase 0.02 M Na_2HPO_4 + 0.03 M citric acid (pH 3.25)–methanol–acetonitrile 55:36:9 with 1 mM sodium dodecyl sulfate. The methanol/acetonitrile ratio was adjusted so as to separate N-n-propylnorapomorphine from O-methyl metabolites, which coeluted when either methanol or acetonitrile was used as the single organic modifier (428).

The role of organic modifier in the control of selectivity was also established when separating ephedrine alkaloids (429) or tertiary and quaternary alkaloids from Corydalis tuber (Corydalis yanhusuo) (430). Whereas with acetonitrile as organic modifier good separations were obtained, with methanol all alkaloids eluted with the same capacity factor. The measured dependencies of retention on pairing reagent (sodium dodecylsulfate) concentration in the mobile phase are typical for ion pair chromatography (Figure 5.23). Maxima on curves and crossing of the dependence for glaucine and tetrahydropalmatine between 0.5 and 1% of sodium dodecyl sulfate are apparent. Increasing the concentration of acetonitrile caused a decrease in retention of ephedrine alkaloids (429). Contrary to the case of reserpine mentioned above (422), a decrease in retention at higher pH of about 5 was reported for tetrahydrocolumbamine, tetrahydropalmatine, and corydaline (430).

The separation systems described above contain rather complex electrolytes. Along with protons in acidic mobile phase, other ions

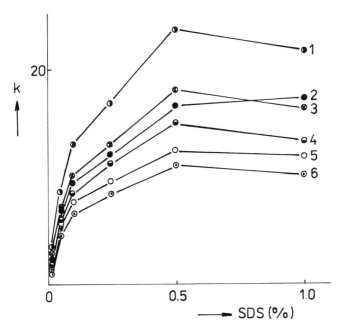

FIGURE 5.23 Effect of sodium dodecyl sulfate (SDS) on capacity factors of tertiary alkaloids from corydalis tuber. Column 150 × 4 mm packed with TSK gel LS-410 (ODS chemically bonded silica gel), 5 μm, mobile phase 0.05 M tartaric acid–acetonitrile 62:38, 1.5 ml min^{-1}, 30°C. 1, Corydaline; 2, glaucine; 3, tetrahydropalmatine; 4, 5, unknown; 6, tetrahydrocolumbamine. (Reproduced from Chemical and Pharmaceutical Bulletin, Ref. 430, by permission of Pharmaceutical Society of Japan.)

as sodium, potassium, tetrabutylammonium (421), triethylammonium (425), and tetramethylammonium (427) act as eluting cations, although simpler systems are sometimes sufficient. Pilocarpine, physostigmine, and rubreserine were analyzed on µBondapak C18 column with mobile phase containing 0.005 M heptanesulfonic acid in 40% methanol at pH 3.6 (431). In such a system with the only pairing reagent in the form of an acid, the hydrogen ions simultaneously ensure protonization of basic solutes and serve as the eluting cations. If the pairing reagent is added in the form of salt of strong base, its cations act as the main eluting cations. Such an unbuffered system hardly yields a satisfactory result for weak bases because their degree of dissociation is not well defined. Only strong, completely ionized bases or weak, completely undissociated solutes can yield good separations. A similar system based on Zorbax ODS column with mobile phase containing 5 mM of sodium lauryl sulfate in 95% acetonitrile was used to determine acrinol and berberine chloride simultaneously in pharmaceutical preparations (432). Sodium octanesulfonate (1 mM) in methanol-water 4:1 was used as the mobile phase for pilocarpine determination in biological fluids after quaternization with p-nitrobenzyl bromide (433).

A strongly hydrophobic pairing reagent, once retained by a reversed phase column, is only slowly washed out with the mobile phase rich in water. The effect of ionic surfactant loaded to the column on solute retention is known to persist for some time even if it is no longer present in the mobile phase (396,434). Chromatographic systems of this type were applied to the separation of Tabernaemontana (62), opium, and other alkaloids (435). A column packed with LiChrosorb RP-18 loaded by recycling overnight 100 ml of 0.01 M dodecyl sulfonic acid in methanol-water 1:1 followed by 100 ml of 0.02 M aqueous cetrimide displayed a substantially improved efficiency compared to untreated column or column treated only with dodecyl sulfonic acid or cetrimide. Even the column loaded with both surfactants but in reversed order possessed poor efficiency. The loaded column could be used for several weeks without a decrease in efficiency with a mobile phase 0.02 M methanesulfonic acid-dioxane-sulfuric acid 98.5:1.5:0.5 (pH 3.5). Some types of reversed phase materials other than LiChrosorb RP-18 were so altered by this treatment that alkaloids were no longer retained (435).

Micellar chromatography was applied to the separation of berberine-type alkaloids in Rhizoma coptidis and Chinese patent medicines (436). The effects of stationary phase, concentration of micelles, organic modifier content, and pH were investigated. Coptisine, jatrorrhizine, palmatine, and berberine were separated on µBondapak Phenyl column with mobile phase containing 3.5% of sodium dodecyl sulfate in 0.1 N tartaric acid-methanol 30:70 (436).

Xanthine alkaloids are very weak bases that are undissociated at the pH values commonly used in liquid chromatography. They behave as neutral compounds and their retention is usually not markedly affected by the presence of an anionic pairing reagent (413). However, some of them dissociate as a weak acid and a cationic pairing reagent can therefore be applied to their separation (437–439). Also the separation of other alkaloids or their derivatives containing an acidic group can be performed using a cationic pairing reagent. Addition of 0.001 M trioctylmethylammonium chloride to the mobile phase acetonitrile–0.01 M ammonium carbonate 7:3 enabled retention of vincaminic acid on Nucleosil 10 C18 (31). Using a mobile phase containing 0.005 M pentanesulfonic acid in a methanol–water–acetic acid–triethylamine 40:53:6:1 mixture, noscapinic acid, noscapine, and papaverine were separated on a Spherisorb S5 ODS column (440). Otherwise in order to suppress the ionization of acidic groups such as glucuronic acid ($pK_a \approx 3.2$), low pH values of the mobile phase are required. The determination of morphine and its glucuronides in plasma and urine was performed at pH 2.1 using 10 mM sodium dihydrogen phosphate containing 1 mM dodecyl sulfate and 26% acetonitrile on an Ultrasphere ODS column (441).

As mentioned above, the pairing reagent in the mobile phase of a reversed phase system has a minor effect on retention on nitrile phases (413). Interesting results were obtained by applying the pairing reagent to normal phase chromatography on columns of this type (287,442). The questionable assumption about the higher polarity of uncharged alkaloid compared to its ion pair with acid added to the mobile phase led to the postulation of another molecular complexation between these species. Di-(2-ethylhexyl)phosphoric acid (DHP) added to the mobile phase hexane–chloroform–acetonitrile 65:20:15 at concentrations higher than 0.5 mM increased the retention of ergot peptide and eburnane alkaloids on μBondapak CN column. The dependence leveled off at a concentration of about 2.5–5 mM and eventually decreased slightly further. For ergocryptinines, ergocorninine, ergocristinine, and ergotaminine, the retention increased approximately linearly up to a concentration of 10 mM. The enhancement in retention was not observed on DHP-loaded column unless DHP was also present in the mobile phase. The example of the separation of eburnane alkaloids is presented in Figure 5.24. Studies of eburnane alkaloids eluted by mixtures of camphorsulfonic acid and diethylamine demonstrated increased retention only if the molar concentration of camphorsulfonic acid was higher than the concentration of diethylamine (287). The separation of optical isomers was studied using (+)- and (-)-camphor-10-sulfonic acid (442). The application to the separation of optical isomers of vincamine is shown in Figure 5.25.

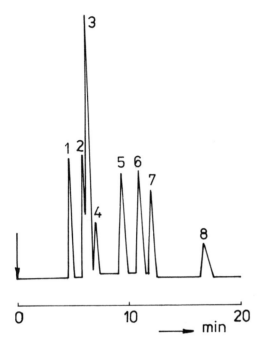

FIGURE 5.24 Ion pair chromatography of some eburnane alkaloids. Column 300 × 3.9 mm packed with μBondapak CN, mobile phase hexane-chloroform-acetonitrile 65:20:15 containing 1 mmol liter^{-1} of di-(2-ethylhexyl)phosphoric acid, 1 ml min^{-1}, UV detection at 280 nm. 1, (+)-cis-Vincamone; 2, (+)-cis-apovincaminic acid ethyl ester; 3, (+)-cis-apovincamine; 4, (+)-cis-vincamenine; 5, (+)-cis-epivincamine; 6, (+)-cis-vincamine; 7, (+)-cis-vincanol; 8, (+)-cis-isovincanol. (After Ref. 287.)

Similar to the case in ion exchange chromatography, the method of indirect detection in reversed phase ion pair chromatography (443-447) has not yet been applied to the separation of alkaloids. In indirect photometric detection the ionic component absorbing on a required wavelength is added to the mobile phase. In contrast to ion exchange chromatography, both pairing reagent and its counterion can be used as an active component since generally the concentrations of both are changed by the presence of solute ions.
The sign and magnitude of the response are reported to be dependent on charge and relative retention of active component and solute. However, if methylpyridinium was the only cation present in the mobile phase and its concentration was low (0.25 mM), the response factor was constant for all separated alkylsulfonates and indirect

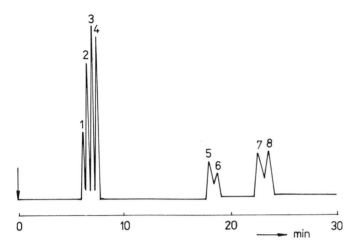

FIGURE 5.25 Ion pair chromatography of optical isomers of vincamine. Column 150 × 4.6mm packed with Nucleosil 5 CN, mobile phase hexane-chloroform-ethanol 80:18:2 containing 2 mmol liter^{-1} of (+)-10-camphorsulfonic acid and 1 mmol liter^{-1} of diethylamine, 1 ml min^{-1}, UV detection at 280 nm. 1, (+)-cis-Epivincamine; 2, (-)-cis-epivincamine; 3, (+)-cis-vincamine; 4, (-)-cis-vincamine; 5, (+)-trans-epivincamine; 6, (-)-trans-epivincamine; 7, (+)-trans-vincamine; 8, (-)-trans-vincamine. (After Ref. 442.)

detection in gradient elution was possible (407). Application of analogous techniques to weakly absorbing alkaloids might be fruitful.

Ion pair chromatography using disposable cartridges can be effectively applied during the sample preparation step. Sep-Pak C18 cartridges were used in the purification of the potato glycoalkaloids (448,449). Also classical ion pair extraction can be applied for this purpose. A wider range of pairing reagents than that used in ion pair chromatography is used for this purpose. As an example, perchlorates (51-53), picrates (448-449), 9,10-dimethylanthracenesulfonates (450), 3,5-tert-butyl-2-hydroxybenzenesulfonates (51), or octylsulfonates were applied to the isolation of purified alkaloids from the sample.

V. DETECTION IN LIQUID CHROMATOGRAPHY

In the liquid chromatography of alkaloids, as in other fields of LC application, the absorbance (photometric) detectors are dominant.

Besides these, the fluorescence detectors due to their selectivity and the amperometric detectors due to their sensitivity are also frequently employed. On the other hand, the refractive index detector has been applied only in instances when the use of another detector was not possible. These four types of detectors are discussed in detail in the following sections.

However, there are other, less widely used detectors that have been tried for alkaloid detection. In the first place the mass spectrometer coupled with a liquid chromatograph must be mentioned. At present, LC-MS interfaces mostly employing thermospray process have become commercially available, and one can expect the usefulness of this technique in chromatography to increase. The latest development in the combination of supercritical fluid chromatography and mass spectrometry (SFC-MS) has demonstrated the potential of this technique for alkaloid analysis (96). Using a direct liquid introduction device and chemical ionization (96) or a moving belt with modified thermospray deposition device and electron impact ionization (452), alkaloids were separated and determined by SFC-MS with greater speed and lower detection limit than by a comparable LC-MS system.

In order to monitor alkaloids, some other detectors were used: capacitance-conductance detector, thermoionic nitrogen-sensitive detector, circular dichroism spectrophotometer, etc., but none of them showed a wider range of application.

A. Absorbance Detectors

Absorbance detection can be classified as the most important and most popular mode of detection in alkaloid chromatography. It allows the detection of minimal amounts of approximately 1 μg to 1 ng. A comprehensive review on the advances in absorbance detector design was published by Barth et al. (453). The principles of absorbance measurement are given in Chapter 2, Section III. It should be emphasized that modern spectrophotometric detectors are convenient for monitoring the great majority of alkaloids.

Three types of absorbance detectors can be distinguished: (a) fixed-wavelength; (b) variable-wavelength; (c) rapid-scan variable-wavelength.

The first type, which is still very popular, operates with a discrete wavelength source such as a low-pressure Hg lamp. Very often detectors of this type operate at 254 nm and employ a flow cell of 1 mm ID with optical pathlength 1 cm. Some of these detectors are equipped with a filter which enables detection at 280 nm. Fixed-wavelength detectors are considerably less expensive than the other types, possess low-noise characteristics, and in comparison with variable-wavelength detectors show higher sensitivity for the given wavelength. With the exception of certain alkaloids, a fixed-wavelength detector that makes it possible to operate at 254 and 280 nm is very suitable for alkaloid detection.

The second type, the variable-wavelength detector, provides a continuum of wavelength from 200 nm to the visible employing a continual source, e.g., deuterium or xenon lamp. A simplified scheme of this detector is given in Figure 5.26A. The polychromatic beam passes through the entrance slit and strikes the grating. Light rising from the grating is allowed to pass through the exit slit and flow cell and to strike the photomultiplier tube (PMT). The selection of wavelength is done mostly by the manual grating adjustment. Commercially available detectors employ the flow cell with a dead volume of a few microliters, and even submicroliter flow cells for microbore HPLC have been developed.

The last development in variable-wavelength detectors was directed at two points: (a) automatic change of wavelengths during an analysis and (b) on-line acquisition of spectral information.

The first problem has been solved by a control module which can be programmed to automatically change wavelengths in order to take full advantage of maximum absorptivity of each sample component. In another device, a programmable detector makes possible to continuously record the ratio of absorbances of two selected wavelengths vs. retention time. This so-called ratiogram makes possible the estimation of two chromatographically unresolved peaks. The ratiogram of a pure compound yields a "square wave" with a height proportional to the ratio of molar absorptivities (Figure 5.27A). When two compounds with different absorptivities at the wavelengths used are chromatographically unresolved, the ratiogram shows the presence of two components (Figure 5.27B). However, to obtain this a certain chromatographic resolution of both components must be achieved. The ratio of absorbances can be also used for the characterization of drugs including alkaloids. To identify isoquinoline alkaloids, opium bases, and the like, the ratios of absorbances measured at 254 and 280 nm (64), or at 220 and 254 nm (419) were used. To obtain a ratiogram by means of a variable-wavelength detector requires a special device, but there is no problem with the use of a photodiode array detector (454,455).

The on-line acquisition of spectral data provides spectral information on eluted components. While the measurement of complete spectra is an obvious operation of a photodiode array detector, a variable-wavelength detector requires cessation of eluent flow when a component of interest is in the flow cell. This is tedious and moreover the spectra are distorted by diffusion.

Photodiode array detectors (PDA) are the most sophisticated absorbance detectors. In principle they are constructed similarly to variable-wavelength detectors, but the grating is fixed and they

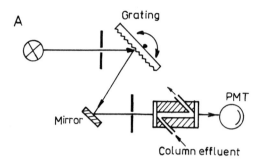
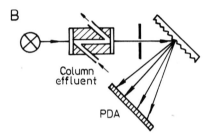
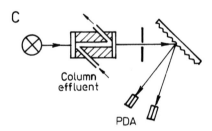

FIGURE 5.26 Simplified schemes of absorbance detectors. (A) Variable-wavelength detector. (B) Linear photodiode array (PDA) detector. (C) Discrete photodiode array detector.

Liquid Chromatography 233

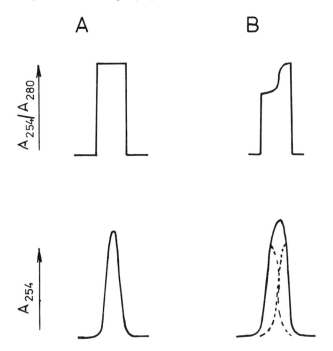

FIGURE 5.27 Ratiogram and chromatogram. (A) Pure compound, (B) two unresolved compounds.

do not employ an exit slit. Instead of a photomultiplier tube they have photodiode elements. There are two arrangements of PDA detectors. Linear (or self-scanned) PDA detectors generally contain more than 100 elements in a row (Figure 5.26B). The white light from the flow cell dispersed by the grating is simultaneously measured by an array of photodiodes. In this way each photodiode receives a different wavelength and an individual signal can be obtained for every 2-5 nm of the wavelength range. The signals enter a computer with facilities for multivariate mathematical treatment. Another arrangement is called a discrete PDA detector and contains only 2-40 diode elements (Figure 5.26C). Each of the discrete diodes is read in a parallel fashion from individual outputs. A discrete PDA detector is less expensive than a linear PDA owing to the reduction in volume of multichannel data. On the other hand, the spectral coverage of the discrete PDA is essentially reduced. At the time of writing there are no discrete PDA detectors available commercially but their further development is confidently expected.

Linear PDA detectors generate a large volume of data, thus extending the possibilities for identification and quantitation. Due to very rapid spectral response across the measured region (e.g., from 200 to 400 nm), absorption spectra of the all-eluted component can be recorded. On this basis three-dimensional plots (absorbance, wavelength, retention time) may be obtained. They allow the determination of the optimal wavelength for detection and a search for unexpected peaks that might not be observed at a single wavelength. A modification of the three-dimensional plot is the graphic representation of the same data as a two-dimensional contour diagram.
There are isoabsorptive contours plotted as a function of wavelength and retention time. This plot can be used in the same manner as the three-dimensional plot as well as for the evaluation of peak homogeneity. Nevertheless the utility of these plots is limited by the character of UV-visible spectra. The spectrum depends mainly on the type of chromophore and variation in structure brings only small changes. If the variation is separated from the chromophore, the spectra of similar compounds may be essentially identical.

The advantage of the ratiogram for the evaluation of peak homogeneity has been already mentioned. For details the reader is referred to the original papers (454,455), in which the methods of data treatment, including multicomponent analysis, are described. It is only necessary to be reminded of the utility of the absorbance ratio for drug identifications (64,419) and the fact that the ratio of absorbances is the form readily adapted to the PDA detector.

The spectral data obtained by making use of the multiple-wavelength detection can be further employed for spectral matching by means of computer-aided archive retrieval if a database of spectra is available. It must be emphasized that the standard spectra must be obtained under conditions identical to those of the sample. The composition and pH of mobile phase are of primary importance, especially for the ionizable compounds as are most alkaloids.

It was already stated that absorbance detectors are the most frequently used detectors in the liquid chromatography of alkaloids. In order to estimate the suitable wavelength for the detection of an alkaloid, see Tables 2.7 and 2.8. It can be ascertained that wavelengths 254 and 280 nm suit the majority of alkaloids perfectly. Small departures from the given values may be neglected owing to the spectral width of the monochromatic beam. Nevertheless, some exceptions exist where other wavelengths or detector types must be employed.

First, some bases possess no chromophore in their structures. They are the products of hydrolysis of tropane and steroid alkaloids. In both cases operation at 200–220 nm makes it possible to detect the bases even when the limit of detection is relatively high, about 1 µg (85). In addition, a refractive index detector permits the detection of these alkaloids (see Section V.B of this chapter). Table 2.7 lists another four alkaloids—arecaidine, arecoline, isopilocarpine, and pilocarpine—which are detected at wavelength 215 nm. This requirement does not concern the above-mentioned alkaloids but also some other bases derived from their structures.

As far as the other bases are concerned, the subgroup of Cinchona alkaloids is of interest. For the detection of quinine, quinidine, cinchonine, and cinchonidine, a first maximum of 231 nm was used (285), but detection at a secondary maximum of 330 is not intercepted by other components. Nevertheless, owing to the strong native fluorescence of Cinchona alkaloids and their dihydro derivatives, fluorescence detection of these bases is often employed (see this chapter, Section V.C).

The alkaloids of the morphine group can be easily detected at 280 nm, but due to the complexity of opium and heroin samples, detection at the two wavelengths 254 and 280 nm is frequently used. A photodiode array detector was successfully applied to the analysis of heroin (456). Morphine and some of its derivatives can be also sensitively detected by fluorescence and electrochemical detectors (see Sections V.C and V.D of this chapter).

The position of ergot alkaloids from the standpoint of absorbance detection is rather exceptional. Their maxima at 240 and 310 nm have sufficiently high molar absorptivities and were detected with high sensitivity at 241 nm (457), but the fluorescence detection of ergot alkaloids is more frequently employed. The same is valid for their dihydro derivatives, which can be determined by UV absorption at 280 nm; however, fluorescence detection is preferred.

Besides quinine and quinidine, Table 2.7 names a few other alkaloids showing maxima above 300 nm: cytisine, dictamnine, α-narcotine, palmatine, and ricinine. There is also berberine which, compared to the other alkaloids in the list, possesses the longest wavelength absorption band (354 nm) with the highest molar absorptivity. It must be emphasized here that bands of longer wavelength (280 nm and higher) should be used for quantitative determination. This is due to limited interference by other compounds, low baseline drift related to the mobile phase or the instrument, and no problem with mobile phase cutoff. Conversely, at low wavelengths (below 220 nm) there are problems with detector linearity and with high relative absorbances of impurities.

The selection of detection wavelength depends on the mobile phase employed, as the solvent cutoff limits the usable range. The values of UV cutoff for some solvents are given in Table 5.1. The operation at the shortest wavelengths (about 200 nm) is allowed by the use of an acetonitrile-water mixture, while the mixtures of aliphatic alcohols and water can be used at wavelengths slightly higher (205–210 nm). Detection up to the same wavelength is possible with

alkanes as the mobile phase, while diethyl ether and methylene chloride shift the mobile phase cutoff to 220 and 245 nm, respectively.
At the end of this chapter are given a few remarks on the derivatization methods applied to enhance the absorbance of some alkaloids. These methods are preferably employed in fluorescence detection but seldom in absorbance monitoring. However, many derivatization reagents contain aromatic structures in their molecules and therefore derivatized alkaloids will show increased absorbance. Another way to increase the absorbance of poorly absorbing bases is to use ion pair separation with a strongly absorbing pairing ion. This method with picric acid as pairing ion was employed for the separation of tropane alkaloids (57).

B. Refractive Index Detectors

The second type of optical detector is based on differential measurement of the refractive index. In principle two methods can be used: the method of the angle of deviation and the critical angle method. The refractive index (RI) detector records a bulk property of effluent, which means that the measured changes in refractive index are contributed by both solvent and solute. The RI detector due to its universal nature makes it possible to determine all molecular types. On the other hand, this property is a source of frustration. The RI detector is sensitive to small changes in the mobile phase composition, is temperature- and pressure-dependent, and requires the use of degassed solvents delivered by pulseless flow. However, the main disadvantage is the low sensitivity of the RI detector. It is claimed that contemporary RI detectors employing built-in electronic thermal control may reach sensitivity in the nanogram range (458).

Referring to Table 2.7, the reader can find several alkaloids that lack a chromophore in their structures and therefore are nondetectable by photometric detectors. However, some of them have boiling points in the range convenient for gas chromatographic determination. In these cases the application of GC should be preferred due to its speed, sensitivity, and reproducibility. Thus, there are only a few alkaloids for which the use of a RI detector should be considered. They are some bases of subgroups II.D, tropane alkaloids, and VIII.B, steroid alkaloids (see Chapter 1).

The major problem in the monitoring of tropane alkaloids creates the detection of tropine, scopoline, and methylecgonine, which are obtained mostly as the products of hydrolysis of parent alkaloids. Other naturally occurring tropane alkaloids (e.g., atropine, scopolamine, etc.) also do not show any outstanding UV absorption (see Table 2.7). Nevertheless, when RI and photometric detectors were tested for the detection of tropane alkaloids (459), the RI detector showed essentially worse results.

Liquid Chromatography

Similarly, there is the problem of the detection of steroid alkaloids, both free bases and glycoalkaloids. Because they possess either no chromophore or slightly absorbing chromophore (carbonyl group), some authors tried to monitor them by RI detector (85,291). However, in both cases the absorbance detector operating at short wavelengths (200 and 215 nm) gave better sensitivity than the RI detector.

The last example of RI detector employment is pilocarpine determination (460). Pilocarpine shows good absorptivity at 215 nm, but the baseline stability is sometimes a problem at this wavelength. As in previous cases, the limit of detection with an RI detector was about 6 µg in comparison with 0.04 µg for the photometric detector at 215 nm.

It can be concluded that up to now RI detectors have not found any significant application for alkaloid detection and this situation is unlikely to change in future.

C. Fluorescence Detectors

Detectors of this type, classified as optical and nondestructive, are suitable for the detection of either fluorescent compounds or nonfluorescent substances which can be converted to fluorescent products.

Fluorescence is the emission of light accompanying the transition of an excited molecule to its ground electronic state. Thus before this phenomenon can occur a substance must absorb radiation of the proper wavelength. Fluorescence is observed mainly for aromatic and highly conjugated molecules, while compounds with a high degree of vibrational freedom lose their energy by vibrational relaxation processes. The number of fluorescing compounds is rather limited and therefore fluorescence detection offers a high degree of selectivity.

The intensity of fluorescence I_f is linear with the solute concentration c according to Eq. 5.29:

$$I_f = 2.3 k \psi_f I_o \varepsilon bc \tag{5.29}$$

where k is the proportionality constant, ψ_f is the quantum yield of fluorescence, I_o is the intensity of the excitation light, ε is molar absorptivity, and b is the length of the cell path. Equation 5.29 is valid for low solute concentrations only, where the term εbc is less than about 0.05. Generally, concentrations of a few parts per million or less will suit for fluorescence measurement. At high concentrations the fluorescence emission may be reabsorbed by other sample molecules.

The sensitivity of fluorescence detection depends on at least two parameters: on the molar absorption coefficient of a solute ε and its quantum yield of fluorescence ψ_f, supposing that the other parameters of Eq. 5.29 are constant and given by the detector construction. A high ε value (10^4–10^5) is advantageous because it represents a large number of absorbed photons. The quantum yield ψ_f is always smaller than unity. For well-fluorescing compounds it reaches values greater than 0.1 and cannot be smaller than 0.01 for fluorescence to be observable. Fluorescence detectors are among the very sensitive detectors. In some cases they permit the determination of a fluorescing solute with a minimal detectable quantity of about 10 pg.

The requirements for mobile phase compatibility are the same as for absorbance detectors. The mobile phase must not absorb UV radiation at the excitation wavelength and therefore acetonitrile, alcohols, and saturated hydrocarbons are the most suitable eluents.

Commercial fluorescence detectors are equipped with the source of excitation radiation and with a device for the selection of excitation and emission wavelengths, flow cell, and photomultiplier tube. As a radiation source the xenon lamp is preferentially used, but the application of lasers is becoming popular. The selection of excitation and emission wavelengths may be realized by either a set of filters or two monochromators equipped with gratings. The combination of excitation monochromator and cutoff filter on the side of emission is frequently employed. The flow cell is usually made of narrowbore quartz tubing. The volume of fluorescence flow cells is about 20 μl, even though cells of 1-μl volume have been developed.

The employment of a fluorescence detector may bring about a new feature in solute characterization. It is possible to characterize a fluorescing compound by excitation and/or emission spectra, but the later ones are more frequently measured. The scanning of emission spectrum in effluent, without any need for stopping the flow, may be carried out by an instrument equipped with a photodiode array. This makes it possible to measure emission spectra of eluate at a speed compatible with the usual elution rates applied in HPLC.

The majority of alkaloids can be determined by absorbance measurement, but only a few of them can be monitored by fluorescence detector. Some alkaloids showing sufficiently high native fluorescence are given in Table 5.7. The fluorescence intensity depends on solvent composition and on pH via protonation of basic functional groups. For these reasons the mobile phase composition and pH are also given in Table 5.7. Besides the alkaloids given there, some other bases, e.g., codeine (exc. 213 nm; em. >320 nm), morphine (290; 340), ellipticine (360; 455), psilocybin (267; 335), psilocin (260; 312), N-propylajmaline (242; 320), and protoberberine alkaloids (350; 520), have a certain native fluorescence to allow for detection.

TABLE 5.7 Fluorescence Detection of Some Natively Fluorescing Alkaloids

Alkaloid	LC mode	Excitation wavelength (nm)	Emission wavelength (nm)	Mobile phase pH	Mobile phase Composition	Ref.
Dihydroergocristine Dihydroergotamine	RP	295	350	7.2	ACN–0.018 M phosphate buffer 6:4	461
Ergotamine Ergotaminine Ergocristine	RP	328	398	—	ACN–0.01 M $(NH_4)_2CO_3$ 3:7	462
Glaucine Boldine	NP	310	340	—	C6–MeOH–THF–DEA 88.5:7.5:4.0:0.15	463
Harmaline Harmalol	RP	396	475	8.5[a]	MeOH–H_2O–formic acid 166:34:1	464
Harmine Harmol		304	355			464
Quinidine Quinine	RP	340	>418		H_2O–ACN–acetic acid 86:10:4	465

[a]pH adjusted with triethylamine; ACN, acetonitrile; C6, hexane; THF, tetrahydrofuran; DEA, diethylamine; RP, stationary phase alkyl-modified silica; NP, stationary phase silica gel.

TABLE 5.8 Fluorescence Detection of Derivatized Alkaloids

Alkaloid	Derivatization method	Reagent	Medium	Ref.
Cephaeline				
Emetine	Pre	Dansyl chloride	0.1 M Na_2CO_3	466
Ephedrine				
Histamine	Pre	o-Phthalaldehyde	2-Mercaptoethanol + borate buffer	467
Morphine	Post	Potassium ferricyanate	4 M NH_4OH	468
Reserpine	Post	Nitrous acid	H_2SO_4 + $NaNO_2$	469

Another way to employ the fluorescence detector is to convert nonfluorescing alkaloids to fluorescing products. This operation can be carried out either before or after chromatographic separation. Precolumn derivatization is a widely used technique whereby a solute is mixed with a selected reagent prior to the separation and left to react for the required time. If the components of the derivatization mixture do not interfere with the detection of fluorescent product, the mixture can be injected directly onto the column. Examples of precolumn derivatization are given in Table 5.8. Dansyl chloride is the most widely used derivatization reagent, reacting easily with primary and secondary amino groups. The reaction can be performed in either aqueous or nonaqueous media. The reaction time varies from 30 min to 24 hr. The reagent is subjected to hydrolysis occurring competitively with an acetylation reaction but the derivatives are relatively stable. o-Phthalaldehyde is used for the derivatization of compounds with a primary amino group. It reacts in an aqueous reducing medium (e.g., 2-mercaptoethanol) at pH 10 and the reaction is completed within 1–2 min. Using this reagent histamine and N-methylhistamine give products which are stable for 60 min. o-Phthaldehyde has no native fluorescence and therefore the reaction mixture can be injected directly onto the column.

Postcolumn derivatization always requires modification of the instrument to facilitate the postcolumn reaction. A device for postcolumn derivatization consists of a reactor placed between the column outlet and detector and a second pump to deliver the derivatizing reagent to the effluent. The reagent and effluent have to be well mixed before they enter a reactor. Three types of reactors have been applied in practice. The first type, the tubular reactor, is a capillary tubing a few meters long which ensures the time necessary for a fast reaction to occur. The second type, the bed reactor, is a short chromatographic column packed with small glass beads. This reactor is suitable for reaction times up to a few minutes. In the third type, the air segmentation reactor, air bubbles are introduced into the column effluent beside the stream of the derivatizing reagent (see Figure 5.28). The segmentation of flow ensures long residence time simultaneously with reduced band broadening. The bubbles are removed in a debubbler before entering the detector.

Table 5.8 shows two examples of postcolumn derivatization. The oxidation of morphine by means of alkaline potassium ferricyanate yields the fluorescent dimer pseudomorphine. The fast reaction allows the use of a tubular reactor, e.g., reaction coil 5 m × 0.3 mm ID. In a similar way some other morphine alkaloids, e.g., normorphine, dihydromorphine, nalorphine, and 6-monoacetylmorphine, can be derivatized to fluorescing products while the others, e.g., codeine, norcodeine, ethylmorphine, acetylcodeine, morphine-3-glucu-

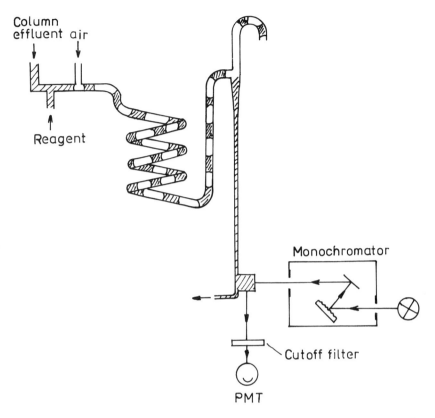

FIGURE 5.28 Postcolumn derivatization with fluorescence detection. Air segmentation reactor.

ronide, and heroin, do not react. In order to increase the fluorescence of 6-monoacetylmorphine, which is rather weak, the modified prederivatization procedure was employed (470). The oxidation was carried out in the presence of morphine to give 6-monoacetylmorphine-morphine dimer. Also the second example of postcolumn derivatization in Table 5.8 is basically alkaloid oxidation, i.e., via reaction of reserpine with nitrous acid as oxidizing agent fluorophore 3,4-dehydroreserpine is formed. In this case the reaction occurs slowly and an air segmentation reactor must be employed.

The use of fluorescence detectors in the liquid chromatography of alkaloids is certainly limited to a narrow range; however, due to their selectivity and sensitivity the importance of these detectors is not in question. The bioanalytical applications of liquid chromatography which have been reviewed (471) deal with these problems in detail.

D. Electrochemical Detectors

The use of electrochemical detectors is gaining in popularity due to their sensitivity. They are able to determine an electrochemically reducible (or oxidizable) compound at the level of 1 pg. The principles of electrochemical detection are given in Section IV of Chapter 2, where the two main types of detector are also introduced. However, it should be underscored here that the electrochemical conversion (and naturally the resulting detector response) of a compound at a given potential does not depend on molecular structure only, but to a certain degree on the mobile phase (composition, pH, etc.) and on the flow cell (design and material of working electrode) as well.

Voltammetric flow-through detectors, often called simply amperometric detectors, may be constructed as either the wall-jet or the channel thin-layer cells. The scheme of a wall-jet cell is depicted in Figure 5.29A. Eluent from the column passes a small jet and impinges on the surface of working electrode which is held at a predetermined potential. The cell body made of stainless steel may act as an auxiliary electrode and the potential of working electrode is measured against a reference electrode, often a silver/silver chloride electrode (SSCE). The description of this cell construction can be found elsewhere (472), and a submicroliter dead volume of cell may easily be obtained by this arrangement. The same dead volume can be reached with a channel thin-layer cell shown in Figure 5.29B. There the working electrode is a part of the channel wall. The channel itself is formed by sandwiching a fluorocarbon gasket between two parts of block made of plastic, e.g., Kel-F. The reference electrode is usually imbedded in the cell outlet. Because of low currents encountered in trace analysis, the amperometric cells may be operated in a two-electrode mode (473). Conversely, when larger samples are injected the employment of three electrodes is preferred. The further improvement of amperometric detectors lies in the use of a dual-working electrode system. This system makes possible the simultaneous determination of reducible and oxidizable species, the improvement of specificity by reelectrolysis at the second electrode of the reaction products produced at the first electrode, etc.

Coulometric detectors are constructed in a similar manner to the thin-layer amperometric detector, but with electrode areas a few square centimeters rather than a few square millimeters. From this fact results a relatively large dead volume of the cell, which suppresses the sensitivity of the coulometric detector. The main disadvantage of this detector lies in the necessity of converting all the solute into a product during its stay in the flow-through cell. This means that this detector may be sensitive to the fluctuations of mobile phase flow rate.

Liquid Chromatography

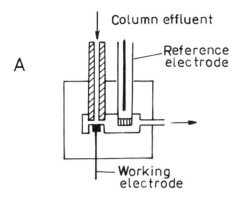

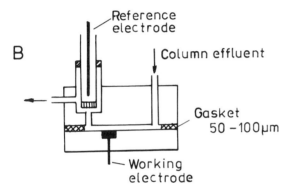

FIGURE 5.29 Voltammetric detectors. A, wall-jet cell; B, channel thin-layer cell.

Among the materials employed for the construction of the working electrode, glassy carbon is the most widely used, but carbon paste, graphite composite, and reticulated vitreous carbon have also found a wide range of applications. The use of static mercury drop in polarographic detectors is rather exceptional.

The application of electrochemical detection in the chromatography of alkaloids shows two features:

1. The use of polarographic detectors has been practically abandoned.
2. The electrochemical oxidation of alkaloids at the anode is the prevalent mode.

No comment is necessary regarding the first point. The second point results from the molecular structure of alkaloids. Detailed studies of this field were carried out by Musch et al. (286,474), who concluded that the compounds containing a phenol, a primary or secondary aromatic amine, an aromatic methoxy or ethoxy, or a thiol function can be detected by amperometric detection using the oxidation mode. The list of functions can be expanded by adding a phenothiazine sulfur, a secondary or tertiary aliphatic amine, piperazine, and dioxazine. The following paragraphs give examples of the most important applications of electrochemical detection.

To give an example of secondary aliphatic amine oxidation, ephedrine can be introduced. The detection was achieved at +1.0 V vs. saturated calomel electrode (SCE) in system of normal phase adsorption chromatography (475). However, the mobile phase used was rather polar (ethyl acetate–methanol–water–ethylamine 78.2:20:0.6:0.2) and moreover lithium perchlorate at a concentration of 0.01 M was added to the mobile phase.

As far as tropane alkaloids are concerned, the amperometric detection of scopolamine has proved to be 20 times more sensitive than absorbance detection (286). Scopolamine was oxidized in a reversed phase buffered system at +1.1 V vs. SSCE. Under identical conditions atropine failed to give any significant response. This means that the electrochemical activity of scopolamine lies in the oxidizable epoxide group.

Some tetrahydroisoquinoline alkaloids (salsoline, salsolinol, tetrahydropapaveroline) possessing one electrooxidizable phenolic substituent were detected at +0.8 V vs. SSCE in a reversed phase buffered system (476). Electrochemical detection of other isoquinoline alkaloids was described by McMurtrey et al. (328).

Morphine is probably the alkaloid most frequently determined by electrochemical detection. According to White (477), morphine gives two anodic waves, the first at +0.44 V, the second at +0.70 V, both measured vs. SSCE. At the lower potential the formation of pseudomorphine is supposed; at higher potential the oxidation may lead to phenoxonium ion. In practice, the potential of +0.60 V vs. SSCE is applied using a buffered system and a wide pH range (472). Even though the response of morphine increases with the potential applied, the selected value +0.60 V appears as most suitable from the standpoint of specificity and low noise level. Codeine, under the identical conditions, gives no significant response. Of the other opium alkaloids, α-narcotine (noscapine) was detected at +1.0 V vs. SCE in normal phase system (475) and apomorphine at +0.50 V vs. SSCE (478).

Purine bases exhibit anodic waves which are pH-dependent. Theophylline was detected at +1.24 V vs. SCE in the mobile phase containing acetate buffer–ethanol 92:8 at pH 4.0 (479). The oxida-

Liquid Chromatography 245

tion potentials of caffeine and theobromine lie at higher values
(see Table 2.10) and their responses at +1.24 V do not disturb the
determination.
  Some indole alkaloids are amenable to electrochemical oxidation.
Physostigmine was detected at +0.80 V vs. SSCE (480), reserpine at
+0.90 V vs. SSCE, and dihydroergotamine at +1.10 V vs. SSCE.
In all cases a buffered mobile phase was used, for reserpine and dihydroergotamine with pH 3.0 (286). Up to a 100 times increase in
sensitivity over that of absorbance detection was confirmed.
  It may be concluded that amperometric detectors have found a
wide range of applications in the chromatography of alkaloids. They
exhibit high sensitivity and, in some cases, high selectivity. On
the other hand, they are best used with buffered mobile phases and
from time to time need the surface of the working electrode to be
renewed. The employment of an amperometric detector is justified
only when the absorbance detector response is insufficient.

E. Quantitative Analysis

Quantitative analysis in liquid chromatography is based on peak area
measurement or peak height determination. Essentially, it can be
stated that all the procedures outlined in Chapter 4 for quantitative
analysis in GC apply equally to LC. However, in the distinction
from GC, liquid chromatography does not possess such a universal
detector as FID which provides a similar response to many organic
compounds. The response of detectors in liquid chromatography
often differs in a few orders for equal amounts of various solutes.
The most common absorbance detector may be a good example of this
phenomenon. Its response (absorbance) is linear with the solute
molar concentration and the molar absorptivity. Owing to the great
differences in molar absorptivities (up to three orders), two solutes
having the same molar concentrations may produce very different
peak areas. The situation is similar with a fluorescence detector.
According to Eq. 5.29, its response depends on the solute concentration, molar absorptivity, and moreover on the quantum yield of
fluorescence. Equal amounts of two fluorescing solutes may differ
more than three orders in their responses. The signal of an electrochemical detector (amperometric or coulometric) depends, besides
the solute molar concentration, on the number of electrons involved
in electrochemical reactions (see Eq. 2.19). This number usually
varies from 1 to 6 and this explains the differences in electrochemical detector responses. The response of a refractive index detector
depends for a given chromatographic system on the solute refractive
index and can in principle be negative. Thus, to summarize, the
determination of a particular solute requires a special calibration for
this solute which can scarcely be applied to another one. For this

reason, in liquid chromatography of alkaloids, quantitative analysis that makes use of the internal standard method is most frequently used.

In the beginning of this procedure, a compound suitable as the internal standard has to be selected. This substance should be pure, similar in nature to the alkaloid to be determined, and must be eluted at a point in the chromatogram where it does not overlap the peaks of other sample components. Afterward the relative response factor f_R of the determined alkaloid B and internal standard S has to be found out. A two-component calibration mixture is made up with the known concentrations $(c_B)_1$ of alkaloid and $(c_S)_1$ of internal standard. If the detector responses are given by the peak areas $(A_B)_1$ and $(A_S)_1$, then the relative response factor is calculated as

$$f_R = \frac{(c_B)_1}{(c_S)_1} \cdot \frac{(A_S)_1}{(A_B)_1} \qquad (5.30)$$

In the next step a known amount of internal standard is added to the analyzed sample. Thus, the sample spiked with a known concentration of the standard $(c_S)_2$ is obtained. Supposing that the chromatogram provides the peak areas $(A_S)_2$ and $(A_B)_2$ for this internal standard and the determined alkaloid, respectively, then the concentration of the alkaloid can be calculated as

$$(c_B)_2 = \frac{f_R (A_B)_2}{(A_S)_2} \cdot (c_S)_2 \qquad (5.31)$$

In the majority of cases where the alkaloid of interest is to be determined in biological matrix or in plant species, the internal standard is added directly to the sample (blood, urine) or to the raw extract if a solid material is to be analyzed. These kinds of samples often require a series of pretreatment steps and the addition of standard to the raw sample eliminates the losses that occur during the process. However, this is only true if a condition of similar nature of internal standard and that of determined alkaloid has been fulfilled. If it has not, then the losses may vary and no conclusion can be drawn for the alkaloid of interest.

The method of internal standard should be applicable for any number of components in the sample provided that the corresponding number of calibration mixtures has been prepared and the respective response factors have been ascertained. Steady-state chromatographic conditions during calibration and analysis are strictly required. From this point of view the steady flow rate of mobile phase is of primary importance because the retention axis is directly related not to eluent volume but to time. The fluctuation of mobile phase flow brings about an inaccuracy in quantitative analysis.

REFERENCES

1. L. R. Snyder, Principles of Adsorption Chromatography, Marcel Dekker, New York, 1968.
2. K. K. Unger, Porous Silica, Its Properties and Use As Support in Column Liquid Chromatography, Journal of Chromatography Library, Vol. 16, Elsevier, Amsterdam, 1979.
3. High-Performance Liquid Chromatography (J. H. Knox, ed.) Edinburgh University Press, Edinburgh, 1979.
4. Z. El Rassi, C. Gonnet, and J. L. Rocca, J. Chromatogr., 125:179 (1976).
5. R. E. Majors, J. Chromatogr. Sci., 15:334 (1977).
6. R. E. Majors, J. Chromatogr. Sci., 18:488 (1980).
7. H. Engelhardt and H. Müller, J. Chromatogr., 218:395 (1981).
8. C. Souteyrand, M. Thibert, M. Caude, and R. Rosset, J. Chromatogr., 262:1 (1983).
9. R. Schwarzenbach, J. Chromatogr., 334:35 (1985).
10. J. J. Kirkland, J. Chromatogr., 83:149 (1973).
11. D. L. Saunders, J. Chromatogr., 125:163 (1976).
12. H. Engelhardt, J. Chromatogr. Sci., 15:380 (1977).
13. A. J. Hsu, R. J. Laub, and S. J. Madden, J. Liq. Chromatogr., 7:599 (1984).
14. M. L. Hair and W. Hertl, J. Phys. Chem., 74:91 (1970).
15. K. Marshall, G. L. Ridgewell, C. H. Rochester, and J. Simpson, Chem. Ind. (London), 1974:775.
16. P. Schindler and H. R. Kamber, Helv. Chim. Acta, 51:1781 (1968).
17. D. N. Strazhesko, V. B. Strelko, V. N. Belyakov, and S. C. Rubanik, J. Chromatogr., 102:191 (1974).
18. G. W. Sears, Jr., Anal. Chem., 28:1981 (1956).
19. K. Hostettmann, M. J. Pettei, I. Kubo, and K. Nakanishi, Helv. Chim. Acta, 60:670 (1977).
20. R. Gimet and A. Filloux, J. Chromatogr., 177:333 (1979).
21. R. A. Heacock, K. R. Langille, J. D. MacNeil, and R. W. Frei, J. Chromatogr., 77:425 (1973).
22. R. Verpoorte and A. Baerheim Svendsen, J. Chromatogr., 100:227 (1974).
23. R. Verpoorte and A. Baerheim Svendsen, J. Chromatogr., 109:441 (1975).
24. D. G. I. Kingston and B. T. Li, J. Chromatogr., 104:431 (1975).

25. O. H. Weddle and W. D. Mason, J. Pharm. Sci., 65:865 (1976).
26. L. Szepesy, I. Fehér, G. Szepesi, and M. Gazdag, J. Chromatogr., 149:271 (1978).
27. L. Szepesy, I. Fehér, G. Szepesi, and M. Gazdag, Magy. Kem. Foly, 84:375 (1978).
28. I. R. Hunter, M. K. Walden, and E. Heftmann, J. Chromatogr., 198:363 (1980).
29. S. Hara, N. Yamauchi, Ch. Nakae, and S. Sakai, Anal. Chem., 52:33 (1980).
30. G. Szepesi and M. Gazdag, J. Chromatogr., 205:57 (1981).
31. M. Gazdag, G. Szepesi, and K. Csomer, J. Chromatogr., 243:315 (1982).
32. J. D. Wittwer, Jr., and J. K. Kluckhohn, J. Chromatogr. Sci., 11:1 (1973).
33. I. R. Hunter, M. K. Walden, J. R. Wagner, and E. Heftmann, J. Chromatogr., 119:223 (1976).
34. A. Yoshida, S. Yamazaki, and T. Sakai, J. Chromatogr., 170:399 (1979).
35. C. van der Meer and R. E. Haas, J. Chromatogr., 182:121 (1980).
36. J. Narkiewicz, M. Jaroniec, M. Borówko, and A. Patrykiejew, J. Chromatogr., 157:1 (1978).
37. L. R. Snyder, Anal. Chem., 46:1384 (1974).
38. E. Soczewiński, Anal. Chem., 41:179 (1969).
39. S. Hara, Y. Fujii, M. Hirasawa, and S. Miyamoto, J. Chromatogr., 149:143 (1978).
40. R. Verpoorte and A. Baerheim Svendsen, J. Chromatogr., 100:231 (1974).
41. D. Volkmann, Dtsch. Apoth. Ztg., 118:501 (1978).
42. Ph. van Aerde, E. Moerman, R. van Severen, and P. Braeckman, J. Chromatogr., 222:467 (1981).
43. P. P. Ascione and G. P. Chrekian, J. Pharm. Sci., 64:1029 (1975).
44. A. Turcaut, P. Caileux, and P. Allain, J. Liq. Chromatogr., 3:1537 (1980).
45. M. Danhof, B. M. J. Loomans, and D. D. Breimer, Pharm. Weekbl., 113:672 (1978).
46. R. D. Hossie, S. Sved, K. McErlane, and I. J. McGilveray, Can. J. Pharm. Sci., 12:39 (1977).
47. R. T. Hunter and R. E. Creekmur, Jr., J. Assoc. Off. Anal. Chem., 67:542 (1984).
48. S. H. Hansen, A. M. Hansen, and B. Poulsen, Arch. Pharm. Chemi. Sci. Ed., 8:181 (1980).
49. S. H. Hansen, J. Chromatogr., 212:229 (1981).
50. P. O. Lagerström and B. A. Persson, J. Chromatogr., 149:331 (1978).

51. B. M. Eriksson, B. A. Persson, and M. Lindberg, J. Chromatogr., 185:575 (1979).
52. S. Sved, I. J. McGilveray, and N. Beaudoin, J. Chromatogr., 145:437 (1978).
53. B. A. Persson and P. O. Lagerström, J. Chromatogr., 122:305 (1976).
54. I. Grundevik and B. A. Persson, J. Liq. Chromatogr., 5:141 (1982).
55. W. Santi, J. M. Huen, and R. W. Frei, J. Chromatogr., 115:423 (1975).
56. R. W. Frei and W. Santi, Fresenius'Z. Anal. Chem., 277:303 (1975).
57. J. M. Huen, R. W. Frei, W. Santi, and J. P. Thevenin, J. Chromatogr., 149:359 (1978).
58. J. C. Gfeller, J. M. Huen, and J. P. Thevenin, Chromatographia, 12:368 (1979).
59. R. Verpoorte, Th. Mulder-Krieger, M. J. Verzijl, J. M. Verzijl, and A. Baerheim Svendsen, J. Chromatogr., 261:172 (1983).
60. M. L. Chan, Ch. Whetsell, and J. D. McChesney, J. Chromatogr. Sci., 12:512 (1974).
61. T. W. Guentert, P. E. Coates, R. A. Upton, D. L. Combs, and S. Riegelman, J. Chromatogr., 162:59 (1979).
62. P. Perera, T. A. van Beek, and R. Verpoorte, J. Chromatogr., 285:214 (1984).
63. Ch. T. A. Chung and E. J. Staba, J. Chromatogr., 295:276 (1984).
64. J. K. Barker, R. E. Skelton, and Ch.Y. Ma, J. Chromatogr., 168:417 (1979).
65. J. Pao and J. A. F. de Silva, J. Chromatogr., 221:97 (1980).
66. D. V. McCalley, J. Chromatogr., 260:184 (1983).
67. M. Bauer and G. Untz, J. Chromatogr., 192:479 (1980).
68. J. P. Thomas, A. Brun, and J. P. Bounine, J. Chromatogr., 172:107 (1979).
69. J. P. Thomas, A. Brun, and J. P. Bounine, Analusis, 8:265 (1980).
70. J. P. Thomas, A. Brun, and J. P. Bounine, J. Chromatogr., 139:21 (1977).
71. M. Smellie, M. Corder, and J. P. Rosazza, J. Chromatogr., 155:439 (1978).
72. K. Sugden, G. B. Cox, and C. R. Loscombe, J. Chromatogr., 149:377 (1978).
73. J. A. Greving, H. Bouman, J. H. G. Jonkman, H. G. M. Westenberg, and R. A. de Zeeuw, J. Chromatogr., 186:683 (1979).
74. R. J. Flanagan, G. C. A. Storey, R. K. Bhamra, and I. Jane, J. Chromatogr., 247:15 (1982).

75. L. Safarík, Z. Stránský, Odmerná analýza v organických rozpoustedlech, SNTL, Praha, 1982.
76. I. Jane, J. Chromatogr., 111:227 (1975).
77. B. B. Wheals, J. Chromatogr., 122:85 (1976).
78. B. B. Wheals, J. Chromatogr., 187:65 (1980).
79. B. Law, R. Gill, and A. C. Moffat, J. Chromatogr., 301:165 (1984).
80. A. L. Christiansen and K. E. Rasmussen, J. Chromatogr., 244:357 (1982).
81. S. U. Shin and Ch.E. Inturrisi, J. Chromatogr., 233:213 (1982).
82. B. A. Bidlingmeyer, J. K. Del Rios, and J. Korpi, Anal. Chem., 54:442 (1982).
83. R. G. Achari and E. E. Theimer, J. Chromatogr. Sci., 15:320 (1977).
84. J. Crommen, J. Chromatogr., 186:705 (1979).
85. S. C. Morris and T. H. Lee, J. Chromatogr., 219:403 (1981).
86. K. E. Bij, C. Horvath, W. R. Melander, and A. Nahum, J. Chromatogr., 203:65 (1981).
87. S. H. Hansen, J. Chromatogr., 209:203 (1981).
88. S. H. Hansen and P. Helboe, J. Chromatogr., 285:53 (1984).
89. A. Wehrli, J. C. Hildebrand, H. P. Keller, R. Stampfli, and R. W. Frei, J. Chromatogr., 149:199 (1978).
90. J. G. Atwood, G. J. Schmidt, and W. Slawin, J. Chromatogr., 171:109 (1979).
91. P. E. Barker, B. W. Hatt, and S. R. Holding, J. Chromatogr., 206:27 (1981).
92. F. M. Rabel, Int. Lab., 1980 (March):53.
93. Ch.Y.Wu and S. Siggia, Anal. Chem., 44:1499 (1972).
94. Ch.Y.Wu, S. Siggia, T. Robinson, and R. D. Waskiewicz, Anal. Chim. Acta, 63:393 (1973).
95. R. Aigner, H. Spitzy, and R. W. Frei, J. Chromatogr. Sci., 14:381 (1976).
96. J. B. Crowther and J. D. Henion, Anal. Chem., 57:2711 (1985).
97. C. J. C. M. Laurent, H. A. H. Billiet, L. de Galan, F. A. Buytenhuys, and F. P. B. van der Maeden, J. Chromatogr., 287:45 (1984).
98. G. L. Smitt and D. J. Pietrzyk, Anal. Chem., 57:2247 (1985).
99. C. J. C. M. Laurent, H. A. H. Billiet, and L. de Galan, J. Chromatogr., 285:161 (1976).
100. V. Quercia, L. Turchetta, V. Cuozzo, and I. Donatelli, Boll. Chim. Farm., 115:810 (1976).
101. H. Huizer, J. Forensic Sci., 28:32 (1983).
102. J. A. Schmit, R. A. Henry, R. C. Williams, and J. F. Dieckman, J. Chromatogr. Sci., 9:645 (1971).

103. J. H. Knox and G. Vasvari, J. Chromatogr., 83:181 (1973).
104. J. H. Knox and A. Pryde, J. Chromatogr., 112:171 (1975).
105. C. H. Löchmuller and D. R. Wilder, J. Chromatogr. Sci., 17: 574 (1979).
106. R. M. McCormick and B. L. Karger, Anal. Chem., 52:2249 (1980).
107. G. E. Berendsen, P. J. Schoenmakers, L. de Galan, G. Bigh, Z. Varga-Púchony, and J. Inczedy, J. Liq. Chromatogr., 3: 1669 (1980).
108. C. R. Yonker, T. A. Zwier, M. F. Burke, J. Chromatogr., 241:257 (1982).
109. C. R. Yonker, T. A. Zwier, M. F. Burke, J. Chromatogr., 241:269 (1982).
110. A. M. Krstulović, H. Colin, and G. Guiochon, Anal. Chem., 54:2438 (1982).
111. P. L. Zhu, Chromatographia, 21:229 (1986).
112. D. C. Locke, J. Chromatogr. Sci., 11:120 (1973).
113. M. J. Telepchak, Chromatographia, 6:234 (1973).
114. D. C. Locke, J. Chromatogr. Sci., 12:433 (1974).
115. K. Karch, I. Sebestian, I. Halász, and H. Engelhardt, J. Chromatogr., 122:171 (1976).
116. Cs. Horváth, W. Melander, and I. Molnár, J. Chromatogr., 125:129 (1976).
117. Cs. Horváth and W. Melander, J. Chromatogr. Sci., 15:393 (1977).
118. O. Sinanoglu and S. Abdulnur, Fed. Proc., Fed. Am. Soc. Exp. Biol., 24:12 (1965).
119. O. Sinanoglu, in Molecular Associations in Biology (R. Pullman, ed.), Academic Press, New York, 1968, p. 427.
120. D. W. Sindorf and G. E. Maciel, J. Am. Chem. Soc., 105: 1848 (1983).
121. R. K. Gilpin and M. E. Gangoda, Anal. Chem., 56:1470 (1984).
122. J. P. Beaufils, M. C. Hennion, and R. Rosset, Anal. Chem., 57:2593 (1985).
123. A. Pryde, J. Chromatogr. Sci., 12:486 (1974).
124. H. Colin and G. Guiochon, J. Chromatogr., 141:289 (1977).
125. D. E. Martire and R. E. Boehm, J. Phys. Chem., 87:1045 (1983).
126. M. Jaroniec and D. E. Martire, J. Chromatogr., 351:1 (1986).
127. M. Jaroniec and D. E. Martire, J. Chromatogr., 387:55 (1987).
128. N. Tanaka, H. Goodell, and B. L. Karger, J. Chromatogr., 158:233 (1978).
129. H. Könemann, R. Zelle, F. Busser, and W. E. Hammers, J. Chromatogr., 175:559 (1979).
130. H. Colin and G. Guiochon, J. Chromatogr. Sci., 18:54 (1980).

131. R. M. Smith, J. Chromatogr., 209:1 (1981).
132. M. J. M. Wells, C. R. Clark, and R. M. Patterson, J. Chromatogr. Sci., 19:573 (1981).
133. W. E. Hammers, G. J. Meurs, and C. L. de Ligny, J. Chromatogr., 247:1 (1982).
134. T. Hanai and J. Hubert, J. HRC&CC, 6:20 (1983).
135. T. L. Hafkenscheid and E. Tomlinson, J. Chromatogr., 264:47 (1983).
136. R. E. Koopmans and R. F. Rekker, J. Chromatogr., 285:267 (1984).
137. A. Kakoulidou and R. F. Rekker, J. Chromatogr., 295:341 (1984).
138. T. L. Hafkenscheid, J. Chromatogr. Sci., 24:307 (1986).
139. N. Funasaki, S. Hada, and S. Neya, J. Chromatogr., 361:33 (1986).
140. J. Stahlberg and M. Almgren, Anal. Chem., 57:817 (1985).
141. R. F. Rekker, The Hydrophobic Fragmental Constant, Elsevier, Amsterdam, 1977.
142. R. F. Rekker and H. M. de Kort, Eur. J. Med. Chem., 14:479 (1979).
143. A. Leo, C. Hansch, and D. Elkins, Chem. Rev., 71:525 (1971).
144. C. Hansch and A. Leo, Substituent Constants for Correlation Analysis in Chemistry and Biology, John Wiley, New York, 1979.
145. R. M. Carlson, R. E. Carlson, and H. L. Kopperman, J. Chromatogr., 107:219 (1975).
146. J. M. McCall, J. Med. Chem., 18:549 (1975).
147. R. Kaliszan, J. Chromatogr., 220:71 (1981).
148. J. L. G. Thus and J. C. Kraak, J. Chromatogr., 320:271 (1985).
149. N. El Tayar, H. van de Waterbeemd, and B. Testa, J. Chromatogr., 320:305 (1985).
150. T. Hanai and J. Hubert, J. Liq. Chromatogr., 8:2463 (1985).
151. J. H. Hildebrand and R. L. Scott, Regular Solutions, Prentice-Hall, Englewood Cliffs, New Jersey, 1962.
152. J. H. Hildebrand and R. L. Scott, The Solubility of Nonelectrolytes, Dover, New York, 1964.
153. J. H. Hildebrand, J. M. Prausnitz, and R. L. Scott, Regular and Related Solutions, Van Nostrand-Reinhold, Princeton, New Jersey, 1970.
154. S. R. Bakalyar, R. McIlwrick, and E. Roggendorf, J. Chromatogr., 142:353 (1977).
155. P. J. Schoenmakers, H. A. H. Billiet, R. Tijssen, and L. de Galan, J. Chromatogr., 149:519 (1978).

156. P. J. Schoenmakers, H. A. H. Billiet, and L. de Galan, J. Chromatogr., 218:261 (1981).
157. P. J. Schoenmakers, H. A. H. Billiet, and L. de Galan, Chromatographia, 15:205 (1982).
158. J. I. Szantó and T. Veress, Chromatographia, 20:596 (1985).
159. R. A. Keller, B. L. Karger, and L. R. Snyder, in Gas Chromatography 1970 (R. Stock, ed.), Institute of Petroleum, London, 1971, p. 125.
160. A. F. M. Barton, Chem. Rev., 75:731 (1975).
161. R. Tijssen, H. A. H. Billiet, and P. J. Schoenmakers, J. Chromatogr., 122:185 (1976).
162. B. L. Karger, L. R. Snyder, and C. Eon, J. Chromatogr., 125:71 (1976).
163. B. L. Karger, L. R. Snyder, and C. Eon, Anal. Chem., 50:2126 (1978).
164. R. B. Hermann, J. Phys. Chem., 76:2754 (1972).
165. S. C. Valvani, S. H. Yalkowski, and G. L. Amidon, J. Phys. Chem., 80:829 (1976).
166. S. H. Yalkowski and S. C. Valvani, J. Chem. Eng. Data, 24:127 (1979).
167. P. Roumeliotis and K. K. Unger, J. Chromatogr., 149:211 (1978).
168. K. Jinno and K. Kawasaki, Chromatographia, 18:90 (1984).
169. A. Bondi, J. Phys. Chem., 68:441 (1964).
170. K. Jinno and K. Kawasaki, Chromatographia, 18:103 (1984).
171. K. Jinno and A. Ishigaki, J. HRC&CC, 5:668 (1982).
172. B. L. Karger, J. R. Grant, A. Hartkopf, and P. H. Weiner, J. Chromatogr., 128:65 (1976).
173. R. J. Hurtubise, T. W. Allen, and H. F. Silver, J. Chromatogr., 235:517 (1982).
174. J. F. Schabron, R. J. Hurtubise, and H. F. Silver, Anal. Chem., 49:2253 (1977).
175. J. F. Schabron, R. J. Hurtubise, and H. F. Silver, Anal. Chem., 50:1911 (1978).
176. K. Jinno and M. Okamoto, Chromatographia, 18:495 (1984).
177. K. Jinno and M. Kuwajima, Chromatographia, 22:13 (1986).
178. S. A. Wise, W. J. Bonnett, F. R. Guenther, and W. E. May, J. Chromatogr. Sci., 19:457 (1981).
179. S. A. Wise and L. C. Sander, J. HRC&CC, 8:248 (1985).
180. M. D'Amboise and M. J. Bertrand, J. Chromatogr., 361:13 (1986).
181. R. Kaliszan, K. Osmialowski, S. A. Tomellini, S.-H. Hsu, S. D. Fazio, and R. A. Hartwick, Chromatographia, 20:705 (1985).
182. A. J. P. Martin, Biochem. Soc. Symp., 3:4 (1949).

183. Gy. Vigh and Z. Varga-Púchony, J. Chromatogr., 196:1 (1980).
184. P. Jandera, Chromatographia, 19:101 (1984).
185. P. Jandera, J. Chromatogr., 314:13 (1984).
186. G. E. Berendsen and L. de Galan, J. Chromatogr., 196:21 (1980).
187. M. C. Spanjer and C. L. de Ligny, J. Chromatogr., 253:23 (1982).
188. M. C. Spanjer and C. L. de Ligny, J. Chromatogr., 20:120 (1985).
189. K. Ogan and E. Katz, J. Chromatogr., 188:115 (1980).
190. J. G. Atwood and J. Goldstein, J. Chromatogr. Sci., 18:650 (1980).
191. H. Englehardt, B. Dreyer, and H. Schmidt, Chromatographia, 16:11 (1982).
192. C. Gonnet, C. Bory, and G. Lachatre, Chromatographia, 16:242 (1982).
193. M. Verzele and P. Mussche, J. Chromatogr., 254:117 (1983).
194. M. C. Hennion, C. Picard, and M. Caude, J. Chromatogr., 166:21 (1978).
195. P. Spacek, M. Kubín, S. Vozka, and B. Porsch, J. Liq. Chromatogr., 3:1465 (1980).
196. R. P. Scott and P. Kucera, J. Chromatogr., 142:213 (1977).
197. M. L. Miller, R. W. Linton, S. G. Bush, and J. W. Jorgenson, Anal. Chem., 56:2204 (1984).
198. M. L. Miller, R. W. Linton, G. E. Maciel, and B. L. Hawkins, J. Chromatogr., 319:9 (1985).
199. D. L. Reynolds, A. J. Repta, and L. A. Sternson, J. Pharm. Biomed. Anal., 1:339 (1983).
200. HPLC Bulletin 823, Supelco Inc., Bellefonte, Pennsylvania, 1985.
201. K. Karch, I. Sebestian, and I. Halász, J. Chromatogr., 122:3 (1976).
202. H. A. Claessens, L. J. M. van de Ven, J. W. de Haan, C. A. Crammers, and N. Vonk, J. HRC&CC, 6:433 (1983).
203. D. V. McCalley, J. Chromatogr., 357:221 (1986).
204. M. Verzele and C. Dewaele, J. Chromatogr., 217:399 (1981).
205. R. Vespalec and J. Neca, J. Chromatogr., 281:35 (1983).
206. M. Verzele and C. Dewaele, Chromatographia, 18:84 (1984).
207. P. C. Sadek, P. W. Carr, and L. W. Bowers, J. Liq. Chromatogr., 8:2369 (1985).
208. H. A. Claessens, C. A. Crammers, J. W. de Haan, F. A. H. den Otter, L. J. M. van de Ven, P. J. Andree, G. J. de Jong, N. Lammers, J. Wijma, and J. Zeeman, Chromatographia, 20:582 (1985).
209. J. L. Glajch, J. C. Gluckman, J. G. Charikofsky, J. M. Minoz, and J. J. Kirkland, J. Chromatogr., 318:23 (1985).

210. J. L. Glajch, J. J. Kirkland, and J. Köhler, J. Chromatogr., 384:81 (1987).
211. L. R. Snyder, J. W. Dolan, and J. R. Gant, J. Chromatogr., 165:3 (1979).
212. P. Jandera, H. Colin, and G. Guiochon, Anal. Chem., 54:435 (1982).
213. P. Dufek, J. Chromatogr., 281:49 (1983).
214. X. Geng and F. Regnier, J. Chromatogr., 332:147 (1985).
215. M. J. M. Wells and C. R. Clark, J. Chromatogr., 235:31 (1982).
216. P. J. Schoenmakers, H. A. H. Billiet, and L. de Galan, J. Chromatogr., 282:107 (1983).
217. W. R. Melander and Cs. Horváth, Chromatographia, 18:353 (1984).
218. R. M. McCormick and B. L. Karger, J. Chromatogr., 199:259 (1980).
219. E. H. Staats, W. Markovski, J. Fekete, and H. Poppe, J. Chromatogr., 207:299 (1980).
220. N. L. Ha, J. Ungvárai, and E. sz. Kováts, Anal. Chem., 54:2410 (1982).
221. R. P. W. Scoot and C. F. Simpson, J. Chromatogr., 197:11 (1980).
222. W. E. Hammers and P. B. A. Verschoor, J. Chromatogr., 282:41 (1983).
223. R. K. Gilpin, J. Chromatogr. Sci., 22:371 (1984).
224. I. Girard and C. Gonnet, J. Liq. Chromatogr., 8:2035 (1985).
225. A. Nahum and Cs. Horváth, J. Chromatogr., 203:53 (1981).
226. P. J. Twitchett and A. C. Moffat, J. Chromatogr., 111:149 (1975).
227. S. Eksborg, J. Chromatogr., 149:225 (1978).
228. N. El Tayar, H. van de Waterbeemd, and B. Testa, J. Chromatogr., 320:293 (1985).
229. B. Pekić, S. M. Petrović, and B. Slavica, J. Chromatogr., 268:237 (1983).
230. L. R. Snyder, J. Chromatogr. Sci., 16:223 (1978).
231. R. M. Smith, J. Chromatogr., 324:243 (1985).
232. C. H. Lochmüller, M. A. Hamzavi-Abedi, and Ch.-X. Ou, J. Chromatogr., 387:105 (1987).
233. J. L. Glajch, J. J. Kirkland, K. M. Squire, and J. M. Minor, J. Chromatogr., 199:57 (1980).
234. W. T. Cooper and L.-Y. Lin, Chromatographia, 21:335 (1986).
235. S. E. Walker, R. A. Mowery, Jr., and R. K. Bade, J. Chromatogr. Sci., 18:639 (1980).
236. W. Melander, D. E. Campbell, and Cs. Horváth, J. Chromatogr., 158:215 (1978).

237. H. Colin, J. C. Diez-Masa, G. Guiochon, T. Czajkowska, and I. Miedziak, J. Chromatogr., 167:41 (1978).
238. E. Grushka, H. Colin, and G. Guiochon, J. Chromatogr., 248:325 (1982).
239. J. Chmielowiec and H. Sawatzky, J. Chromatogr. Sci., 17:245 (1979).
240. A. Opperhuizen, T. L. Sinnige, M. van der Steen, and O. Huitziger, J. Chromatogr., 388:51 (1987).
241. W. R. Melander, J. Stoveken, and Cs. Horváth, J. Chromatogr., 185:111 (1979).
242. W. R. Melander, A. Nahum, and Cs. Horváth, J. Chromatogr., 185:129 (1979).
243. D. Morel, J. Serpinet, J. M. Letoffe, and P. Claudy, Chromatographia, 22:103 (1986).
244. J. C. van Miltenburg and W. E. Hammers, J. Chromatogr., 268:147 (1983).
245. D. Morel, J. Serpinet, and G. Untz, Chromatographia, 18:611 (1984).
246. P. Claudy, J. M. Letoffe, C. Gaget, D. Morel, and J. Serpinet, J. Chromatogr., 329:331 (1985).
247. W. R. Melander, J.-X. Huang, Cs. Horváth, R. W. Stout, and J. J. DeStefano, Chromatographia, 20:641 (1985).
248. P. E. Antle, A. P. Goldberg, and L. R. Snyder, J. Chromatogr., 321:1 (1985).
249. W. E. Hammers, C. H. Kos, W. K. Brederode, and C. L. de Ligny, J. Chromatogr., 168:9 (1979).
250. W. E. Hammers, M. C. Spanjer, and C. L. de Ligny, J. Chromatogr., 174:291 (1979).
251. W. E. Hammers, R. H. A. M. Janssen, A. G. Baars, and C. L. de Ligny, J. Chromatogr., 167:273 (1978).
252. E. L. Weiser, A. W. Salotto, S. M. Flach, and L. R. Snyder, J. Chromatogr., 303:1 (1984).
253. W. T. Cooper and P. L. Smith, J. Chromatogr., 355:57 (1986).
254. M. C. Hennion, C. Picard, C. Combellas, M. Caude, and R. Rosset, J. Chromatogr., 210:211 (1981).
255. R. J. Hurtubise, A. Hussain, and H. F. Silver, Anal. Chem., 53:1993 (1981).
256. S. C. Ruckmick and R. J. Hurtubise, J. Chromatogr., 361:47 (1986).
257. H. Takahagi and S. Seno, J. Chromatogr., 108:354 (1975).
258. Cs. Horváth, W. Melander, and I. Molnár, Anal. Chem., 49:142 (1977).
259. D. J. Pietrzyk, E. P. Kroeff, and T. D. Rotsch, Anal. Chem., 50:497 (1978).
260. R. Matsuda, T. Yamamiya, M. Tatsuzawa, A. Ejima, and N. Takai, J. Chromatogr., 173:75 (1979).

261. M. Popl, L. D. Ky, and J. Strnadová, J. Chromatogr. Sci., 23:95 (1985).
262. K. Aramaki, T. Hanai, and H. F. Walton, Anal. Chem., 52: 1963 (1980).
263. H. Y. Mohammed and F. F. Cantwell, Anal. Chem., 50:491 (1978).
264. W. E. Rudzinski, D. Bennett, V. Garcia, and M. Seymour, J. Chromatogr. Sci., 21:57 (1983).
265. E. Papp and Gy. Vigh, J. Chromatogr., 259:49 (1983).
266. E. Papp and Gy. Vigh, J. Chromatogr., 282:59 (1983).
267. S. G. Weber and W. G. Tramposch, Anal. Chem., 57:1771 (1983).
268. A. Sokolowski and G. Wahlund, J. Chromatogr., 189:299 (1980).
269. Z. Varga-Púchony and Gy. Vigh, J. Chromatogr., 257:380 (1983).
270. R. W. Roos and C. A. Lau-Cam, J. Chromatogr., 370:403 (1986).
271. J. S. Kiel, S. L. Morgan, and R. K. Abramson, J. Chromatogr., 320:313 (1985).
272. R. Gill, S. P. Alexander, and A. C. Moffat, J. Chromatogr., 247:39 (1982).
273. F. F. Wu and R. H. Dobberstein, J. Chromatogr., 140:65 (1977).
274. M. A. Johnston, A. R. Lea, and D. M. Hailey, J. Chromatogr., 318:362 (1985).
275. J. R. Miksic and B. Hodes, J. Pharm. Sci., 68:1200 (1979).
276. B. Pekić, Z. Lepojeviĉ, B. Slavica, and S. M. Petrović, Chromatographia, 21:227 (1986).
277. N. R. Ayyangar and S. R. Bhide, J. Chromatogr., 366:435 (1986).
278. J. Gal, J. Chromatogr., 307:220 (1984).
279. J. Gal and A. J. Sedman, J. Chromatogr., 314:275 (1984).
280. Y. Nobuhara, S. Hirano, and Y. Nakanishi, J. Chromatogr., 258:276 (1983).
281. Y. Nobuhara, S. Hirano, K. Namba, and M. Hashimoto, J. Chromatog., 190:251 (1980).
282. M. G. Lee, J. Chromatogr., 312:473 (1984).
283. V. K. Srivastava and M. L. Maheshwari, J. Assoc. Off. Anal. Chem., 68:801 (1985).
284. U. Lund and S. H. Hansen, J. Chromatogr., 161:371 (1978).
285. A. Hobson-Frohock and W. T. E. Edwards, J. Chromatogr., 249:369 (1982).
286. G. Musch and D. L. Massart, J. Chromatogr., 370:1 (1986).
287. G. Szepesi, M. Gazdag, and R. Iváncsics, J. Chromatogr., 241:153 (1982).

288. G. Hoogewijs, Y. Michotte, J. Lambrecht, and D. L. Massart, J. Chromatogr., 226:423 (1981).
289. L. W. Doner and A.-F. Hsu, J. Chromatogr., 253:120 (1982).
290. P. B. Baker and T. A. Gough, J. Chromatogr. Sci., 19:483 (1981).
291. R. J. Bushway, E. S. Barden, A. W. Bushway, and A. A. Bushway, J. Chromatogr., 178:533 (1979).
292. R. J. Bushway, J. Chromatogr., 247:180 (1982).
293. H. Wagner, U. Neidhardt, and G. Tittel, Planta Med., 41:232 (1981).
294. G. Tittel, H. Hinz, and H. Wagner, Planta Med., 37:1 (1979).
295. M. Wurst, M. Flieger, and Z. Rehácek, J. Chromatogr., 150:477 (1978).
296. M. Wurst, M. Flieger, and Z. Rehácek, J. Chromatogr., 174:401 (1979).
297. M. Flieger, P. Sedmera, J. Vokoun, Z. Rehácek, J. Stuchlík, and A. Cerný, J. Chromatogr., 284:219 (1984).
298. J. C. Gfeller, G. Frey, J. M. Huen, and J. P. Thevenin, J. Chromatogr., 172:141 (1979).
299. M. Verzele, P. Mussche, and S. A. Quereshi, J. Chromatogr., 172:493 (1979).
300. P. G. Grabbe and C. Fryer, J. Chromatogr., 187:87 (1980).
301. P. P. Gladyshev and E. F. Matantseva, Izv. Akad. Nauk Kaz. SSR, Ser. Khim., 1982(5):37; CA, 98:27062g (1983).
302. S. Krýsl and E. Smolková-Keulemansová, Chem. Listy, 79:919 (1985).
303. B. Zsadon, M. Szilasi, F. Tüdös, and J. Szejtli, Chromatogr., togr., 208:109 (1981).
304. B. Zsadon, L. Décsei, M. Szilasi, F. Tüdös, and J. Szejtli, J. Chromatogr., 270:127 (1983).
305. B. Zsadon, M. Szilasi, L. Décsei, A. Ujházy, and J. Szejtli, J. Chromatogr., 356:428 (1986).
306. I. W. Wainer, T. D. Doyle, Z. Hamidzadeh, and M. Aldridge, J. Chromatogr., 261:123 (1983).
307. I. W. Wainer, T. D. Doyle, F. S. Fry, Jr., and Z. Hamidzadeh, J. Chromatogr., 355:149 (1986).
308. F. Helfferich, Ionenaustauscher, Verlag Chemie, Weinheim, 1959.
309. W. Reiman III and H. F. Walton, Ion Exchange in Analytical Chemistry, Pergamon Press, Oxford, 1970.
310. A. W. Kingsburg, A. B. Mindler, and M. E. Gilwood, Chem. Eng. Progress, 44:497 (1948).
311. N. Applezweig and S. E. Ronzone, Ind. Eng. Chem., 38:576 (1946).
312. A. Jindra, J. Pharm. Pharmacol., 1:87 (1949).

313. A. Jindra and J. Pohorský, J. Pharm. Pharmacol., 3:344 (1951).
314. M. Marhol, Ion Exchangers in Analytical Chemistry: Their Properties and Use in Inorganic Chemistry. Academia, Prague, 1982.
315. C. A. Pohl and E. L. Johnson, J. Chromatogr. Sci., 18:442 (1980).
316. J. N. Broich, M. M. de Mayo, and L. A. Dal Cortivo, J. Chromatogr., 33:526 (1968).
317. T. Urbányi, A. Piedmont, E. Willis, and G. Manning, J. Pharm. Sci., 65:257 (1976).
318. E. Murgia, P. Richards, and H. F. Walton, J. Chromatogr., 87:523 (1973).
319. T. Hanai, H. F. Walton, J. D. Navratil, and D. Warren, J. Chromatogr., 155:261 (1978).
320. H. F. Walton, G. A. Eiceman, and J. L. Otto, J. Chromatogr., 180:145 (1979).
321. D. H. Rodgers, J. Chromatogr. Sci., 12:742 (1974).
322. P. P. Gladyshev, E. F. Matantseva, and M. I. Goryaev, Zh. Anal. Khim., 36:1130 (1981); CA, 95:175852t (1981).
323. P. P. Gladyshev, M. I. Goryaev, and A. N. Baigalieva, Izv. Akad. Nauk Kaz. SSR, Ser. Khim.; 1971(4):43; CA, 75: 122380t (1971).
324. P. J. Twitchett, J. Chromatogr., 104:205 (1975).
325. J. H. Knox and J. Jurand, J. Chromatogr., 87:95 (1973).
326. J. J. Nelson. J. Chromatogr. Sci., 11:28 (1973).
327. P. J. Twitchett, A. E. P. Gorvin, and A. C. Moffat, J. Chromatogr., 120:359 (1976).
328. K. D. McMurtrey, J. L. Cashaw, and V. E. Davis, J. Liq. Chromatogr., 3:663 (1980).
329. M. Perkal, G. L. Blackman, A. L. Ottrey, and L. K. Turner, J. Chromatogr., 196:180 (1980).
330. K. C. Cundy and P. A. Crooks, J. Chromatogr., 306:291 (1984).
331. H. F. Walton, J. Chromatogr., 102:57 (1974).
332. V. A. Davankov and A. V. Semechkin, J. Chromatogr., 141:313 (1977).
333. J. C. Wolford, J. A. Dean, and G. Goldstein, J. Chromatogr., 62:148 (1971).
334. E. Murgia and H. F. Walton, J. Chromatogr., 104:417 (1975).
335. H. Small and T. E. Miller, Jr., Anal. Chem., 54:462 (1982).
336. D. R. Jenke, Anal. Chem., 56:2468 (1984).
337. J. H. Knox and G. R. Laird, J. Chromatogr., 122:17 (1976).
338. Paired-Ion Chromatography: An Alternative to Ion Exchange. Bulletin F61, Waters Associates, May 1975.

339. J. C. Kraak, K. M. Jonker, and J. F. K. Huber, J. Chromatogr., 142:671 (1977).
340. C. P. Terweij-Groen, S. Heemstra, and J. C. Kraak, J. Chromatogr., 161:69 (1978).
341. Cs. Horvath, W. Melander, I. Molnar, and P. Molnar, Anal. Chem., 49:2295 (1977).
342. B. A. Bidlingmeyer, S. N. Deming, W. P. Price, Jr., B. Sachok, and M. Petrusek, J. Chromatogr., 186:419 (1979).
343. B. A. Bidlingmeyer, J. Chromatogr. Sci., 18:525 (1980).
344. M. T. W. Hearn, editor: Ion-Pair Chromatography: Theory and Biological and Pharmaceutical Applications. Chromatographic Science Series, Vol. 31, Marcel Dekker, New York, 1985.
345. S. Eksborg, O. Lagerstrom, R. Modin, and G. Schill, J. Chromatogr., 83:99 (1973).
346. S. Eksborg and G. Schill, Anal. Chem., 45:2092 (1973).
347. I. M. Johansson, K.-G. Wahlund, and G. Schill, J. Chromatogr., 149:281 (1978).
348. K.-G. Wahlund and A. Sokolowski, J. Chromatogr., 151:299 (1978).
349. P. T. Kissinger, Anal. Chem., 49:883 (1977).
350. J. H. Knox and J. Jurand, J. Chromatogr., 149:297 (1978).
351. P. Jandera, J. Churácek, and B. Taraba, J. Chromatogr., 262:121 (1983).
352. R. P. W. Scott and P. Kucera, J. Chromatogr., 175:51 (1979).
353. J. L. M. van de Venne, J. L. H. M. Hendrikx, and R. S. Deelder, J. Chromatogr., 167:1 (1978).
354. C. T. Hung, R. B. Taylor, and N. Paterson, J. Chromatogr., 240:61 (1982).
355. T. A. Walker and D. J. Pietrzyk, J. Liq. Chromatogr., 8:2047 (1985).
356. F. F. Cantwell and S. Puon, Anal. Chem., 51:623 (1979).
357. R. S. Deelder and J. H. M. van den Berg, J. Chromatogr., 218:327 (1981).
358. S. Afrashtehfar and F. F. Cantwell, Anal. Chem., 54:2422 (1982).
359. R. A. Hux and F. F. Cantwell, Anal. Chem., 56:1258 (1984).
360. J. Stahlberg, J. Chromatogr., 356:231 (1986).
361. J. H. Knox and R. A. Hartwick, J. Chromatogr., 204:3 (1981).
362. B. F. Nilsson and O. Samuelson, J. Chromatogr., 235:266 (1982).
363. W. R. Melander and Cs. Horváth, J. Chromatogr., 201:201 (1980).
364. A. Bartha, H. A. H. Billiet, L. de Galan, and Gy. Vigh, J. Chromatogr., 291:91 (1984).

365. Á. Bartha, Gy. Vigh, H. A. H. Billiet, and L. de Galan, J. Chromatogr., 303:29 (1984).
366. A. Berthod, I. Girard, and C. Gounet, Anal. Chem., 58: 1356 (1986).
367. C. T. Hung and R. B. Taylor, J. Chromatogr., 202:333 (1980).
368. C. T. Hung and R. B. Taylor, J. Chromatogr., 209:175 (1981).
369. B. F. Nilsson and O. Samuelson, J. Chromatogr., 212:1 (1981).
370. A. Sokolowski, Chromatographia, 22:168 (1986).
371. T. D. Rotsch, W. R. Cahill, Jr., D. J. Pietrzyk, and F. F. Cantwell, Can. J. Chem., 59:2179 (1981).
372. A. Tilly-Melin, Y. Askemark, K.-G. Wahlund, and G. Schill, Anal. Chem., 51:976 (1979).
373. S. O. Jansson, I. Andersson, and B. A. Persson, J. Chromatogr., 203:93 (1981).
374. A. Bartha and Gy. Vigh, J. Chromatogr., 260:337 (1983).
375. P. R. Bedard and W. C. Purdy, J. Liq. Chromatogr., 8: 2417 (1985).
376. M. Dreux, M. Lafosse, P. Agbo-Hazoume, Chromatographia, 18:15 (1984).
377. A. Tilly-Melin, M. Ljungcrantz, and G. Schill, J. Chromatogr., 185:225 (1979).
378. R. L. Smith and D. J. Pietrzyk, Anal. Chem., 56:1572 (1984).
379. R. S. Deelder, H. A. J. Linssen, A. P. Konijnendijk, and J. L. M. van de Venne, J. Chromatogr., 185:241 (1979).
380. R. C. Kong, B. Sachok, and S. N. Deming, J. Chromatogr., 199:307 (1980).
381. S. N. Deming and R. C. Kong, J. Chromatogr., 217:421 (1981).
382. Á. Bartha, Gy. Vigh, H. Billiet, and L. de Galan, Chromatographia, 20:587 (1985).
383. Á. Bartha and Gy. Vigh, J. Chromatogr., 265:171 (1983).
384. R. B. Taylor, R. Reid, and C. T. Hung, J. Chromatogr., 316:279 (1984).
385. J. A. Graham and L. B. Rogers, J. Chromatogr. Sci., 18: 614 (1980).
386. W. R. Melander and Cs. Horváth, J. Chromatogr., 201:211 (1980).
387. D. W. Armstrong, Sep. Purif. Methods, 14:213 (1985).
388. F. J. DeLuccia, M. Arunyanart, and L. J. Cline Love, Anal. Chem., 57:1564 (1985).
389. M. G. Khaledi and J. G. Dorsey, Anal. Chem., 57:2190 (1985).
390. M. A. Hernandez-Torres, J. S. Landy, and J. G. Dorsey, Anal. Chem., 58:744 (1986).

391. D. W. Armstrong, T. J. Ward, and A. Berthod, Anal. Chem., 58:579 (1986).
392. P. A. Perrone and P. R. Brown, J. Chromatogr., 307:53 (1984).
393. N. H. C. Cooke, R. L. Viavattene, R. Eksteen, W. S. Wong, G. Davies, and B. L. Karger, J. Chromatogr., 149:391 (1978).
394. B. L. Karger, W. S. Wong, R. L. Viavattene, J. N. Lepage, and G. Davies, J. Chromatogr., 167:253 (1978).
395. S. J. Constanzo, J. Chromatogr., 314:402 (1984).
396. J. P. Crombeen, J. C. Kraak, and H. Poppe, J. Chromatogr., 167:219 (1978).
397. C. M. Riley and E. Tomlinson, Proc. Anal. Div. Chem. Soc., 17:528 (1980).
398. C. M. Riley, E. Tomlinson, and T. L. Hafkenscheid, J. Chromatogr., 218:427 (1981).
399. E. Tomlinson and C. M. Riley, in Ref. 344, p. 77.
400. J. C. Kraak, H. H. van Rooij, and J. L. G. Thus, J. Chromatogr., 352:455 (1986).
401. C. M. Riley, E. Tomlinson, and T. M. Jefferies, J. Chromatogr., 185:197 (1979).
402. P. R. Bedard and W. C. Purdy, J. Liq. Chromatogr., 8:2445 (1985).
403. R. Gloor and E. J. Johnson, J. Chromatogr. Sci., 15:413 (1977).
404. R. F. Adams, in Ref. 344, p. 141.
405. O. A. G. J. van der Houwen, R. H. A. Sorel, A. Hulshoff, J. Teeuwsen, and A. W. M. Indemans, J. Chromatogr., 209:393 (1981).
406. J. Frenz and Cs. Horváth, J. Chromatogr., 282:249 (1983).
407. G. Eppert and G. Liebscher, J. Chromatogr., 356:372 (1986).
408. G. Winkler, P. Briza, and Ch. Kunz, J. Chromatogr., 361:191 (1986).
409. R. A. Barford and B. J. Sliwinski, J. Chromatogr., 171:445 (1979).
410. I. S. Lurie, J. Assoc. Off. Anal. Chem., 60:1035 (1977).
411. I. S. Lurie and J. M. Weber, J. Liq. Chromatogr., 1:587 (1978).
412. R. G. Achari and J. T. Jacob, J. Liq. Chromatogr., 3:81 (1980).
413. I. S. Lurie and S. M. Demchuk, J. Liq. Chromatogr., 4:337 (1981).
414. I. S. Lurie and S. M. Demchuk, J. Liq. Chromatogr., 4:357 (1981).
415. I. S. Lurie, J. Liq. Chromatogr., 4:399 (1981).
416. S. J. Bannister, J. Stevens, D. Musson, and L. A. Sternson, J. Chromatogr., 176:381 (1979).

417. C. Olieman, L. Maat, K. Waliszewski, and H. C. Beyerman, J. Chromatogr., 133:382 (1977).
418. R. K. Jhangiani and A. C. Bello, J. Assoc. Off. Anal. Chem., 68:523 (1985).
419. I. S. Lurie, S. M. Sottolano, and S. Blasof, J. Forensic. Sci., 27:519 (1982).
420. I. Beaumont and T. Deeks, J. Chromatogr., 238:520 (1982).
421. J. J. L. Hoogenboom and C. G. Rammel, J. Assoc. Off. Anal. Chem., 68:1131 (1985).
422. J. Wang and M. Bonakdar, J. Chromatogr., 382:343 (1986).
423. T. Veress, Magy. Kem. Foly., 92:54 (1986).
424. W. Lindberg, E. Johansson, and K. Johansson, J. Chromatogr., 211:201 (1981).
425. R. A. Pask-Hughes, P. H. Corran, and D. H. Calam, J. Chromatogr., 214:307 (1981).
426. P. Duez, S. Chamart, M. Hanocq, L. Molle, M. Vanhaelen, and R. Vanhaelen-Fastré, J. Chromatogr., 329:415 (1985).
427. M. G. M. de Ruyter, R. Cronnelly, and N. Castagnoli, Jr., J. Chromatogr., 183:193 (1980).
428. R. V. Smith, J. C. Glade, and D. W. Humphrey, J. Chromatogr., 172:520 (1979).
429. K. Sagara, T. Oshima, and T. Misaki, Chem. Pharm. Bull., 31:2359 (1983).
430. K. Sagara, Y. Ito, M. Ojima, T. Oshima, K. Suto, T. Misaki, and H. Itokawa, Chem. Pharm. Bull., 33:5369 (1985).
431. M. Kneczke, J. Chromatogr., 198:529 (1980).
432. Y. Akada, S. Kawano, and Y. Tanase, Yakugaku Zasshi, 100: 766 (1980); CA, 93:245588w (1980).
433. A. K. Mitra, B. L. Baustian, and T. J. Mikkelson, J. Pharm. Sci., 69:257 (1980).
434. D. L. Duval and J. S. Fritz, J. Chromatogr., 295:89 (1984).
435. R. Verpoorte, J. M. Verzeijl, and A. Baerheim Svendsen, J. Chromatogr., 283:401 (1984).
436. X. Qiu, Ch. Wu, and B. Chen., Yaoxue Xuebao, 21:458 (1986); CA, 105:120875x (1986).
437. M. Voelter, K. Zech, P. Arnold, and G. Ludwig, J. Chromatogr., 199:345 (1980).
438. K. T. Muir, J. H. G. Jonkman, D.-S. Tang, M. Kunitani, and S. Riegelman, J. Chromatogr., 221:85 (1980).
439. K. T. Muir, M. Kunitani, and S. Riegelman, J. Chromatogr., 231:73 (1982).
440. M. Johansson, S. Eksborg, and A. Arbin, J. Chromatogr., 275:355 (1983).
441. J.-O. Svensson, A. Rane, J. Säwe, and F. Sjöqvist, J. Chromatogr., 230:427 (1982).
442. G. Szepesi, M. Gazdag, and R. Iváncsics, J. Chromatogr., 244:33 (1982).

443. J. Stranahan and S. N. Deming, Anal. Chem., 54:1540 (1982).
444. B. A. Bidlingmeyer and F. V. Warren, Jr., Anal. Chem., 54: 2351 (1982).
445. L. Hackzell, T. Rydberg, and G. Schill, J. Chromatogr., 282: 179 (1983).
446. T. Takeuchi and E. S. Yeung, J. Chromatogr., 370:83 (1986).
447. A. Sokolowski, Chromatographia, 22:177 (1986).
448. J. C. Gfeller, J. M. Huen, and J. P. Thevenin, J. Chromatogr., 166:133 (1978).
449. J. E. Parkin, J. Chromatogr., 225:240 (1981).
450. J. F. Lawrence, U. A. Th. Brinkman, and R. W. Frei, J. Chromatogr., 185:473 (1979).
451. M. De Smet, S. J. P. Van Belle, G. A. Storme, and D. L. Massart, J. Chromatogr., 345:309 (1985).
452. A. J. Berry, D. E. Games, and J. R. Perkins, J. Chromatogr., 363:147 (1986).
453. H. G. Barth, W. E. Barber, C. H. Lochmüller, R. E. Majors, and F. E. Regnier, Anal. Chem., 58:211R (1986).
454. J. Frank Jzn., J. A. Duine, H. A. H. Billiet, A. C. J. H. Drouen, L. de Galan, B. G. M. Vandeginste, and R. Essens, Fresenius' Z. Anal. Chem., 322:761 (1985).
455. T. Alfredson and T. Sheehan, J. Chromatogr. Sci., 24:473 (1986).
456. H. A. H. Billiet, R. Wolters, L. de Galan, and H. Huizer, J. Chromatogr., 368:351 (1986).
457. R. Fankel and I. Slad, Fresenius' Z. Anal. Chem., 303:208 (1980).
458. W. A. Dark, J. Chromatogr. Sci., 24:495 (1986).
459. M. H. Stutz and S. Sass, Anal. Chem., 45:2134 (1973).
460. A. Noordam, L. Maat, and H. C. Beyerman, J. Pharm. Sci., 70:96 (1981).
461. L. Zecca, L. Bonini, and S. R. Bareggi, J. Chromatogr., 272:401 (1983).
462. P. O. Edlund, J. Chromatogr., 226:107 (1981).
463. J.-P. Fels, P. Lechat, R. Rispe, and W. Cautreels, J. Chromatogr., 308:273 (1984).
464. F. Sasse, J. Hammer, and J. Berlin, J. Chromatogr., 194: 234 (1980).
465. R. Leroyer, C. Jarreau, and M. Pays, J. Chromatogr., 228: 366 (1982).
466. R. W. Frei, W. Santi, and M. Thomas, J. Chromatogr., 116: 365 (1976).
467. Y. Tsuruta, K. Kohashi, and Y. Ohkura, J. Chromatogr., 224:105 (1981).
468. P. E. Nelson, S. L. Nolan, and K. R. Bedford, J. Chromatogr., 234:407 (1982).

469. J. R. Lang, I. L. Honiberg, J. T. Stewart, J. Chromatogr., 252:288 (1982).
470. H. J. G. M. Derks, K. Van Twillert, D. P. K. H. Pereboom-de Fauw, G. Zomer, and J. G. Loeber, J. Chromatogr., 370: 173 (1986).
471. H. Lingeman, W. J. M. Underberg, A. Takadate, and A. Hulshoff, J. Liq. Chromatogr., 8:789 (1985).
472. B. Law and R. Gill, J. Chromatogr., 325:294 (1985).
473. P. T. Kissinger, Anal. Chem., 49:447A (1977).
474. G. Musch, M. De Smet, and D. L. Massart, J. Chromatogr., 348:97 (1985).
475. C. Bollet, P. Oliva, and M. Caude, J. Chromatogr., 149:625 (1978).
476. R. M. Riggin and P. T. Kissinger, Anal. Chem., 49:530 (1977).
477. M. W. White, J. Chromatogr., 178:229 (1979).
478. R. W. Smith and D. W. Humphrey, Anal. Lett., 14(B8):601 (1981).
479. M. S. Greenberg and W. J. Mayer, J. Chromatogr., 169:321 (1979).
480. R. Whelpton, J. Chromatogr., 272:216 (1983).

6
Thin-Layer Chromatography

In contrast to paper chromatography, which is used rarely nowadays due to its low speed and efficiency, thin-layer chromatography (TLC) is well established and widely used in various fields of analytical chemistry (1-6). Because quantitation with this technique is not straightforward, its long-time domain has been qualitative analysis. Besides giving retention data, the chemical reactions with various reagents performed directly on the sorbent layer can easily be used for the identification of unknown samples. Identification can rely on a large amount of data accumulated in the past. An extensive summary of alkaloid TLC was completed by Baerheim Svendsen and Verpoorte (7). For identification purposes the preferred separation systems are those in which retention data for a great number of compounds have been established. Unfortunately, only a few systematic studies cover more than hundreds of compounds. As an example, the important systems for drug identification can be mentioned here. Many TLC data for drugs were accumulated in early studies (8). A compilation of TLC data (9) enabled completion of a computer-based identification system utilizing TLC, GC, and spectral data (10) with 1600 drugs included in this data base. A dissimilarity index was used to match unknown samples with the data file. Another approach utilizes principal component analysis of R_F values for 362 drugs in four separation systems using silica gel layers (11). Quantitative analysis using TLC, although more difficult, is gaining importance. With suitable equipment the accuracy of determination may be comparable to that of other chromatographic methods.

I. TECHNIQUE OF TLC

As far as laboratory equipment is concerned, TLC is the simplest chromatographic technique. Essentially only suitable vessels containing mobile phase and precoated plates are required in order to perform separations. Even in that way TLC possesses numerous advantages over column chromatography. Several samples can be analyzed simultaneously in a single run. Since chromatographic plates are used only once, samples with a relatively complex matrix can be separated, in contrast to column liquid chromatography, in which column contamination is to be avoided. Far simpler purification steps are therefore satisfactory. All components of the sample are detectable on the TLC plate. Strongly retained components eluted as broad peaks, difficult to detect, or not eluted at all in column liquid chromatography are in TLC concentrated in small spots near the start and can be sensitively detected. With certain precautions TLC can be used as a pilot technique for solvent system selection in column chromatography. The possibility of two-dimensional separation multiplies the resolving power of the TLC, making possible the analysis of complex mixtures.

About 10 years ago innovation of the methods of TLC, primarily in the development of high-quality adsorbents and layers, began and the term high-performance thin-layer chromatography (HPTLC) appeared (12–15). Layers with optimized thickness and other properties were prepared using smaller particles with uniform size distribution. Higher number of samples with higher number of components can be analyzed on plates of lower dimension. Simultaneously, more sophisticated modes of development were developed. The high quality of the resulting chromatograms is advantageous in direct quantitative in situ analysis, which is possible using special equipment. In fact, modern instrumentation makes TLC the most versatile and rapid separation technique, unsurpassed with respect to the number of samples which can be analyzed per unit of time.

A. Layers for TLC

A large variety of TLC materials differing in sorbent type, particle size and homogeneity, binder, and support are commercially available. Therefore home-made plates are rarely used at present since they do not approach the quality of the commercial products. Only for special purposes such as preparative layers without binder can individual pouring of layers be substantiated. Procedures and devices for this purpose are described elsewhere (1,4).

Silica gel is the most frequently used adsorbent in TLC. Much less application has been found for alumina. Chromatographic properties of these packings are dependent on the water content, and hence plates are sometimes activated before use at elevated temperatures. However, equilibration of activated plates with ambient atmosphere humidity is rapid. For silica gel layers equilibration times of 1–2 min were reported (16). Without special precautions during sample application and other manipulations, the content of water in the layer is determined to a large extent by the atmospheric humidity. On the other hand, hysteresis of the water adsorption causes changes in chromatographic behavior of silica gel layers equilibrated to the given relative humidity from dryer or wetter conditions (17). Among other classical materials acetylcellulose can be mentioned. Plates containing crystalline cellulose are used primarily as supports for liquid-liquid partitioning TLC.

In addition, chemically bonded reversed phase materials, materials containing polar groups, and ion exchangers are available in the form of TLC layers. Precoated chemically bonded reversed phase plates are supplied in an increasing number of types (18–20). Their properties may vary considerably, especially with respect to the separation of polar compounds. Unfortunately, the nature of the bonded phase, the surface coverage, the type of binder, and the reproducibility of preparation are mostly not known. Unexpectedly certain products cannot be used with certain mobile phases (21–23). Useful information on the stationary phase structure may be obtained by NMR spectrometry (24). Alternatively, reversed phase separations may be performed successfully on silica gel layers impregnated with paraffin (23,25–28) or silicon oil (28,29). The layer can be impregnated by spraying or by development in the solution of nonpolar phase. This technique was applied in two-dimensional TLC with normal phase separation in one direction and reversed phase separation in the second direction (30).

Thin layers of chemically bonded phases containing polar functional groups have also been prepared (20). Some studies on separations were carried out with nitrilo (31) and amino (31–33) layers but further experience concerning the separation possibilities of these materials is required. The layers prepared with various types of optically active substances in many cases enabled the separation of optical isomers (20,34–36). The application of TLC in this area was recently reviewed (37). Commercial materials for this purpose are supplied at present.

Packings for HPTLC are produced from sorbents with narrow size distributions and average particle diameters, typically 5–7 μm compared to 10–15 μm in earlier TLC. The time required for the plate development to a given distance is longer in HPTLC than in TLC. However, lower migrating distances (e.g., lower than 5 cm) are used, a factor which is outweighed by the higher efficiency of these materials. Since spots in HPTLC are small, the limit of detection expressed as the amount of compound per spot is lower than in TLC.

Thin-Layer Chromatography

Of all the supports, glass is the most inert material compatible with almost any mobile phase and detection reagent. For optical quantitative evaluation of TLC chromatograms the exact planarity of glass plates is highly advantageous. Aluminum foils, plastic sheets, and TLC materials based on glass or silica fiber tissues can be easily cut to the desired dimensions. Strongly basic or acidic eluents and detection reagents cannot be applied to aluminum-backed layers because the dissolution of aluminum may disturb the layer.

In order to improve the mechanical stability of the plates, 5-20% of a suitable binder is added to the adsorbent slurry for pouring of the layer. Gypsum is the most frequently used binder with good chemical stability which does not interfere with most detection reagents. Organic binders based on organic polymers are also employed. Highly aggressive reagents such as concentrated sulfuric acid and reagents requiring strong heating of the plate must not be applied to such a layer. On layers fixed with starch the detection with reagents containing or generating iodine is of limited usefulness.

The greatest stability and most universal use is achieved with layers fixed by sintering the sorbent with glass powder. The technological aspects and specific properties of such layers were studied with alkaloids and reviewed in Ref. 38. In the form of covered rods of ceramics, quartz, steel, or titanium, these materials are in widespread use for the quantitative TLC analysis by flame ionization detector (39). In addition, nitrogen- and halogen-specific detection is possible using principles similar to those in gas chromatography (40). The layers in form of rods enable easy detection by dipping the rod in the reagent solution. Layers of silica gel mixed with microcrystalline cellulose on glass sticks were developed in the test tube with low consumption of the mobile phase (about 0.2 ml) when the identification of drugs in pharmaceutical preparations was performed (41).

B. Sample Application

Most simply, samples are spotted to the TLC layer manually using glass capillary pipettes. Upon dipping into the sample solution the empty capillary will be filled by capillary action. When the end of the filled capillary is put in contact with the surface of the adsorbent layer, the solution will soak in the layer. Disposable capillaries with defined volumes in the microliter range are commercially available. Sample solutions may be applied also by microliter syringes equipped with a 90° needle tip. The starting spot should be as small as possible because its diameter contributes to the dispersion of spots during development and may adversely affect the resolution of compounds. If it is necessary to apply sample in a large volume of solvent, it is recommended that an application of low volume be repeated several times and the solvent be evaporated between applications. Also continuous sampling by microsyringe with the simultaneous evaporation of solvent in the stream of heated air or inert gas may be applied if sample components are stable at elevated temperature. However, tailing or in some other way deformed spots may arise during plate development if the concentration of the sample components is too high. Either overloading of the sorbent, slow dissolution of the sample, or local flow rate deviations caused by the passage of the solvent front through the starting spot may be responsible for these anomalies.

Small diameter of the starting spot is even more important in HPTLC where separation is achieved with higher efficiency on a shorter path. Sometimes samples as low as 10 nl are required to fully exploit the possibilities of HPTLC layers. With such low volumes sample solution does not penetrate fully through the layer thickness and "spots" in fact have the shape of a hemisphere whose centre is on the surface at the point of application. At higher volumes the full thickness of the layer is soaked with sample and the spot approaches the shape of a flat cylinder. Low volumes are difficult to apply with conventional sampling devices. Platinum-iridium capillaries with inner diameters of about 0.1 mm are used for this purpose. In praxis higher volumes of 0.1–1 μl are applied to HPTLC plates with good reproducibility, which is required for quantitative work, although some resolution may be lost. Careful handling during the sample application is required. The surface layer withstands only limited pressure without scratching. More or less intricate devices are available for the application of sample which bring the pipette into contact with the layer in a defined and adjustable manner. When the sample is applied by syringe, the danger of surface damage is not so acute since the tip of the needle need not touch the surface. It is only sufficient if the meniscus of the ejected liquid comes in contact with the layer surface.

Higher volumes of the sample without excessive broadening of the spot can be easily applied if the sample is dissolved in a solvent of low eluting strength with low R_F values ($R_F < 0.1$) for compounds of interest. Then the components are concentrated at the middle of the solvent area. In another method, large spots can be concentrated to narrow streaks by partial predevelopment of the layer with strong solvent in which all components are eluted with the solvent front. After drying of the plate, the chromatogram is developed in the mobile phase of desired composition. Some commercially available plates are equipped with a preconcentration zone containing nonabsorbing material for sample application. A large volume of sample solution may be eventually be applied to this zone in the form of a band parallel to the direction of development. During the development the samples are eluted with solvent front in this zone and are introduced to the zone of proper adsorbent in the form of narrow streaks.

For preparative purposes or quantitative evaluation, the sample is sometimes applied in the form of a band parallel to the solvent front. Special instruments are designed for this purpose that combine sample application with movement of the plate. In some applicators sample is sprayed to the layer surface by the stream of nitrogen. In this way effluent from the narrow-bore liquid chromatography column can be deposited to the TLC plate (42). Either analysis by reagents for TLC detection or TLC separation may follow.

Another technique of sample application utilizes "contact spotting" (43), in which sample application is preceded by evaporation of the solvent. A relatively large volume of the sample solution (as much as 100 µl) is evaporated on nonwetting polymer film until volatile components are eliminated. Due to nonwettability of the film the residue is of small dimensions and is transferred to the adsorbent layer by contact. To the samples which uield a solid rather than a liquid residue, a small volume of nonvolatile solvent must be added. The equipment for contact spotting is commercially available.

C. Plate Development

Presently three methods of plate development are in use: linear, circular, and anticircular. The simplest linear development is accomplished in conventional development tanks containing developing solvent at the bottom. The layer margin is dipped into the solvent. Solvent ascending by the layer passes through the area with spotted samples, dissolving them and carrying away their components at a speed determined by their capacity factors. The retention data in TLC are usually presented as R_F values. In the case of linear development R_F is defined as the ratio of the migration distance of the component to the migration distance of the solvent front, both measured from the point of sample application. Often the retention data are transformed to R_M values by Eq. 6.1:

$$R_M = \log\left(\frac{1}{R_F} - 1\right) \tag{6.1}$$

In an ideal case, R_F and R_M values are related to capacity factor k by relationships

$$R_F = \frac{1}{1 + k} \tag{6.2}$$

and

$$R_M = \log k \tag{6.3}$$

In practice, the conditions for the exact validity of these equations are never met. The content of the mobile phase in the layer is lower near the solvent front than at the source of the mobile phase; therefore, during development the mobile phase does not move with the same velocity in the developmental direction. Furthermore, the exchange of solvent between sorbent layer and gas phase contributes to the total mobile phase transport (16). If the atmosphere of the development tank is not presaturated with vapors of the mobile phase, the evaporation of the solvent from the layer will increase the amount of solvent passing through lower parts of the layer. The effect is usually more pronounced at the layer borders. Deviated sample paths and higher R_F values are frequently obtained in unsaturated development chambers. With a mobile phase consisting of several components of different volatility, the composition of the mobile phase in the layer may be significantly different from that of prepared eluent. In order to obtain regular chromatograms and reproducible R_F values, gas phase saturation, e.g., by lining the chamber with filter paper soaked with the mobile phase, is necessary. When a dry plate is placed in an atmosphere containing vapors of solvents, vapor adsorption may occure. This presorbed mobile phase shortens the effective pathlength of solvent and unretained components appear behind the solvent front at $R_F < 1$. In adsorption TLC preferential adsorption of polar components of the mobile phase from the gas phase significantly changes the activity of the adsorbent.

If the mobile phase is formed by the mixture of several components, more polar components will be preferentially adsorbed and solvent demixing will occur. Especially mobile phase containing low amounts of highly polar component tends to form a secondary solvent front by this mechanism. Mobile phase before this front is essentially free from the polar component while behind the front the sorbent is equilibrated with polar component and composition of the mobile phase is the same as that of the original solvent. Recently, these effects were also observed on chemically bonded amino layers using both normal phase and reversed phase eluents (33).

Solvent demixing acquires considerable importance for mobile phase additives devised to control the dissociation of solutes in the separation system. For alkaloids essentially the same approaches as in liquid column chromatography are applied to TLC. On silica gel layers acid additives, most often acetic acid, can be used to ascertain the dissociation of alkaloids.

In most cases dissociation is suppressed by the addition of a basic component such as diethylamine to the mobile phase. These compounds may be retained substantially by the adsorbent layer.

It was emphasized several times in Chapter 5 that in liquid chromatography the columns have to be equilibrated with a sufficient volume of the mobile phase to saturate the stationary phase along all the column length and achieve reproducible retention. During a TLC run, however, these additives may be easily adsorbed near the point at which mobile phase enters the adsorbent layer and most of the plate may be unaffected by their presence. Compared to column liquid chromatography, higher concentrations of these components are applied to ensure saturation of the layer to a sufficient distance. It is advantageous in the respect that there are no corrosion problems in simple developing tanks. In instrumental HPTLC protection measures against instrument corrosion or other damage of, say, scanning densitometers by aggressive vapors have to be observed.

For volatile, strongly retained additives, direct transport to the stationary phase through the gas phase controls the acidity of the stationary phase. If a silica gel plate is placed in the saturated developing chamber with mobile phase containing ammonia, its surface will be basified within seconds along all its length as can be easily visualized by spots of a suitable acid-base indicator.

The third approach to solute dissociation control utilizes layers impregnated with buffer. Very illustrative examples of the effect of stationary phase pH on the separation of ionizable solutes including several alkaloids were given by Stahl and Dumont (44) using silica gel layers in which the pH gradient was produced perpendicular to the development direction. Home-made sorbent layers can be buffered during slurry preparation. For ready-made plates pH can be adjusted either by dipping or by spraying with buffering solution. The considerable buffering capacity of proper silica gel has to be taken into account. In order to basify silica gel layer on aluminum support by spraying it with hydroxide, its concentration in solution has to be high and dissolution of support occurs easily. Sufficiently concentrated buffer solutions with pH near the required value are useful in this case.

Similarly as for silica gels in column liquid chromatography (see Chapter 5, Section I.A), the surface pH values of commercial plates may differ substantially. If they are used with mobile phases without pH control, such as chloroform-methanol mixtures, poor reproducibility of retention data among the plates of various types can generally be expected.

Demixing of the mobile phase components can also be expected in ion pair TLC. With long alkyl chain surfactants the impregnation of reversed phase layers before separation is required, otherwise ion pair reagent, which is added to the mobile phase only, would be retained in the lower part of the layer with no effect on the separation. Such behavior was observed for cetyltrimethylammonium bromide (45,46), adsorption characteristics of which were established on TLC layers. Strongly retained pairing reagent may not necessarily be present in the eluent because the mobile phase is equilibrated by passing through the lower part of the impregnated plate. However, hydrophilic pairing reagents not retained in the separation system may be added to the mobile phase only (47). Retention in ion pair chromatography is not determined solely by concentration of pairing reagent but also by other ionic components in the mobile phase. Their sorption by the layer may again cause the properties of the separation system to be inhomogeneous. The mechanism of reversed phase ion pair TLC is under intensive study at present (48-51). More knowledge about the factors affecting separation is required because the process here is more difficult to control than in column ion pair chromatography, where the equilibrium between stationary and mobile phases is established before the separation. Ion pair TLC may be helpful in identifying the functional groups of unknown compounds (52).

Various experimental approaches have been adopted in order to achieve a development with a better defined composition of the gaseous, stationary, and mobile phases and their mutual equilibration. In sandwich-type development chambers the chromatographic layer is covered by a covering plate. The distance between them is low and the gaseous phase in this gap is readily saturated consuming only negligible amounts of the mobile phase. Substantial solvent savings have been achieved with this type of chamber. Of a similar type are usually the chambers used for linear development in HPTLC. In the Camag developing chamber solvent from the trough is introduced to the horizontal adsorbent layer through the edges of the plate by capillary action. The plate is developed either from the opposite sides simultaneously up to the middle of the plate or from one direction only, with the development distance doubled. Preconditioning of the plate with mobile phase or other vapors or with gas of defined humidity is possible. A similar type of mobile phase introduction by capillary forces (53) was used for gradient elution (54,55). Eluting solvent was added to the chamber equipped with the distributor of the mobile phase in small portions of changing composition. In chambers of the Vario-KS type various preequilibration conditions can be applied to the different lanes, thus enabling a rapid testing of chromatographic conditions.

If mobile phase ascending through the layer evaporates at the upper edge of the layer by venting or heating, the development can continue for unlimited time. This continuous development technique possesses some advantages over conventional development (56). Weaker eluents can be used and better resolution may be obtained on the given plate length at shorter time. Optimization procedures for the selection of mobile phase strength and plate length have been proposed (57-59).

Thin-Layer Chromatography

In some cases the repeated development of the plate may be applied. Predevelopment for zone sharpening has already been mentioned. In some cases the major component of the sample, which can adversely affect mobile phase flow through the starting spot, may be washed out using a mobile phase of low eluting strength for components of interest which remain on the start. After drying of the plate, a mobile phase suitable for their separation is applied. Repeated development with solvents of increasing strength enables successive analysis of components that differ widely in polarity. Multiple development with the same solvent may yield better distribution of the zones on the chromatogram than single run. In the programmed multiple development technique (60-63) the plate is repeatedly developed with the same solvent. Drying and developing is programmed in such a manner that each developing run is longer than its predecessor. In an improved mode of automated multiple development mobile phases of decreasing strength are used in successive runs (63,64). With repeated passages of the solvent front through samples, their spots tend to concentrate into narrow bands. Hence, a greater number of components can be resolved on a single plate.

The velocity of the solvent front for a given layer is dependent on the surface tension, wetting angle, and viscosity of the mobile phase and decreases with the square root of the development time. The time required for development is proportional to the square of the development distance (20,65). Hence, upon doubling the plate dimensions the time of development increases fourfold.

As conventional alkyl-bonded reversed phase packings are not wetted by mobile phases with high water content, the use of reversed phase plates has been limited to mobile phases lean in water. The time required for the plate development increases with the water content of the mobile phase (65). Also for sample application and detection reagents solvents that are able to wet the adsorbent must be used. The development of layers prepared by impregnation is usually possible even with aqueous mobile phases without organic modifier (66). In recent years plates with less hydrophobic chemically bonded reversed phases, compatible with a high water content of mobile phase or sample solution, have appeared on the market (19, 20,67). Incomplete silanization is one of the solutions of a poor wettability problem. In some cases about 3% of sodium chloride or other salt has to be added to the mobile phase (19,21,22,68).

Circular development can be simply accomplished on a Petri dish containing mobile phase and wick which touches the plate above the Petri dish in one point. In commercial instruments for circular TLC the eluent is applied to the center of the plate by capillary using a solvent delivery system with adjustable flow rate. Samples are applied in the usual manner at a certain distance from the center, and during the development the spots, in the form of arches with center at the solvent delivery point, arise. Sample can also be introduced by a sampling valve to the flow of solvent in a way similar to that in column chromatography. Only one sample per plate can be analyzed in this manner and its components will form concentric circles. Due to an increasing perimeter of the solvent front, the circular chromatogram is developed with a constant flow rate of mobile phase. The flow rate may be varied to some extent. The development can be interrupted as desired. By interrupting the development just when the solvent front passes the sample spots, even slowly dissolving samples may be analyzed without tailing. Usually the instruments for circular development enable the preconditioning of the plates before development by controlling the composition of the atmosphere above the plate.

In anticircular chromatography (69) the mobile phase is introduced to the plate through a circular capillary trough. Circular adsorbent layers with a diameter slightly higher than that of the trough are used. Samples are applied to the inner side of the trough. Linear solvent velocity is approximately constant in this type of development and is determined by capillary effects. The development is over when solvent front achieves the center of the plate or may be interrupted before it. The speed of this type of development is high, which can be useful if slowly migrating solvents are to be used. The application of anticircular chromatography for preparative purposes was also studied (70).

The advantages of column chromatography and HPTLC are combined in overpressured TLC (71-74). A chamber for circular overpressured development is depicted in Figure 6.1. The chromatographic plate is covered by a plastic membrane which adheres tightly to the surface of the sorbent layer owing to gas or liquid pressure applied to the opposite side of the membrane. Mobile phase is delivered to the sorbent layer by a suitable pump just as in column liquid chromatography. Pressures of about 1 MPa are applied, but a modified instrument for circular development operating at high pressure up to 8 MPa has been constructed (75). For linear development the edges on three sides of the plate are impregnated. Plates of larger dimensions may be utilized, making possible improved separations in shorter times. A comparison of conventional and overpressured TLC separations of dopping agents is given in Figure 6.2. A review of the technique was recently published (76).

Another means for the acceleration of TLC separation is centrifugal motion. Various materials for centrifugal chromatography are produced commercially. Modifications of centrifugal TLC development for preparative purposes have been described (77,78). Another flexible apparatus for preparative planar chromatography based on a combination of centrifugal and anticircular development was constructed (79,80).

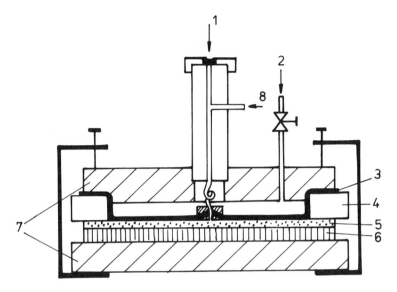

FIGURE 6.1 Cross-section of circular-type overpressurized chamber. 1, Inlet for introducing the sample; 2, pressurized gas inlet; 3, plastic membrane; 4, O-ring seal; 5, sorbent layer; 6, glass or plastic plate; 7, upper and lower support blocks; 8, mobile phase inlet. (After Ref. 71.)

For linear TLC development an apparatus was constructed in which mobile phase can be introduced to any position on the plate at any time, thus enabling so-called sequential development (81). The same principle was applied to preparative separations (82).

In the selection of mobile phase for TLC separations the same strategies may be applied as for column liquid chromatography, taking into account equilibration of the mobile phase and sorbent layer during development. Studies on the relationship between retention and mobile phase composition usually support this approach. In fact, theoretical models for adsorption chromatography are to a large extent based on TLC data. The laws for the effect of composition of binary mixtures on retention have been applied to optimize TLC separations (56,57). A dual-retention mechanism has been proven in TLC on cellulose, silica gel, and alumina, and interpreted using the concept of solubility parameter (83). In reversed phase TLC the retention of solutes is governed mainly by their hydrophobicity or lipophilicity, which in many cases can be approximately estimated from retention data (25,26,28,84,85). The strength of organic modifiers correlates with their lipophilicity and selectivity

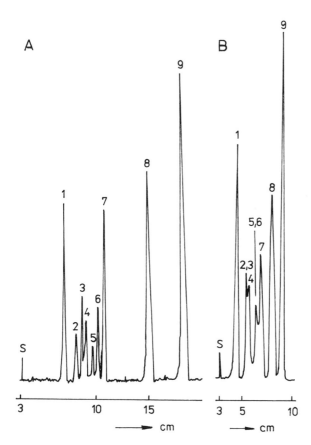

FIGURE 6.2 Densitogram of standard mixture of doping agents. (A) Overpressured linear development, Chrompress-10 chamber. (B) Conventional development. Sorbent HPTLC silica gel 60 F_{254} plate, developing solvent n-butanol-chloroform-methyl ethyl ketone-water-acetic acid 25:17:8:4:6, detection at 210 nm in absorption mode. 1, Strychnine; 2, ephedrine; 3, methamphetamine; 4, phenmetrazine; 5, methylphenidate; 6, amphetamine; 7, Desopimon; 8, Coramin; 9, caffeine. (Reproduced from Journal of Chromatography, Ref. 74, by permission of Elsevier Science Publishers B.V.)

is given by the presence of nitrogen, oxygen, and hydroxyl groups in their structure (27). Recently, mobile phase effects on diphenyl silica layers were studied (86). The Prisma model (87), describing all possible compositions of quaternary, ternary, and binary solvent mixtures, was applied to the optimization of separations in normal and reversed phase TLC (88). Three solvents are used to adjust the selectivity and the fourth one the total elution strength.

D. Detection

After development the visualization of compounds present in the chromatogram is generally required because only colored compounds are directly visible. In order to detect the compounds absorbing UV radiation, fluorescing indicators are added to the adsorbent layer. Spraying of a solution of organic fluorescing compound may be done, but in most cases inorganic luminophores are added directly to the slurry for the layer preparation. Luminophores fluorescing in visible after excitation by UV radiation of mercury lines 254 and 366 nm are commonly used. When irradiated, the chromatographic plate displays uniform fluorescence and the fluorescence of the indicator is only quenched at places where UV-absorbing components are present. Such compounds are therefore visualized as dark spots on a fluorescent background.

Sometimes under UV lamp compounds can be detected exhibiting their own luminescence. In some cases, however, the fluorescence of these compounds is visible on a wetted plate only and vanishes when the plate dries. The impregnation of the layer by nonvolatile solvent may be useful in this case. Some alkaloids are fluorescing in a specific pH range only and the acidity of the sorbent layer has to be adjusted to this range. <u>Cinchona</u> alkaloids (quinine, quinidine) show their typical fluorescence only under strongly acidic conditions. The plates have therefore to be acidified by spraying with diluted sulfuric acid, by dipping the layer into a solution of sulfuric acid in diethyl ether, or by exposing the plate to the vapors of formic acid before examination under UV light.

For compounds which do not absorb visible and UV radiation and cannot be detected by fluorescence, visualization by detection reagents may be used. There are also other special methods such as those utilizing the biological activity of the separated compounds (89), but the application of chemical reactions is most common. Liquid reagents are usually applied by spraying or by dipping the chromatogram into the reagent solution. Volatile reagent may be applied by exposing the plate to its vapors. In this way iodine vapors can reveal many organic compounds. Besides universal detection methods for organic compounds such as charring with concentrated sulfuric acid, there are plenty of detection reagents specific for compounds of certain structural types. Based on the characteristic colors of spots after application of various reagents and R_F values, reliable identification of unknown substances is possible in many cases. Some reagents may be applied in succession on one plate. For example, a very elaborate system for identifying drugs in urine using nine detection reagents has been developed (90).

Elimination of developing solvent from the layer is usually necessary before application of the reagent. In the detection of alkaloids basic components of the mobile phase often disturb the detection. Ammonia is preferred in this respect due to its volatility. Diethylamine and other higher amines are unsuitable as it is nearly impossible to remove them from the layer without destroying alkaloids.

The reader is referred to the literature (1,4) for a list of reagents applied in TLC. Only the most important detection reagents and certain special-purpose reagents for the detection of alkaloids will be mentioned here. In the first place, Dragendorff's and iodoplatinate reagents, which react with most alkaloids, should be mentioned. Both reagents are prepared with many modifications. Frequently, a modification of Dragendorff's reagent according to Munier and Macheboeuf (91) is used. It is prepared by mixing a solution of 1.7 g bismuth subnitrate in 80 ml water and 20 ml acetic acid with a solution of 16 g potassium iodide in 40 ml water. For spraying, 1 volume of this solution is diluted with 2 volumes of acetic acid and 10 volumes of water. With alkaloids Dragendorff's reagent yields orange-red spots on a yellow background with a limit of detection of about 0.1–0.01 µg (92). Overspray with 0.03% potassium hexaiodoplatinate was used to discolor the background of plates developed with mobile phase containing diethylamine (93). Also overspray with aqueous sodium nitrite was sometimes used (94–96). A higher sensitivity of detection was achieved by applying sulfuric acid diluted 1:5 after Dragendorff's reagent followed by exposure to iodine vapors (97). A higher sensitivity (92) was also found for the modification by Munier in which tartaric acid is substituted for acetic acid (1 g for 1 ml) in the above formula (98). Dragendorff's reagent diluted with ethyl acetate was also used (4,93). In addition to alkaloids, other types of compounds such as other amines, polyethylene glycols, ethylene oxides, lactames, lipids, some steroids, etc., yield a positive reaction (4).

Although slightly less sensitive (92), iodoplatinate is sometimes preferred over Dragendorff's reagent because it yields a diversity of colors useful for identification. A mixture of 5 ml of 10% platinum tetrachloride and 250 ml of 2% potassium iodide is one of the formulas for its preparation (4). When used for alkalized plates or plates developed with alkaline buffer, the reagent is acidified with hydrochloric acid. The reagent may be also used as the solution in acetone (4,99), with a lower final concentration of platinate. Iodoplatinate also reacts with some other nitrogen compounds, penicillins, and water-soluble vitamines (4).

Rich chromatograms are obtained with iodine vapors because this nearly universal detection reagent facilitates the appearance of a great number of compounds other than alkaloids. Modifications based on combinations of iodine or bromine vapors with phenothiazine-impregnated plates and vapors of ammonia and pyrrole were compared (100). Iodine can also be applied to the plate as a spray in solution with hexane or aqueous potassium iodide (4,101) eventually acidified with hydrochloric acid (102,103).

A large group of detection reagents utilizes fairly aggressive media. Ceric sulfate (1–2%) in 10% sulfuric acid (104–106) or saturated solution in 65% sulfuric acid (101), 3–5% ferric chloride in about 35% perchloric acid (104–106), or 0.5–1% ammonium vanadate in 50% nitric acid or sulfuric acid (105,107) are typical examples. Recently, an iodic acid (2 g in 10 ml water with 90% sulfuric acid added up to 100 ml) reagent was introduced (108). These solutions are akin to classical spot test reagents used widely in drug identification (8,109), such as Mandelin's, Marquis', Fröhde's, etc., reagents, which can also be used in TLC (105,110). The dangers of handling these reagents, however, undermine the advantages of producing a variety of colors. Some reagents of this type are reserved for special purposes. The increase in fluorescence of some ergot alkaloids after reaction with 0.2% (w/v) o-phthalaldehyde in concentrated sulfuric acid allowed the detection of 1 ng of these bases (111). Chromotropic acid (1 g in 30 ml 50% sulfuric acid) reacts with formaldehyde, which develops at 110–130°C upon hydrolysis from some solutes (112). In this way alkaloids containing a methylenedioxy group can be detected.

Use of Ehrlich's reagent based on p-dimethylaminobenzaldehyde in concentration 1% in ethyl alcohol or, better, ethyl alcohol–carbitol (diethylene glycol monoethyl ether) 7:3 acidified with hydrochloric acid (113) or by exposure of the plate to hydrochloric acid vapors (104) yields blue spots with indoles. With 2% solution in ethanol–hydrochloric acid 80:20 (114) or in concentrated hydrochloric acid (115), reaction was faster and colors were more stable when oversprayed with 1% sodium nitrite, although some sensitivity was lost. Mattocks (113) used Ehrlich's reagent for the detection of pyrrolizidine alkaloids. Upon reaction with peroxide reagents alkaloids were oxidized to their N-oxides followed by reaction with acetic anhydride to pyrroles which were detected by Ehrlich's reagent after heating.

Ninhydrin solution (0.3–0.5%) in acetone, isopropanol–acetic acid 99:1, or butanol (102,116) detects secondary and some primary amines after heating. At a concentration of 1% in ethanol–sulfuric acid 75:25 (105) or 10% in 95% ethanol at elevated temperature up to 160°C (117), other compounds can also be detected.

Chloranil [0.1% in acetonitrile containing 0.2% acetaldehyde (118), 0.5% in dioxane (119), or 1% in toluene or methylene chloride (96,120)] eventually reacts with various alkaloids after heating. Overspraying with 2 N sulfuric acid intensified the spots. By combination with Ehrlich's reagent pyrrolizidine alkaloids were detected (121). Other π acceptors are also useful in detecting alkaloids that contain an aromatic skeleton (106,119,122,123). Benzoyl peroxide (2% in chloroform) can sensitively detect carbazole alkaloids (124). Selective fluorescence detection of some steroidal and glycoalkaloids in the amounts 0.02-0.2 μg was possible after spraying the chromatogram with 0.02% solution of suitable optical brightener in methanol or chloroform (125,126).

Finally, it should be noted that other reagents devised to detect other sample types may yield useful color reactions with alkaloids. For example, Gibbs' reagent, used mostly for the detection of phenols, also reacts with some opium alkaloids (127). The same is true for diazo reagents, e.g., salt Fast Blue B, which is otherwise very useful in the detection of cannabinoids.

E. Quantitation

There are two approaches to quantitative analysis in TLC. First, the separated spot can be extracted from the sorbent with a suitable solvent and determined in the solution. The second possibility lies in the direct evaluation of thin-layer cvhromatograms in situ using suitable equipment. Although such methods as radiometry or conductivity measurement may be useful in special cases, optical methods utilizing either absorption of UV and visible radiation or luminescence of separated compounds are most frequently used (128,129). The schematic diagram of measured light intensities in the optical (densitometric) evaluation of TLC layers is given in Figure 6.3. The intensity of monochromatic radiation which strikes the plate is denoted as I. The radiation is scattered in the layer, and hence reflected (I_R) and transmitted (I_T) radiations spread in all directions above and below the layer, respectively. In the presence of absorbing compounds in the layer the intensities of both transmitted and reflected radiations diminish and the measurement of differences can be used for quantitation. Of course, transmitted radiation can be measured only if a transparent support of the layer is used. For glass plates measurement in the transmission mode is impossible at wavelengths shorter than approximately 330 nm. For this reason the reflection measurement is more universal and more frequently used. In practice, the intensities I_T and I_R in the sample area are compared with the respective background intensities of the layer without absorbing sample I_T^0 and I_R^0. The dependencies of relative reflectance:

Thin-Layer Chromatography

$$R = \frac{I_R}{I_R^o} \tag{6.4}$$

and relative transmission:

$$T = \frac{I_T}{I_T^o} \tag{6.5}$$

on the concentration of absorbing solute in the layer are complex. Only for nonscattering media does the Lambert–Beer law relate log T to concentration in a linear fashion. By impregnating the layer with a liquid of the same refractive index as the layer material, a nearly nonscattering medium can be produced. This technique is sometimes used in paper chromatography. In practical TLC some calibration dependence has to be adopted.

Phenomenological theory of light passage through scattering medium expresses R and T as functions of absorption and scattering coefficient of the medium (130). The famous Kubelka–Munk function has been obtained for the reflectance of an infinitely thick medium and more complex equations are valid for R and T in the case of a layer of final thickness. The absorption coefficient of a medium is proportional to the concentration of solute in the layer and its ab-

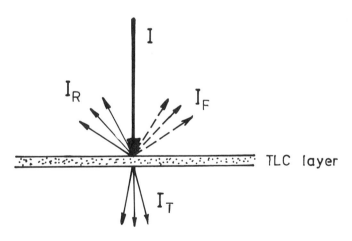

FIGURE 6.3 Optical evaluation of TLC chromatogram. Incident radiation (intensity I), transmitted radiation (I_T), reflected radiation (I_R), and fluorescence radiation (I_F).

sorption coefficient for the measurement in solutions, but the proportionality factor is dependent on the structure and properties of the layer. It may be assessed only by taking these factors into account in discontinuous theories (131). Phenomenological theory after suitable parametrization enables the transformation of the measured data into values, which are proportional to the surface concentration of the solute. In instruments that evaluate the spots point by point with a light beam of small cross-section, integration will be performed after such a transformation. In this manner the integrated area (or "volume") will only be dependent on the amount of compound in the spot and will not be affected by the dimensions and shape of the spot. Moreover, different types of background correction are possible. Microcomputer-controlled "flying-spot" (132,133) or "zig-zag" (134,135) instruments operating on this principle are commercially available.

However, it should be noted that the selection of parameters for linearization requires an empirical approach. Their values have essentially no physical meaning, as the underlying model usually does not consider such effects as absorption of radiation by plain surface layer, the reflective properties of support, and the oriented nature of the incident radiation. During evaporation of developing solvent the solutes may be concentrated at the layer surface and the distribution of the sample along the layer thickness will then be inhomogeneous (136). Moreover, small dimensions of the measuring beam are not considered in the theory. Lateral light scattering in the layer effectively increases the dimensions of the measuring beam and thus limits the achievable resolution.

Point-by-point evaluation of TLC chromatograms by the use of a television camera and digital image processing becomes feasible when costs are acceptable (137–139). A high rate of information can be obtained in this way. Evaluating a large number of spots within a short time is advantageous when large series of samples must be routinely evaluated. The increasing capabilities of computer technology also enables routine evaluation at multiple wavelengths and the production of multidimensional TLC densitograms. In this way the resolving power, identification capability, and accuracy of TLC technique increase dramatically as rigorous mathematical treatment of the data such as factor analysis, least squares or rank anihilation may be used.

On the other hand, there are plenty of cheaper instruments which evaluate the chromatogram using a light beam of dimensions comparable with spot size. In most cases the slit-shaped beam is used and the chromatogram is scanned in one direction only. The measured intensities will then be the averages over the light beam area and it is practically impossible to relate them with any theoretical model. Calibration curves based on peak height or area depend on chromatographic conditions. For any chromatographic plate a new calibration must be assessed. A large number of empirical functions were used in the literature to express the shape of calibration curves (140-143). Some of them are of limited applicability as can be seen from their limiting values and slopes at high and zero concentration (142). Therefore those possessing acceptable behavior in these regions should be preferred. For example calibration in the form of the Kubelka-Munk function, useful in medium concentration range, is

$$m = \alpha \frac{(1 - R)^2}{2R} \qquad (6.6)$$

where m is the amount of compound in the spot, α is constant, and R is the peak value of reflectance in the spot scan. At low concentrations where $R \longrightarrow 1$, the dependence becomes quadratic with zero slope at $R = 1$, which is unrealistic for TLC plates of finite thickness. Polynomial in log R:

$$m = \alpha \log R + \beta (\log R)^2 \qquad (6.7)$$

Treiber's function (140):

$$m = \alpha \log R + \beta \frac{(1 - R)^2}{2R} \qquad (6.8)$$

or an equation based on the Kubelka-Munk function with absorbing background (142):

$$m = \alpha \frac{1 - R}{R} + \beta (1 - R) \qquad (6.9)$$

possess suitable limiting properties. Equations 6.6-6.9 express the linear relationship between the amount m and the coefficients α and β. These coefficients can easily be evaluated using the standard least squares fit. Some other proposed functions, such as that of Michaelis-Menten type (144,145), are not linear in all adjustable parameters, thus requiring a little more complex calculation of nonlinear regression. The evaluation of parameters using linearized forms may lead to misleading results due to error propagation (145).

A modification of densitometric measurement is possible for plates containing fluorescent indicator. In this case incident UV radiation excites the indicator fluorescence and the intensity of the fluorescence signal in the visible wavelength range is scanned. However, the noise is usually higher in this mode of detection due to inhomogeneous distribution of the fluorescence indicator in the layer. Only when photomultiplier tube or other optical detector employed in the instrument is not sensitive to the excitation wavelength can this mode of detection be useful.

The limit of detection by densitometric measurement is adversely influenced by the baseline noise arising from the inhomogeneities of the layer. The quality of the layer is of prime importance in this respect. Two methods have been proposed to diminish this type of noise. The first utilizes simultaneous measurement of transmission and reflection (146). When signals with suitable weights are combined, the baseline noise may to a large extent be reduced. In the second method the measurement is performed on two wavelengths. One wavelength is chosen so as to achieve a maximal signal for the sample. The radiation of the second wavelength is not absorbed by the sample and its changes in intensity will be caused by imperfections in the layer. By recording the ratio of both intensities, the noise can be considerably eliminated (147,148).

Densitometric evaluation of TLC layers can be used directly for compounds which absorb UV or visible radiation. Wavelengths should be used which are not absorbed by the plain layer. The spectral properties of common TLC sorbents can be found in the literature (149). Absorption of radiation by adsorbent impurities or residues of developing solvents may prevent measurement at short wavelengths. In such cases as well as for nonabsorbing compounds, the evaluation may be performed after reaction with a suitable derivatization reagent (150). Unfortunately, many common detection reagents do not react in a sufficiently reproducible and quantitative manner and are therefore inconvenient for quantitative work. Homogeneity of reagent application is of utmost importance. Manual spraying is usually unsuitable in this respect. In practice only by an automatic spraying device can a sufficiently fine mist and homogeneous distribution of reagent on the plate be obtained. The exposition of the chromatogram to the vapors of volatile reagent usually yields better results. However, the wrong color stability and the release of corrosive vapors, which can injure scanning apparatus, can be a problem. Some detection reagents can be applied by dipping the chromatogram in their solution. Solvents which either do not dissolve the sample or are so weak that the sample extraction is negligible should be used for this purpose. Generally, chromatographic quality is worse when the detection is performed after development. Derivatization prior to separation yields better results with respect to reproducibility and limit of detection, and should be done whenever possible. Derivatization in solution before sample application or in situ at the start of the chromatogram can be done with many derivatization reagents in a similar way, i.e., as in column liquid chromatography.

Fluorescence measurement on thin-layer plates is more advantageous than densitometric evaluation for fluorescing compounds. Also phosphorescence at low temperature or room temperature has been applied (151). Fluorescence radiation of intensity I_F (dashed arrows in Figure 6.2) is generated after absorption of incident excitation radiation by the fluorescing compound. Provided that incident radiation is not diminished by the presence of solute, the intensity of emitted radiation is proportional to the amount of solute in the layer. In a low concentration region, linear calibration plots are therefore obtained just as for fluorimetry in solutions. Integration and evaluation of spots can easily be performed using a large cross-section of measuring beam.

Measurement can be performed on the same instrument as that used for densitometry provided that scattered exciting radiation is filtered off and only the wavelengths of fluorescence radiation are detected. The efficient elimination of exciting radiation is decisive in applications near the limits of detection as its presence increases background signal and therefore background noise. In phosphorescence measurements exciting radiation is effectively eliminated by use of a phosphoroscope. As low as picogram amounts of fluorescing compound can be detected in favorable cases compared to nanogram amounts in densitometry. Derivatization procedures giving fluorescing products may be applied for compounds without native fluorescence in a similar way as for column liquid chromatography. Fluorescence emission of compounds on dry silica gel layers is weak in many instances. Enhanced sensitivity is often achieved by impregnation of the layer with liquid paraffin, triethanolamine, glycerol, nonionic surface-active agent Triton X-100, or, recently, poly-(perfluoroalkyl ether) used as vacuum pump fluid Fromblin Y-Vac (150,152–155). The layer can be covered by these agents either by spraying the plate after development or, in certain cases, by adding the agent to the developing solvent.

II. TLC ANALYSIS OF ALKALOIDS

Thin-layer chromatography is still finding a wide range of applications in the field of alkaloid analysis. The main advantage is its simplicity and inexpensive instrumentation, at least for qualitative or semiquantitative analysis. Another advantage is the possibility of separating generally 10–25 samples in one run, according to the size of the plate and the method of development. For qualitative analysis two-dimensional development and color reactions in situ are the important identification tools. When we take into account these advantages, it is not surprising that TLC is a favorite method for the analysis of plant extracts, especially in cases where a large number of samples are to be analyzed. It is not necessary to purify the sample via a complex preparation procedure because the used plates are discarded and the sorbent deterioration is not a problem. The development of new TLC techniques has brought about a new field of applications—preparative separations. It seems that in some respects preparative TLC is superior to preparative column LC.

On the following pages examples of alkaloid separations using TLC are given. Even though different chromatographic systems will be introduced (e.g., reversed phase, ion pair), it has to be emphasized that an absolutely prevailing principle is the adsorption on silica gel with the use of mobile phases similar to those already characterized in Chapter 5, Section I.A.

A. Phenylethylamine Derivatives

Of this group, ephedrine and its derivatives rank among the most frequently analyzed alkaloids. Zhang and Ma (156) determined ephedrine hydrochloride in pharmaceuticals. The ephedrine preparations were dissolved in water and directly applied on a plate with silica gel G and developed using butanol–acetic acid–water (6:3:1). The plate was sprayed with ninhydrin, heated at 90°C for 20 min, and measured by scanning densitometer with measuring and reference wavelengths 520 and 700 nm, respectively. Lepri et al. (157) determined R_F values of ephedrine and related compounds on silanized silica gel (reversed phase system), silanized silica gel impregnated with anionic and cationic detergents (ion pair system), and ammonium tungstophosphate (inorganic synthetic ion exchanger). Some of these results are given in Table 6.1. The data show good separation ability of reversed phase systems when operated in the acidic region without organic modifier. The detection was performed by spraying the plate with 1% ninhydrin solution in pyridine-acetic acid 5:1 and heating for 5 min at 100°C. Hordenine was detected by exposing the plate successively to nitrogen dioxide and ammonia vapors.

In addition, colchicine and related compounds were often analyzed by TLC. Sobiczewska (158) reported a method for the determination of colchicine in colchicine dragees. Ethanolic extract was separated on silica gel 60 F_{254} plates using chloroform-acetone-ammonium hydroxide 25% 5:4:0.2 as developing solvent and evaluated densitometrically. Minor alkaloids in commercial samples of colchicine were investigated by Iorio et al. (159). They separated the samples on silica gel Stratocrom SI F_{254} with two solvents and on alumina Stratocrom AL F_{254} with one solvent. The R_F data are given in Table 6.2. The minor components forming about 5% of the whole alkaloid content were isolated by preparative TLC on a layer 2 mm thick. The spots were detected by UV light and identified by MS with electron impact ionization.

TABLE 6.1 R_F Values of Ephedrine and Related Compounds

Compound	OPTI-UPC$_{12}^a$ S_1	Sil C18-50[a] S_2	Sil C18-50 +4% N-DPC[b] S_3
Norephedrine	0.39	0.55	0.29
Nor-ψ-ephedrine	0.34	0.54	0.21
Ephedrine	0.26	0.53	0.16
ψ-Ephedrine	0.22	0.52	0.15
Synephrine	0.63	0.72	0.52
Metanephrine	0.53	0.69	0.52
Hordenine	0.29	0.60	0.13

[a]Silanized silica gel, reversed phase system.
[b]N-dodecylpyridinium chloride, ion pair system.
Note: S_1, 1 M acetic acid + 3% potassium chloride; S_2, 1 M acetic acid–methanol 80:20; S_3, 0.5 M Na_2CO_3–methanol 70:30.
Source: Data from Ref. 157.

TABLE 6.2 R_F Values of Colchicine and Related Compounds

Compound	Stratocrom SI F$_{254}$		Stratocrom AL F$_{254}$
	S_1	S_2	S_3
C-6-Hydroxycolchicine	0.23	0.07	0.08
C-17-Hydroxycolchicine	0.49	0.26	0.08
N-Formyldeacetylcolchicine	0.55	0.35	0.30
Colchicine	0.60	0.48	0.38
β-Lumicolchicine	0.70	0.70	0.69

Note: S_1, chloroform–methanol–ammonium hydroxide 90:9:1; S_2, chloroform–acetone–diethylamine 50:40:10; S_3, chloroform–methanol 75:25.
Source: Data from Ref. 159.

B. Pyridine and Piperidine Alkaloids

Tobacco alkaloids may be separated by TLC, but this technique cannot be characterized as very convenient for this purpose. Nicotine and other compounds are relatively volatile and amenable to photooxidation. From this point of view, only overpressured TLC seems to fit the separation of tobacco bases, as it has been shown by Nyiredy et al. (160). They separated nicotine, nornicotine, anabasine, nicotyrine, and 2,2'-dipyridyl on silica gel 60 F_{254} plates using ethylacetate-methanol-water 12:35:3 mixture as developing solvent. TLC in combination with MS was used by Henion et al. (161) for determination of caffeine and nicotine in human urine.

Determination of piperine by TLC was reported by Jansz et al. (161). They separated the extract of pepper on silica gel plates and determined piperine quantitatively by scanning densitometry at 342 nm.

Of tropane alkaloids hyoscyamine and scopolamine were often analyzed by TLC. Combaz et al. (163) identified them in Hyoscyamus extract on silica gel 60 with methanol-ammonium hydroxide 98:2 as the developing solvent. Dragendorff's reagent was used for the detection. Grishina et al. (164) analyzed Belladonna tincture in order to determine atropine and scopolamine content. The sample was spotted on silica gel Silufol and developed with methanol-chloroform-25% ammonium hydroxide 30:70:2. The alkaloids were eluted from the plate and determined spectrometrically. Atropine and scopolamine were also identified in some preparations of Pharmacopoeia of Japan. Kawanabe (41) applied thin-layer stick chromatography as the identification test method. The samples were spotted on glass sticks coated with MS with 0.1 mm layer of silica gel and microcrystalline cellulose (5:2). The mixture of chloroform-acetone-methanol-ammonium hydroxide (conc.) 73:10:15:2 was used for development. The eluent was allowed to travel 5 cm. The spots were then visualized by dipping the stick in Dragendorff's reagent. The following R_F values were determined: atropine 0.30–0.35, scopolamine 0.58–0.64 (both alkaloids were identified in six different drugs), papaverine 0.91 (one drug). Atropine, scopolamine, and tubocurarine were used as standards in the TLC investigation of African arrow poison (165). The R_F values of the sample and of three standards were determined for five solvent systems on silica gel F_{254} plates. With chloroform-cyclohexane-diethylamine 3:6:1 the R_F values were as follows: atropine 0.33, scopolamine 0.60, $\underline{d}$-tubocurarine 0.71, and arrow poison 0.0. This separation system gave practically the same R_F values for atropine and scopolamine as that of Kawanabe (41). On the other hand, the lack of similarity between arrow poison and tubocurarine is evident. Cocaine is an important member of the tropane alkaloid family as both local anesthetic and drug of abuse.

Thin-Layer Chromatography

The former problem was solved in the paper of De Spiegeleer et al. (166), dealing with two-dimensional HPTLC applied to local anesthetics. A straightforward and simple strategy for the selection of solvent systems is based on the absolute values of the correlation matrix elements. R_F values of 14 local anesthetics (including cocaine, procaine, isocaine, etc.) are given for 11 mobile phases. The possibilities of this procedure were demonstrated by the separation of 14 anesthetics on silica gel using cyclohexane-benzene-diethylamine 75:15:10 in one direction and chloroform-methanol 80:10 in another. After identification, one-dimensional HPTLC can be used with quantification by reflectance scanning at optimum wavelength.

C. Quinoline Alkaloids

Cinchona alkaloids, which constitute the most important part of this group, have frequently been separated by TLC. Verpoorte et al. (92) investigated all the systems described in the literature for the analysis of Cinchona alkaloids and found 18 of them to be suitable. The R_F values obtained with the four most often applied solvent systems are given in Table 6.3. Further, the sensitivity of a number of detection methods was tested and the colors obtained with some reagents were summarized. Of the various detection methods for Cinchona alkaloids, the most suitable were found to be fluorescence at 366 nm after spraying of the plate with sulfuric acid and colored spots after spraying with potassium iodoplatinate. Quinidine as a component of pharmaceutical preparations must often be determined in blood or serum. In order to assay quinidine and dihydroquinidine in serum, Vasiliades and Finkel (167) extracted it with hexane containing 1.5% of isoamyl alcohol. The organic layer was extracted with 0.05 M sulfuric acid and the acidic extract spotted on silica gel plate type LK5D and developed with ethyl acetate-ethanol (abs.)-n-butanol-ammonium hydroxide (conc.) 56:28:4:0.5. The plate was then sprayed with 10% sulfuric acid and heated at 70°C for 5 min. The scan was made by densitometer in the fluorescence reflectance mode. R_F values were 0.46 and 0.59 for dihydroquinidine and quinidine, respectively. The method is suitable for 3-6 mg liter^{-1} of quinidine. The only disadvantage of this procedure is the narrow range of linear correlation between sample concentration and detection signal.

Of the dihydrofuro(2,3-b)quinoline alkaloids, platydesminium and balfourodinium were separated by TLC technique (see Chapter 7, Section II.C, Refs. 162 and 163).

D. Isoquinoline Alkaloids

The alkaloids belonging to this group have been analyzed by TLC technique using various developing solvents. As an example, the

TABLE 6.3 R_F Values of Cinchona Alkaloids

Compound	Silica gel 60 F_{254}			
	S_1	S_2	S_3	S_4
Quinine	0.17	0.17	0.32	0.18
Dihydroquinine	0.14	0.15	0.32	0.17
Quinidine	0.28	0.26	0.41	0.26
Dihydroquinidine	0.24	0.24	0.41	0.25
Cinchonidine	0.25	0.24	0.39	0.25
Dihydrocinchonidine	0.21	0.22	0.40	0.24
Cinchonine	0.32	0.32	0.44	0.31
Dihydrocinchonine	0.26	0.28	0.44	0.29
Epiquinine	0.52	0.42	0.39	0.30
Dihydroepiquinine	0.51	0.41	0.42	0.31
Epiquinidine	0.55	0.44	0.42	0.35
Dihydroepiquinidine	0.53	0.43	0.43	0.35
Epicinchonidine	0.54	0.44	0.43	0.36
Dihydroepicinchonidine	0.52	0.44	0.45	0.37
Epicinchonine	0.55	0.45	0.45	0.39
Dihydroepicinchonine	0.53	0.43	0.46	0.39

Note: S_1, chloroform-diethylamine 9:1; S_2, chloroform-methanol-ammonium hydroxide (25%) 85:14:1; S_3, kerosene-acetone-diethylamine 23:9:9; S_4, toluene-diethyl ether-diethylamine 20:12:5.
Source: Data from Ref. 92.

conditions of berberine separation are presented in Table 6.4. The applied solvent systems were mostly acidic ones (41,168–170,172) using acetic or formic acid, and two solvents contained alkaline components (173,174). However, if berberine R_F values in these quite different systems are compared, e.g., R_F = 0.40 (174) and R_F = 0.45 (41), they are very much alike. Berberine has received a great deal of attention from chromatographers, especially its determination in pharmaceutical preparations based on plant materials (41, 168–172), or in plant extracts, where berberine has been shown to be present as a minor component (174). For detection and quantitation mostly dual-wavelength densitometry was used for measuring wavelengths from 345 to 400 nm (168,170); in some cases (171,172) berberine was eluted from the plate and determined spectrometrically. Baerheim Svendsen et al. (173) described the analysis of quaternary protoberberine alkaloids, namely, columbamine, palmatine, and jat-

TABLE 6.4 Separation of Berberine on Silica Gel TLC Plates

Solvent system	Material analyzed	Ref.
BuOH-AcOH-H$_2$O 7:1:2	Chinese patent medicine	168
BuOH-AcOH(36%)-H$_2$O 2:1:1	Wuji Wan medicinal powder	169
BuOH-AcOH-H$_2$O 7:2:1	Forsythia suspensa	170
	Berberis root cortex	171
EtAc-MeCOEt-HCO$_2$H-H$_2$O 10:7:1:1	Coptis chinensis	172
iPrOH-1 M (NH$_4$)$_2$CO$_3$-dioxane 15:5:1	Jatrorrhiza palmata, Chasmanthera dependens	173[a]
BuOH-AcOH-H$_2$O 7:1:2	Pulv. Phellod. Comp. ad. Catapl.	41[b]
MeOH-diethylamine 4:1	Glaucium squamigerum	174

[a]Berberine was not present in plant extracts.
[b]Sorbent containing silica gel–cellulose 5:2.

rorrhizine (arranged according to increasing R_F values), with detection in UV light at 254 and 366 nm. Slavík et al. (174) analyzed alkaloids of Glaucium squamigerum containing mainly protopine and protoberberine alkaloid types and isolated the new quaternary alkaloid (−)-β-N-methylisocorypalminium hydroxide. The R_F values of some bases are given in Table 6.5. Along with these, Ref. 174 lists the

TABLE 6.5 R_F Values of Some Alkaloids from Glaucium squamigerum

	Silica gel G		
Compound	S$_1$	S$_2$	S$_3$
Allocryptopine	0.22	0.57	0.64
Protopine	0.35	0.66	0.72
Corydine	0.16	0.50	0.56
Canadine	0.65	0.72	0.76
Scoulerine	0.08	0.23	0.29
Stylopine	0.72	0.75	0.78
Chelidonine	0.23	0.62	0.62

Note: S$_1$, cyclohexane-diethylamine 9:1; S$_2$, cyclohexane-chloroform-diethylamine 7:2:1; S$_3$, cyclohexane-chloroform-diethylamine 6:3:1.
Source: Data from Ref. 174.

R_F values of some other bases by use of various solvent systems. Slavík and Slavíková (175) also analyzed minor alkaloids from Stylophorum diphyllum using the same separation systems as given in Table 6.5 and others. Despite its inclusion in Table 6.5, chelidonine with its structure does not belong to this section dealing with isoquinoline alkaloids but rather is treated in the following section. In nature, however, chelidonine occurs together with protopine-type alkaloids, as can be shown in another example (176) dealing with TLC of extract obtained from the production of chelidonine hydrochloride. Sanguinarine, chelidonine, allocryptopine, and protopine were identified in this material using silica gel 60 TLC plates and three solvents with methanol-diisopropyl ether. The last eluent, which caused protopine and allocryptopine to travel, contained 2 mg $NH_4Cl/10$ ml of methanol. The components were detected by both Dragendorff's reagent and UV light.

Other protoberberine-type alkaloids have been determined in the extracts of Corydalis plant species. Zhu (177) analyzed ethanolic extract of pulverized Corydalis plant. The sample spotted on silica gel plate was developed with ether-cyclohexane-methanol 5:3:0.5 and the spots of tetrahydropalmatine, tetrahydrocoptisine, d-corydaline, bicuculline, and glaucine were quantified by scanning at 280 nm. After subsequent development with a mixture of toluene-chloroform-methanol-diethylamine 7:2:1:0.3, coptisine, palmatine, and berberine were determined at 350 nm. Corydalis bungeana plant extract was analyzed by He and Zhang (178). The plant material was extracted by alkaline chlofororm and the exact separated on silica gel G using hexane-chloroform-methanol in the ratio 8:2:0.25 or 4:5:1 as a developing solvent. To quantify protopine, d-isocorynoline, d-corynoline, tetrahydrocoptisine, acetylcorynoline, and tetrahydrocorysamine, a dual-wavelength TLC scanner was used.

Morphine alkaloids have long been analyzed as the chief components of opium. In early papers as well as in the current literature, their separation has been based mostly on the use of silica gel and an alkaline solvent. In 1961 Waldi et al. (179), by applying silica gel G and a mixture chloroform-acetone-diethylamine 5:4:1, obtained the following R_F values: morphine 0.10, codeine 0.38, thebaine 0.65, papaverine 0.67, and narcotine 0.72. The retention order is entirely in agreement with similar systems of column liquid chromatography (see Chapter 5, Figure 5.4). To analyze morphine alkaloids in opium, as mentioned above, Stahl and Jahn (180) used silica gel G with a mobile phase toluene-acetone-ethanol (96%)-ammonium hydroxide (25%) 45:45:7:3. A detailed study of this problem was carried out by Rajananda et al. (181). In their study 38 TLC systems described in the literature over the last 10 years were evaluated with respect to their suitability for the determination of opiates in urine, opium, and heroin, as well as the adulterants in heroin. Using silica gel plates, a mixture of chloroform-hexane-triethylamine 9:9:4 was found to be the most suitable for the determination of opiates (morphine, 6-monoacetylmorphine, codeine, acetylcodeine, narcotine, papaverine, thebaine) in opium and the adulterants (ephedrine, quinine, methadone, caffeine, cocaine, strychnine) in heroin. Two other solvent systems, hexane-chloroform-diethylamine 50:30:7 and benzene-dioxane-ethanol-ammonium hydroxide 50:40:5:5, were appreciated as being very suitable for the determination of opiates in opium.

Single morphine alkaloids are commonly determined in pharmaceuticals, body fluids, and plant material. Wang et al. (182) reported the assay of morphine in camphor tincture. Directly applied tincture was separated on silica gel G F_{254} plates using ethyl acetate-methanol-ammonium hydroxide 21.5:2.5:1 and quantified by scanning densitometer. For determination of morphine in urine a method using dabsylation was described (see Chapter 7, Section II.D, Ref. 64). Ebel and Rost (183) analyzed pharmaceuticals in order to determine codeine accompanied by a small amount of methylcodeine. The samples were separated on silica gel 60 TLC plates using benzene-acetone-ethanol (70%)-ammonium hydroxide (25%) 35:32.5:35:2.5. The quantitative determination was performed by measuring reflectance at 240 nm. The content of methylcodeine in codeine varied about 0.3%. Codeine in pharmaceuticals was analyzed by TLC using acidic developing solvent (see Chapter 7, Section II.D, Ref. 231), and thebaine in Papaver bracteatum Lindl. by the use of an essentially neutral solvent system (see Chapter 7, Section II.D, Ref. 234).

Jost and Hauck (48) investigated the behavior of some isoquinoline bases (papaverine, codeine, eupaverine) in ion pair TLC using RP-18 F_{254} plates and alkylsulfonates as counterions. They studied the dependence of R_F values on the water content of the developing solvent (water-acetone in ratios from 20:80 to 100:0 with 0.1 mol liter^{-1} of ion pair reagent added) and found the R_F values to decline with an increasing percentage of water. The best selectivity was attained with solvent containing 40–60% water. Further, the relationship between R_F values and alkyl chain length of sodium alkylsulfonate (from 1 to 12 carbon atoms) was evaluated. R_F values decreased with increasing chain length. The spots were detected in UV light at 254 nm. It should be stated that ion pair TLC has thus far found limited use.

Of the remaining isoquinoline alkaloids, emetine and tubocurarine are important. The R_F value of the latter base on silica gel F_{254} plate using ethyl acetate-isopropanol-ammonium hydroxide (25%) 9:7:2 was found to be 0.57 (165) and another separation system was mentioned earlier (see Section II.B of this chapter). Teshima et al. (184) compared TLC and HPLC techniques for the determination of emetine and cephaeline in ipecac root. HPLC was found to be more accurate, more rapid, and simpler than the TLC dual-wavelength densitometric method.

E. Alkaloids Derived from Miscellaneous Heterocyclic Systems

Carbazole alkaloids, namely, makonal, glycozoline, glycozolidine, 3-methylcarbazole, koenidine, and murrayazoline, were separated by TLC on alumina. For details, see Chapter 7, Section II.E, Ref. 290.

Pyrrolizidine (Senecio) alkaloids occurring in a number of plants have often been determined by TLC. Tittel et al. (185) analyzed the bases of this subgroup extracted from Symphyti radix. The samples spotted on silica gel 60 F_{254} plates were developed with chloroform-methanol-ammonium hydroxide (25%) 60:10:1. Symphytine and echimidine were detected in UV light at 254 nm or by Mattocks' reagent (sodium pyrophosphate in 30% H_2O_2). The respective $\underline{N}$-oxides were developed with chloroform-methanol 7:3 and determined using fluorescence measurement at 500 nm. Later Wagner et al. (120) used slightly modified solvent system to separate symphytine, echimidine, lycopsamine, and acetyllycopsamine in their genuine $\underline{N}$-oxide forms. Silica gel plates were developed using chloroform-methanol-$\underline{n}$-propanol-water 70:50:5:10. The detection was performed by spraying the plates with acetanhydride-petrol ether-benzene 10:40:50, drying for 10 min at 95°C, spraying with Ehrlich's reagent (1 g of $\underline{p}$-dimethylaminobenzaldehyde in 100 ml of absolute ethanol with 1.5 ml of concentrated HCl added), and heating at 95°C for 15 min. Alkaloids appeared as blue spots. Bicchi et al. (186) analyzed pyrrolizidine alkaloids from Senecio inaequidens using GC, GC-MS, and TLC techniques. For TLC separation they used precoated silica gel 60 plates and dichloromethane-methanol-ammonium hydroxide (25%) 85:14:1 as developing solvent. For detection either UV light at 254 nm or Dragendorff's reagent and heating for 1 min was applied. The following R_F values were ascertained: senecionine 0.62, retrorsine 0.29. The simultaneous use of GC-MS and TLC techniques for determination of pyrrolizidine alkaloids has also been utilized by other chromatographers (187, 188). In the first case (187), the stability of these alkaloids in hay and silage was evaluated and seneciphylline, platyphylline, jacobine, jacozine, and $\underline{O}$-acetylseneciphylline were determined. The second study (188) describes the determination of senecionine and integerrimine in alcoholic extract of Petasites hybridus. The separation of diastereomeric pyrrolizidine alkaloids can be attained on alkalized silica gel as reported by Mohanraj et al. (189). Ether extract of Heliotropium curassavicum separated on a column with neutral alumina gave a fraction containing curassavine and coromandaline. This fraction was spotted on silica gel TLC plate impregnated with 0.1 M NaOH and developed with chloroform-methanol-ammonium hydroxide (25%) 17:3.8:0.25. The system proved to be successful for the resolution of fraction into two spots; curassavine $R_F = 0.88$ and coromandaline $R_F = 0.85$, however, it only operated at 27–31°C.

Caffeine, one of purine's bases, may be found as a component of doping agents or as a heroin adulterant. Henion et al. (161) applied TLC and MS techniques to analyze caffeine and nicotine in human urine. The sample was separated on a silica gel 60 G_{254} plate using ethyl acetate-methanol-ammonium hydroxide 85:10:5 as an eluent. The visualization of caffeine ($R_F = 0.62$) was made by quenching of UV light. The spots were removed by scrapping out the sorbent layer, and then either the sorbent was washed out and the solution introduced to MS, or silica gel particles were introduced to MS by direct probe inlet. Gulyás et al. (74) analyzed doping agents, among them caffeine, by means of overpressured TLC. For experimental details, the reader is referred to Figure 6.2. Caffeine and strychnine gave the most intense signal at 250–280 nm. Ion pair reversed phase TLC, which was already mentioned (48), was investigated using caffeine as the standard substance. Contrary to other compound behaviors, caffeine did not show any significant change in R_F value along with the increasing chain length of ion pair reagents.

F. Indole Alkaloids

Many alkaloids of this large group, including more than 1500 compounds, have been separated by TLC. However, for the most part the plant extracts containing the bases of several subgroups have been analyzed and hence the classification in Chapter 1 cannot be applied.

Phillipson and Hemingway (104) identified alkaloids of genus Uncaria and by a combination of TLC, GC, UV spectroscopy, and MS 60 compounds were distinguished. They used silica gel plates and eight solvent systems. Some R_F values are presented in Table 6.6. The analyzed material contained simple indole alkaloids of the harmine subgroup, some Rauwolfia bases, and uncarines A and B, pentacyclic oxindoles typical for genus Uncaria. The spots were visualized using UV light at 254 and 365 nm, Dragendorff's and Ehrlich's reagents, and 2% ceric sulfate in 2 N H_2SO_4. Tabernaemontana dichotoma leaf extract was analyzed by Perera et al. (190), who employed TLC and liquid column chromatography. Two solvent systems were used with silica gel 60 F_{254} plates. Some R_F data are given in Table 6.7 (see also Chapter 7, Section II.F, Ref. 407). Later, van Beek et al. (106), in order to identify 100 Tabernaemontana alkaloids, applied four solvent systems:

S_1, hexane-chloroform-diethylamine 6:3:1
S_2, toluene-ethanol (absolute) containing 1.74% (w/v) NH_3 19:1
S_4, chloroform-methanol 9:1
S_4, ethyl acetate-isopropanol-NH_4OH 26% (w/v) 17:2:1

TABLE 6.6 R_F Values of Some Indole Alkaloids

Compound	Silica gel G/GF$_{254}$ (2:1)	
	S_1	S_2
Harman	0.29	0.37
Harmine	0.17	0.26
Harmaline	0.00	0.12
Ajmalicine	0.72	0.71
Isoajmalicine	0.17	0.23
Tetrahydroalstonine	0.15	0.26
Uncarine A	0.70	0.65
Uncarine B	0.62	0.47
Corynantheidine	0.80	0.80
Isocorynantheidine	0.33	0.56

Note: S_1, chloroform-acetone 5:4; S_2, ethyl acetate-isopropanol-ammonium hydroxide (conc.) 100:2:1.
Source: Data from Ref. 104.

All solvent systems were used with silica gel P F$_{254}$ plates. Moreover, the chromogenic reactions of alkaloids with three spray reagents (FeCl$_3$ in HClO$_4$, ceric sulfate in H$_2$SO$_4$, and tetracyanoquinodimethane in acetonitrile) were investigated and utilized for identification. The best results were obtained with FeCl$_3$ as reagent. Thin-layer chromatography was used by Shen and He (191) for the determination of vincine as the major impurity in vincamine. By the use of silica gel G plates with hexane-chloroform-methanol 5:1:1 mixture and dual-wavelength densitometry, the content of vincine ranging from 0.695 to 2.98% was ascertained. Other examples of indole alkaloid separations by means of TLC are given in Chapter 7, Section II.F, e.g., indole alkaloids from the wood of Simaba multiflora (Ref. 370), psilocin, psilocybin, and baeocystin from Psilocybe semilanceata (Ref. 376), ajmaline, reserpine, and rescinnamine from Rauwolfia vomitoria (Ref. 407), and indole alkaloids from cell suspension cultures of Catharanthus roseus (Ref. 392).

TABLE 6.7 R_F Values of Some <u>Tabernaemontana</u> Alkaloids

Compound	Silica gel 60 F_{254}	
	S_1	S_2
Apparicine	0.29	0.21
Dregamine	0.31	0.45
19-Epiiboxygaine	0.28	0.27
19-Epivoacristine	0.34	0.60
Geissoschizine	0.23	0.34
Ibogamine	0.62	0.48
12-Methoxyvoaphylline	0.61	0.72
Methuenine	0.28	0.32
Perivine	0.22	0.32
Stemmadenine	0.11	0.28
Tabernaemontanine	0.36	0.55
Tabersonine	0.62	0.73
Tacamine	0.27	0.33
Vincamine	0.42	0.49
Vobasine	0.32	0.51

<u>Note</u>: S_1, toluene–absolute ethanol saturated with NH_3 19:1. S_2, chloroform-methanol-ammonium hydroxide (25%) 95:5:0.2.
<u>Source</u>: Data from Ref. 190.

G. Imidazole Alkaloids

Of this group mainly pilocarpine and its decomposition products were analyzed by TLC. In an early paper of Waldi et al. (179), R_F values of pilocarpine are given for the systems silica gel G developed with chloroform-acetone-diethylamine 5:4:1 (R_F = 0.41) and alumina developed with chloroform (R_F = 0.32). To evaluate the stability of pilocarpine in aqueous media, Baeschlin et al. (192) used silica gel G plates with chloroform-acetone-water 5:4:1 as developing solvent. The spots were visualized with potassium iodoplatinate reagent. Lactonization of pilocarpic acid ($R_F = 0.0$) to pilocarpine ($R_F = 0.60$), occurring by the acidification of pilocarpic acid solution, was described. Later Durif et al. (193) applied TLC for rapid, semiquantitative determination of pilocarpic and isopilocarpic acids in pilocarpine eye drops. The separation of pilocarpine from both acids was performed using silica gel plates and sufficiently polar mobile phase (ethanol-chloroform-28% NH_4OH 53:30: 17). The regeneration of lactones was accomplished by spraying the plate with p-toluenesulfonic acid in ether which, unlike the other solvents, did not cause distortion of the spots.

H. Diterpene and Steroidal Alkaloids

Using a Chromatotron, a centrifugally accelerated radial TLC instrument, Desai et al. (77) separated diterpene alkaloids on a preparative scale. The rotor was coated with a layer 1 mm thick, containing a mixture of alumina 60 GF-254 neutral and gypsum. The eluting solvent was delivered by gravity feed from a reservoir and the inert atmosphere was maintained by a stream of nitrogen. The separations were achieved using gradient elution by a solvent system selected after trial on a qualitative TLC plate. In this way crystalline "Aconitine Potent Merck," containing aconitine, 3-deoxyaconitine, and mesaconitine, was separated with gradient hexane, hexane-diethyl ether (25:75), diethyl ether, diethyl ether-methanol. In a similar way were isolated two bisditerpenoid C_{20} alkaloids from the seed of Delphinium staphisagria L., staphisine and staphidine, which were eluted with hexane containing 4.5% acetone. A mixture of delsoline and 14-acetyldelcosine was separated using gradient elution of diethyl ether with an increasing percentage of methanol. The separation of veatchine and garryine has also been described.

As far as steroidal alkaloids are concerned, Bushway et al. (194) used TLC to determine the glycoalkaloid metabolite solanidine in milk, but the emphasis has been placed on GC technique (see Chapter 4, Section IV.H, Ref. 91). Thin-layer chromatography was also applied for the assay of conessine in Holarrhena floribunda (see Chapter 7, Section II.H, Ref. 475, Figure 7.30).

REFERENCES

1. E. Stahl, Dünnschichtchromatographie, Springer-Verlag, Berlin, 1967.
2. Pharmaceutical Applications of Thin-Layer and Paper Chromatography (K. Macek, ed.), Elsevier, Amsterdam, 1972.
3. F. Geiss, Die Parameter der Dünnschichtchromatographie, F. Vieweg, Braunschweig, 1978.

4. J. G. Kirchner, Thin-Layer Chromatography, John Wiley, New York, 1978.
5. J. C. Touchstone and M. F. Dobbins, Practice of Thin-Layer Chromatography, John Wiley, New York, 1978.
6. J. C. Touchstone, Thin-Layer Chromatography, John Wiley, New York, 1980.
7. A. Baerheim Svendsen and R. Verpoorte, Chromatography of Alkaloids. Part A. Thin-Layer Chromatography, J. Chromatogr. Library, Vol. 23A, Elsevier, Amsterdam, 1983.
8. E. G. C. Clarke, Isolation and Identification of Drugs, Vols. 1, 2, W. Clowey, Ltd., London, 1974.
9. A. H. Stead, R. Gill, T. Wright, J. P. Gibbs, and A. C. Moffat, Analyst (London), 107:1106 (1982).
10. R. Gill, B. Law, C. Brown, and A. C. Moffat, Analyst (London), 110:1059 (1985).
11. G. Musumara, G. Scarlato, G. Cirma, G. Romano, S. Palazzo, S. Clementi, and G. Giuiletti, J. Chromatogr., 350:151 (1985).
12. HPTLC - High Performance Thin-Layer Chromatography (A. Zlatkis and R. E. Kaiser, eds.), J. Chromatogr. Library, Vol. 9, Elsevier, Amsterdam, 1977.
13. Instrumental HPTLC (W. Bertsch, S. Hara, R. E. Kaiser, and A. Zlatkis, eds.), Dr. A. Hüthig Verlag, Heidelberg, 1980.
14. J. C. Touchstone and J. Sherma, Techniques and Application of Thin-Layer Chromatography, John Wiley, New York, 1985.
15. Planar Chromatography, Vol. 1 (R. E. Kaiser, ed.), Dr. A. Hüthig Verlag, Heidelberg, 1986.
16. F. Geiss, H. Schlitt, and A. Klose, Fresenius Z. Anal. Chem., 213:331 (1965).
17. K. Chmel, J. Chromatogr., 97:131 (1974).
18. U. A. Th. Brinkman and G. de Vries, J. HRC&CC, 5:476 (1982).
19. U. A. Th. Brinkman and G. de Vries, J. Chromatogr., 258:43 (1983).
20. U. A. Th. Brinkman, TrAC, 5:178 (1986).
21. U. A. Th. Brinkman and G. de Vries, J. Chromatogr., 192:331 (1980).
22. U. A. Th. Brinkman and G. de Vries, in Ref. 13, p. 39.
23. I. D. Wilson, J. Chromatogr., 291:241 (1984).
24. J. W. de Haan, L. J. M. van de Ven, G. de Vries, and U. A. Th. Brinkman, Chromatographia, 21:687 (1986).
25. H. J. M. Grünbauer, G. J. Bijloo, and T. Bultsma, J. Chromatogr., 270:87 (1983).
26. T. Cserháti, Y. M. Darwish, and Gy. Matolcsy, J. Chromatogr., 270:97 (1983).
27. T. Cserháti and B. Bordás, J. Chromatogr., 286:131 (1984).

28. G. J. Bijloo and R. F. Rekker, J. Chromatogr., 351:122 (1986).
29. A. M. Barbaro, M. C. Pietrogrande, M. C. Guerra, C. Cantelli Forti, P. A. Borea, and G. L. Biagi, J. Chromatogr., 287:259 (1984).
30. I. D. Wilson, J. Chromatogr., 287:183 (1984).
31. W. Jost, H. E. Hauck, and W. Fischer, Chromatographia, 21:375 (1986).
32. W. Jost and H. E. Hauck, J. Chromatogr., 261:235 (1983).
33. L. Zlatanov, C. Gonnet, and M. Marichy, Chromatographia, 21:331 (1986).
34. I. W. Wainer, Ch. A. Brunner, and T. D. Doyle, J. Chromatogr., 264:154 (1983).
35. U. A. Th. Brinkman and D. Kamminga, J. Chromatogr., 330:375 (1985).
36. A. Alak and D. W. Armstrong, Anal. Chem., 58:582 (1986).
37. K. Günther, GIT Supplement, 1986(3):6.
38. T. Okumura, J. Chromatogr., 184:37 (1980).
39. T. Okumura, T. Kadono, and A. Iso´o, J. Chromatogr., 108:329 (1975).
40. P. L. Patterson, Lipids, 20:503 (1985).
41. K. Kawanabe, J. Chromatogr., 333:115 (1985).
42. J. W. Hofstraat, M. Engelsma, R. J. van de Nesse, C. Gooijer, N. H. Velthorst, and U. A. Th. Brinkman, Anal. Chim. Acta, 186:247 (1986).
43. D. C. Fenimore and C. J. Meyer, J. Chromatogr., 186:555 (1979).
44. E. Stahl and E. Dumont, J. Chromatogr. Sci., 7:517 (1969).
45. G. Szepesi, Z. Végh, Zs. Gyulay, and M. Gazdag, J. Chromatogr., 290:127 (1984).
46. M. Gazdag, G. Szepesi, M. Hernyes, and Z. Végh, J. Chromatogr., 290:135 (1984).
47. S. Lewis and I. D. Wilson, J. Chromatogr., 312:133 (1984).
48. W. Jost and H. E. Hauck, J. Chromatogr., 264:91 (1983).
49. D. Volkman, J. HRC&CC, 6:378 (1983).
50. I. D. Wilson, J. Chromatogr., 354:99 (1986).
51. R. J. Ruane, I. D. Wilson, and J. A. Troke, J. Chromatogr., 368:168 (1986).
52. H. M. Ruijten, P. H. van Amsterdam, and H. de Bree, J. Chromatogr., 252:193 (1982).
53. E. Soczewiński and K. Czapińska, J. Chromatogr., 168:230 (1979).
54. E. Soczewiński, J. Chromatogr., 369:11 (1986).
55. G. Matysik and E. Soczewiński, J. Chromatogr., 369:19 (1986).
56. J. A. Perry, J. Chromatogr., 165:117 (1979).

57. D. Nurok, R. M. Becker, and K. A. Sassic, Anal. Chem., 54:1955 (1982).
58. R. E. Tecklenburg, Jr., B. L. Maidak, and D. Nurok, J. HRC&CC, 6:627 (1983).
59. R. E. Tecklenburg, Jr., R. M. Becker, E. K. Johnson, and D. Nurok, Anal. Chem., 55:2196 (1983).
60. J. A. Perry, T. H. Jupille, and L. J. Glunz, Anal. Chem., 47:65A (1975).
61. J. A. Perry, J. Chromatogr., 110:27 (1975).
62. R. von Wandruska and F. Gottschalk, Rev. Sci. Instrum., 57:119 (1986).
63. D. E. Janechen, Int. Lab., 1987 (March):66.
64. K. Burge, Fresenius' Z. Anal. Chem., 318:228 (1984).
65. G. Guiochon, G. Körösi, and A. Siouffi, J. Chromatogr. Sci., 18:324 (1980).
66. I. D. Wilson, C. R. Bielby, and E. D. Morgan, J. Chromatogr., 242:202 (1982).
67. M. Faupel and E. von Arx, J. Chromatogr., 211:262 (1981).
68. J. Sherma and M. Beim, J. HRC&CC, 1:309 (1978).
69. R. E. Kaiser, J. HRC&CC, 1:164 (1978).
70. A. Studer and H. Traitler, J. HRC&CC, 9:218 (1986).
71. E. Tyihák, E. Mincsovics, and H. Kalász, J. Chromatogr., 174:75 (1979).
72. E. Mincsovics, E. Tyihák, and H. Kalász, J. Chromatogr., 191:293 (1980).
73. E. Tyihák, E. Mincsovics, H. Kalász, and J. Nagy, J. Chromatogr., 211:45 (1981).
74. H. Gulyás, G. Kemény, I. Hollósi, and J. Pucsok, J. Chromatogr., 291:471 (1984).
75. R. E. Kaiser and R. I. Rieder, GIT Supplement, 1986(3):32.
76. Z. Witkiewicz and J. Bladek, J. Chromatogr., 373:111 (1986).
77. H. K. Desai, B. S. Joshi, A. M. Parsu, and S. W. Pelletier, J. Chromatogr., 322:223 (1985).
78. Sz. Nyiredy, S. Y. Mészáros, K. Nyiredy-Mikta, K. Dallenbach-Toelke, and O. Sticher, J. HRC&CC, 9:605 (1986).
79. Sz. Nyiredy, C. A. J. Erdelmeier, and O. Sticher, J. HRC&CC, 8:73 (1985).
80. C. A. J. Erdelmeier, Sz. Nyiredy, and O. Sticher, J. HRC&CC, 8:132 (1985).
81. P. Buncak, Fresenius Z. Anal. Chem., 318:289 (1984).
82. P. Buncak, Fresenius Z. Anal. Chem., 318:291 (1984).
83. S. M. Petrović, E. Lončar, and Lj. Kolarov, Chromatographia, 18: 683 (1984).
84. M. Kuchař, V. Rejholec, M. Jelínková, V. Rábek, and O. Němeček, J. Chromatogr., 162:197 (1979).

85. M. Kuchar, V. Rejholec, B. Brůnová, and M. Jelínková, J. Chromatogr., 195:329 (1980).
86. G. Grassini-Strazza, I. Nicoletti, C. M. Polcaro, A. M. Girelli, and A. Sanci, J. Chromatogr., 367:323 (1986).
87. Sz. Nyiredy, B. Meier, C. A. J. Erdelmeier, and O. Sticher, J. HRC&CC, 8:186 (1985).
88. Sz. Nyiredy, C. A. J. Erdelmeier, B. Meier, and O. Sticher, Planta Med., 1985:241.
89. H. Jork, GIT Supplement, 1986(3):79.
90. K. K. Kaistha and R. Tadrus, J. Chromatogr., 267:109 (1983).
91. R. Munier and M. Macheboeuf, Bull. Soc. Chim. Biol., 33:846 (1951).
92. R. Verpoorte, Th. Mulder-Krieger, J. J. Troost, and A. Baerheim Svendsen, J. Chromatogr., 184:79 (1980).
93. P. Duez, S. Chamart, M. Vanhaelen, R. Vanhaelen-Fastré, M. Hanocq, and L. Molle, J. Chromatogr., 351:140 (1986).
94. I. Sunshine, W. W. Fike, and H. Landesman, J. Forensic. Sci., 11:428 (1966).
95. A. Puech, M. Jacob, and D. Gaudy, J. Chromatogr., 68:161 (1972).
96. H. J. Huizing, F. de Boer, and Th. M. Malingré, J. Chromatogr., 195:407 (1980).
97. J. E. Wallace, H. E. Hamilton, H. Schwertner, D. E. King, J. L. McNay, and K. Blum, J. Chromatogr., 114:433 (1975).
98. R. Munier, Bull. Soc. Chim. Biol., 35:1225 (1953).
99. J. Gasparic and J. Churácek, Papírová a tenkovrstvá chromatografie organických sloucenin. (Paper and TLC of organic compounds.) SNTL, Praha, 1981.
100. R. A. Egli, Fresenius Z. Anal. Chem., 259:277 (1972).
101. K. Schreiber, O. Aurich, and G. Osske, J. Chromatogr., 12:63 (1963).
102. K. K. Kaistha and J. H. Jaffe, J. Pharm. Sci., 61:679 (1972).
103. K. K. Kaistha and R. Tadrus, J. Chromatogr., 135:385 (1977).
104. J. D. Phillipson, and S. R. Hemingway, J. Chromatogr., 105:163 (1975).
105. W. E. Court, and M. M. Iwu, J. Chromatogr., 187:199 (1980).
106. T. A. van Beek, R. Verpoorte, and A. Baerheim Svendsen, J. Chromatogr., 298:289 (1984).
107. M. Malaiyandi, J. P. Barrette, and M. Lanouette, J. Chromatogr., 101:155 (1974).
108. M. Cavazzutti, L. Cagliardi, A. Amato, M. Profili, V. Zagarese, D. Tonelli, and E. Gattavecchia, J. Chromatogr., 268:528 (1983).
109. B. F. Engelke and P. G. Vincent, J. Assoc. Off. Anal. Chem., 62:538 (1979).

110. L. Vignoli, J. Guillot, F. Gouezo, and J. Catalin, Ann. Pharm. Franc., 24:461 (1966).
111. A. Szabó and E. M. Karácsony, J. Chromatogr., 193:500 (1980).
112. S. W. Gunner and T. B. Hand, J. Chromatogr., 37:357 (1968).
113. A. R. Mattocks, J. Chromatogr., 27:505 (1967).
114. K. Genest, J. Chromatogr., 19:531 (1965).
115. H. Sprince, J. Chromatogr., 3:97 (1960).
116. S. J. Mulé, M. L. Bastos, D. Jukovsky, and E. Saffer, J. Chromatogr., 63:289 (1971).
117. M. Ch. Dutt and T. T. Poh, J. Chromatogr., 195:133 (1980).
118. A. M. Taha, G. Rücker, and C. S. Gomaa, Planta Med., 36:277 (1979).
119. S. P. Agarwal and M. A. Elsayed, Planta Med., 45:240 (1982).
120. H. Wagner, V. Neidhardt, and G. Tittel, Planta Med., 41:232 (1981).
121. R. J. Molyneux and J. N. Roitman, J. Chromatogr., 195:412 (1980).
122. S. Roy and D. P. Chakraborty, J. Chromatogr., 96:266 (1974).
123. G. Rücker and A. Taha, J. Chromatogr., 132:165 (1977).
124. P. Bhattacharyya and S. S. Jash, J. Chromatogr., 298:200 (1984).
125. R. Jellema, E. T. Elema, and Th. M. Malingré, J. Chromatogr., 176:435 (1979).
126. R. Jellema, E. T. Elema, and Th. M. Malingré, J. Chromatogr., 189:406 (1980).
127. T. R. Baggi, N. V. N. Rao, and H. R. K. Murty, J. Forensic. Sci., 8:265 (1976).
128. J. C. Touchstone, editor, Quantitative Thin Layer Chromatography. John Wiley, New York, 1973.
129. J. C. Touchstone and J. Sherma, Densitometry in Thin Layer Chromatography. John Wiley, New York, 1979.
130. G. Kortüm, Reflectance Spectroscopy. Springer-Verlag, Berlin, 1969.
131. H. G. Hecht, J. Natl. Bur. Std., 80A:567 (1976); also in K. D. Mielenz, R. A. Velapoldi, and R. Mavrodineanu, editors, Standardization in Spectrophotometry and Luminescence Measurements, NBS Spec. Publ. 466, U.S. Government Printing Office, Washington, D.C., 1977.
132. V. Pollak, J. Chromatogr., 105:279 (1975).
133. V. Pollak and A. A. Boulton, J. Chromatogr., 115:335 (1975).
134. H. Yamamoto, T. Kurita, J. Suzuki, R. Hira, K. Nakamo, H. Makabe, and K. Shibata, J. Chromatogr., 116:29 (1976).
135. H. Yamamota, in Ref. 13, p. 367.

136. V. Pollak, J. Chromatogr., 133:195 (1977).
137. T. Manabe and T. Okuyama, J. Chromatogr., 264:435 (1983).
138. D. D. Rees, K. F. Fogarty, L. K. Levy, and F. S. Fay, Anal. Biochem., 144:461 (1985).
139. D. H. Burns, J. B. Callis, and G. D. Christian, TrAC, 5:50 (1986).
140. L. R. Treiber, J. Chromatogr., 100:123 (1974).
141. R. A. Egli, H. Müller, and S. Tanner, Fresenius Z. Anal. Chem., 305:267 (1981).
142. J. Fähnrich, M. Popl, and E. Výborná, Sci. Pap. Inst. Chem. Technol. Prague, H18:105 (1983).
143. S. Ebel, D. Alert, and U. Schaefer, Fresenius Z. Anal. Chem., 317:686 (1984).
144. G. Kufner and H. Schlegel, J. Chromatogr., 169:141 (1979).
145. S. Ebel, D. Alert, and U. Schaefer, Chromatographia, 18:23 (1984).
146. L. R. Treiber, R. Nordberger, S. Lindstedt, and P. Stöllnberger, J. Chromatogr., 63:211 (1971).
147. A. A. Boulton and V. Pollak, J. Chromatogr., 63:75 (1971).
148. S. Ebel and J. Hocke, Fresenius Z. Anal. Chem., 277:105 (1975).
149. R. R. Goodall, J. Chromatogr., 78:153 (1973).
150. W. Funk, Fresenius Z. Anal. Chem., 318:206 (1984).
151. J. M. Miller, Pure Appl. Chem., 57:515 (1985).
152. S. Uchiyama and M. Uchiyama, J. Chromatogr., 153:135 (1978).
153. S. Uchiyama and M. Uchiyama, J. Chromatogr., 262:340 (1983).
154. S. S. J. Ho, H. T. Butler, and C. F. Poole, J. Chromatogr., 281:330 (1983).
155. K. K. Brown and C. F. Poole, J. HRC&CC, 7:520 (1984).
156. M. Zhang and S. Ma, Yaoxue Tongbao, 20:719 (1986); CA, 104:193261h (1986).
157. L. Lepri, P. G. Desideri, and D. Heimler, J. Chromatogr., 347:303 (1985).
158. M. Sobiczewska, Acta Pol. Pharm., 41:513 (1984).
159. M. A. Iorio, A. Mazzeo-Farina, G. Cavina, L. Boniforti and A. Brossi, Heterocycles, 14:625 (1980).
160. Sz. Nyiredy, I. Tompa, S. Y. Mészáros, E. Tyihák and G. Verzar, 4th Danube Symposium on Chromatography, Bratislava (Czechoslovakia), 1983, Abstracts, Vol. II, C 18.
161. J. Henion, G. A. Maylin and B. A. Thomson, J. Chromatogr., 271:107 (1983).
162. E. R. Jansz, I. C. Pathirana, and E. V. Packiyasothy, J. Natl. Sci. Counc. Sri Lanka, 11(1):129 (1983); CA, 103: 140401b (1985).

163. D. Combaz, J. M. Morand, and J. Alary, Bull. Trav. Soc. Pharm. Lyon, 26:69 (1982), CA; 100:109178y (1984).
164. M. S. Grishina, V. V. Dyukova, L. I. Kovalenko, and D. M. Popov, Farmatsiya (Moscow), 35(2):24 (1986).
165. E. B. Cook, M. Dennis, and R. F. Ochillo, J. Liq. Chromatogr., 4:549 (1981).
166. B. De Spiegeleer, W. Van den Bossche, P. De Moerloose, and D. Massart, Chromatographia, 23:407 (1987).
167. J. Vasiliades and J. M. Finkel, J. Chromatogr., 278:117 (1983).
168. S. Zhu, Y. Ye, and R. Jin, Zhongyao Tongbao, 10(1):36 (1985); CA, 102:137906u (1985).
169. X. Zhang, X. Yuan, and Y. Sun, Yaowu Fenxi Zazhi, 5:364 (1985); CA, 104:75131t (1986).
170. H. Gui, Q. Li, G. Wang, and J. Zhang, Zhongcaoyao, 17(1):13 (1986); CA, 104:193260g (1986).
171. Z. Kalogjera, H. Krnjevic, and J. Petricic, Pharm. Vestn. (Ljubljana), 36:233 (1985); CA, 104:174732w (1986).
172. F. Tang, X. Yang, and G. Yang, Yaowu Fenxi Zazhi, 6:100 (1986); CA, 105:49131u (1986).
173. A. Baerheim Svendsen, A. M. Van Kempen-Verleun, and R. Verpoorte, J. Chromatogr., 291:389 (1984).
174. J. Slavík, L. Slavíková, L. Dolejš, Collect. Czech. Chem. Commun., 49:1318 (1984).
175. J. Slavík and L. Slavíková, Collect. Czech. Chem. Commun., 49:704 (1984).
176. T. Dzido, L. Jusiak, and E. Soczewinski, Chem. Anal., 31:135 (1986).
177. M. Zhu, Yaowu Fenxi Zazhi, 5:139 (1985), CA, 103:166219y (1985).
178. L. He and Y. Zhang, Yaoxue Xuebao, 20:377 (1985), CA, 103:129136y (1985).
179. D. Waldi, K. Snackerz, and F. Munter, J. Chromatogr., 6:61 (1961).
180. E. Stahl and H. Jahn, Pharm. Acta. Helv., 60:248 (1985).
181. V. Rajananda, N. K. Nair, and V. Navaratnam, Bull. Narc., 37:35 (1985).
182. L. Wang, J. Feng, and S. Bi, Yaowu Fenxi Zazhi, 5:109 (1985), CA, 103:27371d (1985).
183. S. Ebel and D. Rost, Arch. Pharm. (Weinheim), 317:933 (1984).
184. D. Teshima, T. Tsuchiya, T. Aoyama, and M. Horioka, Iyakuhin Kenkyu, 15:63 (1984), CA, 100:145055q (1984).
185. G. Tittel, H. Hinz, and H. Wagner, Planta. Med., 37:1 (1979).
186. C. Bicchi, A. D´Amato, and E. Cappelletti, J. Chromatogr., 349:23 (1985).

187. U. Candrian, J. Lüthy, P. Schmid, C. Schlatter, and E. Gallasz, J. Agric. Food Chem., 32:935 (1984).
188. C. Mauz, U. Candrian, J. Lüthy, C. Schlatter, V. Sery, G. Kuhn, and F. Kade, Pharm. Acta. Helv., 60:256 (1985).
189. S. Mohanraj, W. Herz, P. S. Subramanian, J. Chromatogr., 238:530 (1982).
190. P. Perera, T. A. van Beek, and R. Verpoorte, J. Chromatogr., 285:214 (1984).
191. X. Shen and L. He, Yaowu Fenxi Zazhi, 4:356 (1984), CA, 102:84481t (1985).
192. K. Baeschlin, J. C. Etter, and H. Moll, Pharm. Acta Helv., 44:301 (1969).
193. C. Durif, M. Ribes, G. Kister, and A. Puech, Pharm. Acta Helv., 61(5-6):135 (1986).
194. R. J. Bushway, D. F. McGann, and A. A. Bushway, J. Agric. Food Chem., 32:548 (1984).

7
Applications

This chapter is devoted to some examples of alkaloid determinations as they are carried out in medicine, pharmacy, forensic analysis, etc. By no means can it be considered a complete review. So many applications can be found in literature that a sizable book could deal with this subject alone. Therefore, this chapter will focus on the recently published applications, and the older papers will be included only if they describe simple and reliable procedures. Although thousands of alkaloids are known, only a few of them will be treated here. It is the opinion of the authors that apparent similarities exist among the many procedures which have been elaborated in the past 10–15 years in order to determine different alkaloid groups chromatographically. For this reason the most important ones will be discussed in detail in this chapter, assuming that the described methods can easily be adapted to the determination of another alkaloid group. The first part deals with the procedures for sample preparation, which is often the key step in the method. The possibilities of preparing any sample in several ways are demonstrated and the procedures are subdivided according to sample type. In the second part the conditions of chromatographic determination are given. This division makes it possible to combine various clean-up procedures with various chromatographic methods. The emphasis has been placed on the methods of column liquid chromatography that are of primary importance in the chromatography of alkaloids.

I. SAMPLE PREPARATION

The primary aim of sample preparation is to provide a sample free of components that may deteriorate the column and, to certain extent, other parts of the chromatographic system as well. Such components, which have to be removed, are solid particles, proteins, and lipids. In some cases sample preparation can be also aimed toward the increase of the analyte concentration and/or the removal of interfering components. If the column protection may be considered as the main purpose of sample preparation, the conclusion regarding its necessity for individual chromatographic techniques should be drawn.

This step is probably less important for thin-layer chromatography, where the deterioration of the chromatographic system need not be taken into account. Sample preparation should also be less important in gas chromatography using packed columns, even if on-column injection is applied. If the column inlet is contaminated by nonvolatile substances, this part of the packing can be easily replaced. Nowadays the capillary columns used in GC are mostly coated with bonded stationary phases, making it possible to wash them with solvents of various polarity. From the standpoint of the column deterioration, capillary GC columns are less dependent on sample purity. On the other hand, sample preparation is an important and inevitable step for highly efficient LC columns. This operation is of primary importance, especially when biological samples are to be analyzed. Very often in these cases sample preparation is complex, time consuming, and brings about the risk of analyte loss. However, for some sample types the preparation can be effected by a simple procedure, e.g., sample dissolution and filtration, which is often sufficient for the determination of alkaloids in many pharmaceutical formulations.

Any purification or preconcentration step means the risk of an unexpected loss of analyte, and the more steps that are involved, the greater the probability of solute loss. To calculate the loss, the recovery must be evaluated for the whole sample preparation procedure. This is carried out by applying the complete procedure to a blank sample spiked with a known amount of the analyte. Peak area obtained by the chromatographic analysis of this sample (A_1) is then compared with peak area obtained by the direct injection of the same amount of the analyte (A_2).

$$\text{Recovery (\%)} = \frac{A_1}{A_2} \times 100 \qquad (7.1)$$

If the recovery drops below 60%, the procedure should be considered as unreliable and replaced by another method. When the internal standard is added, its recovery must also be determined.

The decision making regarding the sample preparation procedure has to be judged from at least two standpoints: (1) the type of sample and (2) the drug to be determined.

Applications

Sometimes other criteria may also be operating: solute concentration, sample size, sample history, and so forth. However, these factors are usually only of minor importance and their resolution depends on the experience of the analyst. Samples to be analyzed may be mostly classified into three main types: (1) pharmaceutical formulations, (2) plant materials, and (3) biological samples. They are arranged according to the increasing complexity of sample preparation. Each sample type may be then further subdivided, e.g., biological samples range from body fluids (urine, plasma, serum, blood) to tissues, organs, feces, etc. In the same way, a wide variety of plant parts may be submitted to the analysis: leaves, seeds, pods, stems, etc. Pharmaceutical formulations can also differ in form: tablets, water-based preparations, oil-based formulations, solutions for injections, etc. Very often each form requires a different approach to sample preparation.

When considering a drug to be analyzed, the decision between liquid chromatography and gas chromatography must first be made. Furthermore, it is necessary to take into account mainly the dissociation constant of the drug, its solubility in various solvents, and the possibility of selective detection. If gas chromatography is the choice and a capillary column is available, then sample preparation may be simplified due to the great resolving power of this technique. A highly basic drug can be isolated from a complex sample by using acid–base interactions. This cannot be done with a drug that is essentially neutral. If an extraction is to be used for drug purification, the hydrophilic and hydrophobic drugs should be distinguished and the data of Table 2.6 should be consulted. The detection method is very important from the standpoint of selectivity and sensitivity. When a highly selective detector is used, a simple sample preparation procedure can be employed and a drug can be detected at very low concentrations.

This brief introduction shows that sample preparation is a rather complex matter. Various procedures for many alkaloids have been elaborated, especially for those used in the pharmaceutical industry. With slight modifications, these methods can be applied to other drug determinations as well.

A. Pharmaceutical Formulations

The main objective for these analyses is continuous quality control throughout the production process. This includes the control of synthesis, if the alkaloid is prepared by synthesis; the determination of active drug in plant material, if a naturally occurring drug is isolated; stability tests; and the determination of alkaloids in formulation. Pharmacokinetic studies, which are important in the development of a pharmaceutical product, involve the determination of an alkaloid and its metabolites in body fluids. In this chapter the determination of alkaloids in plant materials is discussed in Section I.B and pharmacokinetic analyses are covered in Section I.C. The present section discusses the determination of alkaloids in pharmaceutical formulations.

The content uniformity assays of pharmaceutical products, which are the most frequently done assays, are performed to satisfy regulatory agency compliance and to check the manufacturing process. Oral dosage forms (tablets, capsules, suspensions, and syrups) and nonoral dosage forms (injection solutions, topical ointments, powders, and aerosols) are analyzed and the assay results compared with the label claims. Alkaloids in dosage forms are usually present in concentrations well above detection limits and therefore the assays do not present any problems from the standpoint of sensitivity.

Of all pharmaceutical formulations, the clear liquids (e.g., injectable solutions) can be treated in the simplest manner. Usually, they are only diluted with a suitable solvent, preferably with the mobile phase. Morphine assay in injection formulations (1,2) and the simultaneous assay of morphine and codeine in injections(3) are typical examples.

The simple tablet is the most common oral dosage form. Typically, in a content uniformity assay the analyses are performed on 10 individual tablets. Tablets are ground and the amount corresponding to one average tablet is dissolved in a suitable solvent by sonication. Then the sample is filtered through a low-porosity filter in order to remove nonsoluble excipients, and an aliquot volume is injected onto a column. Either the solvent or the diluent should be spiked with the internal standard. Methanol-water and acetonitrile-water mixtures are frequently the most suitable solvents. For fully soluble tablets the preparation procedure can be further simplified by omitting the filtration. Capsules and coated tablets can be prepared in a similar way. Determination of alkaloids in analgesic tablets are a typical example. Codeine was determined in tablets and capsules (4) using normal phase liquid chromatography, and a simultaneous assay for codeine, acetaminophen, and aspirin in tablets and suppositories was carried out by reversed phase chromatography (5). The assay of caffeine and codeine in analgesic tablets was reported with recoveries of 99.1-100.7% (6). Similarly, caffeine and other analgesic drugs in tablets were determined by gradient elution with acetonitrile-water (7), caffeine and codeine using isocratic elution (6), and caffeine with mobile phase containing acetic acid and aqueous methanol (8). Octadecyl stationary phases were applied in all three cases. Recent developments in HPLC analysis of analgesic from the standpoint of methods applied were reviewed by Wilson (9). The use of HPLC in the analysis of eburnane alkaloids in Canviton tablets containing apovincaminic acid ethyl ester and in Devincan tablets containing vincamine was investigated (10). It can be concluded that in the majority of cases the preparation of tablets was simple, the recovery high, and the reproducibility very good.

The assay of alkaloids in water-based preparations (syrups, drops, etc.) is often possible by injection of a sample onto the column directly or after dilution. Whenever possible, a solvent identical with the mobile phase should also be used for dilution. Dilution with a polar solvent is suitable both for the polarity of a sample and for the subsequent analysis by reversed phase or ion pair chromatography.

The determination of codeine in syrup with a heptanesulfonate pairing ion (11) can be mentioned as an example.

The assay of alkaloids in oil-based formulations (creams, suppositories, etc.) usually requires an extraction procedure. Alkaloids as polar components can be separated from the hydrophobic excipients in two-phase systems, e.g., aqueous methanol-alkane, water-chloroalkane, etc. The extraction of an alkaloid into the aqueous layer is often facilitated by addition of an acid as shown in the following example.

For the determination of ephedrine in an oily formulation (12), the sample (100 mg) was dissolved in chloroform with the addition of acetic acid (0.5 ml) and procaine solution as an internal standard (1 ml) and made up to 50 ml. This solution (5 ml) was extracted with 0.1 M hydrochloric acid (5 ml). After centrifugation the upper aqueous phase was separated and an aliquot volume (10 µl) was injected onto a column.

Among the many kinds of pharmaceutical formulations and raw materials, opium and poppy straw have been analyzed frequently. These materials will be treated in the next section.

It may be concluded at the end of this section that the assay of alkaloids in pharmaceutical formulations usually requires only a simple sample preparation. Owing to the relatively high concentration of solute to be determined and to the known composition of other components present, the problem does not lie in sample preparation but in the degree of precision and the cost per sample. From this point of view, automated sample preparation has an additional benefit (see Section I.D in this chapter).

B. Plant Materials

Extraction is the main procedure that prepares a sample for the determination of alkaloids in a plant material. However, due to the existing variety of alkaloids and plant materials, there are many variations of the extraction procedure. Before extraction the plant material should be reduced to as fine a powder as possible. The effect of milling was investigated in relation to sieved particle size (13). Generally, two modes of extraction can be applied, both utilizing the basic character of alkaloids.

The first mode is extraction with nonaqueous solvent using hydrocarbons (petroleum ether, benzene), chloroalkanes (dichloromethane), or other solvents as extracting agents. Powdered plant material is prewetted by addition of alkaline solution (soda, sodium hydroxide, ammonia) and the solvent is added. The mixture is shaken vigorously for a few minutes and then filtrated or centrifuged in order to remove solid particles. An aliquot volume of the clear organic layer may be used for injection onto a column or, more frequently, the organic layer undergoes another extraction in order to purify the extract. In this case a diluted mineral acid (e.g., 0.1 M HCl) is applied as an extracting agent and the basic components, including alkaloids, are transferred to the aqueous layer. Then both layers are separated, and the aqueous one is made alkaline and extracted with a hydrophobic solvent (e.g., petroleum ether). Separated organic layer may be evaporated, residue reconstituted in a solvent, made up to a certain volume, and this solution is then used for analysis. As a modification of this procedure, the plant material prewetted with alkaline solution can be placed in the shell of Soxhlet apparatus and extracted with a suitable solvent. Then the next part of the preparation procedure will be the same as in the former case.

The second mode of extraction is carried out with a strongly polar hydrophilic solvent using water, water-alcohol, or alcohol as extracting agent. Very often this solvent contains a few percents of an acid, mostly acetic acid. Powdered plant material is shaken with the solvent and the resulting mixture filtered or centrifuged. The clear supernatant is made alkaline and extracted with a chloroalkane or chloroalkane-alcohol mixture. It is necessary to remove water dissolved in the separated organic layer before evaporation. The drying is accomplished by allowing the organic phase to stand for a period of time over anhydrous sodium sulfate or by percolation of organic layer through a column filled with a drying agent. After drying, the solvent will be evaporated at a mildly elevated temperature (or at room temperature) in the stream of nitrogen or under reduced pressure. The residue will be reconstituted in a suitable solvent, often in the mobile phase, if LC is the choice. The solution will either be injected or undergo another extraction step. Often, if the recovery is low, the plant material will be extracted twice or three times, always with fresh solvent. Then the extracts are combined and undergo the procedure described above.

Extraction procedures are frequently completed by other operations in order to remove the components that interfere with determination or to enrich the concentration of alkaloid in the solution.

Applications

Nowadays the sorption columns, packed with suitable sorbents, are used for this purpose. This process, often called solid phase extraction, is gaining in popularity. Sorbents such as silica, silica with bonded alkyls, porous polymeric sorbents, or, less frequently, activated carbon and alumina are placed in a disposable cartridge (usually made of polypropylene) or in a short glass column. The use of packing with particle size about 30 µm makes it possible to operate the column under gravity flow conditions or at elevated pressure. Before sample loading, the column is usually prewetted by a solvent, even though some procedures use dry columns. Loaded column can either be eluted in the forward-flush, i.e., in the same direction as the sample loading, or in the back-flush mode. In the former case loaded column will be washed in order to remove interferences and after that, using a stronger solvent, the analyte will be eluted in the same direction. The latter case requires elution of interferences in the forward-flush mode and after that a change of solvent and its direction of flow in order to obtain the desired analyte. Solid phase extraction is a very convenient process which is amenable to the automation.

There are many other variations of the procedures for purification and preconcentration of the analyte, so that it is practically impossible to give an exhaustive review. Some of them are described in Table 7.1.

C. Biological Samples

The task of determining alkaloids in biological samples is very often met in pharmacokinetics and in forensic analyses. Pharmacokinetic studies, which are important in the development of pharmaceutical products, require the determination of a drug and its metabolites in body fluids over a period of time. The analyzed material is generally plasma or serum where the components to be determined are often present in very low concentrations. Forensic samples include a wide variety of materials from urine to tissue and organs with the analytes frequently present in trace concentrations. In these cases precautions have to be taken to avoid the loss of drug during the sample preparation. Sometimes the preparation must be carried out in silanized glassware in order to minimize the adsorption of drug onto the glass surface. The drugs to be analyzed in forensic samples are often more or less converted to metabolites by the reactions in vivo.

Without any doubt the preparation of biological samples represents a complex problem. Components with large molecules, such as proteins and lipids, are present in the majority of samples, which may be harmful to column life. For the assay of alkaloids in biological samples several preparation procedures have been elaborat-

TABLE 7.1 Sample Preparation of Plant Materials

Material prepared	Alkaloid determined	Extraction			Follow-up procedure	Ref.
		Extractant	Ratio extractant/sample (ml g^{-1} or g g^{-1})	Technique and time		
Colchicum seeds	Colchicine, colchicoside	10 g of sample with 0.15 g of CaCO$_3$ extracted with 90% MeOH	150:10 (v/w)	Soxhlet 6 h	Evaporated and reconstituted in 100 ml of MeOH	14
Pepper	Piperine	EtOH	90:0.5 (v/w)	Boiling 15 min	Made up to 100 ml, filtered, first 5 ml discarded, 0.5 ml of filtrate mixed with 1.5 ml of EtOH and 2.0 ml of I.S. (phenazine) solution, 20 µl injected	15
Tobacco	Nicotine	H$_2$O	8:0.15 (v/w)	Stirring 8 h	Tobacco washed 2x with 4 ml H$_2$O, total volume made up to 20 ml, 2-ml aliquot + 1 ml of I.S. (phendimetrazine) solution + 1 ml of 4% NaOH extracted 3x	16

Applications 317

Glaucium flavum	Glaucine	0.33 g of sample in 1 ml of 10% NH$_4$OH extracted with Et$_2$O	20:0.33 (w/w)	Shaking	10 g of Et$_2$O layer evaporated, residue reconstituted in 50 ml of 0.05 M KH$_2$PO$_4$ and solution loaded on C18 Sep-Pak col. The first 2 ml discarded, the next 1 ml used for analysis with 6 ml of Et$_2$O, evaporated (42°C) to 0.5 ml, 3 μl injected	17
Fumaria species	Phthalido-isoquinoline alkaloids	MeOH			Extracted treated with 10% HCl, extracted with benzene, aq. phase adjusted to pH 8.0, extracted with benzene and subsequent extraction with Et$_2$O+CHCl$_3$	18
Corydalis bungeana	Corynoline-type alkaloids	5% acetic acid	50:0.25 (v/w)	Sonication 30 min	Extract filtered, 25 ml of filtrate adjusted to pH 8.5 and extracted 5x with 50 ml of	19

TABLE 7.1 (Continued)

Material prepared	Alkaloid determined	Extraction			Follow-up procedure	Ref.
		Extractant	Ratio extractant/sample (ml g^{-1} or g g^{-1})	Technique and time		
Opium powder	Opium alkaloids	H$_2$O	1:0.03 (v/w)	Maceration overnight	CHCl$_3$. Pooled extracts evaporated, residue reconstituted in 2 ml of MeOH, 5 µl injected	
Opium powder	Opium alkaloids	H$_2$O	1:0.03 (v/w)	Maceration overnight	Supernatant mixed with equal volume of acetonitrile, 20 µl injected	20
Opium powder	Opium alkaloids	2.5% acetic acid repeated 3×	20:2 (v/w)	Shaking 20 min	Centrifuged, supernatant made up to 100 ml with 2.5% acetic acid, a part 5 ml diluted to 20 ml with MeOH, 6 µl injected	21
Opium powder	Opium alkaloids	2.5% acetic acid repeated 4×	20:1 (v/w)	Sonication 30 min	Centrifuged, supernatant made up to 100 ml with 2.5% acetic acid, filtered	22

Applications

Poppy straw concentrate	Opium alkaloids	Acetonitrile-H₂O (1:19) contg. 40 ml/liter of acetic acid and 0.04 ml/liter of N,N-dimethyl-octylamine, pH 3.5	25:0.05 (v/w)	Sonication 30 min	(0.45 μm), 10 μl injected Filtered (0.45 μm) and 10 μl injected	23
Papaver somniferum plant material	Opium alkaloids	Alkalinized material extracted with CH₂Cl₂		Soxhlet	CH₂Cl₂ solution re-extracted with 1 M HCl, layers separated, aq. layer alkalinized and extracted with CH₂Cl₂-isopropanol (3:1)	24
Papaver rhoeas flowers	Rhoeadine papaverrubine A, etc.	MeOH		Percolation	Extract acidified with 10% H₂SO₄, extracted with CHCl₃, aq. phase adjusted to pH 10, extracted with Et₂O, etheric layer purified on silica gel	18

TABLE 7.1 (Continued)

Material prepared	Alkaloid determined	Extraction			Follow-up procedure	Ref.
		Extractant	Ratio extractant/sample (ml g^{-1} or g g^{-1})	Technique and time		
<u>Senecio anonymus</u> flowering specimens	Pyrrolizidine alkaloids	EtOH			Extract partitioned between CHCl$_3$ and H$_2$O, $\underline{N}$-oxides reduced with Zn, solution made alkaline and extracted with CHCl$_3$, org. layer evaporated and reconstituted in CHCl$_3$	25
<u>Chelidonium majus</u> L. roots	Phenenthridine (quaternary alkaloids	70% EtOH		Soxhlet		26
<u>Psilocybe semilanceata</u>	Psilocybin, psilocin, etc.	MeOH–1 M NH$_4$NO$_3$ (9:1), repeated 2×		Rotation in centrifuge tube 30 min	Centrifuged, supernatant made up to 5.0 ml with 1 M NH$_4$NO$_3$-MeOH (1:9), 10 μl injected	27

Applications 321

Psilocybe semilanceata, Psilocybe cubensis, Panaeolina foenisecii, Amanita citrina	Psilocybin, psilocin, etc.	MeOH, contg. 80 μg/ml of bufotenine hydrogen oxalate	20:0.2 (v/w)	Sonication 15 min	Extract concentrated to 4.0 ml, 10 μl injected	28
Tabernaemontana	Perivine, tabernaemontanine, etc.	MeOH repeated 3×		Grinding 5 min	Extract evaporated, dissolved in 1 ml of MeOH + 1 ml of I.S. (dihydroquinine) solution + 8 ml of 0.05 M Na$_2$HPO$_4$, pH 7.0, filtered, loaded on RP-8 col. Column washed with 0.05 M Na$_2$HPO$_4$, pH 9/5–isopropanol (7:3). Alkaloids eluted with 0.05 M buffer, pH 2.0–isopropanol (6:4)	29
Cultures of Tabernaemontana species	Perivine, tabernaemontanine, etc.				Suspension of culture medium adjusted to pH 7.0, 1.0 ml of I.S. (dihy-	29

TABLE 7.1 (Continued)

Material prepared	Alkaloid determined	Extraction			Follow-up procedure	Ref.
		Extractant	Ratio extractant/sample (ml g^{-1} or g g^{-1})	Technique and time		
					droquinine) solution added, filtered, filtrate applied to RP-8 col. (see above)	
Biomass of Tabernaemontana orientalis, T. elegans	Perivine, tabernaemontanine, etc.	96% EtOH			Extract adjusted to pH 4.0, extracted with Et$_2$O, aq. phase adjusted to pH 9.0, extracted with CHCl$_3$, extract dried over Na$_2$SO$_4$ and evaporated	30
Strain of Ochrosia elliptica Labill	Pyridocarbazole alkaloids	MeOH	100:1 (v/w)	Maceration 2 hr	Centrifuged, 0.5–2.0 µl applied on TLC sheet	31

Applications

ed as shown in Table 7.2. The individual preparation procedures will be discussed briefly and some examples will be described in detail.

Procedure 1 involves a single operation, denoted as "simple treatment," which may include, for example, dilution, filtration, or centrifugation. It can mean also the direct injection of a biological fluid onto a column. In the latter case, saliva, urine, and serum samples have been directly injected in a system operated with an aqueous mobile phase (reversed phase, ion pair, ion exchange) that is compatible with the sample composition. However, repeated direct injections of serum sample usually shorten the life of a column due to the precipitation of plasma proteins on the packing.

Thus, as an alternative, the use of precolumn can solve the problem. The precolumns are from 3 to 5 cm long, often with the same diameter and packing material as the analytical columns. They must be replaced after a number of injections, when the separation starts to deteriorate. Naturally, the direct injection of sample cannot be applied in the case of blood, tissues, and the like.

A precolumn was also used in a more promising system, the so-called precolumn venting technique (32) shown in Figure 7.1. A three-port valve is inserted between precolumn and analytical column in order to effect a preseparation of the analyte from plasma proteins. Endogenous components of plasma are then eluted to waste. With sample volumes of about 10 μl, the precolumn can stand around 100 plasma samples. However, the concentration of organic modifier in the mobile phase should be kept as low as possible in order to solubilize proteins during chromatography.

TABLE 7.2 Preparation of Biological Materials[a]

Operation	Preparative procedure					
	1	2	3	4	5	6
Protein precipitation				x	x	x
Liquid-liquid extraction			x			x
Solid-phase extraction		x			x	
Simple treatment	x			x		
Type of sample	U,P	U,P,B	U,P,B	P,B	P,B	P,B

[a]U, urine, saliva; P, plasma, serum, cerebrospinal fluid, B, blood, tissues, feces, organs, etc.

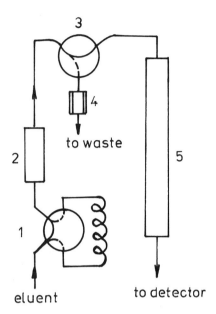

FIGURE 7.1 Precolumn venting technique. 1, sample injector; 2, precolumn; 3, three-port valve; 4, pressure regulator; 5, analytical column. (After Ref. 32.)

This limitation has been removed in the improved system, the so-called precolumn venting plug technique (33), shown in Figure 7.2. Another three-port valve, with a large loop containing 2 ml of buffer, has been inserted in this system. The sample is injected in the center of the buffer plug coming from the plug fluid injector. An acidic (pH 2.0) phosphate buffer was found to be the best fluid for this purpose. In this case plasma proteins will not come in contact with organic modifier and the precolumn can handle more than 20 ml in total volume of plasma (34).

Another approach to direct injection of plasma onto a column in liquid chromatography is by use of a micellar mobile phase, where the partitioning of a solute between a stationary phase, a mobile aqueous phase, and a micellar phase takes place. As surfactants sodium dodecyl sulfate or nonionic Brij 35 are applied, which have the ability to solvate proteins in aqueous solutions. Usually a small amount of organic modifier is added to the mobile phase in order to improve the efficiency. For the stationary phase silica gels modified with CN or C18 groups are used. Cline Love et al. (35) used Supelcosil CN eluted with 0.08 M aqueous Brij 35 with 10% of propanol for the determination quinine and quinidine in serum.

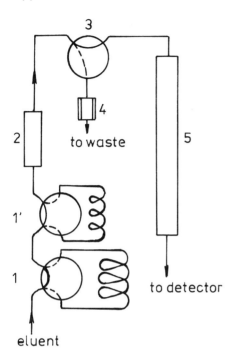

FIGURE 7.2 Precolumn venting plug technique. 1, plug fluid injector (loop 2 ml); 1', sample injector; 2, precolumn; 3, three-port valve; 4, pressure regulator; 5, analytical column. (After Ref. 33.)

There are other possibilities for injecting plasma directly onto a LC column without risking its deterioration. The problem has been reviewed by Westerlund (36).

Preparation procedure 2 involves a single operation—solid-phase extraction—which was mentioned in Section I.B of this chapter. This operation is gaining popularity and in sample preparation competes successfully with liquid-liquid extraction. Body fluids can be applied on the cartridge either directly or after dilution, while tissues, organs, and the like require a simple pretreatment. They are homogenized with distilled water added (sometimes with pH adjusted), filtered or centrifuged, and the filtrates or the supernatants are loaded onto a column. If necessary, the sample clean-up procedure can be repeated on a second disposable column. Solid phase extraction is the universal procedure for preparation of any biological sample. It may also be applied as the follow-up operation after protein removal (procedure 5). A special packing material has been developed for biological sample preparations, consisting of octadecyl groups bonded on silica gel which are coated with protein. The packing preparation is described elsewhere (37).

Preparation procedure 3—liquid-liquid extraction—is still the most frequently used technique, allowing any type of biological sample to be prepared. The aqueous biological samples are shaken with water-immiscible organic solvents, directly in the case of urine, plasma, and blood, or after homogenization in the case of tissues and organs. Due to the low solubility of ionized forms in organic solvent, the alkaloid to be extracted must first be transferred to the nonionized form. This is done by adjusting the pH to a value that allows for quantitative extraction. In most cases the formula $pH = pK_a + 2$ for basic compounds proved to be satisfactory (see Table 2.1). If the drug recovery is low the extraction should be repeated two or three times. Sometimes the preextraction procedure is used in order to remove undesirable components. In this case the samples are acidified and extracted with an organic solvent, the layers are separated, and the organic layer is discarded. Then the sample is made alkaline and the alkaloid is extracted in the usual way.

When the sample is complex, containing compounds that could interfere with an analysis, one single extraction step might not be sufficient. In these cases, the back extraction of the alkaloid from organic solution into an acidified aqueous phase can be used. This phase can be injected onto a column—naturally, when the applied mobile phase is compatible with it. When the sample is to be separated by normal phase chromatography, another extraction step must be carried out to obtain the solution of the analyte in an organic solvent.

The recovery of extraction step can be increased by salting out. This procedure requires the addition of an excessive amount of sodium chloride to the aqueous solution containing the analyte. This is especially so in the case of the extraction of compounds which are essentially neutral and which can only be slightly influenced by pH adjusting.

Among the organic solvents used for the extraction of alkaloids from aqueous solutions, the mixtures of alcohol-chloroalkane or alcohol-hydrocarbon in ratios from 10:90 to 5:95 are the most suitable ones. As a typical example, isopropanol-chloroform 5:95 could be mentioned here, but isobutanol-chloroform, isobutanol-cyclohexane, and n-butanol-toluene have also been applied. From other organic solvents, diethyl ether, ethyl acetate, n-butanol, and others have been used for extraction. In some cases ion pair extraction has proven applicability for the preconcentration of alkaloids from biological samples. To understand the mechanism of separation and for the selection of suitable counterion reagents, the reader is referred to Chapter 5, Section IV. One of the advantages of liquid-

liquid extraction is the ease of analyte preconcentration. The organic solution can be evaporated to dryness, usually in a gentle stream of nitrogen, and the residue reconstituted in a small known volume of the mobile phase or other solvent compatible with the used chromatographic technique.

Preparation procedure 4—protein precipitation—involves protein removal as the main clean-up step. This operation is used extensively when drugs are not protein-bound. The precipitation is followed by centrifugation, denoted as a "simple treatment" in Table 7.2. The proteins can be precipitated by several methods. The addition of water-miscible solvents (e.g., acetonitrile, methanol) is a frequently applied method. The precipitation can be also carried out by the addition of strong acids, such as trichloroacetic acid or perchloric acid. Among the other methods, the salting-out technique using ammonium sulfate or some other salt has been used with success. In order to minimize alkaloid coprecipitation, dilution of the sample with water in a ratio of approximately 1:3 is recommended. The precipitated proteins have to be separated from the mixture. In practice, the only technique used for protein removal is centrifugation. The clear supernatant can then be injected onto a column. Preparation procedure 4 has been elaborated mainly for blood sample clean-up but may be also used for plasma samples. The determination of recovery rates of this procedure is quite inevitable, and is usually accomplished with a blank blood sample spiked with the known amount of the drug to be determined. When the recoveries are poor, another clean-up procedure should be chosen.

Preparation procedure 5 involves the combination of two operations already discussed: protein precipitation followed by solid phase extraction. A precipitation reagent is added to the sample, the mixture is diluted with water, and the precipitated proteins are separated by centrifugation. The supernatant is injected onto a disposable cartridge for further purification. From the beginning endogenous components are eluted from the column and then, using a second solvent, the analyte is eluted. This preparation procedure may be employed for the wide variety of biological samples: plasma, serum, blood, tissues, feces, organs, etc. Tissues and similar types of samples have first of all to be homogenized with water as mentioned previously.

Preparation procedure 6 uses protein precipitation followed by centrifugation and supernatant extraction using a water-immiscible solvent. The range of sample types treated by this procedure is the same as that in the preceding case.

It should be stated at the end of this brief description of procedures used for the preparation of biological samples that in practice many other procedures and modifications have also been tried, but in the main the most important and proven methods are those presented here. Tables 7.3–7.5 give details of actual experiments.

TABLE 7.3 Preparation of Plasma Samples

Alkaloid determined	Preparation procedure	Analytical method and remarks	Ref.
Nicotine and metabolites	Sample (1 ml) + 1 ml of acetonitrile mixed, centrifuged, supernatant injected	LC	38
Cocaine, benzoylecgonine	Sample (1 ml) + 0.1 ml of 5 mM H_2SO_4 contg. I.S. (m-toluylecgonine) vortexed, + 0.5 ml of 1 M K_2CO_3, pH 8.5, saturated with NaCl, extracted with 5 ml of $CHCl_3$-isopropyl alcohol (1:1), vortexed, centrifuged at 0°C, org. layer evaporated (60°C). Residue + 0.1 ml of oxalyl chloride stands for 1 min, 1 ml of n-butanol added and heated at 95°C for 20 min. Sample alkalinized, 2 ml of toluene-t-amylalcohol (9:1) added, vortexed, centrifuged, and aq. layer frozen. Toluene layer separated, 0.5 ml of 0.5 M H_2SO_4 added, vortexed, centrifuged, and frozen as above. Aqueous layer washed with 2 ml of toluene-t-amylalcohol (9:1), 1 ml of 2 M K_2CO_3 + 0.5 ml of butyl acetate added, vortexed, centrifuged, and frozen. Butyl acetate layer injected	GC Extraction and conversion to butyl ester derivatives	39
Quinidine and metabolites	Filtered (0.45 μm) sample loaded on 6 × 0.4 cm col. packed with octadecyl-bonded silica gel coated with protein (37). Proteins eluted within 7.5 ml with phosphate buffered saline, pH 7.4, and alkaloids eluted in back-flush mode	LC Applicable also to urine	40

Applications 329

Quinine, quinidine	Sample (1 ml) + 0.15 ml of I.S. (monoethylglycin-exylidide) solution + 1.0 ml of NH_4OH extracted with hexane-Et acetate (9:1), centrifuged, org. layer evaporated (35°C), residue reconstituted in 0.1 ml of mob. phase, 40 μl injected	LC	41
Morphine	Sample (from 0.1 to 0.4 ml) + 0.4 ml of borate buffer, pH 8.9, + I.S. (nalorphine) solution applied to Clin-Elute CE-1001 col. and eluted with 5 ml of $CHCl_3$-isopropanol (95:5). Eluate evaporated (55°C), residue reconstituted in 0.2 ml of mob. phase, 10–100 μl injected	LC Recovery 89.8 ± 4.0% Applicable also to cerebrospinal fluid	42
Morphine	Sample (2 ml) + 2 ml of borate buffer (0.05 M H_3BO_3 + 0.043 M $Na_2B_4O_7$) extracted with $CHCl_3$-isobutanol (9:1) by shaking for 15 min. Centrifuged, layers separated, org. layer washed 2x with 5 ml of phosphate buffer, pH 9.8, and extracted with 5 ml of 0.5 M HCl. Centrifuged, layers separated, 4.5 ml of aq. phase transferred to test tube, pH adjusted to 8.7 and extracted with Et acetate-isopropanol (9:1). Layers separated, org. layer evaporated, residue reconstituted in 0.1 ml of MeOH, 20–50 μl injected	LC Recovery 79 ± 6.5% Applicable also to serum	43
Morphine	Sample (1 ml) + 3 ml of 0.5 M $(NH_4)_2SO_4$, pH adjusted to 9.3 with NH_4OH loaded on Sep-Pak C18 col., washed with 20 ml of 5 mM $(NH_4)_2SO_4$, pH 9.3, and 0.5 ml H_2O. Drug eluted with 5 ml of acetonitrile-0.01 M phosphate buffer, pH 2.1 (1:9). Eluate mixed with 3 ml of 0.5 M $(NH_4)_2SO_4$ and purified in the same way on second Sep-Pak C18 col. Eluate (0.4 ml) injected	LC Morphine-3-glucuronide and morphine-6-glucuronide were also determined. Applicable also to urine	44

TABLE 7.3 (Continued)

Alkaloid determined	Preparation procedure	Analytical method and remarks	Ref.
Physostigmine	Sample (2 ml) + 1 ml of 0.5% sodium dodecyl sulfate loaded on Sep-Pak C18 col. prewetted with 3 ml of MeOH and 5 ml of H$_2$O. Col. washed with 2 × 5 ml of H$_2$O, 3 ml of 60% MeOH, and eluted with 1 ml of MeOH. Eluate evaporated (30°C), residue reconstituted in 100 μl of 40% acetonitrile	LC Recovery 62 ± 6%. Physostigmine is irreversibly hydrolyzed in alkaline solutions	45
Theophylline	Sample (0.1 ml) + 0.1 ml of 10% trichloroacetic acid contg. I.S. (β-hydroxyethyltheophylline) mixed, centrifuged, 50 μl of supernatant injected	LC	46
Caffeine, theophylline	Sample (0.1 ml) + 0.05 ml of acetonitrile/I.S. (β-hydroxypropyltheophylline) mixture vortexed, centrifuged, and 10 μl of supernatant injected	LC Recovery 104%	47
Caffeine	Sample (1 ml) + 0.2 ml of I.S. (bupivicaine) solution + 60 μl of 12.5 M NaOH extracted with CHCl$_3$-isopropanol (95:5) by shaking for 45 min, aq. layer frozen, both layers centrifuged, org. layer separated and evaporated, residue reconstituted in 0.5 ml of Et acetate, and 2 μl injected	GC Applicable also to breast milk. Recoveries 99.7 and 94.1% for plasma and milk, resp.	48
Purine bases, theophylline, 1,7-dimethylxanthine, caffeine, theobromine	Sample (0.1 ml) extracted with 1.0 ml of CHCl$_3$-isopropanol 95:5 contg. I.S. (β-hydroxyethyltheophylline) by vortex mixing, centrifuged, org. layer evaporated, residue reconstituted in 0.2 ml of mob. phase, 40 μl injected	LC	49

Applications

Compound	Procedure	Method	Ref.
Theophylline	Sample (0.2 ml) + 50 μl of I.S. (3-isobutyl-1-methylxanthine) solution + 0.2 ml of 0.1 M acetate buffer, pH 4.6, extracted with 0.8 ml of $CHCl_3$ by shaking. Centrifuged, org. layer evaporated (57°C), residue reconstituted in 60 μl of CH_2Cl_2, 20 μl injected	LC	50
Caffeine and metabolites	Sample (0.5 ml) + 0.2 ml of 0.01 M HCl extracted with 4.0 ml of Et acetate-isoamylalcohol (98:2) and centrifuged. For LC 3.0 ml of organic layer evaporated, residue reconstituted in 0.3 ml of mob. phase, 15 μl injected. For GC 1.0 ml of org. layer evaporated, residue reconstituted in 0.2 ml of toluene-isoamylalcohol-asolectin (83:14:3), 6 μl injected	GC, LC	51
Theophylline	Sample (0.1 ml) + I.S. (3-isobutyl-1-methylxanthine) solution extracted with 0.5 ml of hexane, centrifuged, layers separated, aq. layer + 50 μl of 1 M NaOH + 50 μl of tetrabutylammonium hydrogen sulfate extracted with 1 ml of CH_2Cl_2 + 0.01 ml of pentafluorobenzyl bromide. Vortexed, centrifuged, layers separated, org. layer washed with 0.4 ml H_2O, evaporated (25°C), residue reconstituted in 50 μl of Et acetate, 10 μl injected	GC	46
Theophylline, 1,7-dimethylxanthine	Sample (1 ml) + 0.05 ml of I.S. (β-hydroxyethyl-theophylline) solution + 2 ml of charcoal suspension (0.5 g of charcoal in 250 ml H_2O) vortexed, centrifuged, supernatant discarded. Residue extracted with isopropanol-CH_2Cl_2-Et_2O (10:65:25), centrifuged, org. layer evaporated (37°C), residue reconstituted in 0.5 ml of mob. phase, 10 μl injected	LC	52

TABLE 7.3 (Continued)

Alkaloid determined	Preparation procedure	Analytical method and remarks	Ref.
Galanthamine and metabolites	Sample (2 ml) + 2 ml of 20% trichloroacetic acid vortexed, centrifuged, 3 ml of supernatant + 0.5 ml of 5 M NaOH to pH 11.0, extracted with 5 ml of $CHCl_3$ contg. I.S. (codeine), centrifuged, 4.5-ml aliquot filtered through Na_2SO_4, filter washed with 1 ml of $CHCl_3$, filtrate evaporated (62°C), residue reconstituted in 0.2 ml of mob. phase, 100 µl injected	LC Applicable also to urine, but without precipitation step	53
Vinpocetine	Sample (1 ml) + 1 ml of 0.2 M glycine, pH 11.0, +10 µl of I.S. (apovincamine) solution + 0.1 ml of 1% aq. solution of NaF extracted with 5 ml of Et_2O. Layers separated, org. layer washed with 1 ml of 0.1 M HCl, aq. layer + 1 ml of 0.2 M glycine extracted with 2 ml of Et_2O. Org. phases combined, evaporated (25°C), residue reconstituted in 0.5 ml of mob. phase, 100 µl injected	LC Recovery 71.2 ± 2.8%	10
Vinblastine, vincristine	Sample (1 ml) + 10 µl of 1.8 M H_2SO_4 mixed and centrifuged. Supernatant (1 ml) loaded on Bond-Elute Diol col., prewetted with 5 ml MeOH and 0.05 M tetramethylammonium bromide in 0.09 M H_2SO_4. Column washed with 2 ml of 0.025 M phosphate buffer, pH 8.5, 5 ml of MeOH–H_2O (1:9), col. dried by air and drugs eluted with 0.5 ml of MeOH. Eluate evaporated, residue reconstituted in 20–100 µl of mob. phase, 10–50 µl injected	LC Applicable also to urine	54

Ergot alkaloids	Sample (3 ml) + 0.1 ml of I.S. (methysergide) solution + 3 ml of 1 M NH$_4$Cl, pH 9.0, extracted with cyclohexane-butanol (9:1) by shaking for 15 min. Centrifuged, layers separated, org. layer reextracted with 2 ml of 0.05 M H$_2$SO$_4$ by shaking for 10 min. Layers separated, aq. layer + 2 ml of 1 M NH$_4$Cl, pH 9.2, extracted with 10 ml of cyclohexane-butanol (9:1). Org. layer separated, evaporated (50°C), residue reconstituted in 0.2 ml of mob. phase, 50–150 μl injected	LC Recoveries from 50 to 99%	55
Dihydroergotamine	Sample (3 ml) + 0.1 ml of I.S. (dihydroergocornine methanesulfonate) solution + 0.3 ml of NH$_4$OH (25%) vortexed, centrifuged, 3 ml of supernatant loaded on dry Extrelut col. After 15 min drugs eluted with 15 ml of Et$_2$O, eluate evaporated, residue reconstituted in 0.2 ml of 0.01 M ammonium carbamate-acetonitrile (1:1), sonicated, centrifuged, 100 μl of supernatant injected	LC	56

TABLE 7.4 Preparation of Whole-Blood Samples

Alkaloid determined	Preparation procedure	Analytical method and remarks	Ref.
Morphine	Blood (10 ml) digested with 1 ml of 25% HCl at 100°C for 30 min, pH to 9.0 with ammonia, extracted with CHCl$_3$-isopropanol (9:1), org. phase separated and extracted with 0.1 M H$_2$SO$_4$. Layers separated, aq. layer made alkaline, extracted with CHCl$_3$-isopropanol (9:1). Layers separated, org. layer evaporated, residue reconstituted in 100 μl of mob. phase	LC	57
Morphine	Blood (0.1 ml) + 0.5 ml of H$_2$O, pH to 10.0 with NH$_4$OH, saturated with NaCl. Solution extracted with 1 ml of toluene-butanol (9:1) by shaking for 15 min. Centrifuged, org. layer evaporated, morphine methylated with tetramethylanilinium hydroxide	GC Extraction and derivatization	58
Codeine	Sample (0.1 ml) + 0.5 ml H$_2$O + 1 ml of 0.15 M HCl + 10 μl of I.S. (caffeine) solution extracted 2x with 5 ml of Et$_2$O. Layers separated, aq. layer + 2 ml of 0.15 M NaOH extracted 2x with Et$_2$O-hexane (7:3). Combined extracts dried, evaporated (40°C), residue reconstituted in 10 μl of MeOH, and injected	LC Applicable also to urine. Recoveries 91 and 94% for blood and urine, resp.	59
Codeine, morphine	Sample (0.1 ml) + 10 μl of I.S. (dihydromorphinone) solution + 0.1 ml of 1 M Na$_2$CO$_3$ extracted with butanol-methyl-t-butyl ether (2:98). Layers	LC Applicable also to plasma (61)	60

Applications

	separated, org. layer extracted with 0.25 ml of 0.025 M H_3PO_4, pH 2.8, aliquot of aq. phase injected		
Codeine, morphine	Sample (1.0 ml) + 50 μl of I.S. solution sonicated for 20 min, then poured on Bond-Elute C18 col. (vol. of 1 ml) prewetted with MeOH and carbonate buffer. Column washed with 50 μl of carbonate buffer and eluted with 3 × 0.5 ml of MeOH–1% H_3PO_4–acetonitrile (5:3:2). Eluate evaporated (40°C), residue reconstituted in 100 μl of MeOH, 20 μl injected	LC Recoveries from 95 to 100%	62
Morphine free and conjugated (3-morphineglucuronide)	Total morphine: sample (0.5 ml) + 5 ml of acetate buffer, pH 5.5 + 0.05 ml of β-glucuronidase incubated at 37°C for 40 hr. After cooling 0.05 ml of 3.5 M NaOH and 4 ml of borate buffer added and sample purified on Sep-Pak C18 col. Free morphine: sample (10 ml) + I.S. (nalorphine) solution adjusted to pH 9.0 with borate buffer, vortexed for 10 min, loaded on Extrelut silica. Eluted with 20 ml of $CHCl_3$-isopropanol (9:1), eluate evaporated, residue derivatized with heptafluorobutyric anhydride	GC Applicable also to plasma. For morphineglucuronide hydrolysis with enzyme see (64), in autoclave (65)	63
Theophylline, dyphylline	Sample centrifuged for 5 min at 1380g, 50 μl of supernatant mixed with 20 μl of I.S. (β-hydroxyethyltheophylline) solution and precipitated with 30 μl of 40% trichloroacetic acid. Centrifuged, 15-μl aliquot of supernatant injected	LC Recoveries from 98.9 to 99.4%	66 67

TABLE 7.5 Preparation of Urine Samples

Alkaloid determined	Preparation procedure	Analytical method and remarks	Ref.
Drugs: ephedrine, morphine, tubocurarine, etc.	Sample (5 ml) acidified to pH 2–3 (HCl), extracted with 10 ml of Et_2O, centrifuged, layers separated. Aqueous layer adjusted to pH 7.0 (NaOH) + 2 ml of 0.1 M Sorensen's phosphate buffer, pH 7.0, extracted with 10 ml of CH_2Cl_2 contg. 0.01 M bis(2-ethylhexyl)phosphate. Centrifuged, org. layer evaporated, residue reconstituted in 0.1 ml of MeOH–1 M HCl (7:3). Centrifuged and 5 μl of supernatant applied to TLC plate	TLC	68
Nicotine and metabolites	Sample filtered through a 0.22-μm filter	LC	38
Nicotine and metabolites	Sample (0.5 ml) + 0.2 ml of 4 M acetate buffer, pH 4.7 + 0.1 ml of 1.5 M KCN + 0.1 ml of 0.4 M chloramine T + 0.5 ml of 50 mM diethylthiobarbituric acid, mixed, incubated for 15 min, 5 μl injected	LC Sample derivatization	69
Opiates: morphine, codeine, nalorphine, pholcodine, etc.	Sample (1.0 ml) + 1 ml of buffer A (1 M NH_4OH, pH 9.0 adjusted with HCl), extracted with 8 ml of CH_2Cl_2-butanol (95:5), org. phase separated and extracted with 1 ml of 0.05 M H_2SO_4. Layers separated, aq. layer + 1 ml of buffer B (5 M NH_4OH added to buffer A until a mixture with an equal volume of 0.05 M H_2SO_4 gave pH 9.0) extracted with 8 ml of CH_2Cl_2-butanol (95:5). Organic layer separated, evaporated, residue + 80 μl of bis(trimethylsilyl)trifluoroacetamide-acetonitrile (1:2) or	GC Sample derivatization. Recoveries from 58% (monoacetylnalorphine) to 86% (codeine)	70

	80 μl of pentafluoropropionic anhydride heated at 60°C for 15 min. Evaporated, residue reconstituted in 0.1 ml of butyl acetate, 1–2 μl injected	
Theophylline, 1,7-dimethylxanthine	Sample (1.0 ml) + 0.1 ml of 1 M phosphate buffer, pH 7.4 + I.S. (β-hydroxyethyltheophylline) solution extracted with 5.0 ml of isopropanol-CH_2Cl_2-Et_2O (10:65:25), centrifuged, org. layer separated, evaporated, residue reconstituted in 0.5 ml of mob. phase, 3 μl injected	LC 52
Theophylline and metabolites	Sample diluted to creatinine concentration of 0.2 g liter^{-1}, spiked with I.S. (8-chlorotheophylline), centrifuged, 6–20 μl of supernatant injected	LC 71
Dihydroergotamine	Sample (5 ml) + 0.1 ml of I.S. (dihydroergocornine methanesulfonate) solution + 0.4 ml of NH_4OH (25%) vortexed, centrifuged, 5 ml of supernatant loaded on Sep-Pak C18 col. prewetted with 5 ml of acetonitrile and 5 ml of 0.01 M NH_4 carbamate. Column washed with 1.5 ml of 0.01 M NH_4 carbamate-acetonitrile (55:45) and eluted with 2 ml of MeOH-acetonitrile (1:9). Eluate evaporated, residue reconstituted in 0.2 ml of 0.01 M NH_4 carbamate-acetonitrile (1:1), vortexed, centrifuged, 100 μl of supernatant injected	LC 56

The most difficult problems are associated with samples of tissues and organs, which usually require a complex preparation procedure. This can be demonstrated on histamine and N-methylhistamine assay in rat brain, as described by Tsuruta et al. (72). A sample of rat brain (100–500 mg) was homogenized with 4.0 ml of 0.4 M perchloric acid, centrifuged, and the supernatant was transferred to a test tube containing 0.5 ml of 5 M sodium hydroxide, 10 ml of n-butanol, and 1.5 g of sodium chloride. After the extraction, the n-butanol layer was separated by centrifugation. To this layer 1 ml of 0.1 M hydrochloric acid and 15 ml of benzene were added, the mixture was shaken and centrifuged, and the aqueous layer (0.4–0.9 ml) was treated in the same manner as urine samples. This means that the solution was mixed with 0.25 ml of 5 M sodium hydroxide, 0.4 g sodium chloride, and 4.5 ml of n-butanol and extracted by shaking for 5 min. After centrifugation the butanol layer (4 ml) was separated and back-extracted with 0.5 ml of 0.1 M hydrochloric acid in the presence of 4 ml of benzene. The organic layer was discarded and to the aqueous layer 2.5 ml of 0.01 M phosphate buffer pH 6.0 was added. This solution was applied to Cellex-P column, a cation exchanger in phosphate buffer cycle, pH 6.0. Histamine and its methyl derivatives were eluted with 0.12 M hydrochloric acid. The eluate pH was adjusted to 8–10 and the solution evaporated to dryness. The residue was dissolved in 0.2 ml of the reaction buffer in order to convert histamine to the fluorescing products (see Chapter 5, Section V.C).

It can be concluded that of the methods introduced here, liquid-liquid extraction occupies the most important position for the preparation of biological samples. Often this method is combined with another procedure, especially protein precipitation or solid phase extraction. Even though the number of introduced procedures is limited, the described methods may be applied either directly or with minor modifications for the assay of most of the alkaloids in biological samples.

D. Automated Sample Preparation

Injection techniques in gas and liquid chromatography have developed significantly in recent years and microprocessor-controlled facilities for automatic injection are now available to chromatographers. Automatic sampler systems in gas chromatography usually employ a simple low dead volume syringe providing unattended injection of a variable sample volume (1–10 µl) and the flushing of the system after injection. On the other hand, in liquid chromatography the sample valves are used regularly. They can be operated either in a syringe loading valve mode for low volumes to be injected (1–10 µl), or in a loop-fill mode, where a sample volume of a few hundred microliters is required. The injection system is usually washed through with solvent between two injections. In both GC and LC automatic samplers the interface module allows for the control of all injection sequences. The samples are loaded into small glass vials, some of them low volume, allowing the injection of, say, 1 µl from the total volume of 10 µl. With the automatic injection system and the computing integrator and data station, it is possible to carry out about 100 analyses in an entirely unattended regime.

It can be seen that the analysis and data treatment steps can nowadays be highly automated, but sample preparation is mostly done manually. For laboratories dealing with a large number of samples, the requirement of a higher productivity without increasing the technical staff arises and from this standpoint sample preparation is the weak element. The only solution of this problem is automated sample preparation via robotics. In such cases, where the sample preparation is less complex, the procedure is more amenable to automation.

The simplest preparation procedure is considered to be sample dilution followed by filtration or centrifugation. This procedure is often applied to urine samples but is seldom fully automated. In the majority of cases the samples are prepared manually and placed in autosampler vials. Very often a selective detector is utilized which screens out the endogenous compounds, leaving only the analyte detectable.

Solid phase extraction makes it possible to prepare the samples in a procedure which, in combination with the autosampler, can be easily automated. Clarke and Robinson (73) reported on equipment for the preparation of urine samples (Figure 7.3). The column for solid phase extraction with silica gel ODS packing is mounted in the loop of the six-port valve. The sample is introduced onto the clean-up column either by autosampler or by manual injection, and the column is then washed with water or buffer. Rotating of the valve brings the mobile phase from the pump to the opposite end of the clean-up column and the purified sample is back-flushed onto the analytical column. By the use of an air-activated six-port valve and autosampler, both controlled by microprocessor, the equipment can be operated as a fully automated system. It is necessary only to change the clean-up column on a regular basis.

A modification of this system is the use of two six-port valves with two clean-up columns. In Figure 7.4 the clean-up column 2 is loaded with sample and flushed with water, while column 1 is back-flushed with the mobile phase. According to Clarke and Robinson (73), no loss of analytical performance has been found as a consequence of the essentially pluglike injection of the drug from back-flushed column and drug recoveries exceed 95%.

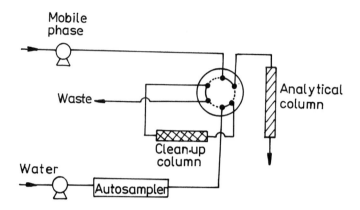

FIGURE 7.3 Automated sample preparation using solid phase extraction with single clean-up columns. (After Ref. 73.)

If the solid samples are to be prepared or a more complex sample preparation is to be carried out, the automation calls for the application of more sophisticated equipment. In these cases the use of laboratory robots can solve the problem. The robots, currently marketed by Zymark or Perkin-Elmer, consist of a robot arm with interchangeable hands which is controlled by a microcomputer and of a set of laboratory peripherals arranged beside the robot arm. These peripherals may consist of an electronic balance, a syringe or pipetting system, a vortex mixer, a centrifuge, a HPLC injec-

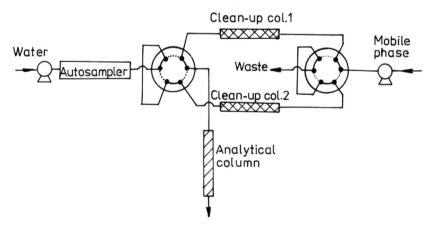

FIGURE 7.4 Automated sample preparation using solid phase extraction with two clean-up columns. (After Ref. 73.)

Applications

tor station, etc. (74). Using a robot, the prepared samples can be interfaced to a liquid chromatograph in two ways: (a) direct injection by robot hand, (b) filling of vials (by robot hand) with utilization of an existing autosampler. The success of robotics in sample preparation depends on the ability of the robot to interact with laboratory peripherals.

Automated sample preparations have been applied to the determination of content uniformity of pharmaceutical formulations, the analysis of alkaloids in food, plasma, etc. by means of a robotic system (75). The content uniformity test of tablets has been carried out as follows: transfer tablet to a tared tube, weigh tablet, add solvent, cap tube, place tube in an ultrasonic bath and then a vortex mixer, take aliquot and dilute, filter and place in autosampler vial. Hurst and Martin (76) reported the assay of caffeine and theobromine in cocoa with robot-prepared samples. The method consisted of weighing, diluting to a required volume, mixing, heating, filtering and injecting into an HPLC. The coefficients of variation were 3.47 and 3.18% for theobromine and caffeine, respectively. Myers et al. (77) described the assays of theophylline and tolazamide in serum. The procedure of sample preparation included preconditioning of a solid phase extraction column, loading of sample and internal standard onto column, elution of the compounds of interest, and placement of eluate in an autosampler vial. The average within-day coefficient of variation was 4.25%.

The use of robotics in sample preparation results in time savings, greater sample throughput, and improved precision of analysis. In this way laboratory robots represent a substantial gain in productivity.

II. SAMPLE SEPARATION

Once alkaloids have been extracted from the sample and a concentrate of sufficient purity has been prepared, various chromatographic procedures can be used for the analysis. The choice of separation system is partially affected by the type of sample because residual contaminants present in the purified samples can interfere with certain chromatographic techniques, whereas in another system these components will be well separated or suppressed by the use of selective detection. Nevertheless, a separation system originally developed for the analysis of, say, pharmaceutical products will probably operate quite well in the analysis of plant materials, cell cultures, biological materials, etc.

The decisive criterion in the choice of separation system is the type of alkaloid to be analyzed. This section summarizes some chromatographic systems which have been used in the analysis of alkaloids. It is subdivided according to type of alkaloids and follows the classification given in Chapter 1. Because extensive reviews of this type can be found in the older literature (9,78-80), only recent data are included here. Hopefully, they give in most cases sufficient insight into the range of available methods and enable comparison of their advantages and drawbacks. Besides column liquid chromatography, which is of primary importance in the determination of alkaloids, thin-layer, gas, and other chromatographic systems are included. Separation conditions are arranged in Tables 7.7-7.23. A list of abbreviations used in these tables is given in Table 7.6. The reader is also referred to Chapters 4-6, where the separation of alkaloids is discussed from the perspective of respective chromatographic techniques and where other examples are given.

A. Separation of Phenylethylamine Derivatives

A RP-LC system was developed for the rapid screening and quantitation of β-phenylethylamines in cactus tissues (81).

Among alkaloids of phenylethylamine-type, ephedrine is one of the pharmacologically most important substances. Along with LC and TLC, GC is appropriate for its separation. In a compilation of retention indices of substances of toxicological interest, the mean value 1363 was reported for ephedrine on nonpolar stationary phases SE-30 and OV-1 (82). This means that ephedrine can be chromatographed at a column temperature of about 150°C or lower. Ephedrine, if present as an adulterant in street drugs, can be identified by capillary GC with SE-54 (83,84) or OV-215 (83) stationary phases (see Figure 4.1, Table 7.15). The separation of ephedrine enantiomers was achieved on the polymeric chiral stationary phase XE-60-L-valine-(R)-phenylethylamide (85). Also in LC chiral stationary phases were applied to the resolution of enantiomers (86-88). Either oxazolidin-2-one of ephedrine and pseudoephedrine (87) or oxazolidins produced by the condensation of alkaloids with aromatic aldehydes (86) were separated. For preparative purposes pure enantiomers of norephedrine were recovered from separated derivatives by alkaline hydrolysis with a total recovery of 60% (88). As mentioned in Chapter 5, Section II.A, enantiomers of ephedrine were separated (89) by RP-LC as diastereomeric derivatives produced by reaction with chiral agent (Figure 5.13).

Enantiomers of underivatized ephedrine were separated on nitrilo stationary phase by the addition of chiral component to the mobile phase. The effect of enantiomeric purity of chiral counterion and of other factors on the stereoselectivity of the separation was studied (90).

Recently, thermospray coupling of LC with mass spectrometry was applied to ephedrine, norephedrine, and other ingredients in

(text continues on p. 563)

Applications

TABLE 7.6 Abbreviations Used in Tables 7.7–7.23

abs.	absolute
acid.	acidified
ACN	acetonitrile
AcOH	acetic acid
AcONa	sodium acetate
AcONH$_4$	ammonium acetate
anal.	analytical
anh.	anhydrous
approx.	approximately
BuNH$_2$	n-butylamine
BuOH	n-butyl alcohol
Bu$_2$NH	di-n-butylamine
Bu$_4$NBr	tetra-n-butylammonium bromide
Bu$_4$NHSO$_4$	tetra-n-butylammonium hydrogensulfate
Bu$_4$NOH	tetra-n-butylammonium hydroxide
Bu$_4$N phosphate	tetra-n-butylammonium phosphate
CCC	countercurrent chromatography
CH	cyclohexane
chem.	chemically
col., cols.	column, columns
conc.	concentrated, concentration
concs.	concentrations
contg.	containing
CV	coefficient of variation
C1SO$_3$H	methanesulfonic acid
C1SO$_3$Na	sodium methanesulfonate
C5SO$_3$H	1-pentanesulfonic acid
C5SO$_3$Na	sodium 1-pentanesulfonate
C6	n-hexane

TABLE 7.6 (Continued)

C6SO$_3$H	1-hexanesulfonic acid
C7	n-heptane
C7SO$_3$Na	sodium 1-heptanesulfonate
C8SO$_3$H	1-octanesulfonic acid
C8SO$_3$Na	sodium 1-octanesulfonate
C8SO$_4$Na	sodium octyl sulfate
C12SO$_4$Na	sodium dodecyl sulfate
deriv., derivs.	deirvative, derivatives
derivat.	derivatization
det.	detector
DMF	N,N-dimethylformamide
EDTA	ethylenediaminetetraacetic acid
electrochem.	electrochemical
em	emission
EtAc	ethyl acetate
EtOH	ethanol
Et$_2$NH	diethylamine
Et$_2$O	diethyl ether
Et$_3$N	triethylamine
ev.	eventually
ex	excitation
fluor.	fluorescence
i.d.	inner diameter
iPrOH	isopropyl alcohol
iPr$_2$O	isopropyl ether
I.S.	internal standard
LOD	limit of detection
LOQ	limit of quantitation
MeOH	methanol

Applications

TABLE 7.6 (Continued)

Me$_4$NBr	tetramethylammonium bromide
mob.	mobile
Na$_2$EDTA	ethylenediaminetetraacetic acid disodium salt
OPTLC	overpressured thin-layer chromatography
PC	paper chromatography
ph.	phase
pharm.	pharmaceutical
PIC B5	sodium 1-pentanesulfonate
PIC B7	sodium 1-heptanesulfonate
precol.	precolumn
prep.	preparative
PrNH$_2$	n-propylamine
rel.	relative
reten.	retention
SFC	supercritical fluid chromatography
SSCE	silver/silver chloride electrode
stat.	stationary
THF	tetrahydrofuran
Tris	tris(hydroxymethyl)aminomethane

[a]Abbreviations used in a single table only will be defined at the end of that table.

TABLE 7.7 Chromatographic Conditions for Phenylethylamine Derivatives[a]

Technique	Compounds separated[b]	Sample	Detection	Col. Stat. phase Col. temp.	Mobile phase flow rate	Remark	Ref.
GC	N,O-Bisheptafluorobutyryl derivs. of E, ψE, norE, and other pharm-		FID	18 m × 0.25 mm XE-60-L-valine-(R)-phenylethylamide, 120°C		Enantiomer separation	85
LC	Tyramine, N-methyltyramine, hordenine, 3,4-dimethoxy-β-phenethylamine, mescaline (I.S.)	Tissues of Mammillaria microcarpa and Mammillaria tetrancistra	UV 280	300 × 3.9 mm μBondapak C18	Linear gradient 5–30% (B) in 40 min. A. H$_2$O-MeOH-PIC B-7 90:10:5 B. MeOH-PIC B-7 100:5, 1.5 ml min^{-1}		81
LC	Oxazolidins of E, ψE, 4-methoxyE, halostachine		UV 254	250 × 4.6 mm Ionically bonded form (R)-N-(3,5-dinitrobenzoyl)phenylglycine, 20°C	C6-iPrOH 99.5:0.5	Enantiomer separation after condensation with various aromatic aldehydes	86
LC	2-Oxazolidones of (±)norE		UV 254	250 × 10 mm Covalently bonded (R)-	C6-(EtOH-ACN 2:1) 97:3, 6 ml	Preparative separation of enantiomers	88

Applications

LC	Oxazolidin-2-one of E, ψE, and derivs. of other pharmaceuticals		N-(3,5-dinitrobenzoyl)-phenylglycine 35°C	C_6-CH_2Cl_2-iPrOH 94:3:1 and 90:8:2 min^{-1}	Enantiomer separation	87
LC	2,3,4,6-Tetra-O-acetyl-β-D-glucopyranosyl isothiocyanate derivs. of E, ψE, norE	UV 254	150 × 4.6 mm Ultrasphere ODS, 5 μm	2.33 g liter^{-1} $NH_4H_2PO_4$-ACN 6:4 1 ml min^{-1}	See Fig. 5.13. Enantiomer separation	89
LC	E and other aminoalcohols	UV	150 × 3 mm Nucleosil CN, 25°C	21.5 mM benzoxycarbonyl-glycyl-l-proline, 0.25 mM Et_3N, 80 ppm H_2O in CH_2Cl_2 1 ml min^{-1}	Enantiomer separation	90
LC	E, norE, phenacetin, guaifenesin,	MS UV	250 × 4.6 mm RP-SCX, 5 μm	1 M ammonium trifluoroacetate, pH 2.5–	Thermospray coupling	91

TABLE 7.7 (Continued)

Technique	Compounds separated[b]	Sample	Detection	Col. Stat. phase Col. temp.	Mobile phase flow rate	Remark	Ref.
	phenyleph- rine				MeOH 75:25 0.5 ml min^{-1}		
LC	E, caffeine, theophylline, codeine, pa- paverine, and other basic, neutral, and acidic drugs		UV 254	250 × 4.6 mm LiChrosorb CN, 5 μm	MeOH-phos- phate buffer of various percentages, pH, and ionic strengths 1 ml min^{-1}	Effect of meth- anol content, pH, and ionic strength on retention was studied	92
LC	E, codeine, morphine	Syrup Solu- camphore	UV 254	250 × 4.6 mm Partisil PXS 5/25, 5 μm	0.3% (v/v) Et$_2$NH in Et$_2$O saturat- ed from 95% with H$_2$O 2 ml min^{-1}	See Fig. 7.5 With 0.4% Et$_2$NH k val- ues for 16 al- kaloids are given. CV 3.9, 2.5, and 4.2%, resp.	93
LC	E	Oily formu- lations	UV 214	300 × 4.5 mm Silica gel 60 C8	0.01 M KH$_2$PO$_4$-ACN 9:1, pH 3.0 with 50% H$_3$PO$_4$	CV 0.8%	12

Method	Analytes	Sample	Detection	Column	Mobile phase	Results	Ref.
LC	a. E b. E, theophylline, caffeine, lobeline, phenobarbital, phenacetin, aminophenazon	Anasthman tablets	UV 258	250 × 4 mm a. LiChrosorb RP-18, 30°C b. LiChrosorb RP-8, 10 μm	MeOH–1% ammonia a. 80:20 2 ml min^{-1} b. linear gradient from 0:10 to 7:3	a. Quantitation (Ret. times 3.6 and 2.93 min for E and lobeline resp., resolution 1.79) b. Qualitative analysis	94
LC	Theophylline, E, guaifenesin (I.S.), phenobarbital	a. Suspensions b. Tablets	UV 215	100 × 8 mm Rad-Pak μBondapak C18, 10 μm. Guard column 25 × 3 mm, Bondapak C18/Corasil, 37–50 μm	0.01 M KH$_2$PO$_4$ a. pH 5.7 b. pH 5.5 with 1 N NaOH–MeOH 67:33 2 ml	Recovery 97.79–103.93% b. CV < 1%. LOD 1.2 and 9.7 mg for theophylline and E, resp.	95
LC	E	Liquid animal dosing formulations	UV 210	250 × 4.6 mm Partisil SCX 10 μm	2 g NaH$_2$PO$_4$ 1 g H$_3$PO$_4$ (85%) 500 ml H$_2$O 500 ml ACN 1.5 ml min^{-1}	CV 0.5–1.4% at 0.6 and 60 mg ml^{-1}	96
LC	NorE, ψE, E, methylE	Ephedra-herbs, oriental pharmaceuticals	UV 210	150 × 4 mm TSK gel LS-410 (ODS-bonded silica, 5 μm, 50°C	H$_2$O–ACN C12SO$_4$Na–H$_3$PO$_4$ 65:35: 0.5:0.1 1 ml min^{-1}	Recovery 99.4–101.6%. CV 0.24–3.92%. LOD 20 ng	100

TABLE 7.7 (Continued)

Technique	Compounds separated[b]	Sample	Detection	Col. Stat. phase Col. temp.	Mobile phase flow rate	Remark	Ref.
LC	Derivatives of ψE, norE, other amines, and α-phenylglycinol (I.S.)	Plasma	Fluor. λ_{ex} 460–500 nm λ_{em} 500 nm	250 × 4.6 mm Alltech silica	CH and CH-Tol-BuOH-MeOH 50:47:2:1 2 ml min^{-1}	Derivatization using 7-chloro-7-nitrobenzo-2,1,3-oxadiazole. Automated clean-up on Waters Guard Pak, CN. Linear calibration 20–500 ng ml^{-1}. Rel. error 1.3–8.3%. CV 2.1–10.6%. LOD 10 ng ml^{-1}	97
LC	a,b. ψE, acebutol (I.S.) c. E, norψE, codeine, caffeine, theophylline, and other drugs	a. Plasma b. Urine	UV 205	a,c. 250 × 4.0 mm b. 125 × 4.0 mm LiChrosorb CN, 5 μm	H$_2$O-ACN-MeOH-C5SO$_3$-Na-C7SO$_3$Na 800:160:20:1:1 (w/w) 1.5 ml min^{-1}	a. Linear calibration up to 1000 ng ml^{-1}. Recovery 95%. Precision 1.1–6.8%. LOQ 25 ng ml^{-1} using 0.25 ml of plasma	98

Applications

LC	E, α-(methylaminomethyl)-benzyl alcohol (I.S.)	Plasma Urine	UV 220 nm	300 × 4.6 mm μBondapak C18, 10 μm	0.03 M C7SO3Na, pH 3.0 with 0.1 M HCl–ACN 77:23 1.5 ml min^{-1}	b. linear calibration 2.5–100 μg ml^{-1}. Recovery 95%. Precision 0.6–6.8% Clean-up using Bond Elut cols. Recovery 85–88%. CV 8.0, 7.4, and 1.8% at 20, 100, and 500 ng ml^{-1} resp. LOQ 10 ng ml^{-1} in 1 ml plasma	99
LC	a,b. Colchicine, N-desacetylcolchicine, β-lumicolchicine, γ-lumicolchicine, isocolchicine b. Colchiceine, trimethylcolchicinic acid		UV 350 nm	Rad-Pak B C18, 10 μm	a. 0.25 M KH2PO4, pH 5.6–ACN–MeOH 65: 26.25:8.75 b. 1% CuSO4, pH 4.0 with 2.5 M H2SO4–ACN-MeOH 65:26.25:8.75	Effect of pH Cu(II) concentration and percentage of ACN in the organic component of the mobile phase was studied. LOD 2–20 ng	102

TABLE 7.7 (Continued)

Tech-nique	Compounds separated[b]	Sample	Detection	Col. Stat. phase Col. temp.	Mobile phase flow rate	Remark	Ref.
LC	Colchicine, colchicoside	Colchicum autumnale seeds	UV 254 nm	500 × 3 mm silanized LiChrosorb Si-60, 30 μm	Sequential linear gradient of ACN in H_2O 10% 0–2 min 10–19% in 2–4 min 19% 4–6 min 19–37% in 6–10 min 37% after 10 min 2 ml min^{-1}	CV 3.77% and 4.63%, resp.	14
LC	Colchicine, morpholinopropylcolchicamide (I.S.)	Human plasma, urine	UV 245 nm	300 × 4.0 mm and precolumn 40 × 4 mm Micropak MCH, 10 μm	ACN-H_2O 50:50 2.0 ml min^{-1}	Linear calibration 5–100 ng ml^{-1}, resp. Recovery 78–98%. CV 1.8–8.73% at 50 and 10 ng ml^{-1}, resp.	103
TLC	norE and its 2-oxazolidone		Visualization with ninhydrin reagent	Layer 20 × 5 cm silica gel GF, 0.25 mm	MeOH-$CHCl_3$-AcOH 65:25:10	Resolution of norE and its 2-oxazolidone	88

Applications

OP-TLC	E, strychnine, and other doping agents	Scanning UV 210 nm ... and iodine vapor	Layer 20 × 20 cm with impregnated three edges silica gel 60 F$_{254}$	BuOH–CHCl$_3$–methyl ethyl ketone–AcOH–H$_2$O 25:17:8:6:4 0.85 cm min^{-1}	See Fig. 6.2 101
TLC	Colchicine	Scanning fluor. λ_{ex}365 nm λ_{em}450 nm after exposure to Et$_2$NH–EtOH 1:1 vapors	Tablets, Iphigena indica, Colchicum autumnale seeds	CHCl–MeOH 80:0.5	Linear calibration 20–100 ng 104

[a]For other examples, see Chapter 4, Section IV.A; Chapter 5, Sections I.A, II.A, II.B, IV.A, V.D; Chapter 6, Section II.A; and Tables 5.8, 6.1, and 6.2.
[b]E, ephedrine; ψE, pseudoephedrine.

TABLE 7.8 Chromatographic Conditions for Pyridine and Piperidine Alkaloids[a]

Technique	Compounds separated[b]	Sample	Detection	Col. Stat. phase Col. temp.	Mobile phase flow rate	Remark	Ref.
GC	Nic, norNic, meophytadiene, anabasine, anatabine, myosmine, 2,3'-dipyridyl, 2,4'-dipyridyl (I.S.)	Cured tobacco, tobacco leaves	FID NPD	a. 35 m × 0.25 mm Carbowax 20M 170–200°C at 2 K min^{-1} b. 15 m × 0.25 mm Superox-4 130–200°C at 4 K min^{-1}	a. Helium 20 cm sec^{-1} b. Helium 25 cm sec^{-1}	See Fig. 4.2. Myosmine and anabasine not resolved. Some NPDs failed to produce linear calibration in the range 0.05–1.0 mg ml^{-1}. CV 0.6–2.5%	105
GC	Nic and phendimetrazine (I.S.)	Cigarette tobacco extract	FID	2 m × 1/8 in. 3% Carbowax 20M and 5% KOH on Chromosorb G AW 100–120 mesh, 170°C	Nitrogen 33 ml min^{-1}	Calibration slightly curved below 80 μg ml^{-1}, linear over 80–250 μg/ml^{-1}. CV 1.9–3.8%	16
GC	Nic and 9 major volatiles	Tobacco green leaves	FID MS	60 m × 0.32 m Supelcowax 10 60°C for 1 min, 60–220°C	Helium 31 cm sec^{-1}	Headspace analysis with trapping on Tenax	107

Method	Analytes	Matrix	Detector	Column/Conditions	Carrier	Notes	Ref
GC	ISTD-nitrosopiperidine, Nic, norNic, 2,3-dipyridyl, ISTD-2,4-dipyridyl, N-nitrosonorNic, 4-(N-methyl-N-nitrosamino)-1-(3-dipyridyl)-1-butanone	Smokeless tobacco products	NPD	30 m × 0.25 mm, OV-17 100–250°C at 4 K min^{-1} at 2 K min^{-1}, 220°C for 30 min	Helium 20 cm sec^{-1}		109
GC	Nic, Cot, modoline (I.S.), N-ethylnorCot (I.S.)	Plasma	Thermo-ionic det.	3% SP 2250 DB or 10% Apiezon L + 2% KOH or 10% Carbowax 20M + 2% KOH		5–40 ng ml^{-1} CV 4.4%. Below 5 ng ml^{-1} environmental Nic limits reliable determination	110
GC	Nic, caffeine	Municipal wastewater	FID	6 ft × 1/8 in. 3% SE-30 on GasChrom Q, 80–100 mesh 50°C for 2 min 50–250°C at 5 K min^{-1}	Helium 50 ml min^{-1}	Dichloromethane extract	106

TABLE 7.8 (Continued)

Technique	Compounds separated[b]	Sample	Detection	Col. Stat. phase Col. temp.	Mobile phase flow rate	Remark	Ref.
GC	Nic, quinoline (I.S.)	Plasma	FID	1.8 m × 2 mm 10% Apiezon L + 2% KOH on Chromosorb W-AW 80, 150°C		Linear calibration up to 100 ng ml^{-1}. Recovery 83.2%. CV 10.5–4.5% at 2.9–19.1 ng ml^{-1}. LOD 0.1 ng ml^{-1}	111
GC	Nic, Cot, $\underline{N}$-ethylnorNic (I.S.), $\underline{N}$-ethylnorCot (I.S.)	Plasma	NPD	30 m × 0.32 mm Carbowax (amine-deactivated), 0.25 μm 80–190°C at 15 K min^{-1}, 190°C for 0.5 min, 190–215°C at 2.5 K min^{-1}	Helium 30 ml min^{-1}	1–100 ng ml^{-1} CV 12.7%	112
GC	a. Nic b. Cot	Urine	FID	2 m SP 2250 DB on Supelcoport 100–120 mesh a. 115°C b. 190°C		CV a. 3.8% at 300 ng ml^{-1} b. 2.73% at 360 ng ml^{-1}. LOQ 50 ng ml^{-1}	113

Method	Analytes	Matrix	Detection	Column/Conditions	Carrier gas	Performance	Ref.
GC	a,c. Nic b,d. Cot a. Methylanabasine (I.S.) b. 1-Methyl-6-(3-pyridyl)-2-piperidone (I.S.) c. Nic d_3 d. Cot-d_3	Mouse tissues (liver, brain)	a,b. NPD c,d. MS	a,b. 12 m × 0.2 mm, (deactivated with Carbowax 20M); dimethylsilicone liquid phase a. 80–150°C at 4 K min^{-1} b. 120–200°C at 4 K min^{-1} c,d. 1.8 m × 2 mm, 3% Ov-22 on Supelcoport, 80–100 mesh c. 170°C d. 240–270°C at 8 K min^{-1}	a,b. Helium c,d. Helium 12 ml min^{-1}	Linear calibration 5–1000 ng g^{-1}. Recovery 81–87%. LOD 2–3 ng g^{-1} (50 pg)	114
GC	Cot, noludar (I.S.)	Plasma	MS	25 m × 0.2 mm, SE-30 80–200°C at 30 K min^{-1}	Helium 35 cm sec^{-1}	Linear calibration 0–1000 nmol liter^{-1}. Recovery 111%. CV 6–16%. LOD 100 nmol liter^{-1}	115
GC	Derivative of trans-3'-hydroxyCot with hepta-	Plasma, urine	ECD-(^{63}Ni)	5.4 m × 2 mm 20% Silicon QF 1 on Chromosorb	Argon-methane 1:1 50 ml min^{-1}	Linear calibration 20–200 ng ml^{-1}. Recovery 75.2 % for	116

TABLE 7.8 (Continued)

Tech-nique	Compounds separated[b]	Sample	Detection	Col. Stat. phase Col. temp.	Mobile phase flow rate	Remark	Ref.
	fluorobutyric anhydride			W-HP, 100–120 mesh 240–340°C at 4 K min^{-1}, 340°C for 8 min		plasma, 88.8% for urine. LOD 1 ng ml^{-1}	
GC	Piperine, tetrahydro-piperine (I.S.)	Pepper, pepper extracts		25 m × 0.5 mm, (silylated at high temperature) OV-1 100°C increased to 250°C immediately after sample introduction	Hydrogen 4.6 ml min^{-1}	See Fig. 4.3. Linear calibration 0.027–0.435 mg. CV 2.5%	128
LC	Nic and eight analogs		UV 254	1000 × 0.25 mm, C1, 5 μm	H$_2$O-ACN 8:2 saturated with β-cyclodextrin 1.3 μl min^{-1}	Enantiomer separation. Elution time about 4 hr	117
LC	a. Nic, Cot b. Nic, 4-chloroaniline (I.S.)	Allergenic extracts of tobacco leaf	UV 254	150 × 4.6 mm Ultrasphere C8, 5 μm	Gradient of B in A A. 0.02 M NH$_4$Cl, 0.02 M NH$_4$OH, pH		118

Applications

LC	Cot, norNic, Nic, Nic-$\underline{N}$'-oxide, $\underline{N}$-methylCot, $\underline{N}$'-methylNic, $\underline{N}$-methylNic, $\underline{N},\underline{N}$'-dimethylNic	Guinea pig urine	UV 254, radioactivity	250 × 4.6 mm Partisil-10 SCX with guard col. 70 × 4 mm CSK-I (cation exchanger)	9.2–MeOH 8:2 B. MeOH a. 5–50% in 2–6 min b. 35–50% in 0–5 min 1.2 ml min^{-1} Linear gradient of B in A A. 0.3 M AcONa, pH 4.5 with AcOH–MeOH 70:30 B. 1% (v/v) Et$_3$N in A, pH 4.5, 0–100% in 12–22 min, 1 ml min^{-1}	See Fig. 5.18. Metabolism of ^{14}C-labeled Nic. Dependence of retention on pH and conc. of Et$_3$N, acetate, and MeOH was measured	119, 120, 121
LC	$\underline{N}$-MethylCot, $\underline{N}$-methylnorNic, $\underline{N}$-methylNic	Urine	UV 254, radioactivity	250 × 4.6 mm Partisil 10 SCX, 10 μm with guard col. 70 × 4 mm CSK-I	1% Et$_3$N in 0.3 M AcONa–MeOH 7:3, pH 4.5 with AcOH and then pH 5.5 with NH$_4$OH 2 ml min^{-1}		122

TABLE 7.8 (Continued)

Technique	Compounds separated[b]	Sample	Detection	Col. Stat. phase Col. temp.	Mobile phase flow rate	Remark	Ref.
LC	a,c. R-N-MethylNic b,c. N-MethylnorNic c. Cot, R-Nic-N'-oxide	Urine	UV 254, radioactivity	70 × 4 mm CSK-I in series with two cols. 250 × 4 mm Partisil 10 SCX, 10 μm	Stepped gradient of B in A A. MeOH B. 1% Et_3N in MeOH, pH 7.0 with AcOH, 0–100% after 20 min, 2 ml min^{-1}	a,b. Preparation	a. 122 b,c. 123
LC	Nic (^{14}C-labeled) and 12 metabolites	a. Plasma b. Urine	UV 254, radioactivity	a. 250 × 4.5 mm, IBM cyano RP, 5 μm b. 150 × 4.5 mm, IBM Optima cyano RP, in series with (preceding) 250 × 4.5 mm IBM cyano RP, 5 μm a,b. Guard col. 50 × 4.5 mm, IBM Opti-	Complex gradient of A and B A. H_2O–MeOH–0.1 M acetate, pH 4.0–ACN 187:5:11:1: 0.5 B. H_2O–MeOH 0.5 M acetate, pH 4.0–ACN 187.5:11:1:0.5 adjusted to pH 5.0 with	CV for Nic in urine 5.1% (within day), 9.4% (inter-day). Effect of pH, buffer, MeOH, and ACN conc. on retention was studied	38

Applications

Method	Analytes	Matrix	Detection	Column	Mobile phase	Notes	Ref.
LC	Colored deriv. of Nic, Cot, and 5 metabolites	Urine of smokers	UV 546	ma cyano RP 300 × 3.9 mm µBondapak C18	0.1% Et$_3$N, 1 or 2 ml min^{-1} 0.02 M C$_5$SO$_3$H in MeOH-H$_2$O 2:1. 2 ml min^{-1}	Precolumn derivatization. Diethylthiobarbituric acid was used as the final color reagent	69
LC	a,b. Nic, N-methylNic b. Cot, Nic-N'-oxide, 3-hydroxyCot		a. Coulometric b. UV 254	300 × 3.9 mm µBondapak C18, 5 µm	2 mM NaH$_2$PO$_4$, 0.25 mM C8SO$_4$Na- (MeOH-ACN 3:1) 92.5:7.5 pH 3.0 with H$_3$PO$_4$ 1.2 ml min^{-1}	a. Linear calibration 0.2–200 and 0.2–5000 ng. LOD 0.1 and 0.2 ng for Nic and N-methyl-Nic, resp.	124
LC	Cot, Nic, amphetamine (I.S.)	Urine	UV 260	125 × 4 mm Nucleosil Si-50, 5 µm	CH$_2$Cl$_2$-iPr$_2$O-MeOH-25% NH$_4$OH 62:30:7.9:0.1 2.0 ml min^{-1}	Linear calibration 0.1–10 µg ml^{-1}. CV 5% or less. LOD 15 ng	125
LC	Cot-N-oxide, N-methyl-N-(3-pyridyl methyl)acetamide-N-oxide (I.S.), N-	Human urine	UV 254	250 × 4.6 mm silica (Alltech), 5 µm	1.5% β-Methoxyethylamine in ACN	Linear calibration 0–2000 ng ml^{-1}. CV 10–18.4% at 100 ng ml^{-1}. LOD 10 ng	126

TABLE 7.8 (Continued)

Technique	Compounds separated[b]	Sample	Detection	Col. Stat. phase Col. temp.	Mobile phase flow rate	Remark	Ref.
	methyl-N-(3-pyridylmethyl(propionamide-N-oxide (I.S.)					ml^{-1}	
LC	a,d,e. Nic-1'-N-oxide b. Cot, Nic. Nic-1'-N-oxide Cot-N-oxide c,f. cis- and trans-Nic-1'-N-oxide	a,e. Brain, liver, blood, feces, urine of mice c,f. liver of mice	a,b,c. UV 260 or 254 d. Electrochemical e,f. Radioactivity of fractions	a,b,e. Guard col. 46 × 3.2 mm, Co:Pell PAC, 250 × 4.6 mm, Alltech NH$_2$, 10 μm c,f. Guard col. 46 × 3.2 mm, Co:Pell PAC, 250 × 4.6 mm, Partisil-10 PAC, 10 μm d. 250 × 4.6 mm, Brownlee RP-2, 10 μm	a,b,e. iPrOH-H$_2$O 75:25, 1 ml min^{-1} c,f. MeOH H$_2$O 95:5 d. Deoxygenated 1 mM EDTA in 0.1 M chloroacetic acid–iPrOH 9:1	Recoveries about 70%. e,f. Mice were injected with (^{14}C)Nic. LOD a. 20–25 ng d. 25–20 ng. LOQ a,d. 30 ng g^{-1} e. 5 ng g^{-1}	127
LC	a,b,c. Capsaicin, nor-	a,b,c. Capsicum oleo-	a,d. MS (electron	a. 250 × 4.6 mm, Spheri-	a. MeOH-H$_2$O-AcOH 70:28:2,	a,b,d. Moving belt inter-	129

Applications

	Compounds	Sample	Detection	Column	Mobile phase	Notes	Ref
	hydrocapsaicin, dihydrocapsaicin, homocapsaicin d,e. Piperine isomers, piperettin isomers, piperylin, piperanine, piperolein A, B, sitosterol, N-isobutyleicosatrienamide, N-isobutyloctadecadienamide, N-isobutyloctadienamide	resin d,e. Black pepper oleoresin	impact) b,d. MS (chem. ionization) c. Fluor. e. UV 343, MS of fractions (field desorption)	sorb ODS, 5 μm b. 150 × 4.6 mm, Hypersil ODS, 3 μm c. 150 × 4.6 mm, Hypersil ODS d. 300 × 4.6 mm, Polygosil 60-5	1 ml min^{-1} b. EtOH-H$_2$O-AcOH 50:49:1, 1 ml min^{-1} c. MeOH-0.1 M AgNO$_3$ 60:40, 2 ml min^{-1} d. C6-EtOH-AcOH 95:4:1 e. CHCl$_3$-C6 70:30, 50% saturation with H$_2$O 1.5 ml min^{-1}	face d. See Fig. 7.6	
LC	Piperine, phenazine (I.S.)	Ground pepper	UV 345	250 × 7 mm o.d. LiChrosorb RP-8, 5 μm	H$_2$O-ACN-THF 345:160:40 2 ml min^{-1}		15
LC	Piperine	Pepper (black, green, and white)	UV 345 or 280	300 × 3.9 mm μBondapak CN	MeOH-H$_2$O 50:50 2 ml min^{-1}	Nearly 8 times greater sensitivity at 345 nm compared to 280 nm	130

TABLE 7.8 (Continued)

Tech-nique	Compounds separated[b]	Sample	Detection	Col. Stat. phase Col. temp.	Mobile phase flow rate	Remark	Ref.
LC	Norhydrocapsaicin, capsaicin, piperine, dihydrocapsaicin	Chillies, black pepper	UV 280	250 × 4.6 mm Altex C18, 5 μm	0.95 g C_5SO_3Na + 160 ml H_2O + 335 ml MeOH + 5 ml AcOH 1 ml min^{-1}	For capsaicinoids recovery 98–101%. LOD 0.5 g (25 ppm in 20 μl of injected solution)	131
LC	Vanillyloctanamide, nordihydrocapsaicin, capsaicin, vanillyldecanamide, homocapsaicin, homodihydrocapsaicin	Capsicum fruit and oleoresin	UV 280	8 mm i.d., Rad-Pak C18, 5 μm	MeOH–H_2O 63:37 3.5 ml min^{-1}	Preparation of sample using C18 Sep-pak filter cartridges takes 5 min. Recovery of capsaicinoids	132

[a]For other examples, see Chapter 4, Section IV.B; Chapter 5, Sections I.A, II.B, IV.A, and V.A; and Chapter 6, Section II.B.
[b]Nic, nicotine; Cot, cotinine.

TABLE 7.9 Chromatographic Conditions for Tropane Alkaloids[a]

Technique	Compounds separated[b]	Sample	Detection	Col. Stat. phase Col. temp.	Mobile phase flow rate	Remark	Ref.
GC	Atropine, mepyramine (I.S.)	Formulations containing cholinesterase deactivators	FID	1.83 m × 4 mm, 3% OV-17 on Gas-Chrom Q, 80–100 mesh, 190°C	Nitrogen 60 ml min^{-1}	Linear calibration 0.14–1.14 mg ml^{-1}. CV 13.4–3.1%	134
GC	Ketamine (I.S.), Atropine, scopolamine	Leaf and cultured cells of <u>Atropa belladonna</u>	FID	10 m × 0.3 mm Sil-5 CB, 0.13 μm, 40–210°C at 19 K	Helium 2.5 ml min^{-1}	Linear calibration 10–100 mg liter^{-1}. Recovery 85.8–98.7%. CV 4.0% and 2.9% for atropine and scopolamine, resp. LOD 5 mg liter^{-1} for 2 μl splitless injections	135
GC	Hygrine, cuscohygrine, apoatropine, hyoscyamine, scopolamine, etc.	<u>Atropa belladonna</u> (roots, leaves, root and cell cultures)	NPD FID MS	15 m × 0.25 mm, DB-1, 70–300°C or 150–270°C at 6 K min^{-1}	Helium 0.7 bar	22 alkaloids were determined	136

TABLE 7.9 (Continued)

Technique	Compounds separated[b]	Sample	Detection	Col. Stat. phase Col. temp.	Mobile phase flow rate	Remark	Ref.
GC	a,b. C, ψC, AlloC, AlloψC c,d. Methyl esters of ψEc, Ec AlloEc, AlloψEc		FID MS	1.8 m × 2 mm a,c. 3% SP 2250-BD b,d. 2% OV-17 a. 210°C or 240°C or programmed 140°C for 10 min, 140–240°C at 10 K min^{-1} c. 140°C d. 130°C	Nitrogen 30 ml/min	AlloC and AlloψC, AlloEc methyl esters decomposed on col. Also their solutions were fairly unstable	148
GC	C Tetracosane (I.S.)	Powders, tablets		3% OV-1		Collaborative study. Recovery 98.7–103%. CV 0.89–1.16%	149
GC	C, m-toluyl-Ec (I.S.)	Plasma	NPD	1.8 m × 2 mm 3% OV-101 + 0.1% KOH on Chromosorb W 100–120 mesh isothermal 200–220°C	Nitrogen 30 ml/min	Rate of hydrolysis is similar for C and (I.S.). Automated GC. Linear calibration 0–1000 ng	150

						Ref
GC	a. C, m-toluylEc methyl ester (I.S.) a,b. BEc butyl ester, m-toluylEC (I.S.)	Plasma, urine	NPD	a. 12 m × 0.2 mm, cross-linked methylsilicone, 0.33 μm, 120–250°C at 32 K min⁻¹, 250°C for 4 min b. 6 m, 3% SP 2100 DB on Supelcoport 100–120, 250°C	Helium a. 1 ml/min b. 30 ml/min	Linear calibration 11–500 and 70–1000 ng ml⁻¹. CV 1.01% at 75 ng ml⁻¹ and 4.18% at 350 ng ml⁻¹ for C and BEc, resp. ml⁻¹. CV 2.0–7.5%	39
GC	MSTFA derivative of Ec	Urine	a. FID b. MS	a. 2 m μ 2 mm 10% OV 101 on Chromosorb W HP 80–100 mesh b. 30 m × 0.32 mm, DB-5 a,b. 150°C for 1 min, 150–300°C at 18 K min⁻¹		Recovery 77% at 150 μg ml⁻¹	151
GC	HFBA derivs. of alcohols obtained by LiAlH₄ reduc-	Illicit cocaine, coca	ECD(63Ni)	30 m × 0.25 mm, DB-1701 0.25 μm, 90°C for 1.6 min,	Hydrogen 40–45 cm sec⁻¹		152

TABLE 7.9 (Continued)

Technique	Compounds separated[b]	Sample	Detection	Col. Stat. phase Col. temp.	Mobile phase flow rate	Remark	Ref.
	tion of C, truxillines, cis- and trans- cinnamoyl C, and 30–40 other alkaloids			90–160°C at 25 K min^{-1}, 160°C for 1 min, 160–275°C at 4.0 K min^{-1}, 275°C for 5.0 min			
LC	Scopolamine, atropine		UV 254	250 × 4.6 mm LiChrosorb RP-18, 5 μm, impregnated with dodecylsulfonic acid and cetrimide	0.02 M ClSO$_3$H in H$_2$O-MeOH-dioxane-H$_2$SO$_4$ 75:23.5:1:0.5 pH 3.5 with 10 M NaOH 1.2 ml min^{-1}	Impregnation improved efficiency of col.	138
LC	Homatropine, scopolamine, methylscopolamine, tropic acid, methylatropine, atropine		UV 230	300 × 3.9 mm μBondapak C18, 10 μm	H$_2$O-MeOH-AcOH-Et$_3$N 83:15:1.5:0.5	Retention data for 150 drugs of pharm. interest in dependence on MeOH concentration	139

LC	Atropine	Tablets, injections, belladonna extract	Fluor. λ_{ex} 255 nm λ_{em} 285 nm	250 × 4.6 mm Alltech C18	1 g C5SO$_3$Na + 50 ml H$_2$O + 950 ml MeOH 1 ml min^{-1}	Recovery 97–100.7%. CV 1.46–3.35%. LOQ 0.12 mg ml^{-1}	140
LC	Scopolamine, atropine, apoatropine, ethylbenzoatropine	Belladonna tincture	Electrochemical	250 × 4 mm LiChrosorb diol, 5 µm	0.0125 M phosphate, pH 7.2–ACN 80:20 1 ml min^{-1}	LOD 10 ng of atropine	141
LC	Scopolamine, l-hyoscyamine, apoatropine, tropic acid	Tinctures belladonna, datura, yusquiame	UV 254	300 × 3.9 mm µBondapak C18, 10 µm	3% AcOH–MeOH 70:30 or 75:25	The dependence of retention on MeOH conc. was measured. CV 1.7 and 0.5% for scopolamine and l-hyoscyamine, resp.	142
LC	Atropine, scopolamine	Root, stem, and leaf of Atropa belladonna, Datura innoxia, Datura stramonium	RI	300 × 3.9 mm MikroPak Si-5 with precol. 50 × 3.9 mm, Porasil B	THF-Et$_2$NH 100:1 1/5 ml min^{-1}	Too strong signals for UV 254 detection. CV 2.5–3.3%	143
LC	Hyoscyamine, scopolamine	Leaves, fruits, and seeds of	UV 259	a. 150 × 4.6 mm, Zorbax TMS, 5 µm	0.2 M NaH$_2$PO$_4$, 0.005 M d-10-camphor-	Composition of mobile phase adjusted dur-	144

TABLE 7.9 (Continued)

Tech-nique	Compounds separated[b]	Sample	Detection	Col. Stat. phase Col. temp.	Mobile phase flow rate	Remark	Ref.
		Datura inoxia; seeds of Datura stramonium		40°C b. 125 × 4 mm, LiChrosorb RP-2, 5 μm 40°C	sulfonic acid-0.2 M H_3PO_4, 0.005 M d-10-camphor-sulfonic acid-ACN-$PrNH_2$ a. 34:60:6:0.6 b. 38:58:4:0.6 (pH 2.5 ± 0.05)	ing col. life CV 2–8%	
CCC	a. Scopolamine, hyoscyamine, norhyoscyamine b. Scopolamine, anisodine, 8-isopropylscopolamine		UV 254	a. 0.07 M phosphate, pH 6.5 b. 0.07 M phosphate, pH 6.3	$CHCl_3$	Droplet countercurrent chromatography	145
CCC		Extract of Anisodus tangulicus	UV 278	a. 0.07 M phosphate pH 6.4 b. $CHCl_3$	a. $CHCl_3$ b. 0.07 M phosphate, pH 6.4 200 ml hr^{-1}	Peaks corresponding to scopolamine and hyoscyamine were detected	146

Applications

LC	C, ψC, AlloC, AlloψC, N,N-dibenzylamide (I.S.)	250 × 4.6 mm Partisil 10 PXS	UV 230	C7-iPrOH-Et$_2$NH 75:25 0.1. Programmed flow rate 0.48–4.0 ml min^{-1} over 12 min	See Fig. 7.7	148
LC	C, Ec, BEc	250 × 4.6 mm Partisil 10 ODS-3, 10 μm 40°C	Electrochemical UV 240	ACN–0.02 M KH$_2$PO$_4$, pH 8.0 with NaOH 50:50	Linear calibration 3.1–12 nmol. CV 1.5%. LOD 2 ng for C (about 15–30 times more sensitive than UV detection)	153
LC	C and some pharmaceuticals	250 × 4.6 mm Alltech C18, 10 μm or 100 × 5 mm, Radial Pak C18, 5 μm	Post col. photolysis followed by electrochem. det.	7 ml liter^{-1} BuNH$_2$ in MeOH–H$_2$O 1:1, pH 3.2 with H$_2$SO$_4$ 1.8 ml min^{-1}	Linear calibration 0.011–20 μg. LOD 50 ppb in 50 μl injection	154
LC	C and impurities (benzoic acid, BEc, cinnamic acids, cinnamoyl cocaines, truxillic acids, truxinic acids, truxillines)	125 × 4.6 mm HS-5 C18	UV 215 and 277	Various gradients of organic components (MeOH, ACN, THF) in the buffer H$_2$O–2 mM NaOH–H$_3$PO$_4$ 3480:120:40	Illicit cocaine Spectral data were obtained by diode array detector	155

TABLE 7.9 (Continued)

Technique	Compounds separated[b]	Sample	Detection	Col. Stat. phase Col. temp.	Mobile phase flow rate	Remark	Ref.
LC	a. cis- and trans-Cinnamoyl C, BEc, C, ephedrine, pseudoephedrine, caffeine, and other amines b. Phenylisothiocyanate derivs. of ephedrine, pseudoephedrine, phenylpropanolamine	Cocaine lookalike samples	UV 254 and 280	300 × 3.9 mm μBondapak C18 preceded by 70 × 2.1 mm, Co:Pell ODS guard col.	a. 9.2 g liter^{-1} NaH$_2$PO$_4$, pH 3 with 2 N H$_3$PO$_4$–MeOH 2:1, 2 ml min^{-1} b. H$_2$O–MeOH–AcOH 54:45:1 2 ml min^{-1} pH 2.0 or in the same buffer with 0.02 m C12SO$_3$Na	Ratios A_{254}/A_{280} are reported	156
SFC	a. C b. Codeine c. Caffeine		MS UV 254	200 × 2.1 mm a,b. Spherisorb NH$_2$ 5 μm, 70°C c. Silica, 5 μm	a,b. 10% (w/w) c. 20% (w/w) MeOH in CO$_2$, 140 bar back pressure	SFC-MS coupling by a modified direct liquid interface. LOD about 20 ng	157

Applications

TLC	Hyoscyamine, scopolamine	Leaves, fruits, and seeds of Datura innoxia; seeds of D. stramonium	Layer 10 × 20 cm, silica gel 60 F 254	1,1,1-Trichloroethane–Et$_2$NH 90:10	p-Dimethylaminobenzaldehyde reagent followed by scanning at 495 nm	CV 1–7%	144
TLC	C and 14 local anesthetics		Layer 10 × 10 cm silica gel 60F 254	11 mobile phases. The best resolution was obtained for CH-benzene-Et$_2$NH 75:15:10 in the first and CHCl$_3$-MeOH 8:1 in the second direction	Scanning at different wavelengths, Dragendorff's reagent	A strategy for mobile phase selection in two-dimensional TLC was applied	160

aFor other examples, see Chapter 4, Sections III.A, III.B, IV.B; Chapter 5, Sections I.A, I.B, II.B, III.A, IV.A, V.A, V.B, V.D; and Chapter 6, Section II.B.
bC, cocaine; ψC, pseudococaine; Ec, ecgonine; ψEc, pseudoecgonine; BEc, benzoylecgonine; MSTFA, N-methyl-N-trimethylsilyltrifluoroacetamide; HFBA, heptafluorobutyric anhydride.

TABLE 7.10 Chromatographic Conditions for Quinoline Alkaloids[a]

Technique	Compounds separated[b]	Sample	Detection	Col. Stat. phase Col. temp.	Mobile phase flow rate	Remark	Ref.
GC	Dictamnine, γ-fagarine, skimmianine	LC fractions of Dictamni radicis cortex	MS	1 m × 3 mm 3% OV-17 on Chromosorb W AW DMCS 80–100 mesh 190–240°C at 5 K min^{-1}	Helium 20 ml min^{-1}	Compound identification	161
GC	a, C, Cd, Qd, Q, piperine (I.S.) b. Q c. Q, Qd	a. Cinchona bark pharm. preparation b. Soft drinks c. Pharmaceuticals		30 m × 0.3 mm, RSL-903 (highly polar aromatic sulfone), 0.15 μm, 280°C. Also OV-1, OV-17, OV-225, Superox-4, and RSL-702 stat. phases were tested.	Hydrogen 5 ml min^{-1}	See Fig. 4.4. Resolution of Q and Qd improves with increasing polarity of stat. phase. CV b. 1.97% c. 1.07% and 0.9%, resp.	128
LC	a. Dictamnine a,b. γ-fagarine a,b. Skimmia-	Dictamni radicis cortex	a. UV 244 b. Fluor. λ_{ex}333 nm λ_{em}438 nm	250 × 4.6 mm TSK gel ODS 120T	H$_2$O-ACN-MeOH 53:30:17, 1 ml min^{-1}	Linear calibration a. up to 0.5 μg	161

Applications

Method	Compounds	Sample	Detection	Column	Mobile phase	Remarks	Ref.
	nine, o-phenylphenol (I.S.)					b. up to 0.05 μg	
LC	a. Platydesminium, balfourodinium, evoxine, skimmianine b. Platydesminium c. Balfourodinium	b,c. Choisya ternata cell cultures	b. UV 313 a,c. UV 254	300 × 4 mm μBondapak C18 and pre-Column μBondapak C18/Corasil	a. Various electrolytes in aqueous MeOH, EtOH, and PrOH of various concentrations. a,b,c. 5 mM C5SO3 or C7SO3 (PIC B5 or PIC B7 reagent, resp.) in H$_2$O-EtOH 70:30, 2 ml min^{-1}	CV 1–3% at 0.044–3.033 μmol g^{-1} of lyophylized material. LOD 10 pmol	162
LC	a. Cd, C, HC+HCd, Qd, Q, HQd, HQ b. Ascorbic, salicylic, and benzoic acids, Q, HQ	b. Paired ion combinations of Q	a. UV 280 b. UV 261	300 × 3.9 mm μBondapak C18, 10 μm	H$_2$O-MeOH-AcOH-Et$_3$N 58:40:1.5:0.5 1.5 ml min^{-1}		139
LC	a,b. Q c. Q, Qd d,e. C, Cd, HC, HCd, Qd, Q, HQd, HQ		UV 220	a,c,d. 300 × 3.9 mm μBondapak C18, 10 μm b,d. 250 ×	a. 1% AcOH, pH 2.8 in H$_2$O-MeOH 75:25 or H$_2$O-ACN 85:15	d. See Fig. 5.11. e. See Fig. 5.12. Testing of columns for	165

TABLE 7.10 (Continued)

Tech-nique	Compounds separated[b]	Sample	Detection	Col. Stat. phase Col. temp.	Mobile phase flow rate	Remark	Ref.
						the separation of basic compounds. Better performance was achieved with mob. phases containing phosphate buffer instead of acetic acid	
LC	a. Qd, Q, C, HQd, HQ, HC. Usual components of Qd and epiQd, epiQ, quininone, quinitoxine	d. Injection of Qd gluconate	UV 254	a. 250 × 4.6 mm Partisil-5, 5 μm b. 300 × 4 mm μBondapak C18 c,d. 300 × 4.0 mm	4.0 mm, Li-Chrosorb RP-8 Select B, 7 μm c,e. 150 × 3.9 mm, Nova-pak C18, 4 μm c. 250 × 4.6 mm Hypersil ODS, 5 μm, Spherisorb ODS I, 5 μm, ODS II, 5 μm	a,c,d. 0.1 M phosphate, pH 3.0–ACN 85:15 b. 0.01 M phosphate pH 3.0 or 0.1 M phosphate pH 3.0, 2.5 or 2.0–MeOH 9:1 e. 7% ACN, 0.05 M hexyl-amine, pH 3.0 with H_3PO_4 1 ml min^{-1} a. Ethylene chloride-MeOH-NH_4OH 96:4:0.25 2 ml min^{-1} b. H_2O-ACN-1 M $ClSO_3H$-	Also μPorasil, Partisil 10, Zorbax-Sil columns and CH_2Cl_2-2-methoxy-ethanol-NH_4OH or	166

Applications

	b. C, Cd, HC, HCd+Qd, Q, HQd, HQ c. Cd, C, HCd, HC, Q, Qd, HQ, HQd, Quinitoxine, epiQd, epiQ, quininone, Qd, HQd d. Cd (I.S.), quininone, Qd, HQd			μBondapak Phenyl	1 M Et$_2$NH 86:10:2:2 c,d. 0.05 M NaH$_2$PO$_4$–ACN–2-methoxyethanol 70:15:15 1.0 or 0.6 ml	ethanolamine mobile phases for NP and ODS-3 column for RP separations	
LC	Theophylline (I.S.), C, Cd, Qd, Q+HQd, HQ	Pharm. dosage forms	UV 254	300 × 4 mm μBondapak C18	H$_2$O–MeOH–AcOH 75:25:1, 80:20:1 1.5 ml min^{-1}	See Fig. 7.8. Usable also Whatman ODS and Partisil ODS. QD and HQd failed to elute from Waters Rad-Pak ODS and Varian CH-10 columns	167, 168
LC	Sodium saccharin, Q, HQ, sodium benzoate	Soft drinks	UV 254	300 × 3.9 mm μBondapak C18, 10 μm	H$_2$O–MeOH–ACN–AcOH 70:20:10:1 1.5 ml min^{-1}	Direct injection after CO$_2$ elimination by sonication and ev. filtration. Lin-	169

TABLE 7.10 (Continued)

Technique	Compounds separated[b]	Sample	Detection	Col. Stat. phase Col. temp.	Mobile phase flow rate	Remark	Ref.
						ear calibration 20–120 µg ml^{-1}. Recovery 100.3%. CV 0.82 and 0.42% for Q and HQ, resp.	
LC	Q	Shampoos, hair lotions	UV 332	250 × 4.6 mm Partisil ODS	H$_2$O–0.1 M Me$_4$NBr–0.1 M KH$_2$PO$_4$–0.1 M H$_3$PO$_4$–ACN 340:100:50:10:88, pH 2.4 1 ml min^{-1}	Linear calibration up to 1 µg. Recovery 97–99%. CV 1.36–1.4% at 0.2–0.8 µg. LOD 5 ng	170
LC	C, Cd, HC, HCd, Qd, Q, HQd, epiQd+ epiQ, HQ	Cinchona bark	UV 220	250 × 4.6 mm Hypersil ODS, 5 µm with guard col.	4 or 5% ACN in 0.1 M KH$_2$PO$_4$–0.05 M hexylamine, pH 3.0 with H$_3$PO$_4$ 1 ml min^{-1}	C, Cd, Qd, Q, HQd, and HQ identified in the sample	164

Applications

LC	Quinidinone, Qd, C, HQd, Cd, Q, HCd, HQ	Cinchona bark	UV 312	250 × 4.6 mm Hypersil, 5 µm, 22°C	C6-CH$_2$Cl$_2$-MeOH-Et$_2$NH 66:31:2:0.65 1 ml min^{-1}	See Fig. 5.3. Effect of MeOH conc., temperature, and sample size was evaluated	171
LC	Q, HQ, Qd, HQd, C, HC, Cd, HCd, quinidinone, isocinchophyllanine	Cinchona bark	UV 325	200 × 4.6 mm LiChrosorb Si-60, 5 µm	a. Toluene-EtAc-Et$_2$NH 7:2:1 b. MeOH-NH$_4$OH (25%) 100:1 c. CHCl$_3$-iPrOH-Et$_2$NH 94:1:2 d. CHCl$_3$-Et$_2$NH 98:2 e. CHCL$_3$-ethanolamine 100:0.5 f. iPrO$_2$-MeOH-Et$_3$N 70:5:5 1 ml min^{-1}	k-Values are reported for a,b,e, and f. Complete separation of 4 parent cinchona alkaloids and their dihydro derivatives was not achieved	172
LC	C, Cd, Qd, HCd, Q, HQd, HQ	Cinchona bark, Cinchona ledgeriana tissue cultures	UV 254	a. 250 × 4.6 mm, Altex Ultrasphere-Si 5 µm b,c. Col.	a. Dioxane-C6-NH$_4$OH 80:20:0.21 b,c. THF-butyl chlo-	See Fig. 7.9. a. Separation of C and Cd incomplete b. HQd and Q	173

TABLE 7.10 (Continued)

Technique	Compounds separated[b]	Sample	Detection	Col. Stat. phase Col. temp.	Mobile phase flow rate	Remark	Ref.
				in series with 250 × 4.5 mm Whatman Partisil PXS 10/25	ride-NH$_4$OH b. 60:40:0.5 c. 60:40:0.25 2 ml min^{-1}	not separated	
LC	Quinamine, 3-epiquinamine, 3α,17β-cinchophylline, cinchonamine, 10-methoxy-cinchonamine, 3α,17β-cinchophylline, Qd, Q, C+Cd, HQd, HQ	Leaves of Cinchona ledgeriana, C. succirubra, and C. succirubra, C. ledgeriana	UV 280	250 × 4.5 mm LiChrosorb Si-60, 5 μm	CHCl$_3$-MeOH-NH$_4$OH (conc.) 500:7:1 1.5 ml min^{-1}	Quinine alkaloids were more retained than indole alkaloids from Cinchona leaves	174
LC	a. Q, Qd Cd, C b. Q, C	b. Cinchona succirubra leaf	UV 231	250 × 4.6 mm Spherisorb CN, 5 μm 50°C	6.8 mM H$_3$PO$_4$, pH 7.0 with 1 M NaOH-MeOH-ACN-THF a. Various ratios b. 50:28.7:	See Fig. 5.15 b. Mobile phase composition was optimized to obtain baseline separation of four major	175

	Analytes	Matrix	Detection	Column	Mobile phase	Notes	Ref.
LC	C, Cd, HC, Qd, Q, HQd, HQ	Cell cultures of Cinchona ledgeriana	UV 231	300 × 3.9 mm μBondapak C18, 35°C	17:3.3, 1.5 ml min^{-1}	Cinchona alkaloids Recovery 85–90%	176
LC	C, Cd, HC, Qd, Q, HQd, HQ	Cinchona ledgeriana cell culture	MS	μBondapak C18	H$_2$O-ACN-AcOH-THF 450:50:5:2 1.8 ml min^{-1}	Thermospray interface	177
LC	Theobromine (I.S.), Qd, caffeine, HQd	Plasma	UV 254	300 × 3.9 mm RP-C18	0.1 M AcONH$_4$-ACN-AcOH-THF 86.5:10:3:0.5, 1.5 ml min^{-1}	LOQ 0.5 mg liter^{-1}. Recovery > 96.6%	178
LC	Qd, loxapine (I.S.)	Serum	UV 254 and 330	300 × 3.9 mm μBondapak C18	0.4% AcOH-MeOH 70:30, 2 ml min^{-1}	Methaqualone and oxazepam interfere. Recovery 100%. CV 3.0% (within run), 6.0% (between runs) at 3 mg liter^{-1}	179
LC	Q	Human serum	Fluor. λ_{ex} 350 nm λ_{em} > 418 nm	250 × 4.6 mm Partisil 10 ODS 2, 10 μm and guard	5 mM C8SO$_3$H in MeOH-H$_2$O 60:40, pH 3.5 with 0.05 M H$_2$SO$_4$ in MeOH, 1.5 ml min^{-1}	Linear calibration 1.25–10 μg ml^{-1}. Recovery 99%.	180
					5 mM C5SO$_3$H in H$_2$O-ACN MeOH 4:3:3 1.5 ml min^{-1}		

TABLE 7.10 (Continued)

Tech-nique	Compounds separated[b]	Sample	Detection	Col. Stat. phase Col. temp.	Mobile phase flow rate	Remark	Ref.
				col. 30 × 4.6 mm, LiChrosorb RP-18, 10 µm		CV 4.2% (within-day), 7.7% (day-to-day). LOD 3 ng. QD interferes	
LC	Q, Qd, mon-oethylglycin-exylidide (I.S.)	Blood of pregnant sheep and fetus	UV 254	100 × 8 mm Rad-Pak µBondapak C18, 10 µm with in-line precolumn	1% Et_2NH in H_2O-ACN 91:9, pH 2.5 with H_3PO_4, 3.5 ml min^{-1}	Linear calibration 0-20 µg ml^{-1}. Recoveries 63 and 35% for Q, Qd, and I.S., resp. CV 5% at 1 µg ml^{-1}, 10% at 10 ng ml^{-1}. LOD 10 ng ml^{-1}	41
LC	10,11-Dihy-drodiol Qd N-oxide glu-curonide, 10,11-dihy-drodiol Qd glucuronide,	Urine, blood	Fluor. λ_{ex} 340 nm λ_{em} 440 nm UV 340	60 × 4 mm TSK gel 120 (ODS packing) anal. col. 5 µm, precol. coated with pro-	Stepwise elution 0.1 M citrate, pH 3.7, 3% ACN 0-15 min, 0.1 M citrate, pH 2.8, 6% ACN	Direct injection onto pre-col. and backflushing to anal. col. Linear calibration 0-0.02	40

Method	Compounds	Matrix	Detection	Column	Mobile phase	Time/flow	Remarks	Ref
	10,11-dihydrodiol Qd oxide, 10,11-Qd, 3-hydroxy Qd, O-desmethyl Qd, Qd N-oxide, Qd, HQ, 2'-quinidinone			tein, 20–30 μm		15–40 min 1.5 ml min⁻¹	mM. Recovery 98–102%. CV < 4%	
LC	a,b. Qd c. Q c,d. Qd, propranolol	Urine, serum	Fluor. a. λ_{ex}336 nm λ_{em}>370 nm b,c,d. λ_{ex}215 nm λ_{em}>300 nm	a,b. 150 × 4.6 mm, Supelcosil LC-CN, 5 μm c,d. 300 × 3.9 mm μBondapak C18, 10 μm a,b,c,d. Pre-col. 125 × 4.6 mm, silica gel 25–40 μm	a. 0.1 M C12SO4Na b. 0.05 M C12SO4Na, 10% PrOH c. 0.03 M C12SO4Na, 10% PrOH d. 0.02 M C12SO4Na 10% PrOH 1.0 ml min⁻¹		Direct injection of urine and serum samples. For urine CV 2.75 and 6.27% at 0.6 and 0.3 μg ml⁻¹ for Q and Qd, resp. LOD 0.6 ng (0.03 μg ml⁻¹)	181
TLC	Balfourodinium, isoptelefonium, ribalinium, kokusaginine, skimmianine		Visualization by fluor.	Layers of silica gel 60, alumina Whatman KC18F	11 mobile phases		R_F values reported	162
TLC	a. Platydesminium b. Balfourodinium	Choisya ternata cell cultures	Fluor. scanning λ_{ex} a. 313 nm	Layers of silica gel 60	EtAc-H2O-formic acid 10:1:1 (15–		Linear calibration up to 60 pmol. CV	163

TABLE 7.10 (Continued)

Technique	Compounds separated[b]	Sample	Detection	Col. Stat. phase Col. temp.	Mobile phase flow rate	Remark	Ref.
	odinium		b. 254 nm Filter no. 1 for emission		min development in sandwich-type chamber)	5-8%. Fluorodensitometer Shimadzu CS 920	
TLC		Cinchona ledgeriana cell cultures (LC fractions)	Visualization under long and short wavelengths, UV, iodine vapor	Layer Si 250 μm GF	CHCl$_3$-MeOH 35:10	22 or more spots partially corresponding to alkaloids and quinones were distinguished	173
TLC	Qd	Serum	Spraying with 10% H$_2$SO$_4$ and heating to 70°C for 5 min before fluor. scanning λ_{ex}364 nm λ_{em}440 nm	Layer 20 × 20 cm LK5D precoated silica gel with preadsorbent area	EtAc-EtOH-BuOH-NH$_4$OH (conc.) 56:28:4:0.5	Linear calibration 0.4-10 mg liter^{-1}. Recovery 95%. CV 5.2% (within run), 9% (between runs) at 3 mg liter^{-1}. 0.46 and 0.59 for HQ and Q, resp.	179

Applications

TLC	Disopyramide, propranolol, Qd, procainamide, N-acetylprocainamide	Serum	Scanning UV 265 or 290	Layer 10 × 10 cm, silica gel 60 (HPTLC plates) prewashed	Benzene-EtAc-MeOH-Et$_3$N 7:6:1:1 with a beaker containing 10 ml NH$_4$OH (25%) in tank	Serum extract was applied to the plate by contact spotting. Calibration 1.5–6.0 μg ml^{-1}. Recovery 90–100%. CV 7.8% and 3.9% at 1.7 and 2.1 μg ml^{-1}. LOD 20 ng per spot for Qd	182

[a] For other examples, see Chapter 4, Sections III.A, III.B, IV.C; Chapter 5, Sections I.A, II, II.A, II.B, V.A; Chapter 6, Sections I.D and II.C; and Tables 5.7 and 6.3.
[b] C, cinchonine; Cd, cinchonidine; Q, quinine; Qd, quinidine; HC, dihydrocinchonine; HCd, dihydrocinchonidine; HQ, dihydroquinine; HQd, dihydroquinidine.

TABLE 7.11 Chromatographic Conditions for Tetrahydro-, Benzyl-, Bisbenzyl-, and Phthalidisoquinoline Alkaloids and Alkaloids of Emetine and Chelidonine Type[a]

Technique	Compounds separated[b]	Sample	Detection	Col. Stat. phase Col. temp.	Mobile phase flow rate	Remark	Ref.
GC	TFAA derivatives of di-hydroxyphenylethanol, dihydroxyphenylacetic acid, norepinephrine, dopamine, salsolinol	Food and beverages	MS	2 m × 2 mm 3% OV-17 on Supelcoport 100–120 mesh 146°C for 1 min, 146–200°C at 26 K min^{-1}	Helium 30 ml min^{-1}	Deuterated compounds used as I.S. Derivatization with TFAA in in conjunction with TFE[b]. Linear calibration up to 1.5 nmol/sample. Recovery 75%. CV 4.2–5.1% at 50 pmol ml^{-1} of beer, 8.2% at 5.4 pmol	185, 186
GC	Allocryptopine	Root cells of Macleaya cordata	MS	0.5 m × 5 mm 3% Silicone GE SE-52 on Chromato base AX (AW), 60–80 mesh, 250°C	Helium 30 ml min^{-1}	Histochromatographic study. Deuterium-labeled tetrahydroberberine was incorporated in the plant	200

Applications

LC	Norepinephrine, dihydroxybenzylamine (I.S.), dihydroxyphenylethanol, dihydroxyphenylacetic acid, dopamine, salsolinol	Food and beverages	Electrochemical	250 × 4.6 mm, Ultrasphere ODS 5 μm, 35°C	75 mg liter^{-1} C8SO$_4$Na, 2 or 1 mM EDTA in 0.15 M monochloroacetate, pH 3.0 1.5 ml min^{-1}	In some matrices interfering peaks near salsolinol. Late eluting peaks up to 2 hr after injection	185, 186
LC	3,4-Dihydroxybenzylamine, dopamine, salsolinol, norcoclaurinecarboxylic acid, 3-methoxynorlaudanosolinecarboxylic acid, tetrahydropapaveroline, 3',4'-deoxynorlaudanosolinecarboxylic acid, higenamine		Electrochemical	250 × 4.6 mm, Ultrasil ODS, 5 μm 26°C	Linear gradient 1–7.5% iProH in 1.25% (v/v) AcOH, 0.2 mM EDTA over 20 min 1.5 ml min^{-1}	Effect of AcOH and iProH concs. on retention was studied. Linear calibration 1.25–200 pmol. CV 11–27% at 1.25 pmol, 3–11% at 2.5–10 pmol	187

TABLE 7.11 (Continued)

Tech-nique	Compounds separated[b]	Sample	Detection	Col. Stat. phase Col. temp.	Mobile phase flow rate	Remark	Ref.
LC	Tetrahydro-papaveroline, 8-methyl-2,3,10,11-tetrahydro-berberine (I.S.)	Rat brain	Electrochemical	250 × 4.6 mm, Supelco-sil LC-18DB 5 μm, 29°C	0.05 M NH_4-H_2PO_4, 0.75 M Et_3N, 0.05 mM C_5SO_3Na, 4.5% (v/v) dioxane adj. to approx. pH 4.5 0.85 ml min^{-1}	Linear calibration 0–1.6 pmol. Recovery 43.4% at 0.25–3 pmol per whole brain. CV 3.5%. LOD 0.03 ng g^{-1} (0.1 pmol g^{-1})	188
LC	Papaverine, laudanosine (I.S.)	Plasma, urine	UV 239	250 × 4.6 mm, C8, 10 μm	MeOH–0.015 M sodium borate, pH 8.5, 58:42 2.7 ml min^{-1}	Linear calibration 0.05–1.0 μg ml^{-1}. Recovery 78.6–87.2% (plasma) 78.57–104.46% (urine). CV 0.8–7.5% at 0.1–10 μg ml^{-1}. LOD 25 ng ml^{-1}	189

Applications

LC	Glaucine	Glaucium flavum	RI UV	300 × 3.9 mm, μBondapak C18 precol. 40 × 3.9 mm μBondapak C18/Porasil B	0.05 M KH_2PO_4-MeOH-ACN 5:4:1 1 ml min^{-1}	Recovery 101.4%. CV 2.11%	17
LC	Glaucine, papaverine (I.S.), dehydroglaucine	Microbial cultures	UV 280	300 × 3.9 mm, μBondapak Phenyl 10 μm	0.05 M KH_2PO_4-MeOH-ACN 5:4:2 2 ml min^{-1}	See Fig. 7.10. Linear calibration up to 500 μg ml^{-1}	190
LC	a. Tetrahydrocolumbamine, glaucine, tetrahydropalmatine, corydaline b. Columbamine, coptisine, palmatine, berberine, dehydrocorydaline	Corydalis tuber., oriental pharm. preparation	a. UV 280 b. UV 345 Fluor. λ_{ex}350 nm λ_{em}520 nm	150 × 4 mm TSK gel LS-410 (ODS chem. bonded silica gel), 5 μm a. 30°C b. 40°C	a. 0.5% (w/v) C_{12}-SO_4Na in 0.05 M tartaric acid, pH 2.3–ACN 62:38 1.5 ml min^{-1} b. 0.5% (w/v) $C_{12}SO_4$Na in Britton-Robinson buffer, pH 3.0–ACN 56:34 1.0 ml min^{-1}	See Fig. 5.23. Linear calibration a. 2–12.5 μg b. 0.15–1 μg Recovery 99.8 and 100.3%. CV 2.78% and 1.91% for corydaline and dehydrocorydaline, resp.	191

TABLE 7.11 (Continued)

Technique	Compounds separated[b]	Sample	Detection	Col. Stat. phase Col. temp.	Mobile phase flow rate	Remark	Ref.
LC	a. Dehydroemerine, dehydroglaucine, roemerine, glaucine, mecambrine, epimurinine, amurine, dihydronudaurine b. Amurinine, epiamurinine c. Epimeres of dihydronudaurine	a. *Papaver pilosum*	UV 280	a,b. 300 × 3.9 mm, μPorasil, precol. silica gel H c. μBondapak C18, precol. Li-Chroprep RP-18	C_6-CH_2Cl_2-EtOH (abs.)-Et_3N a. 30:6:2:2 2 ml min^{-1} b. 30:6:4:2 c. H_2O-ACN-Et_3N 60:40:0.1		192
LC	Diethylbolldine (I.S.), glaucine	Plasma, urine	Fluor. λ_{ex} 310 nm λ_{em} 340 nm	125 × 4 mm LiChrosorb SI-60, 5 μm	C_6-MeOH-THF-Et_2NH 88.5:7.5:4:0.15 1.5 ml min^{-1}	Linear calibration 0–5 μg ml^{-1}. Recovery 79.5%. CV 17.8% at 5–20 ng ml^{-1}, 3.2–8.7% above	193

LC	Enantiomers of tetrahydroprotoberberine, tetrahydropalmatine, stylopine, canadine, capaurine, tetrahydrojatrorrhizine, benzyltetrahydrojatrorrhizine, 13-methyltetrahydrojatrorrhizine, corydaline, tetrahydrocorysamine, thalictricavine, corybulbine		UV 254	250 × 4.6 mm, Chiracel OC-L cellulose tris (phenylcarbamate)	EtOH 1 ml min^{-1}	20 ng ml^{-1}. LOQ 5 ng ml^{-1} of plasma. Enantiomer separation was complete for 8 alkaloids	196
LC	Sanguinarine, chelidonine, chelerythrine,	Chelidonium majus roots	UV 280	250 × 4.6 mm, Alltech silica, 10 μm	5 mM AcONa in MeOH-dioxane-AcOH 88:10:2	Linear calibration 0.01–0.5 mg ml^{-1}. CV 0.93–	26

TABLE 7.11 (Continued)

Technique	Compounds separated[b]	Sample	Detection	Col. Stat. phase Col. temp.	Mobile phase flow rate	Remark	Ref.
	berberine				1.5 ml min^{-1}	4.69%. Effect of AcONa conc. on retention was studied	
LC	Sanguinarine, chelidonine, protopine, allocryptopine	a,b. Extract of Chelidonium majus	a,b. UV 280 c. UV 254	200 × 1 mm LiChrosorb Si-60, 10 μm	a. 0% or 1% H$_2$O in iPr$_2$O-0.0 or 0.2 g liter^{-1} NH$_4$Cl in MeOH 3:1 b. iPr$_2$O-C7-MeOH 12:7:1 after 9 min 0.8 g liter^{-1} NH$_4$Cl in iPr$_2$OH-MeOH 3:2 c. C7-iPrOH 19:1 or 9:1 0.05 ml min^{-1}	c. Check of sanguinarine and chelidonine purity	198
LC	a. Caffeine, colchicine, physostig-	b. Extract of Chelidonium majus	UV 280	100 × 3.8 mm, LiChrosorb RP-	a. MeOH-sodium phosphate buf-	Effect of pH, MeOH content, and	199

Applications

	Compounds	Sample	Detection	Column	Mobile phase	Notes	Ref.
	mine, boldine, allocryptopine, brucine, chelidonine, protopine, papaverine, glaucine b. Allocryptopine, protopine, chelidonine			18, 10 μm	fer-HDEHPb in various concs. b. MeOH-(H_2O-85% H_3PO_4 80: 0.5, pH 7.0 with satd. NaOH)-HDEHPb 60:40:0.16	ion pair reagent conc. on reten. was studied	
LC	Protopine, allocryptopine, sanguinarine, chelerethrine	Colored cells from Macleaya cordata root and Sanguinaria canadensis rhizome	UV 285	100 × 8 mm NV-pack C-18, 5 μm	ACN-0.1 M tartaric acid, 0.125% C12-SO_4Na 55:45 2.0 ml min^{-1}	Histochromatographic study. Content from 200 cells was analyzed	201
LC	Corynoline, 13-epicorynoline, 11-epicorynoline, acetylcorynoline, acetylisocorynoline, acetylcorynoline	Corydalis bungeana	UV 289	250 × 4 mm YWG-C 18 10 μm, 25°C	0.017 M KH_2PO_4 in MeOH-H_2O 7:3 0.8 ml min^{-1}	Linear calibration 0.2–2.0 μg. Recovery 96.4–103.7%. CV 1.23–1.70%	19

TABLE 7.11 (Continued)

Tech-nique	Compounds separated[b]	Sample	Detection	Col. Stat. phase Col. temp.	Mobile phase flow rate	Remark	Ref.
LC	Narcotine	Serum	UV 254 and 230	300 × 3.9 mm, μBonda-pak C18, 10 μm, 25°C	0.6% KH_2PO_4, pH 3 with H_3PO_4-ACN 55:45 0.9 ml min^{-1}	Reten. times for 9 other drugs are reported. Recovery 90% at 10–250 ng ml^{-1}. CV 3% at 75 ng ml^{-1}. LOD 10 ng ml^{-1}	202
LC	a. Noscapine (narcotine), papaverine (I.S.), caffeine b. Noscapine, papaverine c. Noscapine acid, nosca-pine, papa-verine	Plasma	UV 310	a. 250 × 4 mm, LiChro-sorb Si 60 5 μm b,c. 150 × 4.6 mm Spherisorb S 5 ODS	a. C6-MeOH-$CHCl_3$-Et_2NH 86.5:10:1:3.4:0.034 freshly prepared b,c. 5 mM C_5SO_3H in MeOH-H_2O-AcOH-Et_3N 40:53:6:1 1.0 ml min^{-1}	Linear cali-bration 6–400 ng ml^{-1}. Recovery 81% at 92 ng ml^{-1}. CV a. 3.8–9.5% at 89–5.9 ng ml^{-1}. b. 6.6–9.3% at 89–18 ng ml^{-1}. LOQ a. 5 ng ml^{-1}. b. 15 ng ml^{-1}.	203

LC	Emetine, cephaeline	Ipecac root (Cephaelis ipecacuanha)	UV 280	300 × 3.9 mm, μPorasil 10 μm	CHCl$_3$-MeOH-Et$_2$NH 90:10:0.2 0.5 ml min^{-1}	Linear calibration 0.5–5 μg. CV 1.5% and 2.8% for emetine and cephaeline, resp. c. Hydrolysis and lactonization of noscapine was studied	204
LC	a. Emetine, naphthalene (I.S.) b. Emetine oxidized to rubremetine	Plasma	a. UV 280 b. Fluor. λ_{ex} 225 nm λ_{em} >418 nm	300 × 3.9 mm, μBondapak C18	0.5% AcOH, 2.5 mM C8SO$_3$Na in MeOH-H$_2$O a. 56:44 b. 60:40 2 ml min^{-1}	a. See Fig. 7.11. Linear calibration a. 1–100 μg ml^{-1} b. 0.02–1.05 μg ml^{-1} LOQ a. 0.5 μg ml^{-1}. b. 0.01 μg ml^{-1}	205
LC	Tubocurarine (I.S.), alcuronium	Plasma, urine	UV 292	300 × 6.4 mm, μBondapak C18 10 μm, 40°C	0.25% (v/v) AcOH, 5 mM C12SO$_4$Na in MeOH-H$_2$O 8:2 1.5 ml min^{-1}	Recovery 100 and 95% for tubocurarine and alcuronium, resp.	206

TABLE 7.11 (Continued)

Technique	Compounds separated[b]	Sample	Detection	Col. Stat. phase Col. temp.	Mobile phase flow rate	Remark	Ref.
LC	Tubocurarine, metocurine	Plasma	UV 204	Radial-Pak CN, 10 μm with Guard-Pak, 10 μm CN precol. insert	ACN-MeOH-H_2O-1.0 M Bu_2NHPO_4, pH 2.5, 40: 10:10:1 2.4 ml min^{-1}	One compound was used as I.S. for the others. Linear calibration 25–5000 ng ml^{-1}. Recovery 79–82.4%. CV 7.9–3.4%	207
TLC PC	a. 20 alkaloids b. 26 alkaloids	a. Stylophorum diphyllum b. Glaucium squamigerum	Visualization fluor. under UV, iodoplatinate, and Dragendorff's reagents	Layers silica gel G, Silufol UV254, Paper C Whatman No. 1	a,b. CH-Et_2NH 9:1, CH-$CHCl_3$-Et_2NH 7:2:1; 6:3:1, PrOH-H_2O-formic acid 12:7:1, MeOH-Et_2NH 4:1; 1:1 a. MeOH-H_2O-Et_2NH 15:3:1	R_F values reported b. See Table 6.5	a.194 b.195

TLC	Jatrorrhizine, columbamine, palmatine	Jatrorrhiza palmata roots, Chasmanthera dependens	Visualization UV 254, 366	Layer 20 × 20 cm silica gel 60 F$_{254}$ 0.2 mm	EtOH-H$_2$O-Et$_2$NH 15:9:1 b. CH sat. MeOH, CHCl$_3$-EtOH-Et$_2$NH 8:1:1, MeOH-H$_2$O-25% NH$_4$OH 15:3:1 a,b. For PC: BuOH-H$_2$O-AcOH 10:3:1, EtOH-H$_2$O 3:2 iPrOH-1 M(NH$_4$)$_2$CO$_3$-dioxane 15:5:1. Circular development	By preparative TLC (NH$_4$)$_2$CO$_3$ volatilized by solvent evaporization	197
TLC	Sanguinarine, chelidonine, protopine, allocryptopine	Extract of Chelidonium majus	Visualization UV 254, 366	Layers silica gel Si-60	iPrO$_2$-0.0 or 0.2 g liter^{-1} NH$_4$Cl in MeOH 4:1, 3:1, or 3:2		198

TABLE 7.11 (Continued)

Technique	Compounds separated[b]	Sample	Detection	Col. Stat. phase Col. temp.	Mobile phase flow rate	Remark	Ref.
TLC	a. Isorhoeadine, papaverubine A, rhoeagine, rhoeadine b. Dihydrosanguinarine, norsanguinarine, protopine, hydrastine, adlumine c. Alstonine, ajmaline, rescinnamine	b. Fumaria species	TLC anal. of fractions	Layer alumina 60 HF-254 (40 g) + gypsum (10 g) 1 mm	a. CH-toluene-Et$_2$NH 320:80:1 followed by 160:40:3 b. CH-toluene 9:1 (50 ml), CH-toluene-Et$_2$NH 160:40:1 (100 ml) c. CH-toluene-Et$_2$NH 35:1:3, 15:1:3	Centrifugal preparative chromatography. R_F values on conventional alumina plates reported	18

[a] For other examples, see Chapter 4, Sections IV, IV.D; Chapter 5, Sections I.A, I.B, II.A (Fig. 5.7), II.B, III.A, IV.A, V.A, V.C, V.D; Chapter 6, Section II.D; and Tables 5.3, 5.7, 5.8, 6.4, and 6.5.
[b] TFAA, trifluoroacetic anhydride; TFE, trifluoroethanol; HDEHP, ethylhexylorthophosphoric acid.

TABLE 7.12 Chromatographic Conditions for Phenanthrenisoquinoline (Morphine) Alkaloids[a]

Technique	Compounds separated[b]	Sample	Detection	Col. Stat. phase Col. temp.	Mobile phase flow rate	Remark	Ref.
GC	C, norC, and their acetyl derivatives	Microbial cultures	FID	2 m × 4 mm 3% SE-30 Ultraphase on Chromosorb W-HP 100–120 mesh 210°C	Nitrogen 124 ml min⁻¹	Dealkylation study. On-column derivatization with acetic anhydride	210
GC	a. TMS derivs. of 6α + 6β-hydromorphols, hydromorphone, M + 6α + 6β-hydrocodols, α-isoC, hydrocodone, C + norM, norhydrocodone, norC	Guinea pig urine	a–d. FID e. MS	a–d. 1.83 m × 2 mm e. 1.52 m × 2 mm a,e. 3% Silar-5CP 250°C b. 3% Silar 10C c. 3% OV 225 d. 3% OV 17 all on Gas Chrom Q, 100–120 mesh	a–d. Nitrogen 50 ml min⁻¹ e. Methane 20 ml min⁻¹	a,c,d. Retention times for parent and derivatized compds. are given. Following C administration 6α-, and 6β-hydrocodols and hydromorphols, hydromorphone and M were determined.	211

TABLE 7.12 (Continued)

Technique	Compounds separated[b]	Sample	Detection	Col. Stat. phase Col. temp.	Mobile phase flow rate	Remark	Ref.
						Following M administration only 6α- and 6β-hydromorphols were assayed	
GC	BSTFA derivs. of C and M, nalorphine (I.S.)	Urine	FID	1.83 m × 6.4 mm o.d., 3% OV-1 on Chromosorb W-HP 100–120 mesh 240°C	Nitrogen 40 ml min^{-1}	See Fig. 7.13. Glucuronides converted to free bases by acid hydrolysis. Linear calibration 0–6 μg ml^{-1}. Recovery 51.7–54.2% for C, 60.1–67.4% for M at 0.3–0.6 μg ml^{-1}. LOQ 0.02 μg ml^{-1}	65

Applications

GC	a. PFPA derivs. of M, C, norM, ethylM, 6-acetylM, norC, nalorphine, 6-acetylnalorphine, pholcodine b. BSTFA derivs. of norM, norC, ethylM, M, 6-acetylM, nalorphine, 6-acetylnalorpine, pholcodine	Urine	NPD	12.5 m × 0.2 mm, cross-linked methylsilicone, 0.33 μm 120°C for 0.5 min, 120–220°C at 30 K min^{-1}, 220–240°C at 5 K min^{-1}, 240–300°C at 40 K min^{-1}	Helium 1.3 ml min^{-1}	HFBA gave more than one peak for most of opiates. Recovery 58–86%. CV 2.1–4.7%. LOD 0.25 μmol liter^{-1}	70
GC	a. Chlorpheniramine, pyrilamine (I.S.), C b. Chlorpheniramine, mono-+dides-methylchlorpheniramine, pyrilamine (I.S.), C, M+NorC	Plasma	NPD	5 m × 0.32 mm, cross-linked 50% phenylmethylsilicone, 0.25 μm 150°C for 1 min, 150–240°C at 10 K min^{-1}, 240°C for 1 min	Helium 4.2 ml min^{-1}	See Fig. 7.12. Linear calibration 9.2–368 ng ml^{-1}. Recovery 85% at 6.9–27.7 ng ml^{-1}. CV 3.7–7.0%. LOQ 0.7 ng ml^{-1} for C	208

TABLE 7.12 (Continued)

Technique	Compounds separated[b]	Sample	Detection	Col. Stat. phase Col. temp.	Mobile phase flow rate	Remark	Ref.
GC	PFPA derivs. of M, N-ethyl-norM (I.S.)	Plasma	ECD (^{63}Ni)	1.8 m × 2 mm, 3% OV-17 on Gas Chrom Q 100–120 mesh, 210°C	5% Methane in argon 30 ml min^{-1}	Separate linear calibrations 20–160 and 2–20 ng ml^{-1}. CV 5.1–76.9% at 80–2.0 ng ml^{-1}. LOD 2 ng ml^{-1}	213
GC	Methylated M, nalorphine (I.S.)	Blood	NPD	1.22 m × 2 mm 3% OV-17 on Chromosorb W-HP 80–100 mesh 220°C	Helium 40 ml min^{-1}	On-column methylation with trimethylanilinium hydroxide. Linear calibration 3–193 ng. Recovery 90%. CV 5.4% at 0.5–32.2 µg ml^{-1}, 7.9% at 1 µg ml^{-1}. LOD 0.5 µg ml^{-1}	58

GC	HFBA deriv. of M	Blood	ECD(^{63}Ni)	1.5 m × 4 mm, 3% OV-17 on Chromosorb W-HP 80–120 mesh 230°C	5% Methane in argon 45 ml min^{-1}	LOD 0.1 ng. 70 drugs were examined for ECD responses, HFBA reactivities, and ret. times	212
GC	HFBA derivs. of M, nalorphine (I.S.)	Human plasma and blood (in cases of heroin-related deaths)	ECD(^{63}Ni)	25 m × 0.2 mm, cross-linked 5% phenylmethylsilicone 250°C	5% Methane in argon	Thermal decomposition above 265°C. Linear calibration 0.1–1.2 μg ml^{-1}. Recovery 78–90% at 0.1 μg ml^{-1}. CV 6.73% at 0.6 μg ml^{-1} and 5.24% at 1.2 μg ml^{-1} for blood and plasma, resp.	63
GC	PFPA deriv. of M	Urine, body organs from rats	MS	a. 1.30 m × 3 mm, OV-17 250°C b. 25 m ×	Helium a. 20 ml min^{-1}	Deuterium-labeled M used as I.S. Linear	214

TABLE 7.12 (Continued)

Technique	Compounds separated[b]	Sample	Detection	Col. Stat. phase Col. temp.	Mobile phase flow rate	Remark	Ref.
				0.3 mm, OV-17, 270°C		calibration a. 0–200 ng ml^{-1} b. 0–20 ng ml^{-1}. CV 5% at 100 ng ml^{-1}, 10% at 10 ng ml^{-1} for urine, 10.8–2.1% for organ spiked with 1 and 20 ng, resp.	
GC	BSTFA deriv. of M	Serum, cerebrospinal fluid	MS	25 m × 0.32 mm, chem. bonded CP Sil 8, 300°C	Helium 1.2 ml min^{-1}	Deuterium-labeled M used as I.S. Linear calibration 5–200 ng ml^{-1}. Recovery 90% at 10 ng ml^{-1}. CV 5.4% at 50 ng ml^{-1}	215

Applications

GC	Acetylated M, 6-butanoylM, 3,6-dibutanoylM	Rat brain	MS at m/z 268	3 m × 0.75 mm, SPB-1 150–270°C at 12 K min^{-1}	Helium, 35–40 ml min^{-1}	Linear calibration 0–250 ng. CV 10.2, 9.9, and 13%, resp. LOD 0.3, 0.5, and 0.9 ng, resp.	216
GC	T		FID	6 ft × 1/4 in., 2% OV-17 on Gas Chrom Q, 100–120 mesh, 270°C, 225°C	Helium 48 ml min^{-1}	Complete decomposition of T at 270°C; one severely tailing peak at 225°C	232
GC	TMS derivs. of C, T, M	Opium	FID	20 m × 0.25 mm, OV-1, 0.15 μm 150–320°C at 9 K min^{-1} followed by 20 m × 0.25 mm, OV-17, 0.25 μm, 150–240°C at 25 K min^{-1}, 240–270°C at 4 K min^{-1}, 270–290°C at 25 K min^{-1}	Helium	Column-switching technique. T decomposes to four major peaks	233

TABLE 7.12 (Continued)

Technique	Compounds separated[b]	Sample	Detection	Col. Stat. phase Col. temp.	Mobile phase flow rate	Remark	Ref.
LC	Caffeine, M, thiamine, toluene, uracil		UV	150 × 4.6 mm, silica gel and C8-bonded packings with various coverage	Phosphate buffer-MeOH	Effect of MeOH and buffer concentration, pH, and type of packing was studied	218
LC	M, C	Tablets, injections	UV 220	Two columns 120 × 4.6 mm, Nucleosil 5 C8, 5 μm	0.01 M Phosphate, pH 5.0-ACN 6:4	Linear calibration 50-150% label claim	3
LC	C, acetaminophen, aspirin, caffeine, phenacetin, salicylamide	Tablets	UV 254	300 × 4 mm μBondapak C18	0.01 M KH_2PO_4, pH 2.3 or 4.85 in H_2O-MeOH 81:19	Recovery 99.1-100.7%	6
LC	M, ψM	Intrathecal injections	UV 240	100 × 5 mm Radial-Pak μBondapak CN, 10 μm	0.05 M KH_2PO_4, pH 4.5-ACN 8:2 1 ml min^{-1}	Effect of pH on resolution was studied. Linear calibra-	219

Applications

LC	M	Bulk drug, injections	UV 284 and 323	300 × 4.6 mm, μBondapak C18	5 mM C7SO3Na-MeOH-AcOH 72:28:1	Reten. times for M, C, ψM and other solutes given. Linear calibration 3.2–12.2 μg. Recovery 97.2–101.4%. CV 0.23–4.1% tion 5–100 μg ml^{-1}. LOQ 0.1 μg ml^{-1} of ψM in 1% M solution	220
LC	Nicotinamide, norM, M, C (I.S.), norethylM, ethylM	Enzymatic incubation mixtures	UV 254	300 × 3.9 mm, μBondapak C18 10 μm	5 mM C6SO3H in 1% AcOH-ACN 85:15 2.0 ml min^{-1}	Estimation of O-deethylase and N-demethylase activity. Linear calibration 3–300 nmol. Recovery 85%. CV 10% LOD 3 nmol	221
LC	Dabsyl deriv. of M	Urine	UV 436	300 × 3.9 mm, μPorasil 10 μm	10 drops/liter Et3N in CHCl3	Enzyme hydrolysis of urine. Lin-	64

TABLE 7.12 (Continued)

Tech-nique	Compounds separated[b]	Sample	Detection	Col. Stat. phase Col. temp.	Mobile phase flow rate	Remark	Ref.
					95% EtOH 10:1 1 ml min^{-1}	ear calibration 0.26–15.84 µM. Recovery 66%. CV 5.7% at 15.84 µM. LOQ 0.26 µM (75 ng ml^{-1}, 10.5 pmol)	
LC	M, C	Urine	Fluor. λ_{ex}215 nm λ_{em}480 nm	300 × 3.9 mm, µBondapak C18 10 µm	0.03 M C12SO$_4$Na, 10% PrOH 1.0 ml min^{-1}	Direct injection of sample. CV 7.1 and 3.1% at 0.4 and 1.0 µg ml^{-1} for M and C, resp. LOD 0.3 µg ml^{-1}	181
LC	Dansyl derivs. of nalorphine (I.S.), M	Urine	Fluor. λ_{ex}350 nm λ_{em}480 nm	150 × 4.5 mm, Spherisorb S3W 3 µm	C6-iPrOH-NH$_4$OH 95: 4.5:0.5	Recovery 43.9%. CV 4.3% (within run) 5.9% (day-to-day)	227

						at 300 nmol $liter^{-1}$. LOD 1.5 pmol	
LC	NorM, morphinone-2-mercaptoethanol adduct, M, dihydroM, dihydromorphinone	Urine, bile of guinea pig	UV 214	150 × 6 mm YMC AL-312 (ODS with residual silanols), 5 μm, precol. 10 × 4 mm, LiChrosorb RP-18 10 μm	10 mM Na_2HPO_4, pH 6.8–ACN 3:2	Linear calibration 20–400 ng. Recovery 76.2–101.4%. CV 1.5–4.6%. LOD 10 or 20 ng	224
LC	M, NorM (I.S.)	Plasma	Electrochemical	250 × 4 mm μBondapak C18, 10 μm	MeOH-H_2O-NH_4OH 50:50:0.1 1.3 ml min^{-1}	Linear calibration 1.6–400 ng ml^{-1}. CV 4.1–31.5% at 100–2.0 ng ml^{-1}. LOD 1 ng ml^{-1}	213
LC	C, methadon (I.S.)	Plasma	UV 254	300 × 3.9 mm, μPorasil, 10 μm, precol. 100 × 2.1 mm Vydac 101SC	CH_2Cl_2-MeOH-33%NH_4OH 90:10:0.1 1.5 ml min^{-1}	Linear calibration 10–160 ng ml^{-1}. Recovery 99.8%. CV 3.2–9.1% at 160–10 ng ml^{-1}. LOD 5 ng ml^{-1}	222

TABLE 7.12 (Continued)

Technique	Compounds separated[b]	Sample	Detection	Col. Stat. phase Col. temp.	Mobile phase flow rate	Remark	Ref.
LC	a. M-3-glucuronide, M-6-glucuronide, M b. M-3-glucuronide, M-6-glucuronide, norM, M c. M, C, ethylM, heroin	a. Plasma b. Urine	UV 210 Electrochemical (coulometric)	150 × 4.6 mm, Ultrasphere ODS 5 μm	a,b. 1 mM NaC12SO$_4$, 0.01 M NaH$_2$PO$_4$, pH 2.1 with H$_3$PO$_4$, 26% ACN 1.5 ml min^{-1} c. 5 mM NaC12SO$_4$, 0.1 M phosphate, pH 3.3, 36% ACN	Linear calibration 20–100 ng ml^{-1}. Recovery 84% at 100 ng ml^{-1}. CV 7.6% at 22 ng ml^{-1}, 3.8% at 223 ng ml^{-1} plasma for M. LOD 5 ng ml^{-1} for M-3-glucuronide. For electrochemical detector (223): Linear calibration 10–50 nmol liter^{-1}. CV 5.9% at 5 nmol liter^{-1}. LOD 1 nmol liter^{-1} (0.29 ng ml^{-1}) for M	44, 223

Applications

LC	C, N-allyl-norC (I.S.)	Plasma, urine	Fluor. λ_{ex}220 nm λ_{em}355 nm	50 × 4.6 mm, C_{18} Spheralyte 3 μm, 50°C	0.2 M $NaClO_4$, 0.1 M H_3PO_4–MeOH 84:16 1.5 ml min^{-1}	Linear calibration 15–150 ng ml^{-1} and 0.5–20 μg ml^{-1}. CV 5.0–10.6% and 0.4–12.7%. LOD 3 ng ml^{-1} and 0.25 μg ml^{-1} for plasma and urine, resp.	225
LC	Nalorphine (I.S.), M	Serum	Electrochemical	300 × 4 mm ODS bonded to LiChrosorb, 40°C	MeOH–0.01 M KH_2PO_4 85:15 1 ml min^{-1}	Linear calibration 10–200 ng ml^{-1}. Recovery 79%. CV 6.5% at 5 ng ml^{-1}, 11.6% at 1 ng ml^{-1}	43
LC	M, dihydroM (I.S.)	Plasma	Electrochemical (at 0.75 V vs. SSCE)	250 × 4.5 mm, Spherisorb S5W 5 μm, 31°C	0.325 ml $liter^{-1}$ AcOH, 0.55 ml $liter^{-1}$ NH_4OH in MeOH–C7 7:3, app. pH 8.3 2.0 ml min^{-1}	Linear calibration 2–80 ng ml^{-1}. Recovery 91%. CV 4.4% at 2 ng ml^{-1}, 2.7% at 80 ng ml^{-1}. LOD 150 pg.	229

TABLE 7.12 (Continued)

Technique	Compounds separated[b]	Sample	Detection	Col. Stat. phase Col. temp.	Mobile phase flow rate	Remark	Ref.
						At 0.9 V C could also be determined	
LC	M, nalorphine (I.S.)	Plasma, cerebrospinal fluid	Electrochemical	300 × 4 mm µBondapak C18 with guard col. Corasil/C18	0.07 M NaH_2PO_4, 0.5 mM Na_2EDTA-MeOH-ACN 87:8:5, pH 4.5 1.0 ml min^{-1}	Linear calibration 1–200 ng ml^{-1}. Recovery 78–84.8% at 50 and 100 ng ml^{-1}. CV 1.2–9.9% at 174–2.48 ng ml^{-1}. LOD 20 pg	42
LC	Caffeine (I.S.), C	Blood, urine (spiked), tablets, syrups	UV 212	250 × 4 mm LiChrosorb RP 8,10 µm with guard col. RP 8 Spheri-5 25°C	0.5% $AcONH_4$-MeOH 7:3 pH 7.0 1.9 ml min^{-1}	Recovery 90%. LOD 0.5 ng	59
LC	M, 6-acetylM, C	Blood	UV 220	250 × 4.5 mm, Spherisorb ODS	50 mM $C5SO_3H$-ACN 7:3	Recovery 95–100%. LOD 0.2 ng	62

Applications

LC	M, nalorphine (I.S.)	Urine, blood, gastric lavage	a. Postcolumn derivatization with $K_3Fe(CN)_6$ and fluor. det. b. Precolumn derivatization with $K_3Fe(CN)_6$ and fluor. det.	a. 100 × 4.0 mm, Partisil 10 ODS 5 μm b. 200 × 4 mm, silica gel 60 HPLC, 5 μm	pH 2.0 1.2 ml min^{-1} a. 0.1 M KBr–MeOH 87.5:12.5 pH 3 with H_3PO_4 2 ml min^{-1} b. MeOH–2 M NH_4OH–1 M NH_4NO_3 3:2:1 2.0 ml min^{-1}	11 opiates examined for retention and fluorescence after derivatization. Recovery 80%. CV 11%	57
LC	M, nalorphine, dihydroM, norM, 6-acetylM		Postcolumn derivat. with $K_3Fe(CN)_6$ and fluor. λ_{ex}324 nm λ_{em}430 nm	100 × 4 mm Zorbax ODS 8 μm	0.1 M KBr–MeOH 87.5:12.5, pH 3 with H_3PO_4	Fluorescence enhancement by micelle formation	226
LC	Dansyl derivs. of nalorphine (I.S.), M	Blood, plasma	Fluor. λ_{ex}330–380 nm λ_{em}410–500 nm	150 × 4.5 mm, Spherisorb S3W 3 μm	C6-iPrOH-NH_4OH 97:2.7:0.3 1.5 ml min^{-1}	Linear calibration 3.3–1000 nmol liter^{-1}. Recovery 72.8% at 0.15–140 μmol liter^{-1}. CV 3.7% (intra-) 4.5%	228

TABLE 7.12 (Continued)

Technique	Compounds separated[b]	Sample	Detection	Col. Stat. phase Col. temp.	Mobile phase flow rate	Remark	Ref.
LC	C, M, dextrophan (I.S.), dihydroC	Blood	Electrochemical	250 × 4.9 mm, Spherisorb S5W 5 μm	a. (ACN-MeOH 9:1)–0.05 M HClO$_4$, pH 9.0 with 1.0 M NaOH 7:3 1.5 ml min^{-1} b. (ACN-MeOH 8:2)–0.1 M NH$_4$NO$_3$, pH 9.5 with conc. NH$_4$OH 9:1	(interassay) at 150 nmol liter^{-1}. LOD 0.2 pmol b. Effects of ACN-MeOH ratio, pH, and buffer conc. were studied. Retention times for possible interferences are reported. LOD 250 pg for M, C; 500 pg for dihydroC	230
LC	a. M, norC, dihydromorphinone (I.S.), C	a. Human plasma b. Rat blood	Electrochemical	a. 250 mm b. 150 mm Spherisorb Phenyl 5 μm	7% ACN, 2% MeOH, 25 mM H$_3$PO$_4$, pH 2.8 1.5 ml min^{-1}	b. LOQ 2 ng ml^{-1} in 100 μl blood for M and C. Linear cali-	a. 61 b. 60

Applications

	b. M, dihydromorphinone (I.S.), C					
LC	T, Papaver bracteatum straw	UV 285	300 × 4 mm μBondapak C18	MeOH–0.3% $(NH_4)_2CO_3$ 4:1 1 ml min^{-1}	bration 2–100 ng ml^{-1}, 3–250 ng ml^{-1}. Retention times for isoT, orientalidine, M, C are given. For quantitation col. saturation with T was required. Linear calibration up to 12 μg. Recovery 93%. CV 0.7% at 7.91 μg. LOQ 0.05 μg	232
TLC	C, chlorpheniramine, phenylephrine, acetaminophen Syrup, tablets	Scanning UV 285, 260, 273, and 260, resp.	Layer silica gel G-60, F_{254}, 0.25 mm	BuOH–MeOH–toluene–H_2O–AcOH 3:4:1:2:0.1	Direct application of 5 μl of sample solution. Calibration 0–1 mg ml^{-1}. Recovery 97.2–103.5%. CV 1.26–1.64% at 0.16 mg ml^{-1} for C	231

TABLE 7.12 (Continued)

Technique	Compounds separated[b]	Sample	Detection	Col. Stat. phase Col. temp.	Mobile phase flow rate	Remark	Ref.
TLC	Dabsyl deriv. of M	Urine	Zig-zag scanning UV 436	Layer silica gel 60 F_{254} 0.25 mm	$CHCl_3$-EtOH abs.-Et_3N 30:2:0.05	Linear calibration 1.3–15.8 μM. CV 5–7%. LOQ 1.3 μM (0.375 μg ml^{-1}). LOD 7.5 ng (0.075 μg ml^{-1})	64
TLC	T	<u>Papaver bracteatum</u>	Scanning UV 280	Layer 10 × 10 cm silica gel 60	$CHCl_3$-MeOH 9:1. Anticircular development	Calibration 1.5–3.5%	234

[a]For alkaloid mixtures of opium type and heroin, see Tables 7.13 and 7.14. For other examples, see Chapter 4, Sections IV, IV.D, V.B; Chapter 5, Sections I.A, II.A (Fig. 5.7), II.B, III.A, IV.A, V.C, V.D; Chapter 6, Sections I.D and II.D; and Tables 5.3 and 5.8.
[b]C, codeine; M, morphine; BSTFA, bis(trimethylsilyl)trifluoroacetamide; Dabsyl, 4-dimethylaminoazobenzene-4'-sulfonyl; Dansyl, 5-dimethylamino-1-naphthalenesulfonyl; HFBA, heptafluorobutyric anhydride; PFPA, pentafluoropropionic anhydride; T, thebaine.

TABLE 7.13 Chromatographic Conditions for Mixtures of Opium Alkaloids[a]

Technique	Compounds separated[b]	Sample	Detection	Col. Stat. phase Col. temp.	Mobile phase flow rate	Remark	Ref.
GC	Resmethrine (I.S.), C, M, T, P, didecylphthalate (I.S.), Nco	Opium	FID	6 ft × 0.08 in., mixture 1:1 of OV-17 on Varaport 30, 80–100 mesh and 5% SE-30 on Chromosorb W-AW, DMDS, 80–100 mesh 250°C for 5 min, 250–280°C at 48 K min^{-1}	Helium 30 ml min^{-1}	Without derivatization. CV 0.9–3.8%	235
GC	a. C, T, P, triacontane (I.S.), Nco b. BSA deriv. of M, docosane (I.S.)	Opium	FID	6 ft × 4 mm 3% OV-1 on Gas Chrom Q, 100–120 mesh a. 240°C b. 233°C	Nitrogen 70 ml min^{-1}	Linear calibration up to 7 μg. CV less than 4.7%	236

TABLE 7.13 (Continued)

Tech-nique	Compounds separated[b]	Sample	Detection	Col. Stat. phase Col. temp.	Mobile phase flow rate	Remark	Ref.
GC	MSTFA derivs. of C, T, M, P, Nco	Opium, crude M	FID, NPD	25 m × 0.27 mm, cross-linked a. OV-1 b. SE-54, 0.15 μm a. 150–280°C at 9 K min^{-1}, 280°C for 0.5 min b. 150–280°C at 6 K min^{-1}, 280–300°C at 15 K min^{-1}, 300°C for 20 or 30 min	Hydrogen a. 110 cm sec^{-1} b. 65 cm sec^{-1}	a. Quantitation b. Profiling. Effect of acetylation on GC profile was examined	237
LC	M, C, T, Nco		Electro-chemical	250 × 4.6 mm, Partisil 10 ODS-3 10 μm	ACN–0.02 M KH$_2$PO$_4$, pH 6 with NaOH 55:45	Linear calibration 0.42–1.7 nmol. CV 0.9%. LOD 0.3 ng for M	153

LC	M	Poppy straw	UV 285	300 × 4 mm μPorasil, 10 μm, 25°C	C6-CH$_2$Cl$_2$-EtOH-Et$_2$NH 300:30:40: 0.5 2.4 ml min^{-1}	Sample homogenization procedure was examined	13
LC	a. Nco, P, T, C, M b. Nco, P, T, C, M, Nce c. Nce, M, C, T, P, Nco d. M, C, T, kryptopine, Nco, P, Nce		UV 280	a,b. 300 × 3.9 mm, silica gel, precol. silica gel H c,d. 300 × 3.9 mm μBondapak C18, precol. LiChroprep RP-18	a. C6-CH$_2$Cl$_2$-EtOH-Et$_3$N 30:6:6:2 b. Linear grad. B in A A. ACN-H$_2$O-Et$_3$N 60:40:0.1 B. EtOH abs. 2–50% B in 16 min c. ACN-H$_2$O-Et$_3$N 60:40:0.1 d. 5 mM C7SO$_3$H (PIC B7) in H$_2$O-MeOH-AcOH 56:44:1 2 ml min^{-1}	c. See Fig. 7.14. Extractive removal of phenolic alkaloids before RP separation was recommended for samples with high M content	24

TABLE 7.13 (Continued)

Tech-nique	Compounds separated[b]	Sample	Detection	Col. Stat. phase Col. temp.	Mobile phase flow rate	Remark	Ref.
LC	a. Nco, T, C, M b. Caffeine, heroin, acetylM, M, strychnine	a. Opium b. "Chinese heroin"	UV 278	250 × 4.6 mm, fines from Partisil, 6 μm	MeOH–2 M NH$_4$OH–1 M NH$_4$NO$_3$ 27:2:1 1 ml min^{-1}	a. See Fig. 5.4. Retention data for 84 drugs were reported	238
LC	M, C, T, P	Opium, latex from _Papaver somniferum_	UV 280	Rad-Pak Standard C18 packing or Nova C18 (fully end-capped)	Linear gradient B in A A. 10 mM KClO$_4$, 5 mM BuNH$_2$, pH 3 with HClO$_4$ B. ACN 10–70% in 0–5 min 70% in 5–7 min 3 ml min^{-1}	Linear calibration for M 50–250 ng only, other up to 700 ng. LOD 20–40 ng	241
LC	M, C, codeinone, C methyl	(Codeine phosphate drug sub-	UV 254	300 × 3.9 mm μ Bondapak C18	Linear gradient of B in A	Linear calibration up to 1.66 μg	242

Applications 421

	ether, T, cryptopine, Nco, P	stance)		10 μm, 40°C	A. 15 mM KH$_2$PO$_4$, pH 2.5 with H$_3$PO$_4$-ACN 93:7 B. MeOH-ACN 93:7 3% in 0–2 min, 3–57% in 2–17 min, 57% in 17–18 min, 3% in 18–23 min 3 ml min^{-1}	for P, 51.79–99.92 μg for others. CV 0.07–2.14%. LOD 0.6 ng for P, 12–48 ng for the others	
LC	M, C, P, T, Nco	Poppy straw	UV 254	150 × 3.2 mm, LiChrosorb RP-8 5 μm	MeOH-25 mM NaH$_2$PO$_4$ 25 mM Na$_2$HPO$_4$ 6:4 0.5 ml min^{-1}	See Fig. 7.15	243
LC	M, C, T, quinine (I.S.), P, Nco	Poppy straw concentrate	UV 275	250 × 5 mm μBondapak Phenyl guard col. 70 × 2 mm C18/Corasil 37–50 μm	Linear gradient of B in A H$_2$O-ACN A. 95:5 B. 80:20 both contg. 1 ml liter^{-1} AcOH, 0.04	CV 0.6% for M. LOD 0.1, 0.5, 0.1, and 0.2% for C, T, P, and Nco, resp.	23

TABLE 7.13 (Continued)

Tech-nique	Compounds separated[b]	Sample	Detection	Col. Stat. phase Col. temp.	Mobile phase flow rate	Remark	Ref.
					ml liter^{-1} N,N-dimethyloctylamine, pH 3.5 with NaOH, 0–100% in 20 min 1 ml min^{-1}		
LC	M, C, P, T, Nco	Gum opium	UV 254	300 × 3.9 mm, μBondapak Phenyl 10 μm	MeOH–1% AcONa, 7.0 mM Et$_3$N, pH 11, 58:42 apparent pH 10.5 1.0 ml min^{-1}	See Fig. 5.10. AcONa helped to achieve adequate retention of M. Guard col. was recommended	244
LC	M, meconic acid, C, T, laudanosine, cryptopine, Nco, Nce, P	Gum opium	UV 280	300 × 3.9 mm, μBondapak Phenyl 10 μm or 250 × 4.0 mm, MN-Nucleosil 7	Linear gradient of B in A; 1% (v/v) of 10 ml Et$_3$N+ 80 ml H$_2$O, pH 2.2 with	Modification of gradient shape, pH, buffer type, and THF addition to the mobile	22

Applications

	a. M, C, P, T, Nco b. M, P, Nco c. Norephedrine, ephedrine, atropine, strychnine, cinchonine, quinine, and other bases d. Yohimbine, reserpine	a,b. Opium	UV 254	C_6H_5, 7 μm Guard-Pak precol. module with a Resolve CN (10 μm) insert	85% H_3PO_4, diluted to 100 ml in MeOH-H_2O A. 5:95 B. 70:30 0–100% in 0–20 min, 100% in 20–32 min, 100–0% in 32–40 min, 0% in 40–50 min 1.5 ml min^{-1}	phase were examined. Linear calibration up to 20, 5, 4, 8, 3 μg for M, C, T, Nco, and P, resp. CV 1.43–2.5%
LC				220 × 4.6 mm, Hitachi gel 3010 (styrene-divinylbenzene copolymer) 10 μm a,b. 65°C c. 55°C d. 25°C	a. 0.019 M NH_4OH, 48% ACN b. 0.035 M NH_4OH, 60% ACN c. 0.02 M Bu_4NOH, 30% ACN d. 0.02 M Bu_4NOH, 60% ACN 1 ml min^{-1}	See Fig. 5.9. Effect of ACN, NH_4OH, and Bu_4NOH concs. on retention was studied. c,d. k values were reported	20

TABLE 7.13 (Continued)

Tech-nique	Compounds separated[b]	Sample	Detection	Col. Stat. phase Col. temp.	Mobile phase flow rate	Remark	Ref.
LC	M, C, cryptopine, T, Nco, P	a. Gum opium	UV 254	300 × 4 mm a,b. Nucleosil 10 CN c. Nucleosil 10 C18	a. 1% AcONH$_4$, pH 5.8– ACN–dioxane 8:1:1 b. 1% AcONH$_4$, pH 6.3– ACN–dioxane 79:16:5 c. 1% AcONH$_4$, pH 5.8– ACN 65:35 1.5 ml min^{-1}	a. See Fig. 7.16. b. See Fig. 5.14. Retention data were reported for various compositions of the mob. phase	21
LC	M, C, T, P, Nco	Gum opium	UV 254	300 × 3.9 mm, μBondapak CN, 10 μm	1% AcONa, pH 6.78 with AcOH– ACN–dioxane 75:20:5 1.5 ml min^{-1}	Linear calibration up to 1.6–4 μg. Recovery 98–100%	245
LC	a. T, C, M, tyrosine (I.S.)	Latex of <u>Papaver somniferum</u>	UV 286	250 × 4.6 mm, Zorbax NH$_2$	ACN–0.025 M KH$_2$PO$_4$ 75:25	b. See Fig. 5.16. Nco coeluted	246

Applications

	Analytes	Sample	Detection	Column	Mobile phase	Comments	Ref.
	b. P, T, Nce, C, M, tyrosine (I.S.)					a. 1.0 ml min^{-1} b. 2.0 ml min^{-1}	
LC	a. M, C, T, Nco, P b. Caffeine, ephedrine, methamphetamine, phentermine c. LSD	a. Opium b. "Mini-bennie" tablets c. "Microdot" tablets	UV 254	300 × 4 mm μBondapak C18	H_2O–5 mM $C7SO_3Na$ in MeOH-AcOH 59:40:1 pH 3.5 2 ml min^{-1}	See Fig. 5.21. Response was linear in limited concentration range only. Reten. data for 50 drugs were reported	247
LC	M, C, T, Nco, P	Opium from <u>Papaver somniferum</u> × <u>P. setigerum</u>	UV 254	300 × 4 mm μBondapak C18, 26°C	5 mM $C7SO_3H$ (PIC B7) in H_2O-MeOH-AcOH 59:40:1, pH 3.5 2 ml min^{-1}	Linear calibrations 0.75–3, 0.55–2.2, 0.25–1.0, 0.75–3, and 0.2–0.8 μg, resp.	248
LC	M, C, T, Nco, P	Opium	UV 280	LiChrosorb RP-8, 10 μm	500 ml H_2O + 1 g $C7SO_3Na$ + 5 ml 0.05 M H_3PO_4 + 200 ml ACN or H_2O-ACN- with P	CV 0.7–2.6%	249

TABLE 7.13 (Continued)

Tech-nique	Compounds separated[b]	Sample	Detection	Col. Stat. phase Col. temp.	Mobile phase flow rate	Remark	Ref.
LC	M, C, Nce, T, Nco, P	Opium, poppy capsules	UV 220	150 × 4.6 mm, Nucleosil 5 C18 60°C	5 mM C8SO$_3$Na, 50 mM KH$_2$PO$_4$, pH 3.0– ACN 7:3 0.9 ml min^{-1} Et$_3$N 600:400:1	See Fig. 5.22. Linear calibration 0.2–1.0 or 2 μg. CV 3.3–6.0%. LOD 8.5–35.2 ng	250
LC	M, C, T, P, Nco, Nce		UV 280	250 × 4.6 mm, LiChrosorb RP-18 5 μm, impregnated with dodecylsulfonic acid and cetrimide	0.02 M C12SO$_3$H in H$_2$O-MeOH-dioxane-H$_2$SO$_4$ 75:23.5:1:0.5 pH 3.5 with 10 M NaOH 1.2 ml min^{-1}	Impregnation improved efficiency of the column	138

[a]For other examples, see Chapter 4, Sections IV, IV.D; Chapter 5, Sections I.A, II.A, II.B, III.A, IV.A, V.A; and Chapter 6, Section II.D.
[b]C, codeine; M, morphine; Nce, narceine; Nco, narcotine (noscapine); P, papaverine; T, thebaine; BSA, $\underline{N},\underline{O}$-bis(trimethylsilyl)acetamide; MSTFA, $\underline{N}$-methyl-$\underline{N}$-trimethylsilyltrifluoroacetamide.

TABLE 7.14 Chromatographic Conditions for Heroin Analysis[a]

Technique	Compounds separated[b]	Sample	Detection	Col. Stat. phase Col. temp.	Mobile phase flow rate	Remark	Ref.
GC	AcetylC, 6-acetylM, H, P, Nco	Illicit heroin	FID	6 ft, 3% OV 210 on Chromosorb W, 100–120 mesh, 190°C for 15 min, 190–270°C at 8 K min^{-1}, 270°C for 5 min	Nitrogen 25 ml min^{-1}		252
GC	HFBA derivs. of a. p,p'-DDT (I.S.), C, M, 6-acetylM, 3-acetylM a,b. NorC, norM, 6-acetylnorM, 3,6-diacetylnorM, dioctylphthalate (I.S.)	b. Heroin	ECD(^{63}Ni)	a. 12 m × 0.20 mm OV-1 b. 11 m × 0.25 mm OV-17 a,b. 90°C for 1.8 min, 90–160°C at 25 K min^{-1}, 160°C for 1 min, 160–275°C at 4 K min^{-1}	Helium 45 cm sec^{-1}	OV-1 more stable than OV-101. Elution order was different on both columns. Derivatized H was extracted with dil. H$_2$SO$_4$ prior to GC	253

TABLE 7.14 (Continued)

Technique	Compounds separated[b]	Sample	Detection	Col. Stat. phase Col. temp.	Mobile phase flow rate	Remark	Ref.
GC	HFBA derivs. of C, M, 6-acetylM, 3-acetylM, 6-acetylnorM, 6-acetylnorC, 3,6-diacetylnorM, didehydroH, didehydroNco	Illicit heroin	a. ECD (63Ni) b. MS	a. 15 m × 0.25 mm DB-1, 0.25 μm, 90°C for 1.8 min, 90–160°C at 25 K min^{-1}, 160°C for 1 min, 160–275°C at 4 K min^{-1} b. 20 m × 0.2 mm SE-54	a. Helium 40 cm sec^{-1} b. Hydrogen	Derivatization with HFBA in the presence of 4-(dimethylamino)pyridine was followed by acid back-extraction	254
GC	Meconin, caffeine, acetylthebaol, diacetylnorC, 5ADMEP, triacetylnorM, dotetracontane (I.S.)	Heroin acidic and neutral impurities	FID, N-FID	25 m × 0.2 mm, SE 54 150–280°C at 6 K min^{-1}, 280°C for 1 min, 280–300°C at 10 K min^{-1}, 300°C for 5 min	Hydrogen 0.55 bar	Direct analysis or derivatization with either MSTFA or BSTFA	255

GC	MSTFA derivs. of 32 compounds	Heroin acidic and neutral impurities	FID	a. 25 m × 0.25 mm DB-1, 0.25 μm, 200–330°C at 4 K min^{-1}, 330°C for 4 min b. 25 m × 0.27 mm SE-54, 150–280°C at 6 K min^{-1}, 280°C for 1 min, 280–300°C at 15 K min^{-1}, 300°C for 11 min	Hydrogen a. 50 cm sec^{-1} b. 65 cm sec^{-1}	Reten. times and MS data were reported. Nco and norlaudanosine related by-products were studied in detail	256
GC	HFBA derivs. of reduced meconin, 4-methylmeconin (I.S.), AMP (I.S.), p,p'-DDT, ADP, 5ADMEP, diacetylnorC, triacetylnorM,	Heroin neutral impurities	ECD(^{63}Ni)	15 m × 0.25 mm, DB-1, 0.25 μm 80°C for 5.5 min, 80–160°C at 25 K min^{-1}, 160°C for 1 min, 160–275°C at 4 K min^{-1}, 275°C for 15 min	Hydrogen 40 cm sec^{-1}	Extract reduced with LiAlH$_4$ before derivatization with HFBA. Detection at pg level	257

TABLE 7.14 (Continued)

Tech-nique	Compounds separated[b]	Sample	Detection	Col. Stat. phase Col. temp.	Mobile phase flow rate	Remark	Ref.
	tripropion-ylM (I.S.), 8ADMEP						
GC	a. Acetyl-thebaol, 5ADMEP, N-acetyl-norlaudano-sine, N-acetylnor-Nco, (E)-N-acetyl-hydronor-Nce, dotet-racontane (I.S.) b. Acetyl-thebaol	Illicit heroin	FID	20 m × 0.25 mm, SE 54, 0.15 μm 160–240°C at 12 K min⁻¹, 240–280°C at 5 K min⁻¹, 280–330°C at 20 K min⁻¹ followed by 5 m, OV-17 0.25 μm, 150–290°C at various rates	Helium	Col. switch-ing tech-nique. b. Detection of acetyl-thebaol in samples di-luted with rosin or colophony (silylated)	233
LC, GC	H, C, M	Spiked urine	UV 241, FID	100 × 2 mm SPHERI-5 (silica) followed by 20 m × 0.32 mm, OV-1,	Et₂O-MeOH-Et₂NH 91.5: 8:0.5. Hy-drogen 3 ml min⁻¹	LC inter-faced di-rectly to GC	259

Applications

LC	AcetylC, H, 6-acetylM, C, M	Illicit heroin	UV 279	150 or 250 mm, silica gel, 5 μm	Isooctane-Et$_2$O-MeOH-H$_2$O-Et$_2$NH 400:325: 225:15:0.5 2 ml min^{-1}	Sugar diluents were also determined. Effect of Et$_2$NH and H$_2$O contents on retention and peak shape was examined. Retention data for other drugs are given. Linear calibration 0–0.5 mg ml^{-1} for M and H. CV 1.3% for H	260

(Additional column fragment: "0.5 μm with precol. 2.5 m × 0.53 mm 100–200°C at 30 K min^{-1}, 200–300°C at 15 K min^{-1}")

TABLE 7.14 (Continued)

Tech-nique	Compounds separated[b]	Sample	Detection	Col. Stat. phase Col. temp.	Mobile phase flow rate	Remark	Ref.
LC	a,b. P, Nco, T, C, M c. H, acetylC, procaine, 6-acetylM, C, strychnine, M	a,b. Opium c. Illicit heroin	UV	250 × 4.6 mm, Spherisorb A5Y with precol. 75 × 2.1 mm pellicular alumina a. 20°C, home-packed b. 60°C, commercial col.	a. 3 mM, b. 10 mM Me$_4$NOH, citric acid to pH 5, 40% ACN c. 10 mM Me$_4$NOH, pH 6.5 in H$_2$O-MeOH-ACN 75:12.5:12.5	See Fig. 5.5 and Tab. 5.3. Effect of MeOH, ACN, and THF concs. on reten. was studied. Caffeine was not retained	239
LC	a. Dipyron, P, Nco, acetylthebaol, H, acetylC, 6-acetylM b. Meconic acid, meconine, P, Nco, cryptopine, T, C, M	a. Heroin b. Opium	Fluor. λ_{ex}260 nm λ_{em}400 nm UV	250 × 4.6 mm, Spherisorb A5Y (alumina) followed by 450 × 2.1 mm, C18 modif. silica	0.1 M Me$_4$NOH, pH 6.0 with citric acid–MeOH–ACN 55:28:17 1.0 ml min^{-1}	Cols. in series with or without switching. Mob. phase optimization was described. Reten. times for caffeine, strychnine, and 4 other	240

	Compounds	Detection	Column	Mobile phase	Remarks	Ref.
LC	a. M, procaine, C, 3- and 6-acetylM, acetylprocaine, quinine, dihydroquinine, caffeine, acetylC, H, acetylM, piperonal (I.S.), Nco, P b. 3- and 6-acetylM, acetylC, H, Nco, P, propiophenone (I.S.)	UV, programmed wavelength	125 × 4.6 mm, HS-5 C18	Linear gradient of MeOH in phosphate buffer (0.023 M hexylamine in H_2O–2 M NaOH–H_3PO_4 87:3:1, pH 2.2) 5–30% in 0–20 min, 30% in 20–28 min 1.5 ml min^{-1}	compounds are given See Fig. 7.17. Effect of pH, type and conc. of organic and amine modifiers on retention and selectivity was studied. Changes of resolution with col. lifetime were discussed. b. Quantitation. Peak height yielded better precision at low conc., peak area was preferred above approx. 0.2 g liter^{-1} where non-linearity of isotherm was	261

TABLE 7.14 (Continued)

Tech-nique	Compounds separated[b]	Sample	Detection	Col. Stat. phase Col. temp.	Mobile phase flow rate	Remark	Ref.
LC	Nalorphine (I.S.), 6-acetylM, quinine, acetylC, H, Nco, P, propiophenone (I.S.)	Heroin	Dual electrochemical, UV, programmed 218, 228, 240, 305	125 × 4.6 mm, HS-5 C18	Linear gradient of MeOH in phosphate buffer (0.023 M hexylamine, pH 2.2) 5–30% in 0–20 min, 30% in 20–26 min, 30–60% in 26–36 min, 60% in 36–45 min, 5% in 45–55 min 1.5 ml min^{-1}	apparent. CV 0.6–8% Relative reten. times and electrochemical response categories were presented for 50 compounds. CV 4.5%	262
LC	Products of oxidative coupling of 6-acetylM and M	Urine	Fluor. λ_{ex}320 nm λ_{em}436 nm	150 × 4.6 mm a. Hypersil, 5 μm, 30°C b. Hypersil,	a. ACN–2 M NH$_4$OH– 1 M NH$_4$NO$_3$ 85:10:5 b. H$_2$O–	Linear calibration a. 0–160 ng b. 1–100 μg ml^{-1}.	263, 264

				ODS, 5 μm, 30°C	ACN-Et_3N 85:15:0.1 1.5 ml min^{-1}	Recovery a. 69.2% and 81.6% at 10 and 50 μg ml^{-1}, resp. b. 78% and 73% at 5 and 25 μg ml^{-1}. CV b. 6–10% at 94–2 μg ml^{-1}. LOD a. 0.4 ng (6 μg ml^{-1}) b. 0.25 ng	
TLC	Methadone, cocaine, Nco, T, P, acetylC, H, 6-acetylM, caffeine, C, ephedrine, quinine, M	Heroin	Visualization UV 254, 366, Marquis', acidified iodoplatinate, and Dragendorff's reagents	Layer 20 × 20 cm silica gel 60 GF_{254}	C6-$CHCl_3$-Et_3N 9:9:4	35 systems were reviewed. LOD 0.1 μg	252

[a]For other examples, see Chapter 4, Sections IV, IV.D; Chapter 5, Sections I.B, II.B, III.A, IV.A, and V.A; and Chapter 6, Section II.D; and Table 5.3.

[b]5ADMEP, 4-acetoxy-3,6-dimethoxy-5-[2-(N-methylacetamido)ethyl phenanthrene]; 8ADMEP, 4-acetoxy-3,6-dimethoxy-8-[2-(N-methylacetamido)ethyl phenanthrene]; ADP, 4-acetoxy-3,6-dimethoxyphenanthrene; AMP, 4-acetoxy-3-methoxyphenanthrene; C, codeine; H, heroin (diamorphine, diacetylmorphine); M, morphine; Nco, narcotine (noscapine); P, papaverine; BSTFA, N,O-bis(trimethylsilyl)trifluoroacetamide; HFBA, heptafluorobutyric anhydride; MSTFA, N-methyl-N-trimethylsilyltrifluoroacetamide.

TABLE 7.15 Chromatographic Conditions Used for Drug Identification

Technique	Compounds separated	Sample	Detection	Col. Stat. phase Col. temp.	Mobile phase flow rate	Remark[a]	Ref.
GC	570 compounds		NPD	a,c. 6 ft × 2 mm b,d. 3 ft × 2 mm a,b. 3% OV 1 c,d. 3% OV 17 on Chromosorb W-HP, 100/120 mesh a,c. 150°C for 1 min, 150–250°C at 10 K min⁻¹, 250°C for 8 min b. 200°C for 1 min, 200–280°C at 10 K min⁻¹, 280°C for 10 min d. 220°C for	Nitrogen 50 ml min⁻¹	Relative reten. times were reported and correlated with literature reten. indices. Reference substances: a. 2-Amino-5-chlorobenzophenone b,d. Thioridazine c. Methaqualone. LC and TLC were also applied	284

GC	32 drugs	Spiked blood and urine	FID	50 m × 0.32 mm, SE-54 45°C for 1.5 min, 45–300°C at 6 K min^{-1}, 1 min, 220–280°C at 10 K min^{-1}, 280°C for 10 min	Hydrogen 165 cm sec^{-1}	Long-term reproducibility was studied	266
GC	27 drugs		a. FID b. NPD	a. 11 m × 0.2 mm, SE-54, 0.26 μm b. 10 m × 0.2 mm, OV-215, 0.20 μm a,b. 75–280°C at 10 K min^{-1}, 280°C for 5.5 min	Helium 40 cm sec^{-1}	See Fig. 4.1. Both cols. were inserted in one common injection port	83
GC	24 drugs, adulterants, or diluents and their silylated derivs.	Street samples of narcotics	FID	25 m, SE-54, 0.4–0.45 μm 100–170°C at 10 K min^{-1}, 170–230°C at 1	Hydrogen 0.4 kg cm^{-2}	Silylation with HMDS-TMCS-pyridine 3:1:9	84

TABLE 7.15 (Continued)

Technique	Compounds separated	Sample	Detection	Col. Stat. phase Col. temp.	Mobile phase flow rate	Remark[a]	Ref.
GC	187 drugs and metabolites	Clinical and forensic specimens	a. Dual FID and NPD b. MS	K min^{-1}, 230–320°C at 10 K min^{-1}, 320°C for 5 min 25 m × 0.22 mm, crosslinked methylsilicone 250°C	Helium 1 ml min^{-1}	Reten. indices were reported	147
LC	84 drugs	Illicit drug samples	UV	250 × 4.6 mm, silica gel fines	MeOH–2 M NH$_4$OH–1 M NH$_4$NO$_3$ 27:2:1 1 ml min^{-1}		238
LC	About 90 drugs		UV, fluor., electrochemical	250 × 4.9 mm, Spherisorb S5W	0.02, 0.05, 0.1% HClO$_4$ in MeOH-C6 85:15 2.0 ml min^{-1}	Most alkaloids yielded tailing peaks with reten. volume dependent on col. loading (1–100 µg). For some	267

Applications

LC	84 drugs	UV 254	250 × 5 mm Spherisorb S5W, precol. dry packed with silica gel, 40 μm	MeOH–(94 ml 35% NH$_4$OH + 21.5 ml 75% HNO$_3$ + 884 ml H$_2$O, pH 10.1 with NH$_4$OH) 9:1 2 ml min^{-1}	applications comparison with chem. bonded packings Comparison with other cols. (Hypersil, Nucleosil 50-5, and Zorbax BP-Sil) for morphine, ephedrine, and 5 other basic drugs	268
LC	462 drugs	UV, electrochemical	125 × 4.9 mm, Spherisorb S5W	10 mM NH$_4$ClO$_4$, 0.1 mM NaOH in MeOH, apparent pH 6.7 2 ml min^{-1}	Detector response ratios reported. Most of strongly retained alkaloids yielded tailing peaks	269
LC	166 drugs	UV	300 × 3.9 mm, μBondapak C18 10 μm	H$_2$O–MeOH–AcOH–Et$_3$N from 98:0:1.5:0.5 to 8:90:1.5:0.5 1.5 ml min^{-1}		139

TABLE 7.15 (Continued)

Tech-nique	Compounds separated	Sample	Detection	Col. Stat. phase Col. temp.	Mobile phase flow rate	Remark[a]	Ref.
LC	157 drugs	Urine	UV 230	a. Zorbax C18, 31°C b. PRP-1	a. Linear gradient of B in A A. 0.1% (v/v) H_3PO_4 B. 0.1% (v/v) H_3PO_4, 10% H_2O in ACN 0–100% in 30 min b. Linear gradient of B in A A. 1% (v/v) NH_4OH B. 1% (v/v) NH_4OH, in ACN 2.0 ml min^{-1}	a. Acidic and neutral drugs b. Basic and neutral drugs	270
LC	570 sub-stances		UV 220	250 × 4 mm RP-18	a. Convex gradient of ACN in	Relative reten. times were report-	284

Applications

LC	115 drugs	UV, electrochemical	250 × 4 mm LiChrosorb CN	1 mM NaCl in phosphate, pH 3, ionic strength 0.05–ACN 6:4 1 ml min^{-1}		271, 272
				H$_2$O–ACN 688:312 (w/w), 0.1–20% in 0–1 min; 20% in 1–8 min; 20–80% in 8–24 min; 100% in 24–27 min; 0.1% in 27–35 min; b. 4.8 g liter^{-1} 85% H$_3$PO$_4$, 6.66 g liter^{-1} KH$_2$PO$_4$, pH 2.3–ACN 688:312 (w/w) 1 ml min^{-1}	ed. Reference substance 5-(p-methylphenyl)-5-phenylhydantoin. Col. prewashing with H$_2$O before pure ACN was recommended to prevent salt precipitation	

TABLE 7.15 (Continued)

Technique	Compounds separated	Sample	Detection	Col. Stat. phase Col. temp.	Mobile phase flow rate	Remark[a]	Ref.
LC	50 drugs		UV 254	300 × 4 mm μBondapak C18	H_2O–5 mM C_7SO_3Na in MeOH–ACOH 59:40:1 pH 3.5 2 ml min^{-1}		247
LC	43 heroin adulterants and by-products		UV 220 and 254	300 × 3.9 mm, μBondapak C18 or 250 × 4.6 mm, Partisil 10 ODS-3	20 mM $ClSO_3H$ in H_2O–ACN–H_3PO_4 87:12:1, pH 2.2 with 2 M NaOH 3.0 ml min^{-1}	Absorbance ratios were reported	273
LC	a. 78 drugs b. 62 drugs c. 52 drugs		UV 254 and 280	300 × 3.9 mm a. μBondapak C18 b,c. μPorasil	a. 25 mM NaH_2PO_4 in H_2O–MeOH 3:2 pH 7.0 with 5% NaOH b. MeOH–2 M NH_4OH–1 M NH_4NO_3 27:2:1	Absorbance ratios were reported	274

Applications

LC	161 basic drugs	UV	250 × 5 mm a. Silica gel b. Mercaptopropyl bonded phase c. Strong cation exchanger	50 mg liter^{-1} Na$_2$SO$_3$ in MeOH-2 M NH$_4$OH-1 M NH$_4$NO$_3$ 1 ml min^{-1} c. CH$_2$Cl$_2$-NH$_4$OH conc. 1000:2 2.0 ml min^{-1}	Series of 23 chem. bonded packings were tested with 11 basic drugs. Na$_2$SO$_3$ was added to minimize oxidation of SH packing	275
TLC	Cocaine, heroin, and 6 local anaesthetics	Acidic iodoplatinate and p-dimethylaminobenzaldehyde reagents	Layer silica gel (activated), 0.25 mm	EtAc–PrOH–28% NH$_4$OH 40:30:3	Street drugs. Tentative identification of cocaine and heroin in street drugs	159
TLC	43 illicit drugs and similar compounds	Visualization UV 254, 332. Ninhydrin followed by iodo-	Layer 10 × 10 cm Silica gel G 60 F$_{254}$, one-half dipped into 0.1 M	0, 0.01 and 0.1 M KBr in in MeOH	Simultaneous development from both sides of the plate using Camag	278

TABLE 7.15 (Continued)

Tech-nique	Compounds separated[b]	Sample	Detection	Col. Stat. phase Col. temp.	Mobile phase flow rate	Remark	Ref.
			platinate reagent	$KHSO_4$ for 10 sec, dried and activated		linear developing chamber	
TLC	14–17 major drugs of abuse	Urine	Visualization UV, fluor., ninhydrin-fluorescamine, diphenylcarbazone, silver acetate, mercuric sulfate, 0.5% H_2SO_4, iodoplatinate, I_2-KI reagents	Layer 10 × 20 or 20 × 20 cm, pre-coated silica gel microfiber glass sheets 0.25 mm	EtAc-CH-MeOH-NH_4OH conc.-H_2O 70:15:8:2:0.5, EtAc-CH-MeOH-NH_4OH conc. 56:40:0.8:0.4 EtAc-CH-NH_4OH conc. 50:40:0.1 EtAc-CH 50:60	LOD 0.3 μg ml^{-1} of free M for 35-ml urine sample. Detection reagents were applied in succession to the specific area of the same plate and were combined with heating and examination under UV light	158
TLC	34 drugs	Urine	Visualization visible	Layer Silica gel	iPrOH–toluene–		279

TLC	19 drugs	Urine	Visualization UV, a. FPN, Marquis' b. Dragendorff's, acid. iodoplatinate c. Fast Black K salt d. Fast Black K salt, acid. iodoplatinate reagents	Layers 10 × 20 cm a,b,c. Silica gel 60 F_{254} d. RP-18 F_{254}	a,b,c. EtAc-MeOH-NH_4OH 85:10:5 d. MeOH-H_2O-HCl 50:50:1	Drugs were isolated from the sample using ion pair extraction. R_F values were reported also for 6 other TLC systems	68
			light, UV, fluor., and reagents	60 F_{254}	25% NH_4OH 6:3:1		
TLC	794 drugs		Visualization UV 254, 350 ev. one or more reagents (ninhydrin, FPN, Dragendorff's, acidified	Layers 20 × 20 cm Silica gel 60 F_{254} 0.25 mm; dipped in 0.1 M KOH and dried for mobile phases a,b,c,d	a. MeOH-NH_4OH 100:1.5 b. CH-toluene-Et_2NH 75:15:10 c. $CHCl_3$-MeOH 9:1 d. Acetone e. $CHCl_3$-	R_F values were standardized using 4 reference compounds for each system. Systems a–d were used for 594 basic and some	280

TABLE 7.15 (Continued)

Technique	Compounds separated	Sample	Detection	Col. Stat. phase Col. temp.	Mobile phase flow rate	Remark[a]	Ref.
			iodoplatinate, and Marquis')		acetone 4:1 f. EtAc-MeOH-NH$_4$OH 85:10:5 g. EtAc h. CHCl$_3$-MeOH 9:1	neutral drugs and e-h for neutral and acidic drugs	
TLC	55 basic and neutral drugs		Dragendorff's reagent	Layers 10 × 20 cm Silica gel 60 F$_{254}$ in some instances dipped in 0.1 M KOH and dried or exposed to 30% ammonia vapors just before development	40 eluent mixtures	Principal component analysis cf R_F values	282
TLC	362 drugs		10% Sulfuric acid,	Layers 20 × 10 cm Silica	a. EtAc-MeOH-30%	R_F values were stand-	283

Applications

TLC	570 substances	Dragendorff's, and acidified iodoplatinate reagents	gel 60 F$_{254}$ (for eluent d. dipped in 0.1 M KOH and dried)	NH$_4$OH 85:10:5 b. CH-toluene-Et$_2$NH 65:25:10 c. EtAc-CHCl$_3$ 50:50 d. Acetone	ardized using 4 reference compounds. Principle component analysis
		a. Dragendorff's reagent b. Chlorination with Cl$_2$, 0.5% o-tolidine reagent	Layer 10 cm Silica gel F 254	a. MeOH–25% NH$_4$OH 100:1.5; for quaternary nitrogen bases MeOH–32% HCl 99:1 b. CH$_2$Cl$_2$-acetone 4:1	Relative retentions were reported. Reference compounds a. Bupranolol b. Phenobarbital	284

aHMDS, hexamethyldisilazane; TMCS, trimethylchlorosilane.

TABLE 7.16 Chromatographic Conditions for Alkaloids Derived from Miscellaneous Heterocyclic Systems Including Senecio (Pyrrolizidine), Lupin, (Quinolizidine), and Amaryllidaceae Alkaloids[a]

Technique	Compounds separated[b]	Sample	Detection	Col. Stat. phase Col. temp.	Mobile phase flow rate	Remark	Ref.
GC	Carbazole, 3-methyl-carbazole, glycozoline, mukonal, glycozolidine, heptazolidine, koenimbine, murrayazoline, koenidine	Glycosmis pentaphylla	FID	1.5 m × 2.0 mm a. 3% OV-17 b. 3% SE-30 on Gas Chrom Q a. 220°C for 3 min, 220–295°C at 4 K min^{-1} b. 150°C for 3 min, 150–245°C at 4 K min^{-1}	Nitrogen 50 ml min^{-1}	Glycozoline and glycozolidine were found in Glycosmis pentaphylla by this technique	287
GC	TMS derivs. of 9 polyhydroxy pyrrolidine, piperidine, and indolizi-		FID	5 ft × 4 mm a. 3% OV 101 b. 3% OV-17 on Chromosorb	Nitrogen 40 ml min^{-1}	Alkaloids were isolated from Castanospermum australe, Baphia race-	133

Applications 449

	dine alkaloids			W HP a. 135°C for 3 min, 135–175°C at 4 K min^{-1}, 175°C for 5 min b. 125°C for 2 min, 125–200°C at 5 K min^{-1}, 200°C for 15 min	mosa, Morus nigra, Xanthocercis zambesiaca, Lonchocarpus costaricensis, and Swainsona canescens. Silylation with Sigma Sil A reagent (TMCS-HMDS-pyridine 1:3:9)
GC	25 alkaloids	Lycopodium clavatum var. borbonicum, L. deuterodensum, L. australianum, L. fastigiatum	a. FID, NPD (dual det. system) b. MS	30 m × 0.32 mm, DB-1, 0.1 μm, 50–300°C at 10 K min^{-1}	Helium a. 1.0 bar b. 0.2 bar	Reten. indices for 18 identified and 7 unknown alkaloids were reported	291
GC	Senecivernine, senecionine, integerrimine, retrorsine,	Senecio inaequidens	FID MS	20 m × 0.32 mm, OV-1, 0.1 μm, 120°C for 1 min, 120–230°C	Hydrogen 3 ml min^{-1} (helium for MS)	Smaller tailing on soda lime col. was observed compared	293

TABLE 7.16 (Continued)

Technique	Compounds separated[b]	Sample	Detection	Col. Stat. phase Col. temp.	Mobile phase flow rate	Remark	Ref.
	retrorsine analog			at 5 K min⁻¹, 230°C for 20 min		with Duran 50 glass col.	
GC	b. Isomer of seneciomine, senecionine, seneciphylline, platyphylline or isomer, integerrimine, platyphylline or isomer, jacobine, O-acetylseneciphylline + jacozine, jacoline, jaconine, water adduct of jacozine, acetate ad-	Hay and silage. b. Silage from _Senecio alpinus_	b,c. MS	20 m SE-54 a. 200°C for 1 min, 200–250°C at 5 K min⁻¹ b. 100°C for 1 min, 100–220°C at 10 K min⁻¹ c. 140°C for 1 min, 140–260°C at 6 K min⁻¹	a. Hydrogen b,c. Helium	Stability of pyrrolizidine alkaloids in hay and silage was studied	294

	ducts of jacobine and jacozine c. Retronecine				
GC	N-Oxides of senecionine, integerrimine, seneciphylline	Roots and root cultures of Senecio vulgaris	FID, NPD	25 m × 0.25 mm, DB-1, ICT, 120–290°C at 8 K min^{-1}	Helium 0.7 bar	Quantitation	295
GC	b. TMS derivs. of echinatine, 7-acetyl echinatine, β-angelyl/tiglyl trachelanthamine, β-isovaleryl-, β-angelyl/tiglyl-, 7-angelyl/tiglyl echinatine, 7-angelyl/tiglyl di-	a. Eupatorium cannabinum, Anchusa officinalis b. Roots of Eupatorium rotundifolium var. ovatum	MS (negative ion and positive ion chemical ionization)	a. 25 m × 0.5 mm, 25 m × 0.25 mm a,b. 25 m × 0.32 mm C$_p$-Sil 5 150–325°C at 6 K min^{-1}		Silylation with TMSI-BSA-TMCS 3:3:2. Reten. indices were reported	a. 296 b. 297

TABLE 7.16 (Continued)

Technique	Compounds separated[b]	Sample	Detection	Col. Stat. phase Col. temp.	Mobile phase flow rate	Remark	Ref.
	hydroechinatine, acetyl echimidine, echimidine isomer						
GC	a. Epilupinine, sparteine, caffeine (I.S.), 13-hydroxysparteine, cytisine, tetrahydrorhombifoline, angustifoline, α-isolupanine, lupanine, 17-hydroxylupanine, multiflorine, 17-oxolupanine, 13-	b. Lupinseed of <u>Lupinus angustifolius</u>	a,b. FID c. MS	12 m × 0.2 mm, OV-101 a. 60°C for 1 min, 60–120°C at 30 K min^{-1}, 120°C for 0.5 min, 120–260°C at 10 K min^{-1}, 260°C for 2 min c. 90°C for 2.5 min, 90–250°C at 8 K min^{-1}, 250°C for 5 min	Helium 2 ml min^{-1}	a. See Fig. 4.5. b. Quantitation. CV of weight response factors relative to caffeine 1.68–3.44%	305

Applications

Method	Compounds	Source	Detection	Column	Mobile phase/conditions	Remarks	Ref.
	hydroxy-lupanine b. Caffeine (I.S.), angustifo-line, α-iso-lupanine, lupanine, 13-hydroxy-lupanine						
GC	3 alkaloids	Lupinus arbustus subsp. cal-caratus		10 m × 0.25 mm, OV-17 130–230°C at 4 K min^{-1}		Gramoden-drine could not be quan-tified due to thermal de-composition	306
LC	Vasicinone, vasicine	Leaves of Adhatoda vasica	UV 300	300 × 5 mm APEX-ODS (octadecyl silica) 5 μm	MeOH-CH$_2$Cl$_2$-HClO$_4$ 50:50:0.1 1 ml min^{-1}	Linear cali-bration up to 2 and 10 μg, resp. Recovery 94.3–105.4%. LOD 10 and 20 ng, resp.	285
LC	Vasicine, vasicinone, bromovasi-cinone (I.S.)	Leaves of Adhatoda vasica	UV 254	100 × 8 mm Radial-Pak μBondapak C18, 10 μm	5 mM C$_6$SO$_3$H (PIC B-6) in MeOH-H$_2$O 7:3 0.8 ml min^{-1}	Linear cali-bration 0.01–0.1 mg. LOD 5–10 ng. Also analogs of	286

TABLE 7.16 (Continued)

Technique	Compounds separated[b]	Sample	Detection	Col. Stat. phase Col. temp.	Mobile phase flow rate	Remark	Ref.
						vasicine and vasicinone were separated	
LC	Murrayazoline, heptazolidine, 3-methylcarbazole, glycozoline, koenimbine, glycozolidine, koenidine	Root bark of Glycosmis pentaphylla	UV 254	100 × 8 mm Radial-Pak μPorasil 10 μm	C6-CHCl₃ 75:25 0.7 ml min⁻¹	Resolution on μBondapak C18 with MeOH-H₂O 8:2 or 7:3 was not satisfactory. Glycozoline, glycozolidine, murrayazoline, and 3-methylcarbazole were found in plant. LOD 1–2 ng	288
LC	Perloline	Dried grasses	UV 254	250 × 4.6 mm, LiChrosorb Si-100 10 μm	CHCl₃-MeOH-NH₄OH 90:9.5:0.5 1 ml min⁻¹	Linear calibration 50–500 ng. Recovery 97.7%. CV <	292

Applications

Method	Compounds	Source	Detection	Column	Mobile phase	Notes	Ref.
LC	a. N-Oxides of retrorsine, seneciphylline, senecionine + integerrimine b. Parent alkaloids	Root cultures of *Senecio vulgaris*	UV 209	300 × 3.9 mm μBondapak C18	MeOH-K-Pi buffer 15 mM, pH 7.5 a. 1:2 1.5 ml min^{-1} b. 1.5:1	4% at 300–900 mg kg^{-1}	295
LC	Retrorsine + usaramine, spartioidine, seneciphylline, senecionine + integerrimine, narceine (I.S.)	*Senecio vulgaris*	UV 219	300 × 3.9 mm μBondapak C18 10 μm	10 mM KH$_2$PO$_4$, pH 6.3–MeOH 7:3 2 ml min^{-1} for standard mixtures, 1.4–1.8 ml min^{-1} for samples	Z and E geometrical isomers were not resolved. Linear calibrations up to 0.25–0.64 g liter^{-1}. CV 4.4–13.9%	298
LC	Retrorsine, seneciphylline, senecionine	*Senecio vulgaris*	UV 235	300 × 4 mm μBondapak CN, 10 μm	a. 10 mM (NH$_4$)$_2$CO$_3$, pH 7.8–THF 84:16 1.8 ml min^{-1} b. Linear gradient	a. See Fig. 7.18. Whatman Partisil 10 PAC failed to provide adequate separation	300

TABLE 7.16 (Continued)

Technique	Compounds separated[b]	Sample	Detection	Col. Stat. phase Col. temp.	Mobile phase flow rate	Remark	Ref.
					of THF in 10 mM $(NH_4)_2CO_3$, pH 7.8 13–26% in 30 min 1.8 ml min^{-1}		
LC	a. DHP, senecionine N-oxide, senecionine b. DHP, retrorsine N-oxide, retrorsine	Microsomal incubation mixtures	UV 220 or 254	150 × 4.2 mm, PRP-1	Linear step gradient of ACN in 0.1 M NH$_4$OH, 0% in 0–5 min, 0–10% in 5–15 min, 10–40% in 15–30 min, 40% in 30–40 min 1.5 ml min^{-1}	Direct injection of 20–100 μl of supernatant. Linear calibration up to 0.24, 0.24, and 1 mM. CV 5.6, 6.9, and 6.6% at 107, 58, and 640 μM for DHP, senecionine N-oxide, and senecionine, resp.	301

Method	Compound	Matrix	Detection	Column	Mobile phase	Notes	Ref
LC	Otosenine, petasitenine, senkirkine, neopetasitenine	Petasites japonicus	UV 215, RI	150 × 4.6 mm, Cosmosil 5 Ph 5 μm, 25°C	20 mM $(NH_4)_2CO_3$–MeOH 55:45, pH 8.2 1.0 ml min^{-1}	Otonecine was not found in plant. Linear calibration 0–100 μg for RI detection	299
LC	Lycorine	Bulbs and leaves of Sternbergia lutea	UV 290	250 × 4.6 mm, C18 10 μm	10 mM $(NH_4)_2CO_3$–ACN 53:47 2 ml min^{-1}	Linear calibration 0.5–8.0 μg. Recovery 97.9%. CV 0.5%. LOD 5 ng	309
LC	Galanthamine, phenacetin (I.S.)	Serum, urine, bile	UV 235	100 × 4.6 mm, CP Micro Spher Si, 3 μm	C_6–CH_2Cl_2–ethanolamine 500:500:0.25 1 ml min^{-1}	Effects of mob. phase composition, type of amine and col. were examined. Linear calibration 0–100 ng ml^{-1}. Recovery 100.2% at 50 ng ml^{-1}. CV 18.9–2.5% at 10–100 ng	310

TABLE 7.16 (Continued)

Technique	Compounds separated[b]	Sample	Detection	Col. Stat. phase Col. temp.	Mobile phase flow rate	Remark	Ref.
LC	Epigalanthamine, galanthamine, galanthaminone, codeine (I.S.)	Plasma, urine	UV 280	125 × 4 mm Hibar LiChrosorb RP-8, 5 μm	5 mM Bu_2NH, 0.2 ml $liter^{-1}$ 85% H_3PO_4, 400 ml $liter^{-1}$ MeOH, pH 7, 1.2 ml min^{-1}	ml^{-1}. LOD 5 ng ml^{-1} Galanthaminone was not detected in plasma. Linear calibration 0.1–10 μg ml^{-1}. Recovery 97.5%. CV 0.94–4.66% at 6.0–0.5 μg ml^{-1}	53
SFC	Senecionine, integerrimine, senkirkine, neosenkirkine, retrorsine, otosenine, hydroxysenkirkine, anonamine		FID	10 m × 0.05 mm a. Crosslinked methylsilicone, 0.25 μm b. SB-Biphenyl-30 0.25 μm	CO_2 100 atm (1.84 cm sec^{-1}) for 20 min, 100–280 atm at 3 atm min^{-1}	See Fig. 7.19. Alkaloids were isolated from Senecio anonymus	25

Applications

TLC	15 carbazole alkaloids		Visualization 2% benzoyl peroxide in CHCl$_3$	Layers 20 × 10 cm Silica gel G	Benzene-CHCl$_3$ 1:1	R$_F$ values and colors developed were reported. LOD 0.05 µg	289
TLC	Murrayazoline, heptazolidine, 3-methylcarbazole, carbazole, koenimbine, glycozoline, glycozolidine, mukonal, koenidine		Visualization fluor. under UV 254, picric acid	Layer 20 × 20 cm alumina, 0.25 mm	Petroleum ether-AcOH 8:2 in both directions	Separations on silica gel and using other eluents containing CHCl$_3$, C6, petroleum ether, and AcOH were not sufficient for preparative work	290
TLC	Ellipticine, 9-methoxy-, and 9-hydroxy-ellipticine	Cell cultures of Ochrosia elliptica	Visualization fluor. under UV 366. Scanning fluor.[c]	Layers silica gel	EtAc-H$_2$O-octanol 17:2:2 30 min	Conditions for fluorescence measurement in solution and on TLC plates were investigated. Linear calibration 100–5000 fmol.	31

460   Chapter 7

TABLE 7.16 (Continued)

Technique	Compounds separated[b]	Sample	Detection	Col. Stat. phase Col. temp.	Mobile phase flow rate	Remark	Ref.
TLC	Senecionine, retrorsine	Senecio inaequidens	Visualization UV 254, Dragendorff's reagent	Layers Silica gel 60 F254	CH_2Cl_2-MeOH 25% NH_4OH 85:14:1	CV 2-10%. LOD 40 fmol for ellipticine	293
TLC	Senecionine, senecionine N-oxide	Root cultures of Senecio vulgaris	Scanning radioactivity	Layers Silica gel 60 F254	$CHCl_3$-MeOH-NH_4OH 91:15:3 Toluene-EtAc-Et_2NH 5:3:2	^{14}C tracer studies	295
TLC	Lasiocarpine N-oxide, lasiocarpine, heliotridine, europine	Aerial part of Heliotropium ellipticum	Visualization Dragendorff's reagent, iodine vapors	Layers Silica gel G	$CHCl_3$-MeOH-NH_4OH 85:14:1	Separation and isolation. Some spots could not be identified	303
TLC	Doronine, senkirkine	Aerial part of Emilia	Visualization Drag-	Layers Silica gel 60	CH_2Cl_2-MeOH-25%	Preparative TLC. Detec-	304

Applications

	sonchifolia	endorff's reagent	F254, 0.25 mm	NH₄OH 85:14:1	tion with Ehrlich reagent was not possible	
TLC	Gramodendrine, ammodendrine, lusitanine, gramine	<u>Lupinus arbustus</u> subsp. <u>calcaratus</u>	Visualization Dragendorff's and Ehrlich's reagents	Layers a. Silica gel 60 GF 254, 0.25 mm b. Silica gel 60 PF 254, 1 mm	a. CHCl₃-MeOH-18 M NH₄OH 100:10:1 b. CH-Et₂NH 7:3	b. Preparative TLC	306
TLC	Lupanine, 13-hydroxyysparteine, anagyrine, cytisine, calycotomine	<u>Genista anatolica</u>	Visualization UV 254, 366, Dragendorff's reagent	Layers a. Silica gel GF254 b. Silica gel 60 PF 0.5 mm	CH-Et₂NH 7:3, 9:1 CHCl₃-MeOH-25% NH₄OH 85:15:1	b. Preparative TLC. R_F values were reported	307
TLC	Epimers of alangimarckine, emetine, isoemetine		Visualization UV 254, I₂-KI reagent	Layers Silica gel 60 GF254, aluminum oxide GF254	14 mobile phases	R_F values were reported	308

[a]For purine bases, see Table 7.17. For other examples, see Chapter 4, Section IV.E; Chapter 5, Sections II.B, V.A; Chapter 6, Sections I.D and II.E; and Table 4.2.
[b]BSA, bis(trimethylsilyl)acetamide; HMDS, hexamethyldisilazane; TMCS, trimethylchlorosilane; TMS-, trimethylsilyl-; TMSI, trimethylsilylimidazole; DHP, 6,7-dihydro-7-hydroxy-1-hydroxymethyl-5H-pyrrolizidine.
[c]Fluorescence scanning at λ_{ex} = 303 nm, λ_{em} > 440 nm for ellipticine, after impregnation with dimethylsulfoxide and drying at λ_{ex} = 290 nm, λ_{em} > 440 nm for 9-methoxy- and 9-hydroxyellipticine.

TABLE 7.17 Chromatographic Conditions for Purine Bases[a]

Technique	Compounds separated[b]	Sample	Detection	Col. Stat. phase Col. temp.	Mobile phase flow rate	Remark	Ref.
GC	Caf	Coffee, tea, maté, beverages	FID	2 m × 2 mm 3% OV-17 on Chromosorb W-HP 80–100 mesh 220°C			311
GC	Caf	<u>Citrus sinensis</u> flowers and leaves	FID, MS	1% OV-17 on Supelcoport, 150–250°C at 8 K min^{-1}		Caf was found at levels 3–62 µg g^{-1}	312
GC	Caf, mepivacaine (I.S.)	Urine	NPD	2 m × 2.5 mm, 3% OV-7 on Chromosorb W HP a. 195°C b. 180°C	Nitrogen 25 ml min^{-1}	a. 7 metabolites did not interfere. Extrelute extraction b. Fully automated extraction using polystyrene resin col. Linear calibration 0–20	a. 314 b. 315

GC	Caf, bupivacaine (I.S.)	a. Plasma b. Milk	NPD	1.83 m × 2 mm, 3% OV-17 on Gas Chrom Q, 100–120 mesh, conditioned for 36 hr at 280°C and treated with 10% TMCS in toluene a. 240°C b. 250°C	Helium 74 ml min^{-1}	µg ml^{-1}. Recovery a. 90% b. 63.9–66.4% at 2 and 1 µg ml^{-1}. CV b. 2.2–6.3% at 1.55–0.48 µg ml^{-1}. LOD 0.4 µg ml^{-1} (b. for 1 ml urine), 1.75 ng Caf injected	
						Linear calibration 1–10 µg ml^{-1}. Recovery a. 99.7% b. 94.1% CV a. 4.47 and 4.33% b. 2.9 and 1.1% at 1 and 10 µg ml^{-1}, resp. LOD 50 ng ml^{-1} (1 ng)	48

TABLE 7.17 (Continued)

Technique	Compounds separated[b]	Sample	Detection	Col. Stat. phase Col. temp.	Mobile phase flow rate	Remark	Ref.
GC	Caf, 7-(2-chloroethyl)Tp (I.S.)	Plasma	NPD	3 m × 2 mm 3% SP-2250 on Supelcoport, 80–100 mesh 230°C	Helium 30 ml min^{-1}	Col. was primed with soy phosphatides in benzene at the beginning of each working day. CV 1–9.6% at 0.005–10 µg ml^{-1}. LOD 5 ng ml^{-1}	51
GC	PFBB derivs. of Tp, 3-isobutyl-1-methylX (I.S.)	Plasma	MS, NPD	20 m × 0.32 mm, SE 52, 0.52 µm 80°C for 1 min, 80–280°C at 30 K min^{-1}	Helium 2 ml min^{-1}	^{15}N-labeled Tp coadministered with Tp. Calibration 0–6 µg ml^{-1}. Recovery 86.7%. CV 1.39 and 4.42% at 6 and 1 µg ml^{-1}, resp. LOQ 10 ng ml^{-1}	46

Applications

LC	Tb, Tp, Ox, 8-hydroxymethyl-Tp, Caf	UV 273	a. 100 × 4 mm, C18 bonded to silica gel SG-7/G, 8 μm with various densities of coverage b. 150 × 4 mm, LiChrosorb RP-18 5 μm	0.1 M AcOH in H_2O-ACN 9:1, pH 4.05	Effect of surface coverage on Tp retention was studied. LOD 50 μg liter^{-1}	316
LC	26 xanthines	UV 280	250 × 4.6 mm, Spherisorb 5-ODS 5 μm with guard col. 75 × 2.1 mm, pellicular RP (Chrompack)	AcONa in H_2O-0.1 M AcOH-ACN 81.5:12.5:6, pH 5.30 1.5 ml min^{-1}	Contributions of N-methyl and C-hydroxyl substituents to retention were investigated	317
LC	Tb, dyphylline, Tp, Caf, 8ClTp	UV 254	300 × 3.9 mm, μBondapak C18 10 μm	H_2O-MeOH-AcOH-Et$_3$N 83:15:1.5:0.5 1.5 ml min^{-1}		139

TABLE 7.17 (Continued)

Tech-nique	Compounds separated[b]	Sample	Detection	Col. Stat. phase Col. temp.	Mobile phase flow rate	Remark	Ref.
LC	Tb, Tp, Caf		UV	a. 90 × 1 mm b. 61 × 1 mm C18, 5 μm	15% ACN in phosphate buffer 0.085–0.465 ml min^{-1} 0.7 ml min^{-1}	a. Separation in 20 sec. Linear calibration 3–50 μg ml^{-1}. CV 4.4–0.5%	318
LC	3-Hydroxyacetanilide (from aspirin), 3-aminophenol, acetaminophen, Caf	Tablets	UV 240	250 × 4.6 mm, Zorbax ODS, 5 μm	17.7 g KH$_2$PO$_4$ + 700 ml H$_2$O + 300 ml MeOH, pH 3.6 with HCl 1 ml min^{-1}	Linear calibration up to 5 μg. CV 2.1% for Caf. Sample was shaken and refluxed with 3-aminophenol in anh. EtOH	319
LC	Caf, acetylsalicylic acid, salicylic acid, propylphenazone	Three-component analgesic	UV 295	a. 30 × 2.1 mm b. 30 × 4.6 mm Spherisorb RP-8, 5 μm	H$_2$O-MeOH-H$_3$PO$_4$ 70:30:3 a. 1.7 ml min^{-1} b. 8.0 ml min^{-1}	See Fig. 7.20. Very high-speed LC	320

LC	Caf	UV 272	150 × 6 mm, Fine-Pak Sil C18	MeOH-H$_2$O 55:45, 1.2 ml min^{-1}		352
LC	a. Caf, Tp, Tb b. Tb, Tp, Caf	UV 273	150 mm a. Spherisorb 800 5 μm b. LiChrosorb RP-18	a. CH$_2$Cl$_2$-MeOH 9:1 1.25 ml min^{-1} b. 10 mM AcONa, pH 4-ACN 9:1 3 ml min^{-1}	a. Determination of Caf	311
LC	Caf	UV 277	250 × 4 mm Bio-Sil ODS-5S	H$_2$O-MeOH 75:25 1 ml min^{-1}	Linear calibration 10–500 ng. Recovery 93.8–98.3%. CV 0.82–4.65%	321
LC	a,b. Uracil, U, X, Tb, Tp, Caf c. Aspirin, Caf phenacetine d. Caf	c. UV 254 d. UV 280	a,b,c. 250 × 7 mm d. 100 × 7 mm Carboxyl poly-(methyl- or chloromethyl-styrene-divinyl-	a. 0.1 M NaNO$_3$, 0.1 M Tris-MeOH 8:2, pH 2–10 b. 49 mM Na$_2$HPO$_4$, 38 mM citric acid, pH 4.28, 0–25%	a,b. Effects of pH and MeOH percentage on retention were studied. d. Time of analysis about 100 min	322

TABLE 7.17 (Continued)

Tech-nique	Compounds separated[b]	Sample	Detection	Col. Stat. phase Col. temp.	Mobile phase flow rate	Remark	Ref.
				benzene) polymeric packing, 7 μm b. 75°C c. 70°C d. 22°C	MeOH 1 ml min^{-1} c. Phosphate, pH 10–MeOH 7:3 0.5 ml min^{-1} d. 0.1 M NaNO$_3$, 0.1 M citric acid, pH 2.1–MeOH 8:2 0.8 ml min^{-1}		
LC	Caf, 8ClTp (I.S.)	Tea	UV 254	250 × 2 mm Varian MCH-10 with universal 40 × 2 mm guard col.	1% AcOH-ACN 12:3 1.5 ml min^{-1}	Linear calibration 0–12.5 μg. CV 2.0%. LOD 25–50 ng	323
LC	Saccharin, Caf, aspartame,	Cola beverages	UV 214	300 × 4.6 mm, μBondapak C18	1 ml liter^{-1} Et$_3$N, 0.8 ml liter^{-1}	Degassed samples were diluted	324

Applications

	benzoates				with H_2O 5:100 before injection. Recovery 99.1–101.6% at 20–200 mg liter^{-1}. LOD 0.5 mg liter^{-1} for Caf	
LC	Caf	Citrus sinensis flowers and leaves	UV	250 × 10 mm, Nucleosil 10 C18	(10% MeOH, H_3PO_4 pH 3.2)–MeOH 8:2 3 ml min^{-1}		312
LC	Caf, Tp	Paullinia cupana, Cola spp.	UV 280	250 × 4 mm a. Hibar-LichroCART RP-18, 7 µm b. Nucleosil 5 C18	MeOH-20 mM Me$_4$NBr 7:3 1.5 ml min^{-1}		325
LC	Tb, Tp, Caf	Cocoa, coffee, tea, liquid and solid food products	UV 275	150 × 4 mm LiChrosorb RP-8, 5 µm 45°C	H_2O-MeOH-0.2 M KH_2PO_4, pH 5.0 with 0.2 M K_2HPO_4 36:9:5 1.0 ml min^{-1}	Linear calibration 0.05–0.25 µg Tp, Tb, 0.1–0.5 µg Caf. Recovery 92.6–103.3%. CV	326

TABLE 7.17 (Continued)

Technique	Compounds separated[b]	Sample	Detection	Col. Stat. phase Col. temp.	Mobile phase flow rate	Remark	Ref.
LC	Tb, Tp, Caf	Cocoa, milk chocolate	UV 280	250 × 1 mm Micro B ODS-3	H_2O-MeOH-AcOH 68:31:1 0.06 ml min^{-1}	0.37–14.9%. LOD 5 µg g^{-1} Tp, Tb, 10 µg g^{-1} Caf for 2 g of liquid sample	327
LC	a. 7MX + Tb, Tp, adenosine, Caf b,c. 7MX, adenosine, Tb, Tp, Caf	Cocoa	UV 254	a. 250 × 7.6 mm Asahipak GS 32OH (polyvinyl alcohol), 9 µm, with guard col. b,c. 110 × 4.7 mm Partisphere	a. 12 mM C12SO$_4$Na, 5 mM Na$_2$HPO$_4$, pH 11.5 with NaOH 2 ml min^{-1} b. 15% MeOH, 1% AcOH, pH 2.8 and	See Fig. 7.21. CV 4.07% Tb, 1.85% Caf. LOD 100 pg Tb, Tp, 150 pg Caf	328

LC	7-EthylTp (IS), Caf, Tb, Tp, PX, 137-TMU (I.S.)	Plasma	UV 280	Hibar R5 250-4 col. LiChrosorb Si-60 5 μm	CH_2Cl_2-2 g liter^{-1} NH_4 formate, 0.15 ml liter^{-1} formic acid in MeOH 975:25, (965:35 new, 982: 18 old col.) 2 ml min^{-1} (new), 1.2 ml min^{-1} (old col.)	Linear calibration up to 800, 200, 200, and 400 nmol ml^{-1}. CV (interday) 3.7, 7.7, 11.2, and 7.1% at 31.2, 3.7, 2.5, and 14.0 nmol ml^{-1} for Caf, Tb, Tp, and PX, resp. Recovery 85% for [^{14}C]-Caf	329
				C18, 5 μm, with guard col. 50 × 4.6 mm Whatman Co:Pell ODS, 30–38 μm	4.8 1 ml min^{-1} c. Linear gradient of B in A A. 20 mM KH_2PO_4, 0.1% MeOH, pH 5.5 B. MeOH 0–45% in 30 min 1.5 ml min^{-1}		

TABLE 7.17 (Continued)

Tech-nique	Compounds separated[b]	Sample	Detection	Col. Stat. phase Col. temp.	Mobile phase flow rate	Remark	Ref.
LC	Tb, PX, Tp, βHETp, Caf	Plasma	UV 273	150 × 4.6 mm, LC-18 DB 5 μm	1.75 mM $H_3PO_4^-$ ACN-THF 97:2:1 1.2 ml min^{-1}	Linear calibration 0–1.5 μg ml^{-1} Tb, PX, Tp, 0–3 μg ml^{-1} Caf. LOD 25 ng ml^{-1} Tb, PX, Tp, 50 ng ml^{-1} Caf	51
LC	37DMU, 3MX, 13DMU, Tb, PX, Tp, Caf, 8ClTp (I.S.)	Blood, urine, milk, saliva	UV 276	150 × 4.6 mm, C18 5 μm	H_2O-iPrOH-ACN-AcOH 91:4:4:1	CV 12.9, 20.8, 4.8, 4.1, 5.7, and 8.4% for 3MX, 13DMU, Tb, PX, Tp, and Caf at 10 μg ml^{-1}	330
LC	Tb, PX, Tp, βHETp (I.S.), Caf	Plasma	UV 254	300 × 3.9 mm, μBondapak C18 10 μm, with guard col.	10 mM AcONa, pH 5.0– MeOH-THF 95:	Linear calibration 2–10 μg ml^{-1} Caf, 0.5–5 μg/ml^{-1} Tb,	331

Applications

LC	Tb, PX, Tp, βHETp (I.S.), Caf	Plasma	UV 273	Bondapak C18/Corasil	4:1 3 ml min⁻¹	PX, Tp in 3% bovine serum albumin. Recovery 89%. CV 8%. LOD 100 ng ml⁻¹ Caf, 50 ng ml⁻¹ Tb, PX, Tp in 300 μl of plasma	
				Nova Pak C18 with Guard-Pak CN cartridge	1% AcOH-MeOH 83:17 2.7 ml min⁻¹	Reten. times of 25 compounds tested for interference are given. Linear calibration 0–5 μg ml⁻¹ Tb, PX, Tp, 0–10 μg ml⁻¹ Caf. Recovery 82.27–94.93%. CV 2.7–7.46%. LOD 50 ng ml⁻¹ Tb, PX, Tp, 100 ng ml⁻¹ Caf	332

TABLE 7.17 (Continued)

Technique	Compounds separated[b]	Sample	Detection	Col. Stat. phase Col. temp.	Mobile phase flow rate	Remark	Ref.
LC	17DAU, 1MU, 137-DAU, 1MX, 13DMU, 17-DMU, Tp + PX, TMU, Caf	Urine of rats, mice, and chinese hamster	Radioactivity	C18	0.05% AcOH, 8% MeOH	Metabolic disposition of [1-Me^{14}C] Caf	333
LC	U, 7MU + 7MX, 1MU, 3MX, 1MX, 13DMU, Tb, 17-DMU, PX, Tp, 137-TMU, Caf, proxyphylline (I.S.)	Urine	UV 280	250 × 4.5 mm, Hypersil ODS 5 μm	Gradient of B in A A. 1% THF, 10 mM acetate, pH 4.8 B. 15% ACN, 1% THF, 10 mM acetate, pH 4.8, 0% in 0–5 min, 0–25% in 5–10 min, 25–75% in 10–15 min,	Ion pair extraction was compared with direct injection of diluted sample. Linear calibrations 0–200 mg liter^{-1}. Recovery 65.9–97.8% at 50 mg liter^{-1}. CV 4.4–19.7% for direct injection, 2.75–15.81%	334

LC	Tp, βHETp (I.S.), Caf	Serum	UV 280	100 × 1 mm Spherisorb C18, 3 μm	50 mM acetate, pH 5.0-ACN 93:7 0.08 ml min^{-1}	75–85% in 15–20 min, 1.5 ml min^{-1} for extraction. Linear calibration 2–20 mg liter^{-1}. CV 1.5 and 3.5% (within-run), 6.4 and 7.4% (day-to-day). LOD 1.0 and 1.5 mg liter^{-1} for Tp and Caf, resp.	335
LC	Tb, PX, Tp, Caf, enprophylline, D4126 (I.S.)	Serum	UV 280	250 × 4 mm Regis Octyl Hi-Chrom 5 μm	3% THF, 10 mM Na$_2$HPO$_4$, pH 6.5 with diluted H$_3$PO$_4$ 2 ml min^{-1}	Linear calibration 0.5–20 mg liter^{-1}. Recovery 90.7–103%. CV < 7% (between-day), < 5% (within-day) at 0.5 and 5 mg liter^{-1}. LOQ 0.05 mg liter^{-1} Tb, Tp, 0.1	336

TABLE 7.17 (Continued)

Tech-nique	Compounds separated[b]	Sample	Detection	Col. Stat. phase Col. temp.	Mobile phase flow rate	Remark	Ref.
						mg liter^{-1} PX, enprophylline and Caf	
LC	AAMU, 1MX, benzyloxy-urea (I.S.)	Urine	UV 254 or 265	300 × 7.5 mm, Bio-Gel TSK-20 10 μm	0.1% AcOH 0.8 ml min^{-1}	Run time 45 min. Linear calibration 4–100 and 10–40 mg liter^{-1}. Recovery 99.2%. CV 2.4 and 4.4% for AAMU and 1MX, resp. LOD 0.2–0.4 mg liter^{-1} in 50 μl of urine	337
LC	a. Caf, βHPTp (I.S.) b. Tp, Caf, βHPTp	a. Serum	UV 273	150 mm, Supelcosil LC-18, 3 μm with 30 mm guard col.	0.5% AcOH-MeOH 20:5 1.5 ml min^{-1}	Linear calibration up to at least 20 mg liter^{-1}. Re-	47

Applications

	(I.S.)			pellicular LC-18		covery 104%. CV 2.3 and 3.8% at 7.77 and 15.73 mg liter^{-1}, resp. (within-run) 11.0% and 8.6% at 7.63 and 15.55 mg liter^{-1}, resp. (between-day)	
LC	PX, Tp, βHETp (I.S.), Caf	Plasma	UV 214	Radial-Pak C18, 5 μm	10 mM KH$_2$PO$_4$, pH 3.5– MeOH-THF 90:9:1 2.5 ml min^{-1}	Reten. times for other xanthines are reported. Recovery 77–97% dependent on reconstituting solvent. CV < 5.5%. LOD 2 ng	338
LC	Caf, βHETp (I.S.)	Plasma	UV 274	250 × 4 mm LiChrosorb Hibar RP-18	H$_2$O-ACN-AcOH 89.4: 10:0.6 1.4 ml min^{-1}		339

TABLE 7.17 (Continued)

Technique	Compounds separated[b]	Sample	Detection	Col. Stat. phase Col. temp.	Mobile phase flow rate	Remark	Ref.
LC	Caf	Serum, saliva	a. UV 280 b. MS	250 × 4.6 mm, Hypersil ODS 5 μm	0.1 M AcONH$_4$, pH 4.6-ACN 85:15 1.5 ml min^{-1}	a. Linear calibration 0.2–1.6 mg liter^{-1}. Recovery > 95%. CV 2.5% (intraassay) 5% (interassay) at 1.6 mg liter^{-1}. LOD 5 ng ml^{-1} for 0.5 ml of sample b. LOD 10 ng in full-scan mode, 200 pg (1 mg liter^{-1} for 1 μl of sample) in selective ion monitoring mode	340

LC	Tb, PX, βHETp (I.S.), Tp, Caf	Serum	UV 280	150 × 4.6 mm with 3 cm precol. Ultrasphere C18 ion pair, 5 μm	20 mM Bu$_4$NOH, 15 mM Tris in H$_2$O-ACN-MeOH 93:3.5:3.5, pH 7.50 ± 0.02 with conc. HCl 1.2 ml min^{-1}	k values for 31 xanthines and common drugs are reported. Recovery 93.2–111.5%. CV 0.6–6.1% for Tp and Caf at toxic to subtherapeutic levels	49
LC	a,b. 3MX, 1MU, 13-DMU, Tp, dyphylline (I.S.) c. 3MU, 7MX, 3MX, 1MX, 1MU, 13DMU, tryptophan, PX, Tp, dyphylline, 19DMU, Caf	a. Plasma b. Urine	UV 280	150 × 4 mm Ultrasphere IP 5 μm, with guard col. 40 × 2 mm, LiChroprep RP-18 25–40 μm 24°C	0.1 M AcONa, pH 4.0, 0.75 mM decylamine a,c. 2% ACN b. 1% ACN 2 ml min^{-1}	Linear calibration a. 2–25 mg liter^{-1} Tp, 0.05–5 mg liter^{-1} its metabolites b. 5–50 mg liter^{-1} Tp, 10–100 mg liter^{-1} metabolites CV a. < 7.3% b. < 11.4% LOD a. 0.01, 0.01, 0.02, and 0.1 mg liter^{-1} b. 0.5, 0.5,	341

TABLE 7.17 (Continued)

Technique	Compounds separated[b]	Sample	Detection	Col. Stat. phase Col. temp.	Mobile phase flow rate	Remark	Ref.
						0.5, and 1 mg liter^{-1} for 3MX, 1MU, 13DMU, and Tp, resp.	
LC	1MU, 3MX, 13DMU, Tp, 8ClTp (I.S.)	Urine	UV 280	250 × 5 mm Hypersil ODS, 5 μm	Linear gradient B in A A. 1.28 g liter^{-1} AcONa.3 H$_2$O, 0.4% AcOH, pH 4 B. 1.28 g liter^{-1} AcONa.3 H$_2$O, 0.5% AcOH, 20% ACN, 0–80% in 0–22 min, 80% in 22–27 min 0% in 27–37	Reten. times for 14 methyluric acids and methylxanthines were reported. Linear calibrations 0–20 mg liter^{-1}. CV 7.5 and 1.6% (intraassay), 9.7 and 5.3% (interassay) at 1 and 10 mg liter^{-1}, resp. LOD 0.5 mg liter^{-1}	71

LC	1MU, 3MX, 13DMU, Tp, βHETp (I.S.), Caf	a. Urine b. Serum	UV 275	a. 300 × 3.9 mm μBondapak C18, 10 μm b. 250 × 4.6 mm, Ultrasphere ODS, 5 μm	Linear gradient of MeOH in 10 mM NaH_2PO_4, pH 4.5 a. 0% in 0–2 min, 0–6% in 2–5 min, 6% in 5–8 min, 6–26% in 8–18 min, 26% in 18–20 min, 0% in 20–30 min b. 6% in 0–8 min, 6–21% in 8–13 min, 21% in 13–20 min, 6% in 20–30 min 2.0 ml min^{-1}	No Caf was found in samples. Linear calibrations approx. 0–300 ng. Recovery a. 72–89% b. 66–102% CV a. 7.3–11.1% b. 1.6–8.1%	342

min 1 ml min^{-1}

TABLE 7.17 (Continued)

Technique	Compounds separated[b]	Sample	Detection	Col. Stat. phase Col. temp.	Mobile phase flow rate	Remark	Ref.
LC	3MX, 1MU, 13DMU, Tp, βHETp (I.S.), Caf	Serum	UV 273	125 × 4.6 mm, C8 5 μm with precol. μBondapak C18/Corasil 45°C	10 mM AcONa, 5 mM Bu$_4$N phosphate, 4% MeOH pH 4.5 with AcOH 2 ml min^{-1}	See Fig. 7.22. Linear calibrations approx. 0–30 mg liter^{-1}. Recovery 66.6–96.2%. CV < 4.3% (intraday), < 3.8% (interday). LOD 10–30 μg liter^{-1}	343
LC	a. Caf, 3-isobutyl-1-methylX (I.S.), PX, Tp b. 3-isobutyl-1-methylX (I.S.), Tp	b. Serum	UV 254 or 270	125 × 4.6 mm, LiChrosorb Si-60	CH$_2$Cl$_2$-MeOH-25% NH$_4$OH 95.8:4:0.2 1.6 ml min^{-1}	Linear calibration 0–100 μmol liter^{-1} (also for Caf, Tb, and PX). LOD 3.7 μmol liter^{-1} (0.66 mg liter^{-1})	50

Applications

LC	Tp	Blood, serum, plasma	UV 275	250 × 4.6 mm, Silica gel A 10 μm	C6-CH$_2$Cl$_2$-THF-MeOH-AcOH 75:14:7.6:3:0.4	Comparison of Tp content in serum, plasma, and whole blood	344
LC	a. 8ClTp (I.S.), Tp, Tb, PX b. Tp, βHETp (I.S.), PX, Tb, Caf	Plasma	UV 273	a. 250 × 4.6 mm b. 250 × 4.1 mm Spherisorb Phenyl 5 μm	a. 20% MeOH, 20 mM KH$_2$PO$_4$ pH 5.6 with 8% H$_3$PO$_4$, 1.5 ml min^{-1} b. 2% ACN, 8% MeOH, 20 mM KH$_2$PO$_4$, pH 5 1.8 ml min^{-1}	Comparison of two extraction procedures. a. Caf was removed by back-extraction with 0.1 M NaOH. Linear calibration 0.05–50 mg liter^{-1}. Recovery 83%. CV 0–14.3% at 0.08–26 mg liter^{-1}. LOD 0.07 mg liter^{-1}	345
LC	Tp, βHETp (I.S.)	Plasma	UV 280	250 × 5 mm Hypersil ODS, 5 μm	1.28 g liter^{-1} AcONa.3 H$_2$O, 5.3% (w/v) ACN, 0.5% AcOH,	Calibration 0–20 mg liter^{-1}. Run time 15 min. Comparison with GC	46

TABLE 7.17 (Continued)

Technique	Compounds separated[b]	Sample	Detection	Col. Stat. phase Col. temp.	Mobile phase flow rate	Remark	Ref.
LC	a. Tp, βHETp (I.S.) b. PX, Tp, βHETp (I.S.)	a. Plasma b. Urine	UV 276	50 × 4.6 mm Sepralyte C18 3 μm 50°C	pH 4 1.5 ml min^{-1} 50 mM $NH_4H_2PO_4$, 10 mM H_3PO_4, 1% DMF, 4% MeOH 1.5 ml min^{-1}	Linear calibration a. 1–40 mg $liter^{-1}$ b. 1–100 mg $liter^{-1}$ CV a. 1.9–7.8% b. 0.6–8.2% LOD a. 0.25 mg $liter^{-1}$ b. 1 mg $liter^{-1}$	52
LC	Tp, dyphylline, βHETp (I.S.)	Serum	UV 280	100 × 5 mm Radial-Pak Nova-Pak C18, 4 μm with C18 precol. cartridge	10 mM AcONa, 6% ACN, pH 4.0 with AcOH 2.3 ml min^{-1}	16 substances including Caf, PX, 13DMU, 3MX, and 1MU did not interfere. Recovery 97–99.4% at 5–40 mg	66

LC	a. Tp, βHPTp (I.S.) b. MU, 3MX, 13-DMU, PX, Tp, Caf, βHPTp (I.S.)	a. Serum	UV 280	100 × 8 mm Radial-Pak with pre-col. cartridge μBondapak C18	10 mM AcONa, 8% ACN, pH 4.0 with AcOH 4.8 ml min^{-1}	Calibration 0–50 mg liter^{-1}. Recovery 95.4–100%. CV < 6.41%. LOD 0.1 mg liter^{-1}	67
LC	a. Tp b. 7MU, etofylline, Tp, 37-DMU, Caf	a. Serum	UV 270	200 × 4.6 mm, Hypersil ODS 5 μm, 50°C	10 mM KH$_2$PO$_4$/K$_2$HPO$_4$, pH 7.0, 0.25 ml liter^{-1} Et$_3$N, 4.0 g liter^{-1} Bu$_4$NBr, 1% THF 2.5 ml min^{-1}	Linear calibration 1.8–54.5 mg liter^{-1}. Recovery 96–101%. CV 1.6–1.2% at 1.7–38.6 mg liter^{-1}. 85 substances were tested for possible interference with Tp	346
LC	a. Tp, 8ClTp (I.S.) b. PX,	a. Serum	UV 273	32 × 4.6 mm, C18-bonded silica, 3 μm	50 ml liter^{-1} ACN, 30 ml liter^{-1}	Separation completed in 40 sec. Linear calibra-	347

TABLE 7.17 (Continued)

Technique	Compounds separated[b]	Sample	Detection	Col. Stat. phase Col. temp.	Mobile phase flow rate	Remark	Ref.
	Tp, acetaminophen, Caf, 8ClTp				THF, 0.5 ml liter^{-1} AcOH, pH 4.9 with NaOH 2 ml min^{-1}	tion 5–40 mg liter^{-1}. CV 2.7% (within-run), 4.6% (between-day)	
LC	Px, Tp	Serum	UV 275	60 × 4.6 mm, Hypersil ODS 3 μm, with two precol. cartridges (push-pull mode) 20 × 4.6 mm, Hypersil ODS 5 μm	50 mM AcONa, 10 mM Bu$_4$HSO$_4$, pH 4.2, gradient to 10% ACN	Column-switching technique. Linear calibration 2.5–25 mg liter^{-1}. Recovery 99.3% at 11.8 mg liter^{-1}. CV 1.7–2.7% (in series), 3.2% (day-to-day)	348
LC	Tp, βHETp (I.S.)	Serum	UV 273	100 × 4.6 mm, Spherisorb ODS 2 5 μm	H$_2$O-1 M NH$_4$PO$_3$, pH 5.0– THF 96.5:2:1.5	Recovery 50% after 3 min dialysis combined with trace	349

Applications

LC	Tp	Serum	UV 270	125 × 4 mm LiChrosorb RP-18, 5 μm	15% ACN, 0.1% AcOH 0.6 ml min^{-1}	enrichment. CV 2.1–4.6% CV < 5%	350
LC	37DAU, 3MU, 7MU, 7MX, 3MX, 37DMU, Tb	Urine	UV 254, radioactivity	Partisil PXS 10/25 ODS-3 and guard col. Co:Pell ODS	Gradient of MeOH in 60 mM H$_3$PO$_4$/ KH$_2$PO$_4$, pH 2.4, 4% in 0–17 min, 4–15% in 17–18 min, 15% in 18–22 min, 15–4% in 22–23 min 1.5 ml min^{-1}	Comparison of labeled Tb (^{14}C at position 8) metabolism in dog, rabbit, hamster, rat, and mouse	351
SFC	Caf	Coffee	UV 250, 270, and 200–350 (diode array)	150 × 6 mm Fine-Pak Sil C18 40°C	CO$_2$-MeOH (0.1 ml min^{-1}) 5 ml min^{-1} 150 bar	Sample solvent influenced the retention significantly	352
SFC	Caf, Tp, Tb		UV 270 MS	100 × 4.6 mm LiChrosorb 5 μm	12% MeOH in CO$_2$ 2.5 ml min^{-1} 385 bar	Run time 4 min. Deterioration of col. performance was observed	353

TABLE 7.17 (Continued)

Tech-nique	Compounds separated[b]	Sample	Detection	Col. Stat. phase Col. temp.	Mobile phase flow rate	Remark	Ref.
TLC	Caf, Tp, and 15 drugs		Visualization UV 254 and 365	Layers 20 × 20 cm Silica gel GF, 0.25 mm	CH_2Cl_2-MeOH-H_2O 183:27:5, EtAc-toluene-DMF-H_2O-formic acid 75:75:4:4:2, CH_2Cl_2-MeOH 183:27, EtAc-toluene-DMF-formic acid 75:75:4:2	R_F values were reported for adulterants of Chinese herbal preparations	354
TLC	Caf, Tp, PX, TMU, TMA, 1MX, 13DMU, 137DAU, 17DMU, 1MU	Urine	Visualization by autoradiography. Scanning radioactivity	Layers Silica gel, 0.25 mm	Two-dimensional 1. $CHCl_3$-MeOH 4:1 2. BuOH-$CHCl_3$-acetone-NH_4OH conc. 4:3:3:1	Metyluric acids were unstable on silica gel plates	333

| PC | Caf, Tb | Cocoa and cocoa products | Visualization UV 254 | Paper 5 × 20 cm Whatman 3MM | Multiple development 1. Light petroleum (1.5 hr) 2. BuOH satd. with NH4OH (two times 1.5–2 hr and rechromatography 3–4 hr) | Determination by UV spectroscopy after extraction of zones with 1% NH4OH | 355 |

[a] For other examples, see Chapter 4, Sections III.B, IV.E, V.B; Chapter 5, Sections I.A, II.B, III.A, IV.A, V.D; and Chapter 6, Section II.E.

[b] AAMU, 5-acetylamino-6-amino-3-methyluracil; Caf, caffeine (1,3,7-trimethylxanthine); 8ClTp, 8-chlorotheophylline; D4126, 3,7-dihydro-1-ethyl-3-(2-hydroxypropyl)-1H-purine-2,6-dione; 17DAU, 6-amino-5-(N-formylmethylamino)-3-methyluracil; 37DAU, 6-amino-5-(N-formylmethylamino)-1-methyluracil; 137DAU, 6-amino-5-(N-formylmethylamino)-1,3-dimethyluracil; 13DMU, 1,3-dimethyluric acid; 17DMU, 1,7-dimethyluric acid; 19DMU, 1,9-dimethyluric acid; 37DMU, 3,7-dimethyluric acid; βHETp, β-hydroxyethyltheophylline; βHPTp, β-hydroxypropyltheophylline; MU, methyluric acid; 1MU, 1-methyluric acid; 3MU, 3-methyluric acid; 7MU, 7-methyluric acid; 1MX, 1-methylxanthine; 3MX, 3-methylxanthine; 7MX, 7-methylxanthine; PFBB, pentafluorobenzyl bromide; PX, paraxanthine (1,7-dimethylxanthine); Ox, oxyphylline (7-(2-hydroxyethyl)theophylline); Tb, theobromine (3,7-dimethylxanthine); TMA, trimethylallantoin; TMCS, trimethylchlorosilane; Tp, theophylline (1,3-dimethylxanthine); TMU, trimethyluric acid; 137TMU, 1,3,7-trimethyluric acid; U, uric acid; X, xanthine.

TABLE 7.18 Chromatographic Conditions for Simple Indole Alkaloids

Technique	Compounds separated[b]	Sample	Detection	Col. Stat. phase Col. temp.	Mobile phase flow rate	Remark	Ref.
GC	PFBB deriv. of harman	Rat lung	MS	15 m × 0.25 mm, DB-5 100°C for 5 min, 100–250°C at 20 K min^{-1}	Helium, 1–2 ml min^{-1} (0.4 bar)	Proof of LC identification	365
LC	Nitrosation products of gramine		UV 254	250 × 4 mm Spherisorb C18, 10 μm	Gradient 25–40% or 40–45% MeOH 2 ml min^{-1}		356
LC	Phy, Sal	Physostigmine salicylate	UV 261	300 × 3.9 mm, μBondapak C18 10 μm	H$_2$O-MeOH-AcOH-Et$_3$N 63:35:1.5: 0.5 1.5 ml min^{-1}		139
LC	a. Pilocarpine, phenethyl alcohol b. Pilocarpine, Sal, methyl p-hydroxy-	c. Aged Phy solution	a,b. UV 235 c. UV 292	300 × 3.9 mm, μBondapak C18 10 μm	40% MeOH 5 mM C7SO$_3$H, pH 3.6 1.0 ml min^{-1}	Pilocarpine was not resolved from isopilocarpine. LOD 1, 20, and 3 ng for Rub, pilo-	357

Applications

	Compounds	Sample	Detection	Column	Mobile phase	Remarks	Ref.
	benzoate, Phy, propyl p-hydroxybenzoate c. Rub, Sal, Phy					carpine, and Phy, resp. CV of peak height 1.3 and 0.3% for pilocarpine and Phy, resp.	
LC	Phy and metabolites	Rat plasma, muscle, brain, and other tissues	Radioactivity of fractions	300 × 3.9 mm, μBondapak C18	5 mM C8SO$_3$H, 5 mM NaH$_2$PO$_4$, 1% AcOH in H$_2$O–MeOH 6:4, pH 3.1 2 ml min^{-1}	[^{3}H-]Phy administered	358–360
LC	Phy, Ese, N,N-dimethyl analog of Phy (I.S.)	Plasma, whole blood	Fluor. λ_{ex}254 nm λ_{em}346 nm	125 × 4.9 mm Spherisorb 5 ODS 1, 5 μm, precol. 50 × 2 mm, pellicular ODS 37–53 μm	ACN–10 mM AcONa 95:5 2 ml min^{-1}	On fully capped Spherisorb ODS-2 Phy was not retained. Linear calibration 0.1–3 ng ml^{-1}. Recovery 102–93.8%. CV 4.9–0.8% at 0.1–3 ng ml^{-1}. LOD	361

TABLE 7.18 (Continued)

Tech-nique	Compounds separated[b]	Sample	Detection	Col. Stat. phase Col. temp.	Mobile phase flow rate	Remark	Ref.
LC	a. Rub, Phy, Ese b. Phy	b. Plasma	a. UV 254, electro-chemical b. Electro-chemical	250 × 4.5 mm, Spheri-sorb, 5 μm	MeOH–1 M NH$_4$NO$_3$ pH 8.6, 9:1 (degassed to remove oxygen) 1 ml min^{-1}	0.1 ng ml^{-1} of Phy Linear cali-bration 0.5–20 ng ml^{-1}. Recovery 93%. CV 6.3 and 7.3% at 10 and 1 ng ml^{-1}. LOD 0.5 ng ml^{-1} in 3 ml of plasma	362
LC	N,N-Di-methylana-log of Phy, (I.S.) Phy	b. Plasma, blood, urine	Dual elec-trochemical	a. 150 × 4.6 mm, Spheri-sorb, 3 μm b. Two cols. in series	a. MeOH-or b. (MeOH-ACN 1:1)–0.1 M NH$_4$NO$_3$, pH 8.9, 9:1 a. 0.5 ml min^{-1} b. 1 ml min^{-1}	Linear cali-bration 0.1–10 ng ml^{-1}. CV 10.22–1.55% (intra-assay), 13.6% at 2.86 ng ml^{-1} plasma, 16.6% at 2.96 ng ml^{-1} blood (interassay). LOD 25 pg ml^{-1}	363

LC	a. Rub, Ese, Phy b. Phy, 2 major and 6 minor metabolites	b. Mouse liver microsomal incubations	Dual electrochemical	250 × 4.6 mm, Biophase Octyl 5 μm, 35°C	17 mM C12SO$_4$Na in 0.1 M NaH$_2$PO$_4$, 0.1 M H$_3$PO$_4$, pH 3–ACN 6:4 1 or 2 ml min^{-1}	Electrochemistry of standards and major metabolites was studied	364
LC	Phy	Plasma	Dual electrochemical	250 × 4.6 mm, Biophase Octyl 5 μm, 35°C	34 mM C12SO$_4$Na in 0.1 M NaH$_2$PO$_4$, 0.1 M H$_3$PO$_4$, pH 3–ACN 6:4 2 ml min^{-1}	See Fig. 7.23. Recoveries 35, 53, 62, and 63% at 1, 10, 14, and 100 ng ml^{-1}, resp. LOD 0.5 ng ml^{-1}	45
LC	Harman and 16 other β-carbolines		UV 254	300 × 3.9 mm, μBondapak C18	a. 30–80% MeOH in phosphate, pH 7, I = 0.05 M b. 1 g liter^{-1} C12SO$_4$Na, 0.1 M NaClO$_4$ in MeOH–phosphate, pH 3, 65:35	Estimation of lipophilicity	366

494   Chapter 7

TABLE 7.18 (Continued)

Tech-nique	Compounds separated[b]	Sample	Detection	Col. Stat. phase Col. temp.	Mobile phase flow rate	Remark	Ref.
LC	a. Harmol, harmalol, harman, harmine, harmaline b. 6-Hydroxyharman, 6-methoxyharman c. Harmol, 6-hydroxyharman, harmine, 6-methoxyharman	Root of <u>Grewia villosa</u>	UV 260	250 × 4.5 mm, Spherisorb ODS S5 C6, 5 μm	MeOH-H$_2$O-NH$_4$OH a,b. 45:54:1 followed by 60:39:1 after elution of harmine c. 50:49:1 1 ml min^{-1}	Harmaline was not detected in the plant extract	367
LC	Harmol, norharman, 6-methoxyharman, harman, harmine, harmalol, harmaline	7 species of Helico-niini butterflies	Fluor. λ$_{ex}$370 nm λ$_{em}$425 nm	300 × 4 mm C18 col. (Varian) 10 μm	Linear gradient 65–90% MeOH in 5 steps during 33 min, 0.01% Et$_3$N 2 ml min^{-1}	Calibration 1–100 mg liter^{-1}. LOD 10 ng (harman)–0.1 ng (6-methoxyharman). Quantitative results for	368

LC	Harmine, 6-hydroxy-7-methoxy-harman, 3-harmol, 3- or 4-hydroxy-7-methoxy-harman	Mouse liver, microsomal incubation mixtures	UV 254, radioactivity of fractions	250 × 4.6 mm, Spherisorb, 5 μm	CH_2Cl_2-EtOH-NH_4OH Linear gradient 95:5:0.5 to 76:24:0.4 over 20 min 1 ml min^{-1}	norharman, harman, and harmine were reported. Harmalol was not detected in samples. [^{3}H-]harmine and [^{3}H-]harmol were used as substrate	369
LC	a. O-Acetylnorharmol, O-acetylharmol, norharman, 1-harman, 1-propyl-9H-pyrido[3,4-b]indole (I.S.) b. Harman, 1-propyl-	Rat and mouse lung	Fluor. λ_{ex}252 nm λ_{em}430 nm	250 × 4.6 mm, Biosphere C18 5 μm	0.5% Et_3N in MeOH-H_2O 6:4 1.4 ml min^{-1}	Harmols were converted to their acetoxy derivs. by treatment with acetic anhydride. Calibration 0.5–10 ng. CV 2.1, 4.6, 1.9, and 5.6%, resp.	365

TABLE 7.18 (Continued)

Tech-nique	Compounds separated[b]	Sample	Detection	Col. Stat. phase Col. temp.	Mobile phase flow rate	Remark	Ref.
	9H-pyrido-[3,4-b]indole (I.S.)					LOD 750, 500, 150, and 100 pg/sample, resp.	
LC	Psilocybin, psilocin	<u>Psilocybe semilanceata</u>	UV 254, fluor. λ_{ex} 267 nm λ_{em} 335 nm	250 × 4.6 mm, Partisil 5, 6 μm	MeOH-H$_2$O-1 M NH$_4$NO$_3$, pH 9.6 with 2 M NH$_4$OH 22:7:1 1 ml min^{-1}	Linear calibration 5–200 mg liter^{-1}. Recovery > 98%. CV 4.7, 2.4, and 2.5% at 10, 50, and 200 mg liter^{-1} (for standard solutions)	27
LC	Baeocystin, psilocybin, psilocin	<u>Psilocybe semilanceata</u>	UV 254, fluor. λ_{ex} 267 nm λ_{em} 320 nm, electrochemical	250 × 4.6 mm, Partisil 5 μm	1 mM EDTA in MeOH-H$_2$O-1 M NH$_4$NO$_3$, pH 9.6 with 2 M NH$_4$OH	Detector response ratios aided the identification. Extract of <u>Panaeolus</u>	371

LC	Psilocybin, psilocin	Psilocybe bohemica, a,c,d. P. semilanceata c. Inocybe aeruginascens	a,c. UV 267 b. UV 280 a,c. Fluor. λ_{ex}280 nm λ_{em}360 nm b,c. Electrochemical	a,c. 250 × 2 mm, LiChrosorb RP-18,5 μm b. 250 × 4.6 mm Partisil ODS,10 μm or 250 × 4 mm, Separon SI C18 10 μm d. 500 × 8 mm, LiChrosorb RP-18 10 μm	a,c,d. H_2O-EtOH-AcOH 79.2:20:0.8 b,c. (0.1 M citric acid-0.1 M NaH_2PO_4 30:16, pH 3.8)-EtOH 9:1 a,c. 0.333 ml min^{-1} b. 1 ml min^{-1} d. 3 ml min^{-1}	22:7:1 1 ml min^{-1} rickenii was separated for comparison. LOD UV: 10, 7.5 ng Fluor: 5, 20 ng Electrochemical: 5, 75 pg for psilocybin and psilocin, resp. d. Prep. chromatography. CV a. < 3.4% b. UV: 3.02, 10.1%. Electrochemical 2.35, 1.37%. LOD a. UV: 20, 40 ng. b. UV: 80, 93 ng. Electrochemical 2, 12 ng for psilocybin and psilocin, resp.	a,d. 372 b. 373 c. 374

TABLE 7.18 (Continued)

Tech-nique	Compounds separated[b]	Sample	Detection	Col. Stat. phase Col. temp.	Mobile phase flow rate	Remark	Ref.
LC	Psilocybin, psilocin, N,N-dimethyltryptamine (I.S.)	<u>Psilocybe subaeruginosa</u>	UV 267, fluor. λ_{ex}267 nm λ_{em}335 nm	250 × 4.6 mm, Partisil SCX, 10 µm with pre-col. 30 × 2.8 mm pellicular beads, 30 µm, 50°C	0.2% $(NH_4)_3PO_4$, 0.1% KCl in H_2O-MeOH 8:2, pH 4.5 1 ml min^{-1}	See Fig. 5.17. Linear calibration 0–10 mg $liter^{-1}$ by UV, 0–6 ng for psilocybin by fluor. detection. LOD UV: 7 and 150 ng Fluor.: 0.25 and 30 ng for psilocybin and psilocin, resp.	375
LC	a. Psilocin, psilocybin + baeocystin b. Psilocybin, baeocystin	<u>Psilocybe semilanceata</u>	UV 267, 254, or 290	300 mm × 1/4 in. µBondapak Ethylphenyl, 10 µm	5 mM C_7SO_3H in a. H_2O-EtOH 65:35 b. H_2O, pH 3.5 with AcOH 1.0 ml min^{-1}	Linear calibration 0.25–500 mg $liter^{-1}$. CV 3%. LOD 6 ng	376

LC	a. Baeocystin, psilocybin, bufotenine (I.S.), psilocin b. Baeocystin, psilocybin, bufotenine, psilocin, serotonin, 4-hydroxytryptamine, tryptamine, dimethyltryptamine	a. Psilocybe semilanceata, P. cubensis	UV 269 (photodiode array)	250 × 4.6 mm, Spherisorb ODS-1 5 μm	Linear gradient of B in A. 0.3 M AcONH$_4$ pH 8 with conc. NH$_4$OH B. 0.3 M AcONH$_4$ in MeOH. 0% in 0–2 min, 0–95% in 2–14 min, 2.0 ml min^{-1}	Linear calibration 20–200 mg liter^{-1}. CV 2.5–3.7%. LOD 15, 10, 8 ng for baeocystin, psilocybin, resp. Serotonin (0.2%) was found in _Panaeolina foenisecii_ and bufotenine (1.8%) in _Amanita citrina_	28
OP-TLC	a. 10-Methoxycanthin-6-one b. 3-Methoxycanthin-2,6-dione, canthin-2,6-dione		TLC of fractions, fluor.	Layer 20 × 20 cm Silica gel 60 F$_{254}$, 2 mm	a. Ethylene chloride-C6-THF 6:2:2 b. CHCl$_3$-THF-MeOH 70:25:5 followed by 70:20:10 after 60 min 3.0 ml min^{-1}	Prep. OP-TLC. On-plate injection of sample solution. Compounds were isolated from the wood of _Simaba multiflora_	370

TABLE 7.18 (Continued)

Tech-nique	Compounds separated[b]	Sample	Detection	Col. Stat. phase Col. temp.	Mobile phase flow rate	Remark	Ref.
TLC	Harman and 16 other β-carbolines		Visualization UV 254	Layers KC18F	a. 30–80% MeOH in phosphate pH 7, I = 0.05 M b. 1 g liter^{-1} C12SO4Na, 0.1 M NaClO4 in MeOH-phosphate, pH 1–5 65:35	Lipophilicity estimation	366
TLC	Psilocin, psilocybin	Psilocybe semilance-ata, P. bohemica b. Inocybe aeruginas-cens	Ehrlich's reagent	Layers Silufol UV 254	BuOH-H2O-AcOH 24:10:10	b. Psilocin was not found in Inocybe aeruginas-cens	a. 372 b. 374
TLC	a. Psilocin, baeocystin, psilocybin b. Baeo-	Psilocybe semilance-ata	Van Urk's reagent	a. Layer 10 × 20 cm Silica gel 60H-Silica	PrOH-H2O-AcOH 10:3:3	b. Isolation of baeocys-tin	376

Applications

	cystin		gel 60G F_{254} 1:1 (3.2 g) b. Layer 20 × 40 cm Silica gel 60PF$_{254}$ with gypsum-Silica gel 60PF$_{254}$ 1:1 (23 g)		
TLC	Psilocybin	Inocybe aeruginascens, Psilocybe semilanceata	Layer 20 cm, Silufol	Modified Ehrlich's reagent	BuOH-H$_2$O-AcOH 2:1:1	Two unidentified spots were found. Psilocin was not detected. No alkaloid reacting with Ehrlich's reagent was found in Psilocybe callosa	377

aFor other examples, see Chapter 4, Section IV.F; Chapter 5, Sections I.A, III.A, IV.A, V.C, and V.D; and Chapter 6, Section II.F; and Tables 5.7 and 6.6.
bEse, eseroline; Phy, physostigmine; PFBB, pentafluorobenzyl bromide; Rub, rubreserine; Sal, salicylic acid.

TABLE 7.19 Chromatographic Conditions for Indole Alkaloids of Yohimbine Type, Rauwolfia Alkaloids, and Related Bases[a]

Technique	Compounds separated[b]	Sample	Detection	Stat. phase Col. temp.	Mobile phase flow rate	Remark	Ref.
GC	Vca, ethaverine (I.S.)	Plasma	NPD	12 m × 0.5 mm, OV-1, 0.4 μm 130–210°C at 30 K min^{-1}, 210°C for 5 min, 210–250°C at 5 K min^{-1}	Helium 0.7 ml min^{-1}	Linear calibration 0–200 ng ml^{-1}. Recovery 55.6%. CV 5.1% at 100 ng ml^{-1}. LOQ 3 ng ml^{-1}	414
GC	AAEE, AAME (I.S.)	Plasma	MS	12.5 m × 0.2 mm OV-1, 150–270°C at 16 K min^{-1} 270°C for 4.5 min	Helium 0.8 ml min^{-1}	Recovery > 95%. CV 0.31–10.12% at 50–0.1 ng ml^{-1}. LOD 0.1 ng ml^{-1}	415
GC	Hirsutine and 8 oxindole alkaloids	Mitragyna stipulosa	FID	0.5 m, 1% SE-52 on Chromosorb HMDS 80–100 mesh 240°C	Nitrogen 15 psi	Reten. times were reported	423

Applications

LC	Y	Fortified horse serum	UV 280	300 × 3.9 mm μBondapak C18	ACN-MeOH-1.2 mM HCl 6:3:1 1 ml min^{-1}	Linear calibration 1–1000 ng ml^{-1}. Recovery 95.6–105.1%. CV 4.5–1.1% at 1–500 ng ml^{-1}. LOD 1 ng ml^{-1}	378
LC	a. Y, physostigmine (I.S.) b. Physostigmine, corynanthine, Y, R, yohimbic acid	Blood	Fluor. λ_{ex} 280 nm λ_{em} 360 nm	a. 250 × 2.1 mm Partisil 5, 5 μm b. 250 × 2.1 mm or 100 × 2.1 mm, Partisil 5, Novapak C18, Spherisorb C18, 5 μm μBondapak C18, 10 μm	a. MeOH-H$_2$O 95:5 b. MeOH-H$_2$O 8:2 to 10:0	Curved calibration 0.05–1000 ng ml^{-1}. For 5-ml sample: CV 15.7–4.1% at 0.05–30 ng ml^{-1}. LOD 0.05 ng ml^{-1}. For 0.1-ml sample: CV 5.9–2.6% at 10–1000 ng ml^{-1}. LOD 2 ng ml^{-1}. b. Separation was poor on end-capped No-	379

TABLE 7.19 (Continued)

Technique	Compounds separated[b]	Sample	Detection	Col. Stat. phase Col. temp.	Mobile phase flow rate	Remark	Ref.
LC	ABA (I.S.), Y	Plasma	Electro-chemical	300 × 3.9 mm, μBonda-pak C18 10 μm	10 mM NH_4 phosphate in H_2O-MeOH 52:48 1 ml min^{-1}	vapak C18 packing Recovery 82–93%. CV < 6.3% at 50–600 ng ml^{-1}. LOD 10 ng ml^{-1}	380
LC	Y, reser-piline (I.S.)	Plasma	Electro-chemical	300 × 3.9 mm, μBonda-pak C18	MeOH-0.4 M AcONa, pH 6.0 6:4 1.5 ml min^{-1}	Linear calibration 0–250 ng ml^{-1}. Recovery 75%. CV 14.0–12.4% at 57–185 ng ml^{-1}. LOD 10 ng ml^{-1}	381
LC	a,c,d. Clopamide, dihydroer-gocristine, R	Brinerdine tablets	UV 260	a,b. 100 × 2.1 mm c. 30 × 4.6 mm d. 30 × 2.1	a. ACN-0.4 M H_3PO_4 17:3 3 ml min^{-1} b,c,d.	a. See Fig. 7.24. Content uniformity test using high-	320

	b. Clopamide, dihydroergocristine, R 3,4-dehydroR		Fluor. λ_{ex}280 nm λ_{em}360 nm	mm, Spherisorb RP-18 5 µm	ACN-0.2 M H_3PO_4 13:7 b. 1 ml min^{-1} c. 30 ml min^{-1} d. 5 ml min^{-1}	speed LC. CV 0.15, 1.1, and 1.4%. LOD 1.5, 3.0, and 0.9 ng b. Resolution of possible degradation products. Time of analysis a. 18 sec b. 90 sec c,d. 4 sec	
LC	R, 3,4-dehydro-R, 3-iso-R	Tablets	Fluor. λ_{ex}280 nm λ_{em}360 nm	300 × 3.9 mm, µPorasil	MeOH 1.5 ml min^{-1}	Recovery 99–101%	382
LC	a. R b. Rc c. Rn, Ajc, Y, Ajm, S	Rauwolfia serpentina powders and tablets	Fluor. a,c. λ_{ex}280 nm λ_{em}370 nm b. λ_{ex}330 nm λ_{em}370 nm	300 × 3.9 mm, µPorasil	MeOH 1.5 ml min^{-1}	R and Rc coeluted. S eluted at about 225 min	383
LC	Methyl reserpate, R		Fluor. λ_{ex}395 nm λ_{em}470 nm	Bonded phase phenyl col.	ACN-5 mM or 50 mM NaH_2PO_4,	Linear calibration 10–100 ng ml^{-1}.	a. 384 b. 385

TABLE 7.19 (Continued)

Technique	Compounds separated[b]	Sample	Detection	Col. Stat. phase Col. temp.	Mobile phase flow rate	Remark	Ref.
			after post-col. (b. photochemical) derivat.		pH 6 with NaOH 7:3 1 ml min^{-1}	CV a. 6.44 and 5.8% b. 8.0 and 3.8% at 20 and 80 ng ml^{-1}. LOD a. 200 pg b. 80 pg	
LC	a. R b. Rc	Spiked urine	Electro-chemical	250 × 4.6 mm with 3-cm guard col., BAS Biophase ODS, 5 μm	5 mM C7SO$_3$Na in MeOH-50 mM KH$_2$PO$_4$ pH 4.5 with H$_3$PO$_4$ 65:35 1 ml min^{-1}	Dependence of detection response and retention time on pH of the mob. phase was measured. Linear calibration a. 5–40 ng b. 10–50 ng LOD a. 0.9 ng b. 0.8 ng	386

Applications

LC	Ajm, 7-chloroace-tylAjm, pindolol (I.S.)	Tablets, vials	UV 244, 287	300 × 4.6 mm Micro Pack MCH-10 25°C	ACN-5 mM NaClO$_4$, pH 3.0 with HClO$_4$ 4:6 for 2 min, linear gradient up to 6:4 in 2-10 min 1.5 ml min^{-1}	Linear calibration 0-2.5 μg. Recovery 97.1-101.2%. CV 1.0-1.3%. LOD 5 and 10 ng at 244 and 287 nm, resp.	387
LC	a. Ajm b. R, Rc	Bark of <u>Rauwolfia</u> <u>vomitoria</u>	a. UV 292 b. UV 250	250 × 4.0 mm, LiChrosorb RP-18 7 μm a. 45°C b. 40°C	a. 40 mM NaH$_2$PO$_4$, 10 mM Na$_2$HPO$_4$-THF-PrOH 95:4:1 b. 40 mM NaH$_2$PO$_4$, 10 mM Na$_2$HPO$_4$-ACN-PrOH-THF 70:13:13:4 1 ml min^{-1}	a. Frequent washing of col. with MeOH-H$_2$O 1:1 was required. b. PrOH improved selectivity, THF reduced tailing. CV<7%	388
LC	Ajc	<u>Catharan-</u> <u>thus roseus</u>	UV 254	250 × 4.6 mm, Alltech C18, 10 μm	0.5 ml liter^{-1} conc. NH$_4$OH in MeOH-H$_2$O	Direct injection of methanolic extract reduced with	391

TABLE 7.19 (Continued)

Tech-nique	Compounds separated[b]	Sample	Detection	Col. Stat. phase Col. temp.	Mobile phase flow rate	Remark	Ref.
					8:2 0.6 ml min^{-1}	sodium borohydride. Linear calibration 0.8–1.6 μg. Recovery 93.4–101.3%. CV 7.66 and 19.6% within and between assays, resp.	
LC	a. Tryptamine, Vdo, Ajc + epiAjc, Ca, THA, Vbl, S b. Ajc c. S	b,c. Cell cultures of Catharanthus roseus (fractions from silica gel cartridge)	UV 254	300 × 3.9 mm μBondapak C18 10 μm, with a guard col. Bondapak-C18/Corasil 37–50 μm	MeOH–5 mM (NH$_4$)$_2$HPO$_4$ b. 67:33, pH 7.3 c. 77.5: 22.5, pH 7.3 1.0 ml min^{-1}	a. Influences of MeOH content, pH, and ionic strength were investigated. Linear calibration up to about 300 ng. LOD b. 9 ng c. 15 ng	392

Applications

LC	a. Epivindolinine, vindolinine, hoerhammericine, minovincinine, 20-hydroxytabersonine, pleiocorpamine, Ca, vallesiachotamine + isovallesiachotamine, Ajc, 3-iso-Ajm, akuammigine, THA, R (I.S.), tabersonine b,c,d. Y, Ajc, Ca, R e. Y, Ajc, Ca, ibogaine (I.S.) f. Ajc, Ca, R (I.S.)	a. Cell cultures of Catharanthus roseus e,f. Spiked extracts	a,b. LiChrosorb RP-18, 7 μm c. LiChrospher 100 CH-8 d. Nucleosil RP-18	a. MeOH-H_2O-Et_3N gradient from 40:60: 0.5 to 80: 20:0.5 b,c,d,e,f. ACN-10 mM Et_3N formiate, pH 8.5 b,c,d,e. Gradient from 2:8 to 8:2 f. 5:5	b,c,d. Different selectivities of various packings were demonstrated. e. Linear calibrations CV < 4% for Ajc and Y. Extraction to $CHCl_3$, Extrelute, and Bond-Elut SCX procedures were compared	394

TABLE 7.19 (Continued)

Tech-nique	Compounds separated[b]	Sample	Detection	Col. Stat. phase Col. temp.	Mobile phase flow rate	Remark	Ref.
LC	Ajm, S	Cell cultures of Catharanthus roseus	UV 254, 280	100 × 8 mm Radial compression cartridge µBondapak C18	Linear gradient of B in A A. 50 mM C_7SO_3 B. 50 mM C_7SO_3 in 95% MeOH. 50% for 14 min, 50–75% in 14–26 min, 75% in 26–30 min 2 ml min^{-1}		395
LC	5-Methoxy-tryptamine, S, Vdo, Ca, Vcr, Vbl	a. Leaves b. Roots of Catharanthus roseus	UV 280 and 254	220 × 4.6 mm Spheri-5 RP 18, 5 µm with a guard col. 15 × 3.2 mm, Aquapore ODS 7 µm	Linear gradients in two steps MeOH-ACN-25 mM AcONH$_4$-Et$_3$N a. pH 6.8 13:32:55:	Peak homogeneity was tested by absorbance ratio plots and scanning of UV spectra. Linear calibration 30–	396

Applications

LC	Vindoli-nine, Vdo, Vcr, Ca, Vbl, leuro-sine, Co	Catharan-thus ro-seus (Vin-ca rosea)	UV 290	250 × 4.6 mm, R-Sil-C 18 HL-D with a pre-col. 100 mm, R-Sil, R-Sil-C18 HL-D or their mix-ture, 20 μm 5–50°C	0.1% etha-nolamine in MeOH–H_2O, gra-dient from 50:5 to 85:15 2 ml min^{-1}	0.2 to 19: 46:35:0.2 b. pH 7.5, 18:27:55: 0.2 to 26: 39:35:0.2 1 ml min^{-1}, after 17 min 1.5 ml min^{-1}	110 mg liter^{-1} for Ca, 52–156 mg liter^{-1} for Vdo. CV < 3.52 mg liter^{-1}. LOD 10 mg liter^{-1}. a. Ajc co-eluted with Ca	Temperature influence on reten. and plate num-ber was evaluated under iso-cratic condi-tions. In-ferior col. efficiency for dimeric alkaloids was ob-served	397
LC	Lochner-ine, Ajc,		UV 222, fluor.	250 × 4.6 mm, LiChro-		Electro-lytes	Effect of in-organic buf-	398

TABLE 7.19 (Continued)

Tech-nique	Compounds separated[b]	Sample	Detection	Col. Stat. phase Col. temp.	Mobile phase flow rate	Remark	Ref.
	Ca, Vdo, S, Vcr, Vbl		λ_{ex} 226 nm λ_{em} >340 nm	sorb RP-18 10 μm or LiChrospher SI-100	(KH_2PO_4, K_2HPO_4, $NH_4H_2PO_4$, H_3PO_4, H_2SO_4) in MeOH-H_2O 85:15 or 70:30 3 ml min^{-1}	fer conc. on reten. was examined. Use in isolation of dimeric alkaloids was proposed	
LC	a. Leurosidine, Vbl, leurosine b,c. Vbl d,e. Vdo, Ajc, Ca	Catharanthus roseus b,c. callus culture with differentiated roots d,e. plants and multiple shoot culture	UV 214	a,b,d. 150 × 6 mm Shim-Pak CLC ODS c,e. 250 × 20 mm Shim-Pak Prep ODS 60°C	ACN-MeOH-5 mM $(NH_4)_2$-HPO_4, pH 7.3, 4:3:3 a,b,d. 1 ml min^{-1} e. 3 ml min^{-1}	c,e. Prep. purification	a,b, c. 399 d,e. 393
LC	a. Vindesine, Vcr, (I.S.), b. MO_4	a. Plasma, urine b.	UV 220	250 × 4 mm with guard col. 30 ×	ACN-Na phosphate, pH 3, I =	a. Stability of alkaloids was exam-	a. 400 b.

	DAVbl, Vbl b. Vcr (I.S.), Vb	mouse fibrosarcoma cells	4 mm LiChrosorb CN, 5 μm	0.08 M a. 65:35 b. 60:40 1.5 ml min^{-1}	ined. Retentions for 31 other drugs were reported. Recovery 97.3–103%. CV 4.0–5.9% at 25 and 100 ng ml^{-1}. LOD 6 ng ml^{-1} b. Recovery 101.3% at 100 ng ml^{-1} and 90.6% at 200 ng ml^{-1} for Vbl and and Vcr, resp.	401
LC	Vcr (I.S.), DAVbl, vindesine, Vbl	a. Plasma b. Urine	150 × 3.9 mm, Hypersil ODS 5 μm, with a guard col. 20 × 3.9 mm, LiChrosorb RP-8, 5–20 μm, or	Electrochemical	MeOH-10 mM phosphate, pH 7.0, 65:35 or in the range 65:35 to 55:45 depending on col. used.	Vbl was used as I.S. in the Vcr analysis. Linear calibrations 1–1000 ng ml^{-1}. Recovery and CV were de-	54, 402

TABLE 7.19 (Continued)

Technique	Compounds separated[b]	Sample	Detection	Col. Stat. phase Col. temp.	Mobile phase flow rate	Remark	Ref.
				100 × 4.0 mm, Chromsep Hypersil ODS 5 μm	1 ml min^{-1}	pendent on Bond-Elut Diol batch. For Bond-Elut CN: a. 69.7–105.6% b. 76.5–98.5% at 10 and 100 ng ml^{-1}. LOD 0.1 ng for Vbl, 0.25 ng for Vcr	
LC	DAVcr, DAVbl, Vcr, LY 119863, Vbl	Blood, tissues	UV 254, radioactivity of fractions	Partisil 10/25 ODS-3	Linear gradient from 20% MeOH (contg. 10 mM KH_2PO_4, pH 4.9) to to 64% MeOH (pH 4.9) over a	Recovery > 93%. Metabolism of [^{3}H-]Vcr and [^{3}H-]-Vbl was studied	403

Applications

LC	4-Succinyl-4-DAVbl-N-oxide, 4-succinyl-4-DAVbl, DAVbl-N-oxide, DAVbl, Vcr (I.S.)	Plasma	UV 270, radioactivity of fractions	300 × 3.9 mm, µBondapak C18	period of 60 min, 80% MeOH 16 65 min, 1 ml min^{-1}	Disposition of KSI/4-[^{3}H-]DAVbl was studied. Recovery of free DAVbl 80%	404
LC	a. DAVbl, 19'-hydroxy-3',4'-dehydroVbl, Vbl, isomer of Vbl, 19'-oxo-Vbl, 3',4'-dehydro-19'-oxoVbl b. DAVcr, isomer of DAVcr, Vcr, iso-	Incubated a. Vbl b. Vcr	UV 254	300 × 4.6 mm, µBondapak C18 10 µm	Linear gradient of MeOH in 20 mM K phosphate, pH 7.4. 10% in 0–4 min, 10–65% in 4–20 min 1 ml min^{-1}	Tentative assignment of degradation products	a. 405 b. 406
				10 mM KH$_2$PO$_4$, pH 4.5-MeOH 1:1 1.2 ml min^{-1}		

TABLE 7.19 (Continued)

Technique	Compounds separated[b]	Sample	Detection	Col. Stat. phase Col. temp.	Mobile phase flow rate	Remark	Ref.
	mer of Vcr, 3-deoxy-DAVcr, N-formylleurosine						
LC	a. Vob, P, 19-epiiboxygaine, 19-epiVoac, App, 12-methoxyVoap, isomethuenine c. 12-MethoxyVoap, 10-epiVoac, Vob, P, 19-epiiboxygaine, App	a,c. Tabernaemontana dichotoma leaf extract	UV 280	a,b. 300 × 4.6 mm, LiChrosorb RP-18, 5 µm, subsequently loaded with 10 mM C12SO3H in MeOH-H2O 1:1 and 20 mM cetrimide in H2O c,d,e. 300 × 3.9 mm µPorasil	a,b. 20 mM C12SO3H in H2O-dioxane-H2SO4 a. 94.5:5: 0.5, pH 4.0 with NaOH b. 98.5:1: 0.5, pH 3.5 with NaOH c,e. CHCl3-MeOH-25% NH3 c. 99:1:0.2 e. 98:2:0.4 d. CHCl3-10% NH3 100:0.2	a. See Fig. 7.25 a,b,c,d,e. k values were reported for 21 Tabernaemontana alkaloids	407

LC	P, tabernaemontanine, Vob, Voap, dihydroquinine (I.S.), App, tubotaiwine, Co	Cell cultures of *Tabernaemontana divaricata*	UV 275, 313	300 × 3.9 mm, μBondapak Phenyl	50 mM NaH$_2$PO$_4$, pH 3.9 with H$_3$PO$_4$-ACN-2-methoxyethanol 80:15:5 2 ml min^{-1}	Y and Vca were also tested as (I.S.). k values and absorbance ratios for 27 indole alkaloids and related compounds were reported	29
LC	Enantiomers of a. vincadifformine b. aspidospermidine c. vincadine d. quebrachamine e. N-methylquebrachamine f. Vca g. AVca h. eburnamonine		UV 277 or 254	a–h. 800 × 16 mm a,d. 900 × 50 mm, CDP-25 (β-cyclodextrin bead polymer) a–h. 63–90 μm a,d. 90–125 μm	a–h. Phosphate, pH 5.5, 0.67–1 ml min^{-1} a,d. Phosphate, pH 9.5	a. Effect of pH and buffer type on reten. were studied a,d. Prep. chromatographic resolution was achieved	416
LC	17 Eburnane and 15		UV 280	300 × 3.9 mm, μBonda-	0–5 mM DHP or CSA,	See Fig. 5.24. Effect	417

TABLE 7.19 (Continued)

Technique	Compounds separated[b]	Sample	Detection	Col. Stat. phase Col. temp.	Mobile phase flow rate	Remark	Ref.
	ergot peptide alkaloids			pak CN	0–25 mM Et$_2$NH in C6-CHCl$_3$-ACN 65:20: 15, 60:23: 13, or 70: 17:13 or in C6-iPrOH 8:2 1 ml min^{-1}	of mob. phase composition on reten. was investigated	
LC	a. 18 Eburnane alkaloids. Enantiomers of b. EpiVca, Vca c. AAEE, AVca e. Vca		UV 280	a,b,c. 250 × 4.6 mm, Nucleosil 10 CN d,e. 150 × 4.6 mm, Nucleosil 5 CN	2 mM (+)-CSA, 1 mM Et$_2$NH in a. C6-CHCl$_3$-MeOH or EtOH or iPrOH 80: 18:2, 70:27: 3, 60:36:4 1 ml min^{-1} b,c. C6-dioscane BuOH 70:25:5 1.5 ml min^{-1} d. C6-CHCl$_3$-EtOH 80:12:2	Influence of the type of hydrophobic component, moderator, and polar component of the mob. phase on the separation of optical isomers was studied. b,c,d. Check of optical pu-	418

LC	22 stereo-isomers of eburnane alkaloids	UV 280	300 × 3.9 mm, μBondapak C18 or 250 × 4.6 mm, LiChrosorb RP-8 10 μm	2 ml min^{-1} e. C6-CHCl$_3$-EtOH 70:27:3 1 ml min^{-1} rity. e, See Fig. 5.25		419
LC	a,b. 11 eburnane alkaloids c. 21 eburnane alkaloids d. 20 eburnane alkaloids e. 13 eburnane alkaloids	UV 280	a. 250 × 2 mm, Micropak SI-10 10 μm b,c,d. 250 × 4.6 mm, LiChrosorb Si-60, 5 μm e. 300 × 3.9 mm, μBondapak CN 10 μm	ACN-10 mM (NH$_4$)$_2$CO$_3$, various ratios from 8:2 to 4:6 1 ml min^{-1} a,b. CHCl$_3$-MeOH 92:8, 95:5, 98:2 c. C6-CHCl$_3$-MeOH, 4 ratios d. C6-CHCl$_3$-ACN-MeOH in 11 ratios e. C6-CHCl$_3$-ACN 65:20:15, 7:2:1, 75:20:5	Effects of col. type and percentage of ACN in the mob. phase were studied b. See Fig. 5.2	420
LC	a. Vca, vincine	a. Vinca minor	a,b,c,f. UV 280	a. 250 × 4.6 mm LiChro-	a. C6-CHCl$_3$-	a. On RP col. Vca	10

TABLE 7.19 (Continued)

Technique	Compounds separatedb	Sample	Detection	Col. Stat. phase Col. temp.	Mobile phase flow rate	Remark	Ref.
	b. Toluene, vincamone, vincamenine, isovincanole, vincanole	b. Mother liquor	d,e. UV 275	sorb SI-60 5 μm	MeOH 8:1:1	was not separated from vincine	
	c. Vincamone	c,d. Cavinton tablets	g. UV 254	b,d,e. 250 × 2 mm Micro-Pak SI-10, 10 μm	b,d,e. CHCl$_3$-EtOH 95:5	d,e. Content uniformity test	
	d. Diazepam (I.S.), AAEE	e. Devincan tablets		c,f. 250 × 4.6 mm, Nucleosil 10 C18	c,f,g. ACN-10 mM (NH$_4$)$_2$CO$_3$	f. With 1 mM trioctyl methyl ammonium bromide in the mob. phase also vincaminic acid could be determined.	
	e. Diazepam (I.S.), Vca	f. Heat-loaded Devincan injections		g. 250 × 4.6 mm, LiChrosorb RP-8	c. 9:1 f. 7:3 g. 8:2	Linear calibrations	
	f. EpiVca, Vca	g. Plasma			a,f,g. 1 ml min^{-1}	d. 0.05–2.5 μg	
	g. AVca (I.S.), AAEE				b,d,e. 0.33 or 0.5 ml min^{-1}	e. 0.2–10 μg CV d. 1.26% e. 1.17%	
					c. 1,5 ml min^{-1}	g. Recovery 71.2% LOD 2 ng ml^{-1}	

LC	Pindolol (I.S.), Vca	Tablets, vials, oral solutions	UV 232 or 273	300 × 4.6 mm, Micro-Pak MCH 10 25°C	5 mM NaClO$_4$, pH 3 with HClO$_4$-ACN 55:45 2 ml min^{-1}	Linear calibration 0.05–0.25 g liter^{-1}. Recovery 96.9–101.3%. CV < 1.5%. LOD 5 and 10 ng at 232 and 273 nm, resp.	421
LC	a. Vca b. Vca, R	a. Spiked plasma	Electrochemical	250 × 4.6 mm, LiChrosorb RP-8	50 mM LiClO$_4$ in ACN-H$_2$O 1:1 1.2 ml min^{-1}	Linear calibration 5–500 ng ml^{-1}. Glassy carbon electrode was more reproducible than gold electrode	422
CCC	a,b. 20 Alkaloids c. 23 Alkaloids d. Voap, App, and 20 unspecified alkaloids	Suspension cultures of a,b. Tabernaemontana divariatica c. Tabernaemontana elegans d. Tabernaemontana orientalis	TLC of fractions	300 capillaries 400 × 2 mm in series	CHCl$_3$-MeOH-McIlvaine buffer (25 mM citrate, 50 mM phosphate, pH 4.2 with H$_3$PO$_4$) 5:5:3. Gradient of	Unresolved compounds were further separated by TLC	30

TABLE 7.19 (Continued)

Technique	Compounds separated[b]	Sample	Detection	Col. Stat. phase Col. temp.	Mobile phase flow rate	Remark	Ref.
TLC	a,c. Ajm, Ajc, corynantheine, corynanthe-idine, R, Rc, reserpiline, Rn, corynanthine, Des, dihydrocorynantheine, S, tetraphyllicine, Y, αY b. Rc, R, corynanthine		Visualization UV, modified Dragendorff's reagent	Layers a,b. 10 × 20 cm c. 20 × 20 cm Silica gel 60, 0.25 mm	a,c,d. NaClO$_4$ b. KSCN a. EtAc-MeOH, i-Pr$_2$O-MeOH CCl$_4$-MeOH AcOH-acetone b. 16 mob. phases c. Two-dimensional iPr$_2$O-MeOH 85:15, methyl ethyl ketone-AcOH 9:1	a. The dependence of R$_F$ values on the composition of mob. phase was measured b. Selectivity effects were studied c. 15 components resolved	390
TLC	a. Aricine, Ajc, Des, R + Rc, Y b. Y, Ajm,	c. *Rauwolfia serpentina* powders and tablets	Visualization UV (long and short wave-	Layers 20 × 20 cm Silica gel H, 0.25 mm (prep.	CHCl$_3$-MeOH a,c. 7:3 b. 8:2	c. S was determined after extraction from	383

Applications

	reserpiline, S c. R + Rn, S, and unknowns				plate by by fluorimetry	
TLC	a. Ajm b. R, Rc	Bark of Rauwolfia vomitoria	Drying at 105°C a. 2 hr b. 15 hr followed by scanning a. UV 292 b. Fluor. λ_{ex} 365 nm λ_{em} >480 nm	Layers 20 × 10 cm Silica gel 60 F 254	a. $CHCl_3$-CH-Et_2NH 7:2:1 b. Light petroleum-acetone-Et_2NH 7:2:1 Development in unsaturated tanks	α-Tocopherol acid succinate was added to standard solutions to prevent degradation. Linear calibrations: R: 20–200 ng, Rc: 20–100 ng, Ajm: 50–500 ng. CV < 6%. LOD 5 ng for R and Rc, 20 ng for Ajm	388
TLC	Y, R, Ajc, S, Ajm, 17-O-acetyl-Ajm, 17-O-acetylnor-	Tissue cultures of Rauwolfia serpentina	Visualization UV	Layers Silufol 254	BuOH-AcOH-H_2O 4:1:1 $CHCl_3$-MeOH-am-	Quantitation after elution from the plate by UV	389

TABLE 7.19 (Continued)

Technique	Compounds separated[b]	Sample	Detection	Col. Stat. phase Col. temp.	Mobile phase flow rate	Remark	Ref.
	Ajm, perakine, vomilenine				monia 90:10: 0.2, $CHCl_3$-acetone-Et_2NH 5:4:1, light petroleum-acetone-Et_2NH 7:2:1		
TLC	Ajc, Ca, S	Cell cultures of Catharanthus roseus	Visualization ceric ammonium sulfate followed by UV	Layers Silica gel 60 F	$CHCl_3$-MeOH 98:2, acetone-toluene-MeOH-27%NH_3 9:8:2:1	Identification	392
TLC	Ajc, Vdo, Ca	Catharanthus roseus plants and multiple shoot cultures		Layers Silica gel 60 F_{254}, 0.25 mm	EtAc-EtOH 1:1	R_F 0.4–0.9 scrapped for analysis by HPLC	393
TLC	Ajc, S	Cell cultures of Catharanthus roseus		Layer Silica gel G	EtAc-MeOH 3:1, $CHCl_3$-acetone-Et_2NH 5:4:1,	Prep. separation	395

Applications

TLC	a. 100 alkaloids c. 21 alkaloids d. 7 alkaloids e. 7 alkaloids f. 8 alkaloids g. 11 alkaloids h. 6 alkaloids	Tabernaemontana species b. T. divaricata T. elegans, T. orientalis cell cultures c,d,e. T. dichotoma leaf and seed f. T. psorocarpa g. T. chippii h. T. undulata	Visualization UV 254, 366, FCPA, CSSA, and TCNQ reagents[b]	Layers a. 20 × 20 cm Silica gel FF254, 0.25 mm b,f. Silica gel F254 c. Silica gel 60 F254, 0.25 mm e. Silica gel 60, 0.5 mm	CH-CHCl3-Et2NH a,d,e,g,h. 6:3:1 f,g. 8:10:3, toluene-abs. EtOH saturated (1.74%) with NH3 (gas) (plate equilibrated with NH3 vapor prior to development) a,c,g,h. 19:1 e,f. 98:2 f,g. 9:1, CHCl3-MeOH a,b,g,h. 9:1, EtAc-iPrOH-26% NH3 a,g. 17:2:1 b. 9:7:4 and other 11 mob. phases	a. R$_F$ values, fluorescence properties and chromogenic reactions were reported b. Analysis of fractions from CCC g,h. Data for other alkaloids were included in (413) h. Prep. separations with CHCl3-MeOH 13:2	a. 413 b. 30 c. 407 d. 408 e. 409 f. 410 g. 411 h. 412

TABLE 7.19 (Continued)

Technique	Compounds separated[b]	Sample	Detection	Col. Stat. phase Col. temp.	Mobile phase flow rate	Remark	Ref.
TLC	Hirsutine and 8 oxindole alkaloids	*Mitragyna stipulosa*	Visualization UV and daylight, Dragendorff's, vanillin, iodoplatinate, Ehrlich's, and $FeCl_3$-$HClO_4$ reagents	Layers Silica gel	10 mob. phases	R_F values were reported	423

[a]For other examples, see Chapter 4, Section IV.F; Chapter 5, Sections I.A, I.B, II.B, III.A, IV.A, V.C, V.D; Chapter 6, Section II.F; and Tables 5.8, 6.6, and 6.7.
[b]AAEE, apovincaminic acid ethyl ester (vinpocetin); AAME, apovincaminic acid methyl ester; ABA, 4-amino-3,5-dichloro-α-(tert-butylaminomethyl)benzyl alcohol; Ajc, apovincaminic acid; Ajm, ajmaline; AVca, apovincamine; App, apparicine; Ca, catharanthine; Co, coronaridine; CSA, camphor-10-sulfonic acid; CSSA, 1% ceric sulfate in 10% H_2SO_4; DAVbl, desacetylvinblastine; DAVcr, desacetylvincristine; Des, deserpidine; DHP, di-(2-ethylhexyl)phosphoric acid; FCPA, 0.2 M (3.25%) $FeCl_3$ in 35% $HClO_4$; P, pervine; PrOH, n-propyl alcohol; R, reserpine; Rc, rescinnamine; Rn, reserpinine (raubasinine); S, serpentine; TCNQ, 0.2% 7,7,8,8-tetracyanoquinodimethane in acetonitrile; THA, tetrahydroalstonine; Vbl, vinblastine; Vca, vincamine; Vcr, vincristine; Vdo, vindoline; Voac, voacangine; Voap, voaphylline; Vob, vobasine; Y, yohimbine; αY, α-yohimbine.

Applications 527

TABLE 7.20 Chromatographic Conditions for Strychnos Alkaloids[a]

Tech-nique	Compounds separated[b]	Sample	Detection	Col. Stat. phase Col. temp.	Mobile phase flow rate	Remark	Ref.
GC	Caffeine, imipramine, nitrazepam, Str		NPD	Capillary cols. a,b,d. 0.32 mm c. 0.22 mm a. CP-Sil 5 chem. bonded, 0.14 μm b. CP-Sil 5, 0.48 μm c. CP-Sil 5, 0.45 μm d. OV-1701, 0.10 μm 60°C for 2.5 min, 60–190°C at 30 K min^{-1}, 190–295°C at 7 K min^{-1}, 295°C for 5 min	Helium a,b,d. 3 ml min^{-1} c. 1 ml min^{-1}	The dependence of reten. times and indices on the amount of sample injected was studied	424
LC	33 derivs. of Str		UV 254	300 × 4.5 mm, Porasil	CHCl$_3$–2% NH$_4$OH in	k values and relative	425

TABLE 7.20 (Continued)

Technique	Compounds separated[b]	Sample	Detection	Col. Stat. phase Col. temp.	Mobile phase flow rate	Remark	Ref.
				8 μm	MeOH 93:7 3 ml min^{-1}	retentions were reported. Substitution effects were evaluated	
LC	Bru, Str	Nux-vomica extract, tablets	UV 254	150 × 4.6 mm, Shodex ODS	5 mM C7SO$_3$H in H$_2$O-ACN 75:25, pH 3.0 0.7 ml min^{-1}	Purification using Sep-Pak C18 cartridges. Recovery > 98%. CV 2.5–2.6%	426
LC	Str, Bru	Nux-vomica tincture	UV 254	125 × 4.6 mm, Hypersil 5 μm	MeOH-2M NH$_4$OH-1 M NH$_4$NO$_3$ 27:2:1 2 ml min^{-1}	Separation in less than 6 min. CV 0.491% (peak area), 0.403% (peak height measurement)	427
LC	Str, Bru, quinine (I.S.)	Nux-vomica extract	UV 254	Micro-Pak Si-10	MeOH-28% NH$_4$OH 99.25:0.75		428

Applications

LC	Str	a. Strychnos piersiana b. Strychnos pierriana	UV 250	250 × 4 mm silica gel 3–5 μm	Et$_2$O-MeOH-NH$_4$OH 80:20:1 2 ml min^{-1}	a. Linear calibration up to 0.75 μg. Recovery 99.3%. CV 1.29%	a. 430 b. 431
LC	Quinine (I.S.), Str	Spiked plasma and urine	UV 254	250 × 2.6 mm, silica gel 10 μm	MeOH-28% NH$_4$OH 99.25:0.75 1.1 ml min^{-1}	Separation in less than 6 min. Linear calibration 20–0.625 mg liter^{-1}. Recovery 83%. CV 3–5% (intra-assay), < 10% (inter-assay). LOD 0.625 mg liter^{-1}	429
LC	Str, ergonovine (possible I.S.)	Biological specimens (liver, kidney, urine, stomach content of domestic animals)	UV 254	250 × 4.6 mm, LiChrosorb Si-60 10 μm, with guard col. 30 × 4.7 mm MPLC silica 10 μm	CH$_2$Cl$_2$-MeOH-H$_2$O-formic acid-Et$_2$NH 72.3:25:2.5:0.1:0.1 1 ml min^{-1}	Linear calibration 0.083–4.0 μg. Recovery 94–104% at 0.75 and 7.5 ppm. CV 3.58% at 3.9 ppm.	432

TABLE 7.20 (Continued)

Tech-nique	Compounds separated[b]	Sample	Detection	Col. Stat. phase Col. temp.	Mobile phase flow rate	Remark	Ref.
LC	Str, tryp-tamine	Stomach content	UV 254	100 × 4.6 mm, MPLC Spheri-5 RP-8 protected by New-Guard RP-8	10 mM C8SO$_3$Na in 5 mM NaH$_2$PO$_4$, pH 3.0– ACN–THF 750:135:115 1.0 ml min^{-1}	LOD 0.05 ppm Reten. times and extraction recoveries for other 11 alkaloids were reported. Effects of ACN conc. and the presence of C8SO$_3$H were examined. Linear calibration 0–370 ng. Recovery 92.0% at 7.0 µg g^{-1}. CV 0.9% at 73 ng. LOD 4.5 µg per sample	433

Applications

Technique	Compounds	Sample	Detection	Column/Stationary phase	Mobile phase	Remarks	Ref.
LC	a,b. Str-N-oxide, 21,22-dihydroStr, Str-21,22-epoxide, 11,12-dehydroStr, Str b. Strychnone, 16-hydroxyStr, 18-oxoStr	a. Str metabolized in vitro with rat and rabbit liver	UV 254	100 × 8 mm µBondapak C18	ACN-2M NH$_4$OH 75:25 1 ml min^{-1}	a. Metabolites were isolated using Sep-Pak C18 cartridge and prep. TLC a,b. Retention times were reported	434
CCC	Quinine, dihydroquinine, Str, Bru, berberine		UV 254 and 277	300 capillaries 400 × 2 mm in series. Organic layer, 20°C	Aqueous layer of CHCl$_3$-MeOH-1 M NaClO$_4$ in pH 4 McIlvaine's buffer 5:5:3, gradient with decreasing conc. of counterion. 0.17 ml min^{-1}	Various buffers and counterions were tested. Effect of counterions on the partition of alkaloids was examined	435
TLC	Str	Chinese drugs	Visualization iodine va-	Layer Silica gel G	Toluene-acetone-	Determination by pho-	a. 436

TABLE 7.20 (Continued)

Technique	Compounds separated[b]	Sample	Detection	Col. Stat. phase Col. temp.	Mobile phase flow rate	Remark	Ref.
		(pills and powders)	b. Scanning by dual wavelength densitometry		NH₄OH-EtOH 8:6:2: 0.5	tometry after elution from the plate b. Linear calibration up to 5 µg. Recovery 95.9–97.49%. CV 1.86–2.86%	b. 437
TLC	a. Str b. Bru	Chinese preparations containing components of *Strychnos nux-vomica*	Dual wavelength scanning a. UV 255 b. UV 295 (reference 400 nm)	Layer Silica gel GF₂₅₄	MeOH-4 M NH₄OH 9:1	Recovery > 95%. CV<5%	438
TLC	Str, Bru, Str-N-oxide, Bru-N-oxide, α-columba-	Nux vomica preparations	Visualization UV 254, modified Dragendorff's	Layer Silica gel 60 F₂₅₄	180° two-dimensional development 1. Et₂O-EtOH-20%	Identification of alkaloids	439

532 Chapter 7

Applications 533

				reagent	NH$_4$OH 90:5:5 2. CHCl$_3$-acetone-EtOH-28% NH$_4$OH 40:15:5	
		mine, β-columbamine, vomicine, 4-hydroxyStr				
TLC		Str	Biological specimens (liver, kidney, urine, stomach content of domestic animals)	Visualization UV 254, iodoplatinate reagent	Layer Silica gel 60 F-254 0.25 mm	CH$_2$Cl$_2$-MeOH-H$_2$O-formic acid-Et$_2$NH 72.3:25:2.5:0.1:0.1	Preliminary screen prior to HPLC analysis	432
TLC	a,b. Str-N-oxide, 21,22-dihydroStr, Str-21,22-epoxide, 11,12-dehydroStr, Str b. Strychnone, 16-hydroxyStr, 18-oxoStr	a. Str metabolized in vitro with rat and rabbit liver	Visualization UV, Dragendorff's reagent	Layers 20 × 20 cm Silica gel G 60 F 254, 0.3 mm	CHCl$_3$-Et$_2$NH 4:1, CHCl$_3$-MeOH 7:3; benzene-EtAc-Et$_2$NH 7:2:1	a. Isolated metabolites were identified by co-chromatography with authentic samples a,b. R$_F$ values were reported	434

[a]For other examples, see Chapter 4, Section IV.F; Chapter 5, Sections I.A, I.B, II.B, IV.A; Chapter 6, Section I.C; and Tables 5.3, 7.7, 7.13, and 7.14.
[b]Bru, brucine; Str, strychnine.

TABLE 7.21 Chromatographic Conditions for Ergot Alkaloids[a]

Technique	Compounds separated[b]	Sample	Detection	Col. Stat. phase Col. temp.	Mobile phase flow rate	Remark	Ref.
GC	Peptidic part of HECo, HECp, HECs	Dihydroergotoxine product		1 m × 3.2 mm, 2% Dexsil 300 on Gas-Chrom Q, 80–100 mesh, 180–280°C at 5 K min^{-1}	Nitrogen 11.5 ml min^{-1}	See Fig. 7.26. Dihydroergotoxine components decomposed upon injection at 235°C. CV 1.5%	440
GC	Dihydroergotoxine components derivatized to N-TMS-ethyldihydrolysergiate	Human plasma	MS (high or low resolution)	25 m × 0.31 mm, cross-linked methyl silicon		Clin-Elut extraction was followed by hydrolysis, esterification, and derivat. with MSTFA. Linear calibration 0–2.345 ng. Recovery 96.1–119.6% at 1.177 and 0.235 ng.	441

Applications

GC	LSD, LAMPA	Street samples	MS	a. 15 m × 0.25 mm b,c. 5 m × 0.25 mm, SE 30, 0.25 μm a. 100–295°C at 5 K min^{-1} b. 240°C c. 260°C	a,c. Helium b. Hydrogen Linear velocities at 100°C a. 44 cm sec^{-1} b. 75 cm sec^{-1} c. 36 cm sec^{-1}	LOD 2 pg (high resolution SIM). [^{2}H$_3$-]HET was used as (I.S.) b,c. Separation in less than 3 min	453
LC	a. EM, EMn, ES, ET, ESt, ECo, αECp, βECp, ECs, ESn, ETn, ECon, EStn, αECpn, ECsn		UV 310	a. 250 × 4 mm LiChrosorb, RP-18 5 μm b. 250 × 4.6 mm, LiChrosorb RP-18, 5 μm c. 120 × 4.6 mm, Nucleosil C18, 5 μm	ACN-20 mM phosphate, pH 7, 1:1 1.0 ml min^{-1}	Reten. indices (referred to 2-ketoalkanes) on various ODS cols. differed considerably, but the differences were nearly constant	442

TABLE 7.21 (Continued)

Technique	Compounds separated[b]	Sample	Detection	Col. Stat. phase Col. temp.	Mobile phase flow rate	Remark	Ref.
LC	a,b. 18 Peptidic and simple ergot alkaloids c. ESn, ETn, Es, ET d. ECpn, ECsn, ECp, ECs e. ECon, ECsn, ECo, ECs, ES, ET		UV 310	250 × 2 mm a. Micro-Pak NH$_2$, 10 μm b,c,d,e. LiChrosorb NH$_2$, 10 μm 25°C	a. CHCl$_3$-iPrOH 9:1, Et$_2$O-iPrOH 6:4, Et$_2$O-EtOH a. 84:16, 88:12 a,b,c. 93:7 d. 97.5:2.5 e. Linear gradient of EtOH in Et$_2$O. 2.5% in 0–7 min, 2.5%–5% at 0.3%/min, 5% thereafter a. 1 ml min^{-1} b,c,e. 0.67 ml min^{-1}	a,b. Reten. volumes and absorption coefficients at 225, 240, 254, 282, and 310 nm were reported. Effects of molecular configuration and of alkyl and aryl substitution on reten. was discussed e. CV 1.15–3.05%	444
LC	a. CCIS (I.S.), ECpn,	Pure substances	a. UV 312 b,c. UV 280 d. UV 305	250 × 4.6 mm, LiChrosorb Si-60	C6-EtAc-Et$_3$N-formamide	Impregnation of the col. with	445

Applications

LC	ECon, βECp, αECp, ECo, ECs, ET, EM b. CCIS (I.S.), αHECp, HECo, HECs c. CCIS (I.S.), ET, HET d. 2-Bromo derivs. of ergot alkaloids a,b,c,d,e,f. 14 Ergot alkaloids g. ETn, PDDP (I.S.), ET h. ECsn, ECs, PTCP (I.S.) i. ECon + ECpn, ETn, ECo + ECp, EMn, EM j. ECo, ECp, ECs	g,h. Plant extracts i,k. Samples of fermentation products l. Dihydroergotoxine product	g–k. UV 320 l. UV 280	a–i. 250 × 2 mm j,k,l. 250 × 4 mm a,c,g,h,i. LiChrosorb Si-60, 10 μm b. Silica SI-60, 10 μm d. LiChrosorb RP-2, 5 μm e. LiChrosorb RP-8, 10 μm 5 μm	a,i. C6-CHCl$_3$-EtOH 4:4:1 b,g,h. CHCl$_3$-MeOH 95:5 c. CHCl$_3$-EtOH 95:5 d,e,f,i,j,k,l. 10 mM (NH$_4$)$_2$CO$_3$-ACN 3:2 g,h. 0.167 ml min^{-1} or 0.25 ml a. 50:48:1:1 b. 40:58:1:1 c. 30:70:1:1 d. 60:38:1:1	k. See Fig. 7.27 l. See Fig. 7.26. CV 1.3% formamide required about 45 min at 2 ml min^{-1}. Linear calibration 1–15 μg. CV 1.8%. LOD 0.1 μg (ET), 0.5 μg (HET)	440

TABLE 7.21 (Continued)

Technique	Compounds separated[b]	Sample	Detection	Col. Stat. phase Col. temp.	Mobile phase flow rate	Remark	Ref.
	k. ECo, αECp, βECp l. HECo, HECp, HECs			f. Silica RP-18, 12 μm f,j,k,l. Silica RP-18, 10 μm	min^{-1} i. 0.33 ml min^{-1} j,k. 2.17 ml min^{-1} l. 3.17 ml min^{-1}		
LC	11 ergot alkaloids	Ergot sclerotia	Fluor. λ_{ex}360 nm λ_{em}425 nm	Radial-Pak μBondapak RP-18, 10 μm	Linear gradient of B in A A. 0.2 g liter^{-1} (NH$_4$)$_2$CO$_3$-ACN-MeOH 55:30:15 B. ACN-MeOH 2:1 0–40% in 20 min		447
LC	EM, EMn, ET, ECp, ECs, ETn, ECpn, ECsn	Spiked wheat	Fluor. λ_{ex} variable λ_{em}>418 nm	150 × 4.6 mm, crosslinked polystyrene-divinylben-	ACN-50 mM (NH$_4$)$_2$HPO$_4$ 55:45 pH 10.0 0.3 ml min^{-1}	Recovery 60.3–100% at 16–760 ng g^{-1}	448

Method	Compounds	Sample	Column	Detection	Mobile phase	Comments	Ref
LC	a. MEM, methysergide (I.S.) b,c. ET, ECs (I.S.), ETn	a,b. Plasma, serum	zene resin 10 μm 250 × 4.6 mm, Hypersil ODS, 5 μm	Fluor. λ_{ex}328 nm λ_{em}>389 nm	10 mM $(NH_4)_2CO_3$-ACN a. 65:35 b. 50:50 c. 400 ml liter^{-1} ACN, 1.9 g liter^{-1} C7SO$_3$Na, 10 ml liter^{-1} AcOH 1.5 ml min^{-1}	k values were reported. Calibration a. 0–75 ng ml^{-1}. Recovery a. 53% b. 79–92% CV a. 10% at 0.5 ng ml^{-1} b. 7.7% LOD 100–200 pg ml^{-1}	55
LC	ET, ECs	Plasma	250 × 4.6 mm, Aquapore RP-300	Fluor. λ_{ex}320 nm λ_{em}400 nm	ACN–0.2 M NH$_4$NO$_3$, pH 7.5, 1:1 1.2 ml min^{-1}		339
LC	a,b. HECo, αHECp, HECs, βHECp c. Aci forms + MECA + MECA amide, dihydroergop-	Codergocrine a,b. Ampoules and oral solutions c. Spiked with degradation products	a,c,d. 100 × 4.6 mm Spherisorb RP-18, 5 μm b. 250 × 4.6 mm, LiChrosorb RP-18, 10 μm a,b. with precol. 30 ×	UV 280 d. Fluor. λ_{ex}290 nm λ_{em}360 nm	H$_2$O-ACN-Et$_3$N a. 645:336:19 b. 800:200:25 a. 2.2 ml min^{-1} b. 1.0 ml min^{-1}	Col. lifetime was about 500 injections. a. See Fig. 7.28 d. Interference of diethylphthalate with	449

TABLE 7.21 (Continued)

Technique	Compounds separated[b]	Sample	Detection	Col. Stat. phase Col. temp.	Mobile phase flow rate	Remark	Ref.
	tine, B-seco-HECs, HECo, αHECp, HECs, βHECp d. HECo, αHECp, diethylphthalate + HECs, βHECp	d. Tablets		4.6 mm, LiChrosorb Si-100, 5 μm	c. 2.0 ml min^{-1} d. 1.5 ml min^{-1}	HECs was eliminated by use of fluor. detection. Linear calibration a. 0–2.7 μg LOD a. 6–18 ng b. 50–100 ng	
LC	a. m-Hydroxybenzoic acid (I.S.), HLA, HLA amide, aciHECo, aciαHECp, aciHECs, HECo, αHECp + βHECp,	Liquid pharmaceuticals	UV 281	250 × 4.6 mm, LiChrosorb RP-18 7 μm	Curved gradient of B in A a. A. 10 mM (NH$_4$)$_2$CO$_3$ B. ACN. 43% B for 6 min, 43–52% in 6–12 min b. H$_2$O-ACN-Et$_3$N A. 220:50:	a,b. Reten. times were reported. Recovery a. 96.3–101.1% b. 98.2–101.3% CV a. 4.0% b. 3.2%	450

	HECs b. HLA, aci-αHECp, aci-HECo, aci-HECs, HLA amide, papaverin (I.S.), HECo, αHECp, HECs, βHECp			0.2 B. 290:120: 12.3% B for 1.5 min, 3–99% in 1.5–5.5 min, 99% in 5.5–16.5 min 2.5 ml min^{-1}	
LC	8'-Hydroxy-HET, HET, HECo (I.S.)	a. Plasma b. Urine	Fluor. λ_{ex} 296 nm λ_{em} 355 nm	Preseparation: 30 × 2.1 mm Aquapore RP-300, 10 μm. Analysis: 83 × 4.6 mm, HS3 C18, 3 μm 40°C	A. 10 mM NH$_4$ carbamate, pH 8.5 B. ACN. Ratio of A:B programmed by washing and equilibrating cols. For separation the ratio 6:4 was used. 1.8 ml min^{-1}	Sample pretreatment with Extrelute (plasma) and Sep-Pak (urine) cols. Analysis using col. switching technique. Cols. changed every 150 (analytical) and 300 (preseparation) injections. Lin-	56

TABLE 7.21 (Continued)

Technique	Compounds separated[b]	Sample	Detection	Col. Stat. phase Col. temp.	Mobile phase flow rate	Remark	Ref.
LC	a. HET, HECs (I.S.)	Plasma	UV 223, fluor.	250 × 4 mm RP-8 Hibar	ACN-9 mM NaH$_2$PO$_4$,	Calibration 5–50 ng ear calibration a. 0.25–2.5 ng ml^{-1} b. 0.1–5.0 ng ml^{-1} Recovery a. 85.2–89.1% b. 94.6–113.2% CV within day a. 6.2–8.9% b. 2.4–8.1% day-to-day a. 9.2% b. 12.0% LOD a. 80 pg ml^{-1} b. 100 pg ml^{-1}	451

	b. HET (I.S.), HECs		λ_{ex} 295 nm λ_{em} 350 nm	col., 10 μm	9 mM Na$_2$HPO$_4$, pH 7.2, 6:4 1 ml min^{-1}	ml^{-1}. Recovery a. 93% b. 91% CV a. 2.8% (5.2% interassay) b. 3.4% (4.8% interassay)	
LC	Dihydroergotoxine, HES (I.S.)	Plasma	Fluor. λ_{ex} 285 nm λ_{em} 345 nm	150 × 4 mm LiChrosorb Si-60, 5 μm	ACN-MeOH-conc. NH$_4$OH 12:1:0.008 0.5 ml min^{-1}	Curved calibration 0.1–1 ng ml^{-1}. Recovery 53.5%. CV 6.1–19.4%. LOD 100 pg	452
LC	LA, lysergamide, LSD, iso-LSD		UV 320	250 × 4.6 mm, fines from Partisil 6 μm	MeOH-0.2 M NH$_4$NO$_3$ 3:2 1 ml min^{-1}		238
LC	a. 22 ergot alkaloids b. LA, lysergamide, isoLA, LSD, methysergide, ES, ET,	d. Illicit LSD preparations	b,c,d. UV 220 d. Fluor. λ_{ex} 312 nm λ_{em} 400 nm	a,b,d. 100 × 5 mm, Hypersil ODS 5 μm, with precol. 50 × 4.5 mm coarse silica, 40 μm	MeOH-22 mM NaH$_2$PO$_4$, 28 mM Na$_2$HPO$_4$ 6:4, pH 8.1 2 ml min^{-1}	Col. washed with MeOH-H$_2$O 6:4 at the end of each working day. a. k values were report-	443

TABLE 7.21 (Continued)

Technique	Compounds separated[b]	Sample	Detection	Col. Stat. phase Col. temp.	Mobile phase flow rate	Remark	Ref.
	ECp c. LSD, LAMPA, methysergide d. LSD					ed c. Separation of compounds eluting closest to LSD	
LC	LA, LSD, LAMPA		UV 254, 280, 313	300 × 3.9 mm, μBondapak C18	Phosphate, pH 3-MeOH 2:1 1.5 ml min^{-1}	Absorbance ratios were compared	454
LC	LSD	Spiked urine	UV 325, fluor. λ_{ex}308 nm, λ_{em}370 nm, electrochemical	250 mm Spherisorb S5W silica	10 mM NH$_4$ClO$_4$, 0.1 mM NaOH in MeOH	Fluorescence detection provided the best selectivity and sensitivity	269
LC	a,c. LSD b,d. 24 ergot alkaloids	a,c. Blood, urine, stomach washings	Fluor. λ_{ex}320 nm λ_{em}400 nm b,d. UV 280	a,b. 100 × 4.6 mm Spherisorb 5-ODS c,d. 150 × 4.6 mm Spherisorb	a,b. MeOH-25 mM Na$_2$HPO$_4$ 65:35, pH 8.0 with 10% H$_3$PO$_4$ c,d. MeOH-	b,d. Retention volumes were reported a,c. Recovery 60 and 70% at 10 and 20 ng	446

				S5W	0.2 M NH$_4$NO$_3$ 60:40 1 ml min^{-1}	ml^{-1} urine, up to 90% above 10 ng ng ml^{-1} plasma. LOD 0.1 ng (above 0.5 ng ml^{-1} in biological fluids)	
LC	a. 15 Clavine alkaloids and simple derivs. of LA b. IsoSC, lysergene, SC, AC, PyC, FC, ergine, EC, CC	b. Submersed culture of Claviceps purpurea	a. UV 225, 240, 254 b. UV 225	250 × 2 mm Micro-Pak NH$_2$, 25°C	a. CHCl$_3$–iPrOH 9:1, 8:2, Et$_2$O–iPrOH 7:3, 6:4, Et$_2$O–EtOH 84:16, 80:20 1 ml min^{-1} b. Linear gradient of EtOH in Et$_2$O. 21% in 0–15 min, 21–29% in 15–17 min, 29% in 17–40 min 0.2 ml min^{-1}	a. Reten. volumes and absorption coefficients at 225, 240, 254, 282, and 310 nm were reported. Missing natural fluorescence was indicated. Effect of molecular structure on reten. was discussed b. CV 1.9–7.9%	455

TABLE 7.21 (Continued)

Technique	Compounds separated[b]	Sample	Detection	Col. Stat. phase Col. temp.	Mobile phase flow rate	Remark	Ref.
LC	a. AC, SC, FC, PaC, isomer of PaC, N-norAC, EC, PeC, isoCC-I, norCC, CC-I, CC-II d. ECp	a,b,c. Claviceps purpurea	MS, UV 280	a,b,c. 250 × 5 mm a. Spherisorb 5W b. ODS type, 5 μm c. Aminophase, 5 μm d. 100 × 5 mm, Silica, 5 μm	a. CH_2Cl_2-MeOH-NH_4OH 95:5:0.1 b. MeOH-H_2O-NH_4OH 60:40:0.1 c. Isooctane-CH_2Cl_2-MeOH 5:4:1 d. C6-$CHCl_3$-EtOH 72:24.25: 0.75 1 ml min^{-1}	LC was coupled to MS using a moving belt interface. MS spectral data were reported. The best separation was obtained on the Spherisorb 5W col.	456
SFC	a. PyC, AC, FC, isoSC, SC, EC, PeC, norAC, isoCC, N-norisoSC, CC-I, CC-II,	a. Claviceps purpurea	UV 280 a. MS	140 × 4.6 mm, Amino-bonded Spherisorb 5 μm, 75°C	a. 10% MeOH in CO_2, 15% after 2.8 min, 20% after 5 min, 3 ml min^{-1}, 5 ml min^{-1} after 2.5 min	Moving belt interface with modified thermospray deposition. Deterioration of col. performance was	353

Applications

	norCC-II b. Bromocryptine, ECp			b. 20% methoxyethanol in CO_2 4 ml min^{-1}	observed. a. Separation within 10 min. Nine unknown components were detected by MS
TLC	HECo, HECp, HECs	Dihydroergotoxine product	Visualization UV 254 Layers Polygram Sil G_{254}	0.1 M $(NH_4)_2CO_3$–acetone–EtOH 67.5:32.5:1	440 See Fig. 7.26. Quantitation after scraping, color reaction, and colorimetry. CV 3.5%

[a] For other examples, see Chapter 4, Section IV.F; Chapter 5, Sections I.A, I.B, II.A, II.B, IV.A, V.A, V.D; Chapter 6, Section I.D; and Table 5.7.

[b] AC, agroclavine; CC, chamoclavine; CCIS, chlorocarbonyliminostilbene; EC, elymoclavine; ECo, ergocornine; ECon, ergocorninine; ECp, ergocryptine; ECpn, ergocryptinine; ECs, ergocristine; ECsn, ergocristinine; EM, ergometrine (ergonovine, ergobasine); EMn, ergometrinine (ergonovinine, ergobasinine); ES, ergosine; ESn, ergosinine; ESt, ergostine; EStn, ergostinine; ET, ergotamine; ETn, ergotaminine; FC, festuclavine; HECo, dihydroergocornine; HECp, dihydroergocryptine; HECs, dihydroergocristine; HES, dihydroergosine; HET, dihydroergotamine; HLA, dihydrolysergic acid; LA, lysergic acid; LAMPA, lysergic acid methylpropylamide; LSD, lysergic acid diethylamide (lysergide); MECA, 6-methylergoline-8-carboxylic acid; MEM, methylergometrine; MSTFA, N-Trimethylsilyl-N-methyltrifluoroacetamide; PaC, palliclavine; PeC, penniclavine; PDDP, 1-phenyl-2,3-dimethyl-4-dimethylamino-5-pyrazolone; PTCP, 2-propylthiocarbamoyl-4-pyridine; PyC, pyroclavine; SC, setoclavine; SIM, selective ion monitoring; TMS, trimethylsilyl.

TABLE 7.22 Chromatographic Conditions for Imidazole Alkaloids[a]

Technique	Compounds separated[b]	Sample	Detection	Col. Stat. phase Col. temp.	Mobile phase flow rate	Remark	Ref.
GC	Aminophenazone (I.S.), P + IP	Degraded solutions of P	FID	1 m × 3.1 mm, 3% OV-101 on Gas Chrom Q, 80–100 mesh 180°C	Nitrogen 35 ml min^{-1}		457
LC	a. 124 drugs b. 74 drugs c. P	b. Various dosage forms c. Drops	UV 214, 254, 280 c. UV 214	300 × 3.9 mm, μBondapak C18, 10 μm, 25°C	a,b,c. 15% a,b. 30, 45 or 70% ACN in 50 mM KH$_2$PO$_4$, ev. acidified with 2 ml liter^{-1} H$_3$PO$_4$ 2 ml min^{-1}	Atropine, quinidine, quinine, codeine, and colchicine were also included. a. k values were reported b. Recommended conditions for the analysis of pharm. dosage forms were reported	458

LC	IP, P		UV 220	100 × 8 mm Radial-Pak C18, 5 μm	0.15 M Na$_2$SO$_4$-2 M H$_3$PO$_4$-MeOH-2-aminopropane 54.7:22:14.6:8.7 3.2 ml min^{-1}	Reten. data for eight mob. phase compositions were reported. Optimized separation within 8 min	459
LC	P, IP, nitrate	Ophthalmic solutions	UV 220	250 × 4.6 mm, Silica gel Si-60 Hibar, 5 μm	C6-(20 ml liter^{-1} conc. NH$_3$ in iPrOH) 7:3 2.0 ml min^{-1}	Linear calibration P: 1.12–2.08 g liter^{-1} IP: 0.037–0.184 g liter^{-1} at 1.61 mg liter^{-1} of P. Recovery 98–101.5%. Mob. ph. containing MeOH, EtOH, or 2-methyl-1-propanol did not give a comparable separation	460
LC	a. Nitrate, PA, IPA, a. P a,b. Esters		UV 214 or 215	a. 250 × 4.6 mm, LiChro-	a. 3% Na$_2$SO$_4$-2 M	a. Esters of PA were	a. 461

TABLE 7.22 (Continued)

Technique	Compounds separated[b]	Sample	Detection	Col. Stat. phase Col. temp.	Mobile phase flow rate	Remark	Ref.
IP, P b,c. P + IP, esters of PA h. P, diesters of PA d,e,f,g. P esters and diesters of PA	of PA a,d,e,f,g. Diesters of PA degraded in: a,b, d,e,f. Aqueous solution d,e. Plasma g. Ocular tissue homogenate			sorb Si-60 5 μm b,c,d,e,h. 250 × 4 mm LiChrosorb RP-8, 7 μm f. 250 × 4 mm, LiChrosorb RP-18 5 μm g. 250 × 4.6 mm Whatman Partisil PXS 10/25	H_3PO_4-MeOH 92:5:3 1.2 ml min^{-1} b,d. MeOH-20 mM KH_2PO_4 3:1 1.2 or 1.6 ml min^{-1} c,h. MeOH-20 mM KH_2PO_4, pH 4.5, 1:3 e. 60–70% ACN in 20 mM phosphate, pH 7 f. 50 mM phosphate, pH 2.5-MeOH 19:1 1.6 ml min^{-1} g. 75 mM KH_2PO_4-MeOH 83:17 1.5 ml min^{-1}	strongly retained. b. PA and IPA eluted with the solvent front. c,h. Lipophilicity estimation. k values were reported for P and: c. 10 esters of PA h. 12 diesters of PA	a,b, c. 462 a,d, e,f, g,h. 463

Applications

LC	IP, P, PA, IPA		RI	300 × 4 mm LiChrosorb RP-18, 10 μm 22°C	5% KH_2PO_4-MeOH 97:3, pH 2.5 with H_3PO_4 1.5 ml min^{-1}	With ACN separation of P and IP was poor. Selectivity was improved by lowering pH and increasing salt conc. LOD 5.5–9.1 μg	464
LC	IP, P, PA, IPA		UV 216	250 × 4.6 mm, LiChrosorb RP-18 10 μm	5% KH_2PO_4, pH 2.5-MeOH 97:3 1.5 ml min^{-1}	LOD 300 ng. IP/P ratio 1:100	465
LC	IP, P, PA, IPA	Ophthalmic formulations	UV 215	150 × 4 mm ODS packing	5% KH_2PO_4-MeOH 97:3, pH 2.5 with H_3PO_4 1.5 ml min^{-1}	LOD 0.04 μg	466
LC	IP, P, PA, IPA	Degraded solutions of P	UV 220	250 × 4 mm LiChrosorb RP-18, 10 μm, 25°C	50 g liter^{-1} KH_2PO_4, pH 2.5 with H_3PO_4, 30 ml liter^{-1} MeOH 2 ml min^{-1}	LOD 0.4–1.25 μg	467
LC	PA, IPA, IP, P	Eye drops	UV 215	300 × 3.9 mm, μBonda-	5% KH_2PO_4, pH 2.5 with	Baseline separation of	468

TABLE 7.22 (Continued)

Technique	Compounds separated[b]	Sample	Detection	Col. Stat. phase Col. temp.	Mobile phase flow rate	Remark	Ref.
				pak Phenyl 40°C	H_3PO_4 1.0 ml min^{-1}	PA and IPA was not achieved. Linear calibration P: 70–110% of labeled content IP: 1–5% of P contamination	
LC	PA, IP, P	Ophthalmic solutions	UV 220	300 × 3.9 mm, reversed phase phenyl, 10 μm 26°C	5% KH_2PO_4-ACN 97:3, pH 2.5 with 85% H_3PO_4 0.3 ml min^{-1}	Collaborative study. CV 3.2–4.1, 31.3–37.2, and 19.7–27.4% for P, IP, and PA, resp.	469
LC	PA, IPA, IP, P	Degraded solutions of P, eye drops, ointments, and	UV 215	300 × 3.9 mm, μBondapak Phenyl 25°C	5% KH_2PO_4, pH 2.5 with H_3PO_4 1 ml min^{-1}	Linear calibration 0–0.1 g liter^{-1} for for IP, PA, and IPA	470

Method	Compounds	Sample	Detection	Stationary phase	Mobile phase	Remarks	Ref.
LC	IP, P, PA, IPA	Degraded solutions of P / conjunctival insert preparations	UV 215	250 × 4.6 mm, Nucleosil 10 C18	1 mM cetrimide in 10 mM $(NH_4)_2CO_3$-ACN 85:15 1.67 ml min^{-1}	Reten. data were reported for 8 mob. phases with different composition and for 5 methanolic mob. phases	457
TLC	P + IP, PA + IPA	Eye drops	Visualization, p-toluenesulfonic acid in Et_2O followed by Dragendorff's reagent	Layers 20 × 20 cm Silica gel 60 F254	EtOH-$CHCl_3$-28% NH_3 53:30:17	Preliminary semiquantitative screening	471

[a] For other examples, see Chapter 4, Section IV.G; Chapter 5, Sections III.A, IV.A, V.A, V.B; and Chapter 6, Section II.G.
[b] IP, isopilocarpine; IPA, isopilocarpic acid; P, pilocarpine; PA, pilocarpic acid.

TABLE 7.23 Chromatographic Conditions for Diterpene and Steroidal Alkaloids[a]

Technique	Compounds separated	Sample	Detection	Col. Stat. phase Col. temp.	Mobile phase flow rate	Remark	Ref.
GC	Solanidine	Spiked milk	NPD	1.2 m × 2 mm 3% OV-13 on Chromosorb W HP, 80–100 mesh	Helium 25 ml min^{-1}	Linear calibration 0.07–0.28 μg. Recovery 80.8–110.2%. CV 8.1–12.4% at 0.28–1.12 ppm. LOD 0.14 ppm	472
GC	5-α-Cholestane (I.S.), solanthrene, solanidine, demissidine, solasodiene, solasodine, tomatidine	Tuber samples incl. household and industrial varieties	FID	a. 1 m × 2 mm, 10% SE-30 on Chromosorb W, 80–100 mesh, 260–300°C at 5 K min^{-1}, 300°C for 10 min b. 50 m × 0.22 mm CP-Sil 5, 0.12 μm 290°C	a. Nitrogen 15 ml min^{-1} b. Helium 24.5 cm sec^{-1}	Quantitative anal. Retention times are given. Aglycones were obtained by hydrolysis with HCl	473

Applications

GC	Solanidine, demissidine, solasodiene, solasodine, tomatidine	Tuber samples of Solanum vernei, S. sucrense, S. brevicaule, S. oplocense, S. spegazzinii, S. leptophyes	FID and FPD simultaneously	c. 50 m × 0.22 mm CP-Sil 19 CB 50 m × 0.25 mm, CP-Sil 5 CB, 0.12 μm 270 or 280°C	Hydrogen 49 cm sec	Samples were extracted and hydrolyzed. Reten. times and reten. indices are given. CV of reten. times 0.037–0.068%	474
GC	Codeine (I.S.), conessine	Holarrhena floribunda stem bark	FID	1 m × 2 mm 3% SE 30 on Chromosorb W AW DMCS 100–120 mesh, 230°C	Helium 30 ml min^{-1}	See Fig. 7.29. LOD 20 ng. CV 3–10% for 0.09–0.76% conessine in plant	475
LC	a. Mesaconitine, acetylaconitine, aconitine, delphinine, deoxyaconitine, 8-O-	a. Standards b. Aconitum crassicaule c. Products of aconitine oxidation	a–d. UV 254	a–d. 150 × 4.1 mm, Resolve C18 5 μm	a. 1.7 mM NH$_4$ carbonate-THF 1:1 b. 1.8 mM NH$_4$ carbonate-THF 1:2 c. 0.9 mM	b. See Fig. 7.31	476

TABLE 7.23 (Continued)

Tech-nique	Compounds separated	Sample	Detection	Col. Stat. phase Col. temp.	Mobile phase flow rate	Remark	Ref.
	methyl-14-benzoylaco-nine, ben-zoylaconine b. Yuna-conitine, crassicau-line, crass-icautine, foresaconi-tine, crass-icausine c. Oxoni-tine, N-ace-tyl-N-des-ethylaconi-tine d. Falcon-erine ace-tate, falcon-erine	d. Aconitum falconeri			NH$_4$ carbon-ate -THF 3:1 d. 1.8 mM NH$_4$ carbon-ate-THF 1:1 a-d. 1 ml min^{-1}		
LC	Hokbusin A, mesaconi-tine, aljes-	Aconitum japonicum Thunb.	UV 254	300 × 5 mm Nucleosil 5 C18, 40°C	50 mM phos-phate-ACN 71:29, pH	Recovery 96-96.5%	477

Applications

Method	Compounds	Sample	Detection	Column	Mobile phase	Notes	Ref.
	aconitine A, aconitine, jesaconitine				2.5 1.1 ml min^{-1}		
LC	Nudicauline, methyllycaconitine	Delphinium nuttallianum Pritz	UV 210	300 × 4 mm MCH-5 (monomeric ODS on silica), 5 μm, MCH-10 guard col.	4 mM C6SO3Na in ACN-0.1% H$_3$PO$_4$ 92:8 0.4–0.6 ml min^{-1}	Recovery 92–97% at 2 mg g^{-1}. CV < 8% for methyllycaconitine	478
LC	Cathedulines E5, -E3, -E4	Catha edulis (Vahl) Forsk ex Endl (Celastraceae), leaves, twigs	UV 254	500 × 25 mm JAI-GS 320 (polyvinyl-alcohol)	MeOH 3 ml min^{-1}	Recycling prep. GPC. Polystyrene col. failed to separate the mixture	479
LC	a. α-Chaconine + α-solanine, β-chaconine b. β-Chaconine, α-chaconine, tomatine, α-solanine c. α-Chaconine, α-solanine	a,b. Standards b. Potato blossoms c. Potato peels and sprouts	a,b. UV 208, RI c. UV 200 or 208 or 215, RI	a–c. 300 × 4 mm a. μBondapak C18, 10 μm b. μBondapak NH$_2$, 10 μm c. Carbohydrate col. (Waters)	a. THF-H$_2$O-ACN 5:3:2 b. THF-3.4 g liter^{-1} KH$_2$PO$_4$-ACN 5:3:2 or 50:25:25 1 ml min^{-1} c. THF-H$_2$O-ACN 56:14:30 or	b. See Figs. 7.32 and 7.33. CV 11.1 and 10.4% for chaconine and solanine, resp.	480

TABLE 7.23 (Continued)

Technique	Compounds separated	Sample	Detection	Col. Stat. phase Col. temp.	Mobile phase flow rate	Remark	Ref.
LC	α-Chaconine, α-solanine	Tater meal	UV 215	250 × 4.5 mm, amino col., 5 μm	6:1:3, ACN-H_2O 85:15 1 or 2 ml min^{-1}		472
LC	α- and β-Solamargine, α-solasonine, α-chaconine, α-solanine, nicotine (I.S.)	<u>Solanum ptycanthum</u>	UV 200	Col. 8 mm i.d., radially packed silicic acid, guard col. Porasil μBondapak	THF-H_2O-ACN 55:15: 30 1,1 ml min^{-1} 0.01% Ethanolamine in ACN-H_2O 77.5:22.5, pH 4.0 with H_3PO_4	Recovery 100–104.6%	482
LC	α-Solanine, α-chaconine	Potatoes	UV 220	250 × 4.6 mm, octyl (C8) spherical silica gel	1.16 g $liter^{-1}$ NH_4 phosphate in H_2O-ACN 1:1	Recovery 90–96%	483
LC	a. Cevadine, vera-	a,b. Veratrine	a,b. RI, UV 245	a. 300 × 3.9 mm, μPorasil,	a. Light petroleum-	a. Recovery 95% for cev-	485

Applications   559

	tridine b. Veracevine, cevadine, veratridine			10 μm, 30°C b. 100 × 8 mm, CN-8 Radial-Pak	EtOH-Et$_2$NH 950:50:4 2 ml min^{-1} b. MeOH-10 ml liter^{-1} dibutylamine phosphate in H$_2$O 4:6 2 ml min^{-1}	adine	
LC	Jervine, veratramine, 11-deoxojervine	Veratrum californicum roots	UV 254	300 × 3.9 mm, μBondapak C18 10 μm	ACN-H$_2$O-trifluoroacetic acid 32:67.9:0.1 or 28:71.9:0.1 1 ml min^{-1}	Other cols. were also tested	486
LC	Veratridine	Veratrine	UV 220	a. 500 × 9.4 mm, Partisil M 9/C 8 10 μm b. 300 × 3.9 mm μBondapak C18	MeOH-0.1 M AcONH$_4$, pH 5.5, 6:4 a. 3 ml min^{-1} b. 2 ml min^{-1}	a. Semipreparative separation	487
LC	Jervine	Veratrum root and tincture	UV 254	250 × 4.6 mm, Spherisorb silica 5 μm	CHCl$_3$-MeOH-Et$_2$NH 99:1:0.1 1.2 ml min^{-1}	Average conc. of jervine in root 0.0369%	488
CCC	Veratridine, ceva-	Veratrine	UV 250	CHCl$_3$, ratio to mob.	0.5 M phosphate buf-	Droplet countercur-	485

TABLE 7.23 (Continued)

Technique	Compounds separated	Sample	Detection	Col. Stat. phase Col. temp.	Mobile phase flow rate	Remark	Ref.
	dine			phase 1:1	fer, pH 5-H$_2$O-MeOH-PrOH 20:20:40:8	rent chromatogr. Preparative separation	
TLC	a. Aconitine, 3-deoxyaconitine, mesaconitine b. Staphisine, staphidine c. Delsoline, 14-acetyldelcosine d. Veatchine, garryine	a. Aconitine Potent Merck b. Delphinium staphysagria L. seeds	a,b,d. Visualization UV 254 c. Iodine vapors	a–d. Alumina rotor alumina 60 GF 254 and gypsum 1 mm	a. C6-Et$_2$O 1:3, Et$_2$O-MeOH 99.9:0.1, Et$_2$O-MeOH 98.5:1.5 b. C6-acetone 95.5:4.5 c. Et$_2$O-MeOH 99:1, Et$_2$O-MeOH 99.5:0.5 d. Et$_2$O-MeOH 99:1, MeOH	Preparative centrifugal TLC	489
TLC	a. 14-Acetyldelcosine, delso-	e. Delphinium barbeyi Huth.	a–e. Visualization UV	a–e. Alumina rotor alumina 60	a. Et$_2$O + 0.5% MeOH followed by	Preparative centrifugal TLC	490

Applications

	line b. Ezochasmanine c. Heterophyloidine d. Lycoctonine e. Delphatine, browniine			PF 254 + 366, 1 mm	Et_2O + 0.5% $MeOH$ + 0.5% Et_2NH b. Et_2O c. CH_2Cl_2 d. Gradient elution involving C6, CH_2Cl_2, $MeOH$ e. Et_2O + 1% $MeOH$ followed by Et_2O + 5% $MeOH$ + 0.3% Et_2NH		
TLC	Methyllycaconitine	<u>Delphinium nuttallianum</u> Pritz		Rotor Silica gel 60F 254 1 mm	$CHCl_3$, $CHCl_3$–$MeOH$ 8:1	Preparative centrifugally accelerated circular TLC	478
TLC	α- and β-Solamargine, α-solasonine	<u>Solanum ptycanthum</u>	Visualization UV 360 after spraying with 8-anilinonaphthalene-1-sulfonate	Layer Silica gel GF 254 2 mm	$CHCl_3$–$MeOH$–1% NH_4OH 2:2:1, two ascensions	Preparative TLC	482
TLC	Solanidine	Spiked milk	Visualization Dragen-	Layer 10 × 10 cm silica	Lower layer of $MeOH$–		472

TABLE 7.23 (Continued)

Tech-nique	Compounds separated	Sample	Detection	Col. Stat. phase Col. temp.	Mobile phase flow rate	Remark	Ref.
			dorff's reagent	gel, 0.2 mm	$CHCl_3$-1% NH_4OH 2:2:1		
TLC	Codeine (I.S.), conessine	Holarrhena floribunda stem bark	Absorption densitometry 500 nm after application of Dragendorff's and iodoplatinate reagents	Layers 10 × 20 cm Silica gel 60F 254	EtAc-C6-Et_2NH 75:24:6	See Fig. 7.30. LOD 40 ng. CV 2-15%	475
TLC	Jervine, veratroylzy-gadenine, rubijervine, isorubijer-vine, vera-mine	Veratrum root and tincture	Fluor. densitometry λ_{ex} 254 nm λ_{em} 546 nm. Dragendorff's reagent	Layers 20 × 20 cm a. Silica gel 60 F 254 b. Alumina	a. Benzene-EtOH-Et_2NH 80:16:4 b. Benzene-EtOH 95:5		488

[a]For other examples, see Chapter 4, Section IV.H; Chapter 5, Sections I.A, II.A, IV.A, V.A, V.B; Chapter 6, Sections I.E and II.H.

cold remedies (91). Factors affecting the separation of ephedrine, other alkaloids, and some acidic, basic, and neutral drugs on nitrilo column was evaluated using RP mobile phases (92).

Due to its pharmacological importance ephedrine is frequently determined in pharmacological products (12,93-95). Normal phase separation of ephedrine, codeine, and ethylmorphine contained in the syrup Solucamphre in concentrations of about 0.35 g liter^{-1} is shown in Figure 7.5 (93). Pseudoephedrine was determined in animal-dosing formulations using separation on cation exchange column (96). For its determination in plasma a sensitive method based on derivitization and normal phase LC with fluorimetric detection was developed (97).

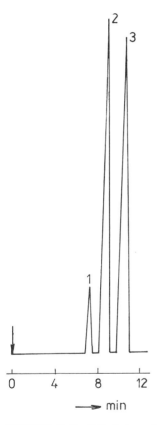

FIGURE 7.5 LC determination of the main alkaloids in syrup Solucamphre. Column 250 × 4.6 mm packed with Partisil PXS 5/25, mobile phase diethyl ether saturated from 95% with water and containing 0.3% (v/v) diethylamine, 2 ml min^{-1}, UV detection at 254 nm. 1, ephedrine; 2, ethylmorphine; 3, codeine. (After Ref. 93.)

When determining pseudoephedrine in plasma and urine without derivatization (98,99), a short wavelength is preferred for photometric detection to achieve higher sensitivity. The same holds for the method used by Sagara et al. (100) for the analysis of Ephedra herbs and oriental pharmaceutical preparations. Both RP and silica gel layers were used for qualitative TLC screening of urine for basic drugs including ephedrine and some other alkaloids (68). As can be seen from data given in Table 7.15, various spray reagents and mobile phases were compared. Overpressured TLC was used for the separation of ephedrine and other doping agents (101) (see Figure 6.2).

A RP-HPLC method which enables the analysis of degradation products formed by colchicine in solution was developed (102). Strong retention of tropolones was eliminated by the addition of copper(II) ions to the mobile phase, which enabled their direct analysis. Colchicine was determined simultaneously with colchicoside in the extract of Colchicum autumnale seed by applying gradient elution on silanized silica gel column (14). A HPLC method for colchicine determination in human plasma and urine (103) was not sensitive enough for pharmacokinetic studies but was suitable for the relatively high levels found in cases of overdose. Quantitation of colchicine in tablets and plant materials was performed by TLC using in situ fluorescence measurement (104).

B. Separation of Pyridine, Piperidine, and Tropane Alkaloids

Owing to a low boiling temperature, nicotine can be easily eluted from GC columns and analyzed by GC. Using a flame ionization (16,105,106) or nitrogen-phosphorus-sensitive (105) detector, nicotine was determined in tobacco extract (16,105) (see Figure 4.2) or municipal wastewater (106). Nicotine was identified (107) in volatiles isolated by a head space using capillary GC and GC-MS with retention index 1925 on Supelcowax 10 (cf. Table 4.3). Capillary columns have been widely applied to the analysis of tobacco smoke (108). Commercial smokeless tobacco products were analyzed for alkaloids, tobacco-specific nitrosamines, and other components (109). The GC determination of nicotine in plasma (110-112), urine (113), and tissues (114) has been possible in low concentration ranges. The limit of quantitation is given by contamination of samples with nicotine from the environment rather than by poor sensitivity of chromatographic detectors. Cotinine, which is retained more strongly than nicotine on GC columns, can also be directly determined in body fluids and tissues by GC (112-115). However, some other nicotine metabolites require derivatization (116).

Enantiomers of nicotine and nicotine analogs were successfully separated on RP-LC microcolumn (117). Chiral resolution was achieved by saturation of the mobile phase with β-cyclodextrin, which is expected to form inclusion complexes with separated compounds. In contrast to GC, LC is less frequently used for the determination of nicotine in tobacco and tobacco products. Allergenic extracts of tobacco leaves were analyzed for nicotine by LC in order to assess their allergenic potency (118). On the other hand, detailed analyses of biological samples for nicotine metabolites have been performed by LC rather than GC. Ion exchange chromatography was used in the study of urine N-methylated metabolites of nicotine for both analytical (119-122) (see Figure 5.18) and preparative purposes (122,123). In a radiometric study using ^{14}C-labeled nicotine, 12 nicotine metabolites were assayed in urine and plasma of rats using RP gradient HPLC (38). In another study on nicotine metabolites the colored derivatives, produced by colorimetric assay, were separated by LC (69). In this way sensitive determination of nicotine and cotinine was possible. Sensitive and selective detection of nicotine and N-methylnicotinium is also possible using electrochemical detection (124). Simple LC method was developed for the determination of nicotine and cotinine in human urine (125). Shulgin et al. used a silica gel extraction column for sample cleanup by LC determination of cotinine-N-oxide in human urine (126). Several LC methods were compared for the determination of nicotine-1'-N-oxide in mouse tissue (127). In this study sample extracts were purified using C18 Sep-Pak cartridges.

Due to the volatility of nicotine, TLC is not suitable for its analysis.

One example of piperine analysis in black pepper by capillary GC (128) was given in Chapter 4, Section IV.B (Figure 4.3). However, because of the thermal lability and low volatility of many piperidine alkaloids, GC is not very suitable for their analysis and LC is usually preferred. Games et al. (129) analyzed components of capsicum and black pepper oleoresins using LC coupled to a mass spectrometer with a moving belt interface. Electron impact and chemical ionization were used in these experiments. For separation both NP-LC and RP-LC were applied. Separation of the black pepper oleoresin by the former mode is given in Figure 7.6. In simple assays of piperine in pepper (15,130), UV detection at 345 nm was more sensitive than at shorter wavelengths. By applying ternary mobile phase to the RP column, piperine was well separated from its isomers produced by exposure to light (15). Weaver et al. (131) resolved piperine from capsaicinoids on C18 column and were able to detect the presence of chilies in black pepper. The RP analysis of capsicum fruits and oleoresins was described by Attuquayefio and Buckle (132).

Some polyhydroxy piperidine alkaloids were separated by GC after trimethylsilylation (133) (for conditions, see Table 7.16).

Conditions for GC determination of many tropane alkaloids are favorable. Average values of their retention indices measured on packed columns with SE-30 and OV-1 stationary phases (82) range from 2050 to 2303 (see Table 4.3). Gas chromatography was used for the determination of atropine in pharmaceutical formulations containing cholinesterase deactivators, which are useful in the treatment of organophosphate poisoning (134). Leaf samples and cell cultures of Atropa belladonna were analyzed for atropine and scopolamine on capillary column (135). Detailed capillary GC-MS analysis of alkaloids from plant and cell cultures revealed the presence of 22 alkaloids in A. belladonna (136) and more than 30 alkaloids in Datura innoxia (137). Nevertheless, LC chromatography is used more frequently, especially for the simple analysis of tropane alkaloids in plant materials. Conditions for their separation on RP column were reported by Verpoorte et al. (138) and Ross and Lau-Cam (139). Small quantities of atropine in tablets, injections, and belladonna extract were determined using fluorescence (140) and electrochemical detection (141). Alkaloids isolated from A. belladonna, Datura stramonium, D. innoxia, and Hyoscyamus niger were determined using RP (142) and adsorption (143) chromatography. Hyoscyamine and hyoscine were determined in leaves, fruits, and seeds of Datura using ion pair RP column chromatography with camphorsulfonic acid as pairing reagent and also using densitometric evaluation of TLC plates (144). Some tropane alkaloids occurring in medicinal herbs (145) and extract from Anisodus tangulicus (146) were separated by droplet countercurrent chromatography.

The GC retention indices of cocaine and pseudococaine are within the range given above for other tropane alkaloids. Only ecgonine was retained too strongly, whereas the value 1193 was reported for tropine (82). Slightly higher retention indices were found by Lora-Tamayo et al. (147) on capillary columns (Table 4.3; for conditions, see Table 7.15). However, some ecgonine esters are not sufficiently stable to allow for GC analysis. Lewin et al. (148) separated four cocaine isomers (cocaine, pseudococaine, allococaine, and allopseudococaine) by GC and observed decomposition products only for allo and allopseudo configurations. The four ecgonine isomers were separated as methyl esters and two peaks were observed for alloecgonine methyl ester. Mass spectrometric investigation revealed that pseudoecgonine methyl ester was also partially decomposed in the injection chamber at 210°C (148).

Interlaboratory study on the GC quantitation of cocaine in pharmaceutical powders and tablets showed precisions ranging from 0.9 to 2.3% (149). By GC analysis cocaine has been identified in samples of street drugs (83,84) (see Figure 4.1 and Table 7.15). Since it is a widely abused drug, the methods for determining cocaine and its metabolites in plasma and urine are of importance. An automated

GC procedure designed for the treatment of a large number of plasma samples was developed (150). Losses on the surface of the borosilicate glass vials were eliminated by washing with a chelating agent (EDTA). In a GC method for the simultaneous determination of cocaine and benzoylecgonine, esterification of benzoylecgonine to its butyl ester was done (39). Goenechea et al. (151) reported two GC procedures for the determination of ecgonine in urine. Ecgonine extracted from urine was separated as MSTFA ($\underline{N}$-methyl-$\underline{N}$-trimethylsilyltrifluoroacetamide) derivative. In an extensive study by Moore et al. (152) on impurities in illicit cocaine samples and alkaloids of South American coca leaf, five diphenylcyclobutanedicarboxylic acids (truxillic and truxinic) and their alkaloidal precursors were identified, determined, and characterized by spectral methods. Strong presumptive evidence for the presence of the remaining possible truxillines was offered by the application of a capillary GC-ECD after reduction with $LiAlH_4$ and derivatization with heptafluorobutyric anhydride.

All four diastereomers of cocaine were separated by LC on silica gel column (148). Using flow programming the separation was completed in 12 min. Figure 7.7 shows the excellent separation of all the components. Electrochemical oxidation of tertiary amines was found to proceed best at pH 6-8 and was applied to the detection of cocaine, ecgonine, and benzoylecgonine (153). A RP-LC assay of cocaine and some pharmaceuticals was described by Selavka et al. (154). By the use of a postcolumn photolytic step followed by oxidative amperometric detection, a high sensitivity was achieved.
The LC analysis of minor components of illicit cocaine was reported by Lurie et al. (155). A procedure for the identification of samples suspected of containing cocaine was developed (156). Derivatization with phenylisothiocyanate prior to separation turned out to be useful for the identification of primary and secondary amines (e.g., ephedrines). Cocaine with codeine and caffeine were also separated by supercritical fluid chromatography coupled to mass spectrometry (157).

Thin-layer chromatography may be useful for the screening of street drugs (158) and local anaesthetics (159) including cocaine (see Table 7.15). Two-dimensional TLC has been used for the identification of local anesthetics (160) via application of a general strategy for the selection of mobile phases.

C. Separation of Quinoline Alkaloids

Some furoquinoline alkaloids occurring in the Rutaceae family exhibit mutagenic properties. Three alkaloids of this type were determined in the roots of Dictamnus albus subsp. dasycarpus. GC-MS was used for their identification in LC fractions (161). Some

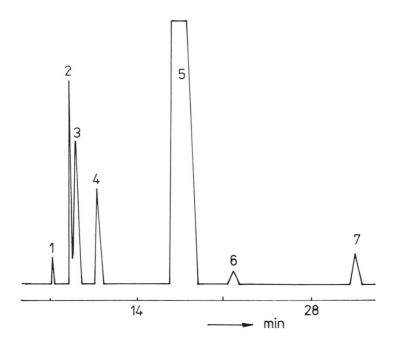

FIGURE 7.6 LC/MS-EI analysis of black pepper oleoresin (computer-reconstructed total ion current). Column 300 × 4 mm packed with Partisil 5 μm, mobile phase ethanol-hexane-acetic acid 4:95:1. 1, sitosterol; 2, N-isobutyleicosatrienamide; 3, N-isobutyloctadecatrienamide; 4, piperolein B; 5, two piperine isomers + piperettin; 6, piperine isomer; 7, piperylin isomer. (After Ref. 129.)

tertiary furoquinoline and quaternary dihydroquinoline alkaloids were separated by TLC on silica gel, alumina, and octadecyl-bonded silica gel layers and conditions suitable for RP-LC determination of platydesminium and balfourodinium in the cell cultures of Choisya ternata were found (162). Later, a TLC procedure with in situ fluorometry was developed for the same purpose with substantially lower limits of detection (163).

The analysis of Cinchona alkaloids received considerable attention because of their wide therapeutic use, particularly as antimalarial and antiarrhythmic agents. As discussed in Section IV.C of Chapter 4, GC separation of diastereomers of Cinchona alkaloids is poor on nonpolar stationary phases. Their retention indices are very similar as can be seen from the average values compiled by Ardrey and Moffat (82), who reported values of 2583 and 2598 for cinchonine and cinchonidine and 2784 and 2803 for quinidine and quinine,

Applications

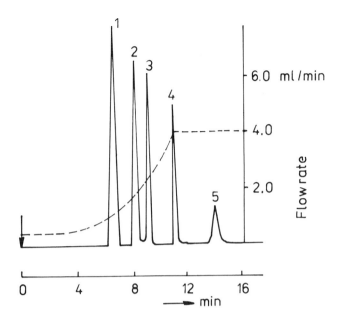

FIGURE 7.7 LC separation of isomeric cocaines. Column 250 × 4.6 mm packed with Partisil PXS 10, mobile phase isopropanol-heptane-diethylamine 25:75:0.1, programmed flow rate from 0.4 to 4.0 ml min^{-1}, UV detection at 203 nm. a, $\underline{N},\underline{N}$-dibenzylamide; 2, cocaine; 3, allococaine; 4, ψ-cocaine; 5, allo-ψ-cocaine. (After Ref. 148.)

respectively (see Table 4.3). Similar values were reported for capillary column (147). Improved separation using capillary column with a polar stationary phase (Figure 4.4) was achieved by Verzele et al. (128), who also analyzed pharmaceutical preparations and soft drinks for Cinchona alkaloids. However, a high column operating temperature renders the separation system less useful and thus LC prevails in practical situations.

The complete resolution of quinine, quinidine, the related desmethoxy alkaloids cinchonine and cinchonidine together with their dihydro analogs is a difficult task. Successful separation was achieved on silica gel (164), alkyl- (135,165) (see Figures 5.11 and 5.12), and alkylphenyl-bonded (166) columns. Such a resolution is desirable for the analysis of plant materials as well as for quality control of pharmaceuticals. In addition to the eight alkaloids mentioned above, Smith (166) resolved epiquinidine, epiquinine, quinone, and quinitoxine from quinidine and dihydroquinidine on alkylphenyl column. This separation system was then used for the analysis of cross-contaminated or decomposed pharmaceuticals.

In many cases of pharmaceutical analysis, however, the requirements for the separation are not so stringent, and simple, more rapid procedures may be satisfactory. Johnston et al. (167) elaborated a RP-LC method for the control of dihydro derivatives and other impurities in antiarrhythmic preparations containing quinidine and quinine salts. Figure 7.8 gives the analysis of a tablet containing quinidine sulfate. Later this procedure was evaluated by interlaboratory studies with encouraging results (168). Only the participating laboratories utilizing Waters Radpak ODS and Varian CH-10 columns were unable to obtain a workable separation. The necessity of using well-defined commercial columns, including criteria for column performance in any official method, was stressed. Roos and Law-Cam (139) analyzed quinine in paired ion drugs simultaneously

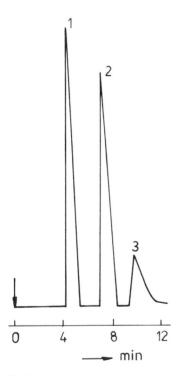

FIGURE 7.8 RP-LC analysis of a tablet containing quinidine sulfate. Column 300 × 4 mm packed with µBondapak C18, mobile phase methanol-water-acetic acid 25:75:1, 1.5 ml min^{-1}, UV detection at 254 nm. 1, theophylline (internal std.); 2, quinidine; 3, dihydroquinidine. (After Ref. 167.)

Applications

with acidic components; RP-LC was used for the determination of quinine and dihydroquinine in beverages (169) and of quinine in hair preparation (170).

Plant materials, primarily the bark of Cinchona and Remija species, are important source of Cinchona alkaloids. For the analysis of the Cinchona bark silica gel columns were used under conditions that enabled favorable separations (164,171–173) (see Figure 5.3). Figure 7.9 shows the analysis of Ecuadorian Cinchona bark extract and good resolution of the couple quinine-dihydroquinidine, which is usually separated with difficulty (173). Silica gel column was

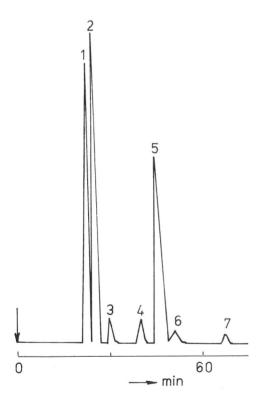

FIGURE 7.9 LC analysis of crude extract of Ecuadorian Cinchona bark powder. Two columns in series, 250 × 4.6 mm packed with Ultrasphere Si, 5 μm and 250 × 4.5 mm packed with Partisil PXS 10, mobile phase tetrahydrofuran-n-butyl chloride-ammonia 60:40:0.25, 2 ml min^{-1}, UV detection at 254 nm. 1, cinchonine; 2, cinchonidine, 3, quinidine; 4, dihydrocinchonidine; 5, quinine; 6, dihydroquinidine; 7, dihydroquinine. (After Ref. 173.)

also used for the comparison of alkaloid patterns of Cinchona leaf samples containing both indole and quinoline alkaloids (174). Optimized separation of four major Cinchona alkaloids (Figure 5.15) on nitrilo column was applied to the analysis of Cinchona succirubra leaf extract (175). With the development of biotechnologies for the production of Cinchona alkaloids have come methods for analysis of their cell cultures (173,176). As an example for the application of LC connected to a double-focusing MS through a thermospray interface, the analysis of alkaloids extracted from cell cultures of Cinchona ledgeriana was reported (177).

The use of Cinchona alkaloids as pharmaceuticals requires methods for their assay in biological materials. These analyses were performed by RP-LC (178) or ion pair LC (179,180). Quinine and quinidine were determined in samples of maternal and fetal sheep plasma following drug administration to the mother (41). A more detailed RP-LC study of quinidine and its metabolites in blood and urine was reported by Tamai et al. (40). Using micellar chromatography on nitrilo or C18 column, the direct injection of urine and serum samples to a LC column was possible (181).

Thin-layer chromatography has been widely applied to the separation of Cinchona alkaloids (see Section II.C in Chapter 6 and Table 6.3) and silica gel layers have been used in most cases. As an example of qualitative analysis, the application of TLC to the evaluation of fractions from microcolumn purification of alkaloid extracts from Cinchona ledgeriana cell cultures can be mentioned (173). Quinidine was determined in serum using TLC separation followed by in situ fluorescence measurement (179). Simultaneous assay of quinidine and four other antiarrhythmic drugs is possible by HPTLC with a work output comparable to that of HPLC (182).

Quinine chemically bonded to silica gel was used as the chiral stationary phase for the separation of enantiomers by LC (183,184).

D. Separation of Isoquinoline Alkaloids

Frequently, various subgroups of isoquinoline alkaloids occur simultaneously in plant materials and therefore the conditions used for their separation are grouped in Table 7.11. Some chelidonine-type alkaloids have also been included here. Voluminous data for phenanthrenisoquinoline alkaloids, to which most opium alkaloids belong, are arranged and discussed separately. Chromatographic conditions for those applications, where single phenanthrenisoquinoline alkaloids or their metabolities were determined, are summarized in Table 7.12. Separations of alkaloid mixtures isolated from Papaver somniferum or standard mixtures with similar composition are separately arranged in Table 7.13. In that table data for papaverine, laudanosine, narcotine, naceine, and cryptopine, when analyzed

Applications

simultaneously with phenanthrenisoquinoline alkaloids found in P. somniferum plants, are also included. Samples of heroin, which is not considered an alkaloid, usually contain a complex mixture of alkaloids related to the raw material used for their preparation. Some examples of the analysis of heroin impurities and metabolites can be found in Table 7.14. Finally, some data are included concerning chromatographic systems employed in the identification of drugs. Such systems are used for toxicological or forensic purposes and many alkaloids of several structural types generally appear in the set of studied substances. Because some of these alkaloids possess isoquinoline rings, such chromatographic applications are arranged in Table 7.15.

The GC separation of a few isoquinoline alkaloids is possible without derivatization (cf. Section IV.D of Chapter 4). Some retention data can be found in Table 4.3 and other data in the original review (82). However, except for morphine alkaloids, GC has been little used in practice. Duncan et al. (185) developed GC-MS procedure for the determination of salsolinol and compared it with LC equipped with electrochemical detection (186). The method was combined with a simple one-step extraction which was free from artificial salsolinol formation. Whereas sensitivity was similar for both methods, the GC-MS method was shown to be more versatile. For complex sample matrices the LC method might require modification of chromatographic conditions or a sample extraction procedure to eliminate interferences. The results for food and beverages suggested that diet is a potential source of "mammalian alkaloids" such as salsolinol.

Six different dihydroxytetrahydroisoquinolines including salsolinol and tetrahydropapaveroline were separated by ion pair RP-LC with gradient elution and electrochemical detection (187). To verify the possibility of endogenous formation of tetrahydropapaveroline in experimental animals, HPLC separation with sensitive electrochemical detection was combined with a gentle multistep isolation technique (188).

The study of Guatam et al. (189) can be introduced as an example of the LC determination of papaverine as a single alkaloid in body fluids. Retention behavior of papaverine, codeine, ephedrine, and 13 other drugs on a nitrilo column using phosphate buffer in the mobile phase was studied (92) (see Table 7.7). Much attention was paid to the expressive antitussive effect of glaucine. When studying the microbial oxidation of glaucine to dehydroglaucine by Fusarium solani, Davis (190) developed a method for the rapid determination of these compounds in microbial cultures. Separation by RP-LC is illustrated in Figure 7.10. Pekić et al. (17) applied a similar mobile phase with phenyl column for the determination of glaucine in Glaucium flavum. Ion pair chromatography was used for the separation of glaucine and other aporphine and protopine tertiary and quaternary alkaloids present in oriental pharmaceutical preparation (191). Aporphines and morphinans including glaucine were isolated from Papaver pilosum and separated on the silica gel column (192). A sensitive method for glaucine determination in plasma and urine after simple extraction on kieselguhr microcolumn was developed by Fells et al. (193).

Alkaloids of aporphine, protoberberine, protopine, and chelidonine types were identified with the aid of TLC and paper chromatography in complex mixtures isolated from Stylophorum diphyllum (194) and Glaucium squamigerum (195) (cf. Table 6.5).

Separation of racemic tetrahydroberberine alkaloids to their respective enantiomers on Chiracel OC-L column was reported by Tagahara et al. (196). Quaternary protoberberine alkaloids from Jatrorrhiza palmata and Chasmanthera dependens were separated and isolated using circular TLC (197). Berberine and three chelidonine-type quaternary alkaloids from Chelidonium majus roots were separated on silica gel column with acidified methanolic mobile phase (26). Wastes from industrial production of chelidonine hydrochloride from C. majus were separated on silica gel using TLC and microcolumn LC (198). Alkaloids of various structural types were used as test compounds in RP ion pair chromatographic study (199) with di(2-ethylhexyl)orthophosphoric acid as pairing reagent. The system was used for the separation of C. majus extract.

Histochemical chromatography, in which intracellular components of living cells isolated by micromanipulator are subjected to chromatographic analysis, was applied to the determination of alkaloids in cells from Macleaya cordata (200,201) and Sanguinaria canadensis (201). [^{2}H$_3$]-Allocryptopine was determined by GC-MS after incorporation of deuterium-labeled tetrahydroberberine to the plant (200). Two protopine- and two chelidonine-type alkaloids were determined in nanogram per cell levels by ion pair chromatography (201).

Five chelidonine-type alkaloids were assayed in Corydalis bungeana by RP-LC (19). Preparative centrifugal chromatography on alumina layer was used for the isolation of some isoquinoline alkaloids from Fumaria species (18). Alkaloids of other structural types were also separated. Noscapine (narcotine) was determined in serum (202) and plasma (203) using both RP and NP-LC. Hydrolysis of the lactone ring and lactonization of noscapinic acid back to noscapine was studied in aqueous buffers (203). In the presence of albumin or proteins from blood, noscapine is protected against opening of the lactone ring (202) and noscapine acid is transformed slowly to noscapine (203). Hence, noscapine can be extracted from alkalinized blood samples without losses due to hydrolysis.

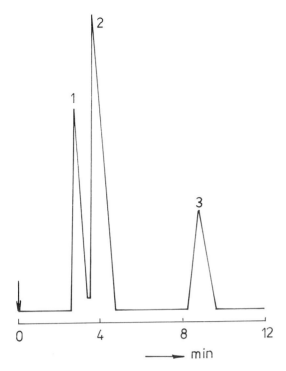

FIGURE 7.10 RP-LC separation of glaucine and dehydroglaucine. Column 300 × 3.9 mm packed with μBondapak Phenyl, 10 μm, mobile phase acetonitrile-methanol-0.05 M KH_2PO_4 2:4:5, 2 ml min^{-1}, UV detection at 280 nm. 1, glaucine; 2, papaverine (internal std.); 3, dehydroglaucine. (After Ref. 190.)

Analysis of two major alkaloids, emetine and cephaeline, in ipecac root by a simple and rapid HPLC method was described by Sahu and Mahato (204). Bannister et al. (205) determined emetine in plasma by ion pair LC with photometric detection (see Figure 7.11). To increase the sensitivity, emetine was converted by oxidation with the mercuric acetate solution into rubremetine, which was detected after LC separation by a fluorescence detector.

Bisbenzylisoquinoline quaternary alkaloid tubocurarine can be separated by ion pair chromatography. This technique was applied to the determination of alcuronium in plasma and urine using tubocuratine as internal standard (206). Avram and Shanks (207) determined either tubocurarine or its trimethyl derivative metocurine in plasma by LC on nitrilo column.

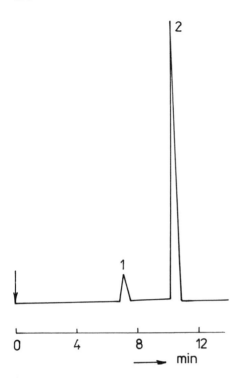

FIGURE 7.11 Analysis of emetine in plasma by ion pair chromatography. Column 300 × 3.9 mm packed with µBondapak C18, mobile phase 0.5% acetic acid and 2.5 mM sodium octanesulfonate in methanol-water 56:44, 2 ml min^{-1}, UV detection at 280 nm. 1, emetine; 2, naphthalene (internal std.). (After Ref. 205.)

Although LC methods prevail in the analysis of phenanthrenisoquinoline (morphine) alkaloids (Table 7.12), their separations by GC are not rare. Practical analyses may be performed in some cases even without derivatization. Masumoto et al. (208) used separation on a short capillary column for the direct determination of codeine and chlorpheniramine in plasma extracts. The chromatogram is shown in Figure 7.12. In most cases, however, derivatization has been used because narcotics, and morphine in particular, may suffer adsorption losses within GC column and, moreover, suitably chosen derivatization may improve the sensitivity and selectivity of the determination. The analysis of derivatized alkaloids requires caution to minimize decomposition of derivatives. Edlund (209) determined the decomposition rate constants for the morphine

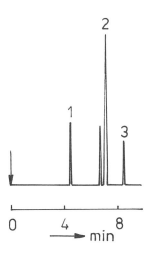

FIGURE 7.12 GC analysis of codeine in plasma. Column 5 m × 0.32 mm with crosslinked 50% phenylmethyl silicone; He 4.2 ml min^{-1}; 150°C for 1 min, then 150–240°C at 10 K min^{-1}, 240°C for 1 min; N-P selective detector. 1, chlorpheniramine; 2, pyrilamine maleate (internal std.); 3, codeine (6.5 ng ml^{-1} of plasma). (After Ref. 208.)

and codeine derivatized with pentafluoropropionic anhydride on capillary columns (C_p-Sil 5, SE-30, OV 1701). At temperatures below 210°C less than 5% decomposition was observed on most of the tested columns.

Methylation after injection of the sample extract dissolved in trimethylanilinium hydroxide into the GC injection chamber was used by Kim (58) for the fast analysis of morphine in blood. Sewel et al. (210) screened various <u>Streptomycetes</u> and species of the fungus <u>Cunninghamella</u> for <u>N</u>-dealkylation activity against a variety of drugs including codeine. Extracts of cultures were injected onto GC column together with acetic anhydride. Incomplete on-column acetylation permitted better identification of separated compounds because at least two species were eluted for each analyte.

Cone et al. (211) analyzed morphine, codeine, and their metabolites in rat urine. Ten potential metabolites were separated on various stationary phases and retention times for parent compounds and trimethylsilyl derivatives were reported. The urinary codeine-to-morphine ratios of 15 volunteers administered codeine tablets were determined by Dutt et al. (65) in order to establish criteria for the recognition of codeine consumption in screening for opiate drug abuse. The analysis performed by GC employing trimethylsilyl de-

rivatives is shown in Figure 7.13. The method developed by
Christophersen et al. (70) is suitable for identification of opiates
in urine and discrimination between use of heroin or morphine and
of legal opiates. Capillary GC was combined with the use of two
derivatization reagents.

Morphine was determined in blood after heptafluorobutyration
by GC with ECD using packed (212,213) and capillary (63) columns.
Analysis of morphine in various biological materials was performed
using a GC coupled to mass spectrometer (214—216). Persistent tissue bonding of morphine in rats was demonstrated 22 days postwith-

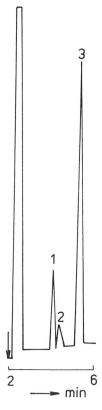

FIGURE 7.13 GC analysis of codeine and morphine in urine. Column 1.83 m × 6.4 mm o.d. packed with 3% OV-1 on Chromosorb W
HP, N_2 40 ml min^{-1}, 240°C, FID. 1, codeine; 2, morphine; 3, nalorphine (internal std.). Silylated derivatives with bis(trimethylsilyl)trifluoroacetamide. (After Ref. 65.)

Applications

drawal (214). Morphine and 6-butanoylmorphine as metabolites of 3,6-dibutanoylmorphine were separated on a very short methylsilicone capillary column after acetylation and were detected using selected ion monitoring (216).

Separation of opium alkaloids by HPLC was recently reviewed (217). Retention of basic compounds on silica gel eluted with methanolic phosphate buffer was studied using morphine as one of the test compounds (218). When LC is applied to the determination of morphine alkaloids in pharmaceuticals, sufficient sensitivity is achieved with photometric detection (3,6,59,93,219,220). Various separation systems are applicable in these cases. Separation of components of the syrup Solucamphre (93) has already been mentioned in Section II.A (see Figure 7.5 and Table 7.7). Even for the analysis of morphine, codeine, and their metabolites in biological materials, the sensitivity of UV detection at the mercury line 254 nm was sufficient in some cases (221,222). With variable-wavelength detectors, shorter wavelengths in the range 210–220 nm have been used in order to improve sensitivity (3,44,59,62,223,224). Detection at short wavelengths is not very selective and limits the range of usable mobile phases. Derivatization of morphine with dabsyl chloride (4-dimethylaminoazobenzene-4'-sulfonyl chloride) followed by detection of derivatives at 436 nm substantially increased sensitivity and selectivity of detection (64). The derivative of morphine was separated and quantitated by both HPLC and TLC. The native fluorescence of morphine and codeine was also used for detection (181,225). With a micellar mobile phase (181), the direct injection of urine samples was possible. Using a short RP column (225) the separation of codeine and internal standard was completed within 5 min. Pre- or postcolumn oxidation of morphine to fluorescent pseudomorphine may be used (57). The presence of some surfactants and micelle formation in the reaction mixture was shown to increase the fluorescence signal (226). Derivatization of morphine with dansyl chloride was also used (227,228). With a simple extraction proeedure the method was suitable for the determination of free morphine. For samples which were hydrolyzed in order to determine total morphine, a more specific sample treatment would be required.

Sensitive detection of some morphine alkaloids may be provided by oxidative electrochemical methods. In coulometric detection with two flow-through electrodes in series (223), more easily oxidizable components prereact completely at low potential at the first electrode and then components of interest are detected at higher potential at the second electrode. This results in a lower background current and consequently in a lower noise level and a cleaner chromatogram. Morphine is detected easily due to the presence of a phenolic group (42,43,213,229) whereas codeine and metabolites require a higher potential of the working electrode (60,61,229,230).

Background current is elevated under these conditions and limits of detection are usually higher for codeine than for morphine detected at a lower potential. The background level varied with the purity of the reagents (acetonitrile) (230). Supercritical fluid chromatography coupled to MS (157) was examined using codeine (see Table 7.9).

Analysis by TLC may yield useful quantitative results at the much lower cost per sample than that by HPLC (64). The determination of codeine and three drugs commonly encountered in analgesics and in cough and cold preparations was described by Al-Kaysi and Salem (231).

Thebaine is prone to decomposition when separated by GC. Complete decomposition was observed on packed columns at 270°C (232) and four major decomposition products were observed by column switching capillary GC (233). Therefore RP-LC was proposed for the determination of thebaine in Papaver bracteatum (232). However, for exact quantitative work, the saturation of the column by repeated injection of thebaine standard was inevitable. Large series of thebaine determinations in P. bracteatum samples were performed by computer-controlled anticircular TLC (234).

A large group of separations has been devoted to the analysis of the major alkaloids present in opium or Papaver somniferum plants (Table 7.13). The GC analysis of opium samples is not frequently applied due to the necessity of sample derivatization, although earlier studies (235) reported the chromatography of underivatized sample. Morphine in particular must be derivatized in order to obtain reproducible results (236). Sperling (236) therefore proposed the preliminary separation of opium extract into two fractions by LC. In the first fraction codeine, thebaine, papaverine, and narcotine were determined by GC without derivatization whereas in the second fraction morphine was determined as a trimethylsilyl derivative. Neumann (237) determined the major opium alkaloids in opium and crude morphine after trimethylsilylation of the total extract by capillary GC. Moreover, GC profiles of impurities were compared and the effect of acetylation was examined.

Liquid chromatographic methods applied to the separation of major opium alkaloids involve a large variety of separation systems. Since detection sensitivity usually poses no problem, the use of photometric detection at 254 nm or approximately 280 nm prevails. However, electrochemical oxidation at higher oxidation potentials was shown to be an universal detection method for tertiary nitrogen compounds (153).

Major opium alkaloids differ substantially in their polarity, ranging from slightly polar papaverine to highly polar morphine and narceine. When a simple isocratic elution is used, the retentions of either papaverine or morphine and narceine are usually unsatisfactory and some compromise has to be accepted. Isocratic elution on silica gel column was used by Engelke and Vincent (13) for the determination of morphine in poppy straw. Hutin et al. (24) were able to resolve five alkaloids by the use of a similar mobile phase. Morphine was eluted as the last component, which is advantageous for the quantitation of samples from P. somniferum where morphine is usually present as the main component. The elution of narceine required gradient elution (24). Jane (238) separated opium alkaloids on silica gel column with methanolic mobile phase containing ammonium buffer (see Figure 5.4). The elution order was the same as for hexane-based mobile phase used by Hutin (24). A roughly similar elution order was observed for ion exchange chromatography on alumina, but it was partially dependent on the content of organic modifier in the mobile phase (239,240). The conditions for the separation of opium and heroin are reported in Table 7.14. The excellent separation of six alkaloids was obtained in a RP system with octadecyl column (24) as shown in Figure 7.14. The elution order was reversed from that of the silica gel system. In order to eliminate interferences due to excessive concentration of morphine in opium samples, the use of preliminary separation of phenolic alkaloids by liquid-liquid extraction was proposed (24). Another solution is represented by better resolution of weakly retained components using gradient elution (241). Using gradient elution, Sisco et al. (242) were able to resolve and determine eight alkaloids in a codeine phosphate drug derived from poppy straw or opium concentrate. Alkyl-bonded RP columns eluted with methanolic phosphate buffer have seldom been applied to the analysis of opium. Separation of the poppy straw extract in such a system is shown in Figure 7.15 (243).

Pettitt and Damon (23) applied a phenyl column and gradient elution to the analysis of poppy straw concentrate. Ayyangar and Bhide (244) used the same packing with isocratic elution (see Figure 5.10). The mobile phase was alkaline and contained sodium acetate, which helps to achieve adequate retention of morphine. Later, using gradient elution and an acidic mobile phase, the separation of eight alkaloids and meconic acid was achieved (22). The alkaloids from brown opium powder were also separated on a macroporous styrene-divinylbenzene copolymer (20) (cf. Figure 5.9). Nitrilo- (21,245) and amino- (246) (see Figure 5.16) bonded packings were used for the analysis of alkaloids with mobile phases of the RP type. Separation of the extract of gum opium on nitrilo column is shown in Figure 7.16 and may be compared with Figure 5.14, which shows the separation of standard mixture (21). Compared to alkyl-bonded phases morphine is usually eluted far enough from the solvent front on nitrilo column. On amino column the same elution order as for NP chromatography was obtained (246).

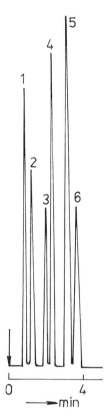

FIGURE 7.14 RP-LC separation of the main alkaloids of Papaver somniferum. Column 300 × 3.9 mm packed with µBondapak C18, mobile phase water-acetonitrile-triethylamine 40:60:0.1, 2 ml min^{-1}, UV detection at 280 nm. 1, narceine; 2, morphine; 3, codeine; 4, thebaine; 5, papaverine; 6, α-narcotine. (After Ref. 24.)

Ion pair chromatography was successfully applied to the analysis of opium alkaloids (247–250) (see Figures 5.21 and 5.22). Chromatographic conditions are usually not very different from those used by Lurie (247). Column efficiency was inferior to that with the mobile phase containing triethylamine (24). On a column impregnated with dodecylsulfonic acid and cetrimide, six opium alkaloids were separated (138).

Heroin does not belong to the alkaloid family but several alkaloids of different polarities commonly occur in it as impurities or adulterants, and hence the analysis of heroin represents a complex

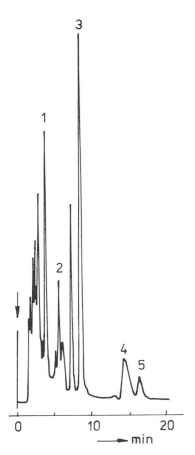

FIGURE 7.15 RP-LC separation of poppy straw extract (243). Column 150 × 3.2 mm packed with LiChrosorb RP-8, 5 μm, mobile phase methanol-(0.025 M NaH_2PO_4 + 0.025 M Na_2HPO_4) 6:4, 0.5 ml min^{-1}, UV detection at 254 nm. 1, morphine; 2, codeine; 3, papaverine; 4, thebaine; 5, α-narcotine.

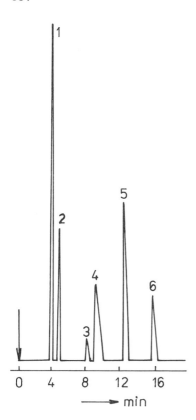

FIGURE 7.16 RP-LC separation of the extract from gum opium. Column 300 × 4 mm packed with Nucleosil 10 CN, mobile phase 1% CH_3COONH_4 (pH 5.8)-acetonitrile-dioxane 80:10:10, 1.5 ml min^{-1}, UV detection at 254 nm. 1, morphine; 2, codeine; 3, cryptopine; 4, thebaine; 5, α-narcotine; 6, papaverine. (After Ref. 21.)

problem. All three chromatographic techniques (LC, GC, and TLC) have been employed to solve it (Table 7.14). By direct GC determination of 6-acetylmorphine in heroin in the presence of morphine, higher contents than those with LC were frequently found. A more detailed study (251) confirmed transesterification occurring on injection into the gas chromatograph. This phenomenon was observed earlier for diacetylmorphine and other alcoholic or phenolic substances. Other pairs of morphinans and the effect of injection temperature were examined. The blocking of susceptible hydroxy groups by sample silylation was recommended (251). Nair et al.

(252) performed semiquantitative analysis of some heroin impurities and adulterants without derivatization on packed columns but morphine detected by TLC was not observed in GC. Moore (253) used derivatization with heptafluorobutyric anhydride and separated several impurities on a capillary column. Prior to analysis the reaction mixture was reextracted with diluted sulfuric acid to eliminate column overloading caused by the bulk of heroin. This treatment also removed pyridine and tertiary amines (morphine, codeine, 3- and 6-acetylmorphine). By pH adjustment tertiary amines can be identified in the chromatogram along with derivatized secondary amines and neutral and acidic impurities (254). It was also demonstrated that reaction yields not only O and N derivatives but, introduces a heptafluorobutyryl group at some carbon sites too (254).

The analyses of neutral and acidic impurities are important for forensic and intelligence purposes, especially in sample comparison and geographic origin studies. The impurities extracted by organic solvent from acidified sample are frequently analyzed. The high resolving power of capillary columns makes them ideally suited for both identification of impurities and sample "fingerprinting" (254–257). In order to identify signals obtained during such chromatographic runs, the acetylation rearrangement products of pure opium alkaloids were analyzed and isolated either by GC (256) or by LC (258).

Neumann and Meyer (233) applied column switching capillary GC to achieve enhanced resolution of heroin trace impurity profiles and resolution of acetylthebaol from rosin acids. The method represents a powerful tool in special cases but requires complex instrumentation for routine applications. Munari and Grob (259) recently described automated on-line HPLC–capillary GC coupling and analyzed a urine extract spiked with codeine, morphine, and heroin.

Using LC separation on a silica gel column, White et al. (260) determined sugar content and opiate concentration in illicit heroin. The separation of heroin components by Jane (238) was mentioned above (see Table 7.13). Laurent et al. (239) applied LC on alumina to the analysis of illicit heroin (see Figure 5.5, Table 5.3). In the further development of the method Billiet et al. (240) used a combined system consisting of alumina column, six-port valve, and C18 column. Sample is injected onto the alumina column. Solutes with low pK_a values (caffeine, papaverine, narcotine), which elute rapidly from the alumina column, can be separated on the C18 column. As soon as the compounds separated on alumina start to leave the alumina column, it can be connected directly to the detector via the switching valve. Alternatively, both columns can be used in series.

Lurie and Carr (261) chose gradient RP elution to overcome the difficulties posed by a wide polarity of heroin impurities. Separation, which was performed using an acidic mobile phase, is shown in Figure 7.17. The hydrolysis of morphine to 6-acetylmorphine on

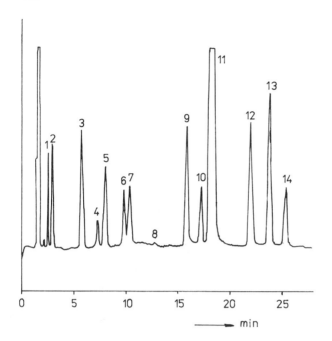

FIGURE 7.17 Standard mixture heroin, selected basic impurities, and the most common adulterants (from Ref. 261). Column HS-5 C18, 125 × 4.6 mm. Gradient elution 5% methanol-95% phosphate buffer (0.023 M hexylamine, pH 2.2), 20 min linear gradient to 30% methanol-70% phosphate buffer, hold 8 min; flow 1.5 ml min^{-1}. 1, morphine; 2, procaine; 3, codeine; 4, 3-monoacetylmorphine; 5, 6-monoacetylmorphine; 6, acetylprocaine; 7, quinine; 8, dihydroquinine; 9, caffeine; 10, acetylcodeine; 11, heroin; 12, piperonal; 13, α-narcotine; 14, papaverine.

standing in solution was also studied. Later, the gradient range was extended and highly lipophilic adulterants could also be separated (262). Dual electrochemical and photometric detection with programmed wavelength selection was applied. The relative response ratios contributed to the reliability of compound identification.

In an attempt to detect the possible continuation of heroin use by drug addicts, Derks et al. (263,264) developed a sensitive method for the determination of 6-acetylmorphine in urine. The precolumn derivatization procedure with morphine added to the sample was later automated (264). The formed mixed dimer (analog of pseudomorphine) was separated either in NP (263) or in RP (264) and detected with a fluorescence detector.

Nair et al. (252) reviewed 35 TLC systems reported in the literature for opiate analysis according to their resolving power. The classification of heroin samples was made possible by using one mobile phase and silica gel layer which was capable of separating eight opiates and five potential adulterants.

The chromatographic methods have wide application in drug identification either in illicit samples of unknown composition or in biological materials of medical, toxicological, or forensic interest. Since several isoquinoline alkaloids or related compounds are important components encountered in samples of this type, a brief comment on this topic is included in this section. Some separation systems applied to drug identification are given in Table 7.15.

The retention data are primary identification parameters in chromatography. For GC columns packed with the nonpolar stationary phases SE-30 or OV-1, the average values of retention indices were computed from published data by Ardrey and Moffat (82) for 1318 substances of toxicological interest. These data were included in the computerized data base of GC, TLC, and UV spectroscopic information containing over 1600 compounds (265). At present, however, capillary columns are substituted for packed columns in drug-screening applications because their excellent efficiency makes the identification more reliable. Plotczyk and Larson (266) included the retention data for 32 drugs in their study on the durability of capillary columns. Alm et al. (83) increased the identification power of the experiment by inserting two different capillary columns in common injection port and using parallel detection with FID and NPD (see Figure 4.1). Barni Comparini et al. (84) separated samples both before and after silylation. The changes in retention indicated which components were amenable to silylation; moreover, silylation enabled simultaneous GC analysis of common organic diluents in narcotics (fructose, glucose, etc.). The retention indices on capillary column were accumulated during the analyses of clinical and forensic specimens by Lora-Tamayo et al. (147) for 184 drugs and metabolites. A capillary GC coupled to a mass spectrometer represents a powerful tool for drug identification. Mass spectra measured during chromatographic runs can be directly compared with a computer-based spectral library.

The LC retention data for drugs of forensic and pharmaceutical interest were determined on a silica gel column with a methanolic mobile phase (238, 267-269) and on an octadecyl column with a mobile phase containing methanol in concentrations from 0 to 90% (139). Reversed phase gradient elution was applied to toxicological drug analysis by Hill and Langner (270). An acidified mobile phase was used for acidic and neutral drugs and an alkaline one for basic and neutral drugs. Some data were published for nitrilo columns (271, 272) and for ion pair chromatography on C18 columns (247,273).

Identification may be improved by repeating an analysis in a different chromatographic system (274). Alternatively, Wheals (275) recommended analysis on several columns of different types with the same mobile phase.

Spectral data obtained by a photodiode array detector aid in identification (270). An archive retrieval algorithm for spectral data acquired by rapid scanning photodiode array detector was proposed and sensitivity enhancement strategies applicable to this type of detection were discussed (276). In the past, simple simultaneous detection at two wavelengths and comparison of signal ratios with tabulated values were used for identification purposes (273,274).

Electrochemical detection is a very sensitive method for compounds of certain structural types (271,272). The response ratio of UV and an electrochemical detector was used as the additional characteristic of substances (269). The use of electrochemical detection in the forensic analysis of drugs of abuse (morphine and related narcotics, cocaine, LSD, hallucinogens from the mushroom Psilocybe semilanceata, and other abused drugs) was recently reviewed (277).

Thin-layer chromatography combined with appropriate detection reagents is ideally suited for use as a preliminary, inexpensive screening method in drug identification. As an example, a rapid method that tentatively identifies cocaine and heroin in street drugs can be mentioned (159). Sundholm (278) proposed the economic use of HPTLC plates by acidification of half of the plate with potassium hydrogen sulfate and development of both parts with the same mobile phase in a Camag linear developing chamber. The type of the unknown substance can be assessed from the effect of layer pH on retention. A very elaborate system for the identification of drugs in urine was reported by Kaistha and Tadrus (158). The basic drugs were isolated from urine either by cation exchange resin-loaded paper which was soaked in urine or by liquid-liquid extraction. After separation on precoated silica gel glass microfiber sheets, a series of seven detection reagents was applied. Other methods for drug screening were described by Moll and Clerc (279) and by Ojanperä and Vuori (68). Extraction of basic drugs as bis(2-ethylhexyl)phosphate ion pairs was used for the isolation of basic drugs from the sample (68).

In a review of drug identification using TLC, Stead et al. (280) reported retention data for 794 basic, neutral, and acidic drugs in eight systems. R_F values were standardized using reference compounds. These data formed the TLC part of the drug identification computer file mentioned before (265). Data for 594 basic drugs in four solvents taken from the same source (280) were analyzed by Musumarra et al. (281) using a principle component analysis. In a two-dimensional model, drug identification can be visualized with

Applications

only a minor loss of information content. Applying the same approach to the R_F values for 55 drugs in 44 solvents, a partially different set of four eluent mixtures was selected (282). Later, 362 substances were evaluated with these eluents (283). Corrected R_F values were again transformed to a two-dimensional space, which enabled the graphic representation of compounds. Standardized retention data alone were not sufficient to achieve unambiguous identification.

A large collection of relative retention data was accumulated by Daldrup et al. (284). About 570 substances were examined in two TLC, two LC, and four GC separation systems.

E. Separation of Alkaloids Derived from Miscellaneous Heterocyclic Systems

Besides Senecio, lupin, and Amaryllidaceae alkaloids and purine bases, some other types of alkaloids not mentioned in Chapter 1 will be discussed in this section.

The quinazoline alkaloid vasicine and its oxidation product, vasicinone, were determined in leaves of Adhatoda vasica using RP-LC (285) and ion pair RP-LC (286). The stability of vasicine in solution was also examined (285, 286) and the separation of some vasicine analogs was studied (286).

Carbazole alkaloids can be analyzed by LC and TLC, with at least some of them by GC as well. The separation of nine carbazole alkaloids was achieved on packed columns with OV-17 and SE-30 stationary phases (287). Application of RP-LC to carbazole alkaloids did not yield acceptable results, but separation on silica gel was found to be suitable for the detection of these alkaloids in plant extracts (288). Thin-layer chromatography was used for the separation of carbazole alkaloids on both silica gel (289) and alumina (290) layers.

The pyridocarbazole alkaloid ellipticine and its derivatives were assayed in cultured plant cells using fluorodensitometry after TLC separation (31). About 200 daily assays could be performed easily.

Nash et al. (133) reported the GC separation of nine polyhydroxy alkaloids after trimethylsilylation. Alkaloids were of pyrrolidine, piperidine, and indolizine types occurring in various plants. The GC-MS method was applied to the identification of bases in Lycopodium species (291), in which alkaloids derived from various ring systems occur. Identification was based on retention indices obtained for 33 compounds and on a reference library of mass spectral data.

Perloline, an alkaloid of some fodder grasses with a diazaphenanthrene ring system, was determined in plant tissues by a simple, one-step clean-up procedure and rapid LC separation on silica gel (292).

Senecio alkaloids, which are derived from a pyrrolizidine ring system, constitute a large class of compounds that are widespread in plants occurring throughout the world. Due to their toxicity, which causes severe intoxication in humans and cattle, the determination of pyrrolizidine alkaloids has attracted considerable attention. Some of these compounds can be subjected to GC analysis without derivatization. Using capillary columns, the analysis of parent alkaloids in Senecio inaequidens (293) and in hay and silage with various amounts of Senecio alpinus (294) was performed. N-oxides, the primary products of biosynthesis and the form in which pyrrolizidine alkaloids are accumulated in plants, were determined in Senecio vulgaris roots and root cultures (295). However, pyrrolizidine alkaloid N-oxides are thermally labile and partially decompose on GC column (293). Because of unsuccessful attempts to perform GC without derivatization, the silylation was applied before separation and tentative structure elucidation of pyrrolizidine alkaloids using GC-MS with positive and negative ion chemical ionization (296,297).

The LC analysis of pyrrolizidine alkaloids requires photometric detection in most cases. However, it has to be performed at a short wavelength due to the absence of useful absorption bands at longer wavelengths for most pyrrolizidine alkaloids. RP-LC separations were applied to the analysis of plant materials (295,298,299). The separation of Senecio vulgaris alkaloids was carried out on nitrilo column, and isocratic and gradient elutions were compared (300). A simple isocratic mode of elution (Figure 7.18) is preferred. Styrene-divinylbenzene polymeric packing was used for the analysis of major metabolites of senecionine, retrorsine, and seneciphylline produced in vitro by microsomal enzyme systems (301).

HPLC techniques for the analysis and isolation of various pyrrolizidine alkaloids were reviewed by Segall (302).

Supercritical fluid chromatography of the mixture of alkaloids which were isolated from Senecio anonymus were performed on capillary columns using FID (see Figure 7.19). Pressure programming was used to allow for the separation of all alkaloids at relatively low column temperatures compared to those of GC (25).

Thin-layer chromatography of pyrrolizidine alkaloids can be used as a preliminary or confirmatory test for the presence of these alkaloids in plant extracts (293,303) or on a preparative scale for the isolation of pure bases (303,304). The biosynthesis of senecionine N-oxide in root cultures of Senecio vulgaris was studied using ^{14}C-labeled tracers (295). Products were analyzed by TLC separation followed by the scanning of layer radioactivity.

The properties of quinolizidine (lupin) alkaloids are favorable for their separation by GC and the underivatized species can be analyzed in most cases. Relative GC retention times for some quinolizidine alkaloids are reproduced in Table 4.2. Priddis (305) used

Applications

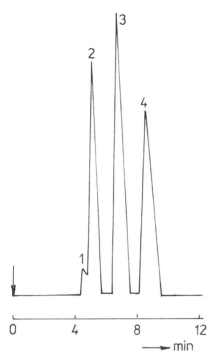

FIGURE 7.18 RP-LC analysis of pyrrolizidine alkaloids from Senecio vulgaris. Column 300 × 4 mm packed with Bondapak CN, 10 μm, mobile phase tetrahydrofuran-0.01 M $(NH_4)_2CO_3$, pH 7.8, 16:84, 1.8 ml min^{-1}, UV detection at 235 nm. 1, unknown; 2, retrorsine; 3, seneciphylline; 4, senecionine. (After Ref. 300.)

capillary GC to separate quinolizidine alkaloids (see Figure 4.5) and to determine some of them in lupin seed of Lupinus angustifolius. The method is suitable for the monitoring of these bitter-tasting toxic components in plants, e.g., by the breeding of sweet lupin varieties. Some alkaloids of Lupinus arbustus subsp. calcaratus were separated by capillary GC and TLC, and some preliminary pharmacological data in mice were ascertained (306). Only one of the four alkaloids found in this species was of the quinolizidine type. Six alkaloids, mostly of quinolizidine type, isolated from Genista anatolica were characterized by their R_F values (307). Preparative scale TLC was used for the clean-up of isolated alkaloids (306,307). Differentiation of synthetic alangimarckine epimers was supported using TLC by comparison with the retention behavior of emetine and isoemetine (308).

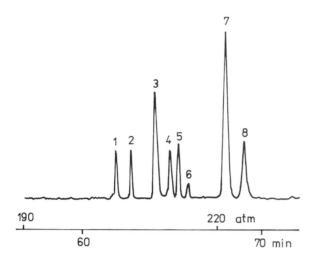

FIGURE 7.19 Chromatogram of alkaloids isolated from Senecio anonymus using SFC. Column 10 m × 0.05 mm coated with SB-Biphenyl-30, temperature 130°C; pressure program, 100 atm for 20 min, from 100 to 280 atm at 3 atm min^{-1}. 1, senecionine; 2, integerrimine; 3, senkirkine; 4, neosenkirkine; 5, retrorsine; 6, otosenine; 7, hydroxysenkirkine; 8, anonamine. (After Ref. 25.)

In contradistinction to the lupin alkaloids, Amaryllidaceae alkaloids are difficult to analyze using GC. Liquid chromatography was used for the determination of lycorine in Sternbergia lutea (309) and galanthamine, a long-acting anticholinesterase drug, in biological fluids (53,310). In a recently elucidated RP method (53), the two major metabolites of galanthamine were also quantified.

A considerable number of chromatographic methods have appeared in the literature for the analysis of purine bases (Table 7.17), due mainly to the pharmacological effect of theophylline and caffeine, i.e., as bronchodilators of narrow therapeutic range. Also, knowledge regarding the caffeine and theobromine content of foodstuffs and beverages is frequently required.

Caffeine in which methyl groups are substituted for all three hydrogens bonded to nitrogen atoms of the xanthine ring can be easily analyzed using GC. Its separation from other drugs on capillary columns (83) is shown in Figure 4.1. Caffeine has been determined in coffee, tea, maté, and beverages (311), in flowers and leaves of Citrus sinensis (312), and has been identified in municipal wastewater (106) (see Table 7.8) and fly ash (313) by this method. Delbeke and Debackere (314,315) used GC to detect caffeine

Applications

abuse by sportsmen. Stavchansky et al. (48) used GC for the determination of caffeine in breast milk and blood plasma. The GC method was more sensitive and required a shorter preparation and run time than LC when caffeine alone was determined in plasma (51). However, LC was essential when caffeine metabolites also had to be determined.

Some other purine bases are amenable to GC whose derivatization is not strictly necessary. Nevertheless, GC is used for analysis in special cases. Bailey et al. (46) used N-pentafluorobenzyl derivative to improve the chromatographic properties of theophylline. Their method was capable of establishing both the absorption properties and the bioavailability of theophylline from sustained-release formulations. A conventional formulation containing isotopically labeled drug was coadministered and the two forms were distinguished in plasma by GC-MS with selected ion monitoring.

It appears that most separations of purine bases are performed using LC. Caffeine and other purine bases are sometimes used as test compounds in studies of retention mechanisms. Silica gel under pseudo-reversed phase conditions (218) (see Table 7.12), alumina (239) (see Tables 5.3 and 7.14), RP packings differing in surface coverage (218,316), and nitrilo packing (92) (see Table 7.7) were studied. The effects of substituents were examined in a RP system for 26 xanthines (317). Xanthines were separated on a RP column using triethylamine as a competing base (139). Dezaro et al. (318) used an algorithm for the time optimization of HPLC separations in order to shorten the analysis time of purine bases by high-speed chromatography on microbore columns.

The majority of samples (pharmaceuticals, foodstuffs, beverages, biological materials) are analyzed by LC with photometric detection at about 270–280 nm, approximately the absorption maxima of purine bases. The sensitivity of detection is usually sufficient because the content of purine in samples is generally relatively high. Separation systems of various types have been applied. Due to their weak basicity, xanthines behave to a large extent as neutral compounds in the common pH range. Caffeine remains neutral even at alkaline conditions, which are sometimes used for its selective extraction from other xanthines to organic solvents. Weak acidity of other xanthines (pK_a about 8–10) may affect their chromatographic behavior in the alkaline region.

Caffeine and other components of tablets were assayed using RPLC on µBondapak C18 (6) (see Table 7.12) and Zorbax ODS (319) columns. Gfeller et al. (320) showed the advantage of very high-speed and superspeed HPLC in the analysis of pharmaceutical preparations including mixtures of analgesics containing caffeine (Figure 7.20). Three drug substances and the degradation product (salicylic acid were) separated within 40 sec. The short run

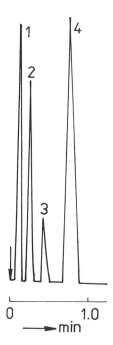

FIGURE 7.20 Separation of three-component analgesic by very high-speed HPLC. Column 30 × 2.1 mm packed with Spherisorb-RP-8 5 μm, mobile phase methanol-water-concentrated H_3PO_4 30:70:3, 1.7 ml min^{-1}, UV detection at 295 nm. 1, caffeine; 2, acetylsalicylic acid; 3, salicylic acid; 4, propylphenazone. (After Ref. 320.)

time enables handling of a large number of samples as required in content uniformity tests. Very high-speed HPLC is especially advantageous in those instances where either no or simple sample preparation is required and where relatively pure samples are to be chromatographed. The determination of theophylline in tablets and suspension formulations by LC (95) was already mentioned in Section II.A of this chapter (see Table 7.7).

Caffeine was determined in coffee (311,321,322), tea (311,321, 323), and beverages (311,321,324) using LC separation on silica gel (311), with RP packings (321,323,324), and on carboxyl poly(methyl- or chloromethylstyrene-divinylbenzene) polymeric packing (322). Along with GC, RP-LC was used to demonstrate the presence of caffeine in citrus flowers and leaves (312). Caffeine and theophylline were determined in Paullinia cupana and Cola spp. samples on C18 columns (325). Reversed phase packings were used for the simultaneous determination of theobromine, theophylline, and caffeine in cocoa, tea, coffee, and various food products (326) and in cocoa (327). Analysis of cocoa extract on microbore columns as shown in Figure 7.21 allowed, along with the lowering of solvent consumption, a decrease in sample size down to 10 mg (327). Theobromine was separated from adenosine in cocoa using a micellar mobile phase with polyvinylalcohol packing and under isocratic or gradient conditions on a C18 column (328).

A large number of LC separations of purine bases are concerned with the analysis of caffeine, theobromine, and their metabolites in biological fluids. Even in cases when only caffeine or theophylline is to be determined, their metabolites have to be well separated from the compounds of interest in order to eliminate interferences. That is why several purine bases have been separated in most of these studies. Because caffeine is metabolized to theophylline, usually

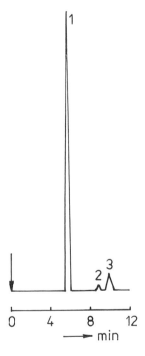

FIGURE 7.21 RP-LC analysis of cocoa extract using microbore column. Column 250 × 1 mm packed with Whatman Micro-B ODS-3, mobile phase water-methanol-acetic acid 68:31:1, 60 µl min^{-1}, UV detection at 280 nm. 1, theobromine; 2, theophylline; 3, caffeine. (After Ref. 327.)

both compounds are separated in assays of caffeine. On the other hand, caffeine is only included in a few studies involving theophylline determination for therepautic purposes. However, caffeine may be present from dietary sources.

Silica gel (329), C18 (51,330–335), and C8 (336) stationary phases were used in the LC determination of caffeine and its metabolites including theophylline. Gradient elution (334) facilitated the separation and quantitation of 13 metabolites within a 20-min period. Reduced solvent consumption in microbore LC (335) was relatively insignificant in the total operating budget. The radioactivity of LC eluent as well as that of TLC chromatograms was monitored (333) in a study on comparative metabolic disposition of [1-Me^{14}C]caffeine in rats, mice, and chinese hamsters. Rather exceptional is the use of exclusion chromatography for the determination of acetylated caffeine metabolites (337). Some of these compounds were either too stronly retained by NP or not retained by RP columns.

Caffeine alone was determined by separation on C18 columns (47,338–340). Theophylline (47,338) was also separated and paraxanthine was determined simultaneously with caffeine (338). Extremely sensitive detection was possible by a direct coupling of MS to LC through a thermospray interface (340). Limits of detection as low as 1 µg ml^{-1} in 1 µl of serum or saliva was achieved, which was advantageous for an oral caffeine load test of liver function. The measurement of salivary caffeine may serve as an alternative to the assay of its level in blood.

Theophylline and its metabolites were determined mostly by C18 packings with isocratic (49,341) or gradient (71,342) elution. A microassay utilizing a C8 column (343) is illustrated in Figure 7.22. A mobile phase with a low content of organic modifier was applied as in many other RP systems for the separation of purine bases.

When theophylline alone is to be determined, considerable attention must generally be paid to the separation of metabolites and other possible interfering compounds. Paraxanthine in particular can be resolved from theophylline with some difficulty. Besides silica gel (50,344) and phenyl-bonded columns (345), C18 packing prevails in these assays (46,52,66,67,346–350). Direct injection of 10 µl serum to the RP column was found less suitable for routine analysis due to the relatively short lifetime of the column (about 40 injections) (346). With application of a column switching technique, the practical lifetime of precolumns was about 150 injections of 6-µl serum samples (348). High-speed LC separations on a short 3-µm RP column were completed in 40 sec (347). However, after about 200 injections (of aqueous standards) the pressure on the column increased and deterioration of resolution became noticeable. Resolution of theophylline and paraxanthine was achieved by the incorporation of dimethylformamide into the mobile phase (52).

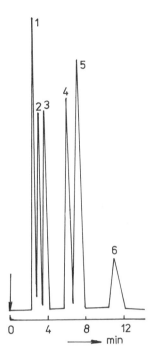

FIGURE 7.22 RP-LC analysis of theophylline and its major metabolites in serum. Column 125 × 4.6 mm packed with C8 packing, 5 µm, 45°C, mobile phase 10 mM CH_3COONa and 5 mM tetrabutylammonium phosphate with 4% methanol, pH 4.5, 2.0 ml min^{-1}, UV detection at 273 nm. 1, 3-methylxanthine; 2, 1-methyluric acid; 3, 1,3-dimethyluric acid; 4, theophylline; 5, β-hydroxyethyltheophylline (I.S.); 6, caffeine. (After Ref. 343.)

Metabolism of ^{14}C-labeled theobromine in five mammalian species was compared (351). Urinary metabolites were analyzed by RP-LC with UV and radioactivity detection.

Coffee extract isolated by supercritical fluid extraction was directly separated by RP-LC or supercritical fluid chromatography (352). Caffeine was identified in the extract with the aid of an absorption spectrum measured by a photodiode array detector. A supercritical fluid chromatograph coupled to a mass spectrometer through a moving belt interface with modified thermospray deposition was applied by Berry et al. (353) to the separation of caffeine, theophylline, and theobromine. Another interface type was used by Crowther and Henion (157) to caffeine (see Table 7.9).

Caffeine was separated with other doping agents by overpressured TLC (101) under conditions given in Table 7.7 (see Figure 6.2).

Thin-layer chromatography has rarely been applied to purine base separation. Caffeine, theophylline, and 15 other drugs could be detected in Chinese herbal preparations (354). As mentioned above, metabolites of ^{14}C-labeled caffeine were determined in urine using two-dimensional TLC (333).

The application of paper chromatography to the isolation of caffeine and theobromine from cocoa was described by Sjöberg and Rajama (355). Cocoa powder was streaked directly to the paper, and caffeine and theobromine were separated using multiple development. Separated compounds were determined after extraction using absorption spectrometry. Compared to the rapid determinations made possible by RP-LC, this method seems to be lengthy even when the sample preparation time required for RP-LC is taken into account.

F. Separation of Indole Alkaloids

The indole alkaloid group comprises a very large number of alkaloids that occur mainly in the plant families Apocyanaceae, Loginaceae, and Rubiaceae. The description of the chromatographic analyses given in this section essentially follows the classification in Chapter 1, Section VI. The chromatographic conditions are summarized in Tables 7.18—7.21; for simple indole alkaloids, see Table 7.18; for yohimbine-type alkaloids, Rauwolfia alkaloids, and related bases, see Table 7.19; for Strychnos alkaloids, see Table 7.20; and for ergot alkaloids, see Table 7.21.

As can be seen from the tables, indole alkaloids are presently rarely separated by means of GC and analysis have been performed by LC in the great majority of instances. This is true even for simple indole alkaloids for which GC can be expected to be highly favorable.

Mangino and Scanlan (356) used HPLC for the analysis of products obtained by nitrosation experiments on gramine. Gramine and hordenine are considered potential precursors of carcinogenic N-nitrosodimethylamine in barley malt.

Physostigmine (eserine) has long been used in ophthalmology, but has more recently found additional applications, such as the treatment of some neurological disorders and poisoning by anticholinergic compounds. This underscored the need for analytical methods applicable in pharmacokinetic or metabolic studies, for which LC appears to be indispensible.

Analysis of physostigmine salicylate as salicylic acid and physostigmine base is possible on a RP column using triethylamine as a competing base (139). Physostigmine is easily hydrolyzed enzymatically or in alkaline medium to eseroline, which is readily oxidized to reddish rubreserine. Kneczke (357) developed a RP ion pair LC method suitable for the analysis of eye drop solutions containing physostigmine and pilocarpine. Some preservatives and rubreserine were also present in separated mixtures.

When biological materials are to be analyzed, the sensitivity of UV detection is usually insufficient. The pharmacokinetics and metabolism of physostigmine in rats were examined with ^{3}H-labeled substrate (358,359) using an ion pair separation method by Somani and Khalique (360) and radioactivity measurement of the collected fractions. Brodie et al. (361) utilized the fluorescence of physostigmine for sensitive detection. Electrochemical detection was used by Whelpton (362) for the determination of physostigmine in plasma using a silica gel column eluted with methanolic ammonium nitrate buffer. Eseroline was not completely resolved from physostigmine in this system but could be distinguished by changing the oxidation potential. Later, an approximately 10-fold improvement in sensitivity was achieved using a dual-electrode detector (363). The effect of pH on electrochemical response and extraction efficiency was examined. The same detection technique was used by Isaksson and Kissinger (45,364) with a RP ion pair separation system. The effects of flow rate and spacer thickness on detector response were evaluated. The method is suitable for the analysis of plasma, as can be seen from Figure 7.23. Also the metabolism of physostigmine in mouse liver microsomal incubations was studied (364).

For the determination of physostigmine in biological materials, the sample treatment must be chosen in such a way as to minimize decomposition processes. The decomposition rates of physostigmine were evaluated under various conditions (361,363) and neostigmine (362,363) or pyridostigmine (361) was added to the plasma samples to prevent enzymatic decomposition. Ion pair extraction from acidified solution using a C18 cartridge was applied by Isaksson and Kissinger (45) in order to avoid the decomposition of physostigmine in alkaline solution.

According to Table 4.3, some harman alkaloids can be easily separated using GC. Bosin and Faull (365) proved the presence of harman in rat lung by capillary GC-MS after derivatization with pentafluorobenzyl bromide. For the quantitation, however, a LC procedure with fluorometric detection was used. The technique was developed in order to examine the possibility of harman and norharman accumulation in mammalian lung, which could be harmful due to comutagenicity of these compounds. LC and TLC were used to determine the lipophilicity of harman and 16 other β-carbolines (366). Bashir et al. (367) ssparated harman alkaloids isolated from roots of Grewia villosa using RP-LC. Cavin and Rodriguez (368) analyzed

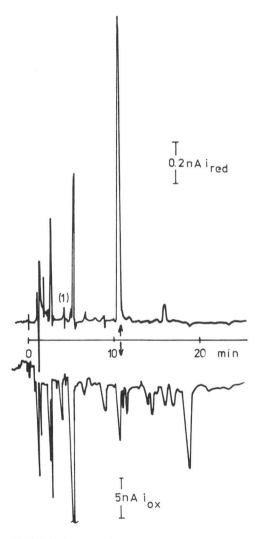

FIGURE 7.23 Chromatogram of plasma spiked with 14 ng ml^{-1} of physostigmine (from Ref. 45). Column 250 × 4.6 mm packed with a Biophase Octyl, 5 μm, mobile phase 0.1 M phosphate buffer, pH 3.0-acetonitrile-sodium dodecyl sulfate 60:40:1, 2 ml min^{-1}, amperometric detector with dual electrodes in a series mode. Column temperature set at 35°C. The retention time of physostigmine is indicated with an arrow.

Applications

specimens of several species of Heliconiini butterflies fed as larvae on plant material containing β-carboline alkaloids. Tweedie and Burke (369) used silica gel column in order to study the metabolism of ^{3}H-labeled harmine and harmol by mouse liver microsomes.

Indole alkaloids of the canthinone subgroup obtained from the wood of Simaba multiflora were purified by preparative overpressured TLC (370).

Several years ago the increasing abuse of hallucinogenic mushrooms containing indole alkaloids of the tryptamine type brought about a need for adequate analytical tools. Psilocybin and psilocin were most frequently determined in mushrooms of Psilocybe and related genera. Although GC can be used for their separation (see Table 4.3), LC has prevailed in more recent applications. Isocratic separations on silica gel (27,371), RP packings (372-374), ion exchanger (375) (see Figure 5.17), and in a RP ion pair system (376) were used along with UV, fluorescence, and electrochemical detection. The simultaneous determination of baeocystin was made possible by isocratic separation on silica gel (371) or by gradient elution in RP-LC (28). In ion pair RP-LC (376) psilocybin and baeocystin were separated using the mobile phase free of organic modifier.

Thin-layer chromatography served as a preparative tool for the isolation of baeocystin from mushrooms (376) as well as for the identification of extracted alkaloids (372,374,376). Gartz (377) compared the TLC pattern of indole alkaloids (reacting with Ehrlich's reagent) in Psilocybe semilanceata, P. callosa, and Inocybe aeruginascens.

Numerous indole alkaloids of the yohimbine type, Rauwolfia alkaloids, and alkaloids occurring in plants of Catharanthus, Tabernaemontana, and other genera are characterized by a more complex structure than the simple indole alkaloids mentioned above. Some of these alkaloids have been found in plant species of various genera and so chromatographic conditions used for their separation are summarized in Table 7.19.

Ardrey and Moffat (82) reviewed GC retention indices for yohimbine, ajmaline (see Table 4.3), and rescinnamine (I = 2180). Reserpine was not eluted in a reasonable time. In practice, the use of GC for the analysis of Rauwolfia and yohimbine-type alkaloids is rare.

Liquid chromatographic methods were developed for the determination of yohimbine in biological fluids using UV (378), fluorometric (379), or electrochemical (380,381) detection. Fluorometric detection seems to be the most sensitive. The separation of yohimbine and reserpine on the polymeric packing Hitachi gel 3010 (20) occurred under the conditions given in Table 7.13 (see Figure 5.9).

Reserpine, which is used for the treatment of hypertension among the applications, can be detected after LC separation by its native fluorescence. This technique was used by Cieri for the analysis of tablets (382) and for the identification of other components of <u>Rauwolfia</u> <u>serpentina</u> preparations (383). A quite simple separation system was used in these instances. Better sensitivity was achieved by postcolumn oxidation of reserpine to a proposed 3,4-dehydroreserpine fluorophore (384). The reaction proceeded faster under UV radiation. The use of a photochemical postcolumn air segmentation reactor provided the decrease in band broadening and the improvement of sensitivity (385). Slightly higher limits of detection were reported for electrochemical detection of reserpine and rescinnamine in RP ion pair LC (386). Gfeller et al. (320) demonstrated the application of fast HPLC in the pharmaceutical industry using the analysis of tablets containing reserpine as an example (see Figure 7.24).

Ajmaline, an antiarrhythmic drug, was determined simultaneously with its semisynthetic derivative 17-monochloroacetylajmaline in

FIGURE 7.24 Content uniformity test of Brinerdine tablets by very high-speed HPLC. Column 100 × 2.1 mm packed with Spherisorb RP-18, 5 µm, mobile phase acetonitrile-0.4 M H_3PO_4 17:3, 3 ml min^{-1}, UV detection at 260 nm. 1, clopamide; 2, dihydroergocristine; 3, reserpine. (After Ref. 320.)

Applications

pharmaceutical preparations by gradient RP-LC (387). Owing to the large polarity difference between ajmaline and the pair reserpine and rescinnamine, a two-step analysis had to be applied for their determination in the bark of Rauwolfia vomitoria using both LC and TLC (388). Preparative centrifugal TLC separation of ajmaline, rescinnamine, and alstonine was reported (18) (see Table 7.11). Cieri (383) applied TLC to the analysis of Rauwolfia serpentina pharmaceutical preparations. Recently a number of indole alkaloids in the callus culture of Rauwolfia serpentina were identified by TLC and quantified after elution from the plate by spectrophotometry (389). Munier et al. (390) separated 15 indole alkaloids of the yohimbine, ajmaline, and alstonine types in order to elucidate solvent selectivity effects in TLC.

Ajmalicine and serpentine, Rauwolfia alkaloids of the alstonine type, predominate in alkaloid mixtures isolated from the periwinkle Catharanthus roseus (Vinca rosea). Recently, their total content in the roots of three pure lines of C. roseus was determined using RP-LC (391). Serpentine was reduced to ajmalicine by treatment with sodium borohydride prior to analysis. Also for the analysis of alkaloids from C. roseus cell and tissue cultures RP-LC (392-394) or RP ion pair LC (395) was used. Gradient elution was required to cover the wide polarity range of alkaloids occurring in cultures (394,395) as well as in C. roseus or V. rosea plants (396,397). In these studies TLC was sometimes used for preparative (393,395) and identification (392) purposes.

Vincristine, vinblastine, leurosine, and leurosidine are dimeric indole alkaloids which accompany monomeric alkaloids in C. roseus (V. rosea) (396,397). Because of the antineoplastic properties of vincristine, vinblastine, and some of their semisynthetic derivatives, these compounds have gained considerable attention. For their isolation Drapeau et al. (398) proposed RP-LC, based on the fact that the retention of dimeric alkaloids depends more on the buffer concentration in the mobile phase than the retention of monomeric alkaloids. Verzele et al. (397) noticed a lower column efficiency for dimeric alkaloids than for monomeric ones and used this fact to aid in identification. Recently, the formation of vinblastine in callus culture with differentiated roots of C. roseus was proved using RP-LC (399).

The determination of vinblastine or vincristine in biological materials was performed using either nitrilo (400,401) or ODS (54,402) columns. Sample preparation using solid phase extraction with a diol cartridge (54) showed variable recoveries for some batches and thus nitrilo Bond-Elut CN columns were used in the later experiments (402). Gradient RP-LC was used to separate metabolites of vincristine and vinblastine (403) and of the monoclonal antibody-desacetyl vinblastine conjugate (404). An acidic medium was used for sample extraction and analysis because vincristine in the form of free base can undergo conversion (403). The degradation products of vinblastine (405) and vincristine (406) produced by incubation in glycine buffer containing bovine serum albumin were analyzed using RP-LC with isocratic elution.

A large number of indole alkaloids have been isolated from plants of the genus Tabernaemontana. The separations of Tabernaemontana dichotoma leaf extract were performed by Perera et al. (407) in either the RP ion pair (see Figure 7.25) or the NP mode. Indole alkaloids from Tabernaemontana divariatica suspension cultures were isolated by means of C8 cartridge and analyzed using RP-LC in a phenyl column (29). Van der Heijden et al. (30) applied ion pair droplet countercurrent chromatography to the alkaloid extracts from cell cultures of three Tabernaemontana species. Most Tabernaemontana alkaloids, however, were isolated, separated, and identified by means of TLC (407-412). Results on R_F values and chromogenic reactions for 100 Tabernaemontana alkaloids were summarized by van Beek et al. (413).

Vincamine, the main indole alkaloid of Vinca minor, is commonly used in the treatment of cerebrovascular diseases. A capillary GC method for its determination in plasma was developed by Michotte and Massart (414). Also vinpocetine (apovincaminic acid ethyl

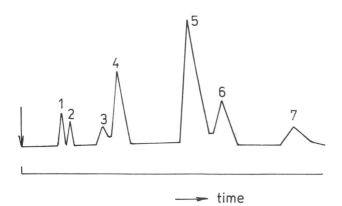

FIGURE 7.25 Ion pair chromatography of alkaloids of Tabernaemontana dichotoma leaf extract. Column 300 × 4.6 mm packed with LiChrosorb RP-18, mobile phase 0.02 M methanesulfonic acid in water-dioxane-H_2SO_4 94.5:5:0.5, pH 4.0 with NaOH, 1 ml min^{-1}, UV detection at 280 nm. 1, vobasine; 2, perivine; 3, 19-epiiboxygaine; 4, 19-epivoacristine; 5, 1-apparicine; 6, 12-methoxyvoaphylline; 7, isomethuenine. (After Ref. 407.)

ester), when determined in human plasma by Hammes and Weyhenmeyer (415), was separated as an underivatized substance and under similar conditions. Most separations of vincamine and other alkaloids of the eburnane type have been performed using LC. Enantiomer separation of eight alkaloids (three of eburnane, two of aspidospermine, and three of quebrachamine structure) on cyclodextrin polymer packing was studied by Zsadon et al. (416). The separation of eburnane alkaloids on nitrilo columns with typically NP mobile phases containing acidic and basic additives (417,418) was mentioned in Section IV.A of Chapter 5 (see Figures 5.24 and 5.25). Earlier, Szepesi and Gazdag studied the separation of eburnane alkaloids in both RP (419) and NP (420) modes (cf. Figure 5.2). Based on these detailed studies, several applications of LC in the analysis of eburnane alkaloids were developeed (10). The RP-LC methods were used to determine vincamine in pharmaceutical products (421) and in plasma (422) with UV and electrochemical detections, respectively. Lala (423) separated hirsutine and eight oxindole alkaloids by GC and TLC when comparing alkaloidal profiles of Mitragyna stipulosa from Ghana and Zaire. The GC retention times and R_F values were reported.

Strychnine is the most important Strychnos alkaloid (Table 7.20). Because of its toxic properties as well as its occurrence as an adulterant in street drugs, strychnine has frequently been included in GC, LC, and TLC data bases devoted to drug identification. Some of these systems (84,147,247,269,270,282,283) are presented in Table 7.15. Gas chromatography is well established in these applications and data for strychnine and brucine are usually reported, even though their retention indices exceed 3000 (see Table 4.3). The values 3025-3145 for strychnine were reported by Bogusz et al. (424) on capillary columns in studies of the dependence of retention on solute concentration. The use of GC for the specific determination of strychnine or brucine, however, is not frequently encountered. Several examples are given in Section IV. F of Chapter 4, but LC prevails in similar instances.

Some LC data for strychnine can be found in papers on the separation of heroin-related mixtures (238-240,260,262) (see Tables 5.3, 7.13, 7.14, and Figure 5.5). Iskander et al. (425) studied the retention of 33 strychnine derivatives on silica gel and established the effect of various substituents. An ion pair RP-LC method for the determination of strychnine and brucine in Nux-vomica and derived pharmaceutical preparations was developed by Hayakawa et al. (426). In other cases (427,428) separation on silica gel was used for this purpose. Dennis (427) used methanolic mobile phase developed by Jane (238) (cf. Table 7.13). Wahbi et al. (428) used a methanolic mobile phase whose composition was the same as that used previously by Alliot et al. (429) for the determination of strychnine in biological media. Li and Zhou (430,431) determined strychnine in Strychnos piersiana and Strychnos pierriana and in their preparations using silica gel column. For the determination of strychnine in biological specimens (domestic animals, stomach content), both NP (432) and RP (433) methods were developed. Mishima et al. (434) analyzed metabolites of strychnine formed in vitro with rat and rabbit livers by both TLC and RP-LC. Ultraviolet detection was used in all the LC methods.

Droplet countercurrent chromatography and partition equilibria of strychnine, brucine, and three other alkaloids were studied in detail by Hermans-Lokkerbol and Verpoorte (435).

Thin-layer chromatographic methods for the determination of strychnine (436–438) and brucine (438) in Chinese herb mecidine preparations containing Strychnos nux-vomica were described. Quantitation was carried out either by spectrophotometry after spot elution and treatment with bromthymol blue (436) or by in situ dual-wavelength densitometric scanning (437,438). Gaudy and Peuch (439) applied a new two-dimensional TLC method involving opposite solvent flow to the identification of alkaloids in pharmaceuticals based on S. nux-vomica. Preliminary TLC screen was used prior to LC quantitative analysis of strychnine in domestic animals (432).

Gulyás et al. (101) determined strychnine in samples of urine using overpressured TLC (see Table 7.7 and Figure 6.2).

Some indole alkaloids isolated from ergot (Table 7.21) and some of their semisynthetic and synthetic analogs exhibit considerable pharmacological activity. These compounds can mostly be found among amides of lysergic acid with various cyclic peptidic moieties. Moreover, lysergic acid diethylamide (LSD) is a potent hallucinogenic substance, and hence analyses of LSD and related compounds are usually required for forensic purposes. Simple ergot alkaloids (clavines) have been analyzed less frequently.

Although some GC retention indices for ergot peptide alkaloids have been reported (82), the values are rather unreliable since these compounds are too labile and of low volatility to be analyzed by GC without decomposition. On the other hand, the separation of peptide moieties formed on injection of dihydroergotoxine to the gas chromatograph was applied to the determination of dihydroergotoxine composition (440) (Figure 7.26). Irie et al. (441) described GC-MS microdetermination of total dihydroergotoxine in human plasma which was sufficiently sensitive for the study of pharmacokinetics of dihydroergotoxine in humans. The method involved hydrolysis of dihydroergotoxine components to dihydrolysergic acid followed by esterification to the ethyl ester of dihydrolysergic acid and trimethylsilylation at indole nitrogen. Detection limit was lower in the high-resolution mode than with conventional low-resolution selective ion monitoring.

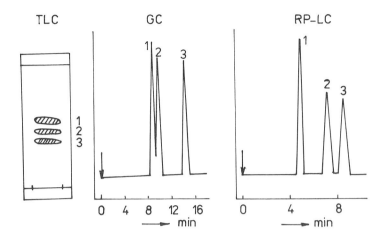

FIGURE 7.26 Analysis of dihydroergotoxine methanesulfonate components. Comparison of different chromatographic techniques. 1, dihydroergocornine; 2, dihydroergocryptine; 3, dihydroergocristine. TLC: Polygram Sil 6 sheet, mobile phase acetone-0.1 M $(NH_4)_2CO_3$-ethanol 32.5:67.5:1. GC: Column 1 m × 3.2 mm packed with 2% Dexsil 300 on Gas Chrom Q, nitrogen 11.5 ml min^{-1}, 180–280°C at 4 K min^{-1}, FID. LC: Column 250 × 4 mm packed with LiChrosorb RP-18, 10 μm, mobile phase acetonitrile-10 mM $(NH_4)_2CO_3$ 2:3, 190 ml hr^{-1}, UV detection at 280 nm. (After Ref. 440.)

The retention behavior of ergot peptide alkaloids in column LC was studied for ODS (442,443), nitrilo (417) (see Table 7.19), and amino (444) packings. Very successful separations were achieved on LiChrosorb RP-18 column (442). Although the separation of major ergot peptide alkaloids on silica gel is difficult, Pötter et al. (445) performed noteworthy separations using mobile phase saturated with formamide. Some LC retention data on ergot alkaloids can be found in papers concerning forensic and clinical analyses of LSD and other preparations (247,269,446). Szepesy et al. (440) investigated the separation of ergot peptide alkaloids on various silica gels and RP packings and reported some applications (Figure 7.27). Silica gel proved to be useful for the separation of groups of indole alkaloids and for the detection of contaminants. Active compounds, stereo- and structural isomers could be separated by RP-LC. When studying the effects of chemical and physical treatment on the alkaloid content of ergot sclerotia, Young et al. (447) applied gradient elution on an RP column. For the assessment

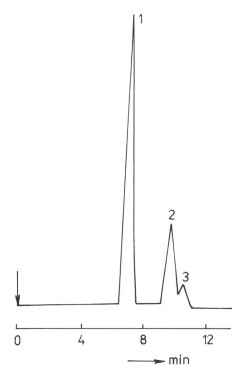

FIGURE 7.27 RP-LC analysis of ergocornine-rich samples. Column 250 × 4 mm packed with LiChrosorb RP-18, 10 μm, mobile phase acetonitrile-10 mM $(NH_4)_2CO_3$ 2:3, 130 ml hr^{-1}, UV detection at 320 nm. 1, ergocornine; 2, α-ergocryptine; 3, β-ergocryptine. (After Ref. 440.)

of wheat contamination by ergot alkaloids, Ware et al. (448) used isocratic separation on polystyrene-divinylbenzene resin which allowed for complete resolution of the major pharmacologically active components. In these applications the native fluorescence of ergot alkaloids was used for detection. The same detection method was used in the determination of ergot alkaloids in plasma (55,339). In the analysis of ergot alkaloids the possibility of their changing to -inine isomers, as well as other degradation pathways which occur in solutions, must be considered (55,448). Edlund (55), for example, used calibration solutions containing ergotamine and ergotaminine in the equilibrium ratio 6:4.

Dihydro derivatives of ergotoxine alkaloids have been determined in various pharmaceutical products (320,440,449,450). In comparison (Figure 7.26) with GC and TLC, the column LC method of dihydroergotoxine analysis proved to be the most precise and practicable (440). The RP-LC method of dihydroergotoxine separation was further improved by Chervet and Plas (449), who applied a short and efficient high-speed column. Figure 7.28 shows the excellent separation of dihydroergotoxine components in Co-dergocrine ampoules and oral solutions. The analysis time was five times shorter than for the conventional method. Another example of high-speed LC analysis of pharmaceuticals (320) is given in Figure 7.24 for tablets containing dihydroergocristine (see Table 7.19). Spiegel

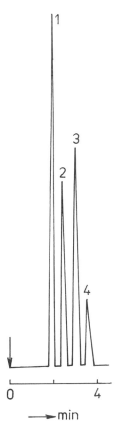

FIGURE 7.28 High-speed RP-LC of dihydroergot alkaloids in pharmaceuticals. Column 100 × 4.6 mm packed with Spherisorb 5 ODS, mobile phase acetonitrile-water-triethylamine 336:645:19, 2.2 ml min^{-1}, UV detection at 280 nm. 1, dihydroergocornine; 2, dihydro-α-ergocryptine; 3, dihydroergocristine; 4, dihydro-β-ergocryptine. (After Ref. 449.)

and Viernstein (450) developed two gradient RP-LC methods capable of separating dihydroergotoxine components and some of their degradation products, and applied them to the analysis of liquid pharmaceuticals.

Liquid chromatography with fluorescence detection was used for the determination of dihydroergotoxine alkaloids in biological fluids (56,451,452). The sensitivity of the methods allowed monitoring of plasma kinetics and urinary excretion after parenteral administration but was not sufficient to follow plasma levels in humans after therapeutic oral doses. Dihydroergotoxine was eluted as a single peak on silica gel column (452).

Properties of simple amides of lysergic acid are a litttle more appropriate for GC analysis, but many GC procedures still suffer from the thermal instability of these compounds. The value I = 3445 was reported for the retention index of LSD (82). Capillary GC was used to discriminate between LSD (I = 3082) and LAMPA (lysergic acid methylpropylamide, I = 3131) in street drug samples (453). Mass spectra demonstrated that both compounds were eluted unchanged from the column but the quantitative evaluation was not carried out.

Various procedures using ion pair RP-LC (247) (see Table 7.13) and separation on silica gel (238) and RP columns (443,454) were described for the analysis of illicit lysergides. When comparing UV, fluorescence, and electrochemical detections of LSD in spiked urine, the best selectivity was provided by a fluorescence detector (269). An analytical scheme for the identification of LSD in biological samples from persons suspected of having taken LSD was described by Twitchett et al. (446). The samples positive in radioimmunoassay were further analyzed using LC both on silica gel and ODS columns. A reversed phase system was shown to provide better discrimination for the identification of LSD.

Clavines have attracted less attention than other ergot alkaloids. Wurst et al. (455) studied the separation of clavines and simple derivatives of lysergic acid on chemically bonded amino packing. Some clavines exhibit useful fluorescence. Eckers et al. (456) tested various packings for the qualitative analysis of clavine alkaloids in ergot fermentation broth using LC coupled to MS. Supercritical fluid chromatography coupled to mass spectrometry was applied by Berry et al. (353) to the analysis of xanthines (see Table 7.17) and alkaloids of the clavine type.

G. Separation of Imidazole Alkaloids

The most important imidazole alkaloids, pilocarpine and isopilocarpine, have been isolated from the plant Pilocarpus jaborandi. Various chromatographic methods have been used for their separa-

Applications

tion (Table 7.22). Very little is known about the chromatography of other imidazole alkaloids. Ergothioneine was included in the series of ergot alkaloids studied by Twitchett et al. (446), but it was retained too strongly on both Spherisorb 5-ODS and Spherisorb S5W columns using mobile phases suitable for indole ergot alkaloids (Table 7.21).

Pilocarpine is used in ophthalmology to lower the intraocular pressure. Analyses of eye drops as well as studies on pilocarpine degradation by epimerization to inactive isopilocarpine and hydrolysis to pilocarpic and isopilocarpic acids usually employed chromatographic methods. The average value of 2014 for the GC retention index of pilocarpine was reported (82) and some applications of GC have appeared. For example, Szepesi et al. (457) determined the sum of pilocarpine and isopilocarpine in aqueous solution on OV-101 column since these compounds had not been resolved. For complete analysis of degradation products RP ion pair LC method was developed (457).

A general LC method for the analysis of pharmaceuticals (458) was also applied to some alkaloids including pilocarpine in eye drops. The simultaneous determination of pilocarpine and physostigmine in eye drops (357) was mentioned in the preceeding section (Table 7.18). Pilocarpine was not resolved from isopilocarpine. The optimized separation of pilocarpine and isopilocarpine on Radial-Pak C18 cartridge within 8 min was developed by Dunn and Thompson (459). Pilocarpic and isopilocarpic acid were eluted with the solvent front. On the contrary, by NP separation on silica gel (460) the reversed elution order of pilocarpine and isopilocarpine was observed and the acids were not eluted from the column. Using an aqueous mobile phase with considerable ionic strength, Bundgaard and Hansen (461) were able to resolve pilocarpine and its degradation products on silica gel column in 12 min. Later this method was applied to the evaluation of lactonization of esters (462) and diesters (463) of pilocarpic acid, which were considered possible prodrug forms for pilocarpine with improved ocular bioavailability. Reversed phase systems were used for the determination of these esters and pilocarpine and for assessment of their lipophilicity. Pilocarpine was not resolved from isopilocarpine and the acids eluted with the solvent front.

Complete RP-LC separation of pilocarpine, isopilocarpine, pilocarpic acid, and isopilocarpic acid, originally described by Noordham et al. (464) and later supplemented with UV detection (465, 466), is still in use (467). Because of peak tailing and difficulties in quantitation, Kennedy and McNamara (468) modified this method by employing a phenyl-type column. The response factor of pilocarpine was slightly dependent on the age of the mobile phase, and the extensive flushing of the system prior to and after use was necessary. Also in a collaborative study based essentially on this method (469), some instrumental problems with a high salt concentration were noted and again the flushing of the system after use was required. Recently, this method was used for the analysis of ophthalmic formulations and for the evaluation of degradation rate constants (470). The separation took about 24 min but all components were more equally separated than in the shorter separation of Kennedy and McNamara (468), which was performed at an elevated temperature within 15 min.

Another RP separation system suitable for similar purposes was developed by Szepesi et al. (457). Cetrimide was shown to change the the retention of alkaloids and acids essentially in opposite directions.

The detection of pilocarpine and its degradation products in LC may cause some difficulties. Due to the lack of long-wavelength absorption, UV detection below 220 nm has to be used, resulting in certain requirements for the choice of mobile phase components and for their purity. Refractive index detection is, of course, less sensitive.

Thin-layer chromatography is suitable for the separation of lactones from the acids but the resolution of epimers has not been achieved. The recent modification (471) of previously reported TLC methods for the preliminary assessment of pilocarpine and isopilocarpine acids in eye drops utilized a mobile phase polar enough for the elution of the acids. The shape of the spots, however, is indicative of their elution with a secondary solvent front. Prior to the detection with Dragendorff's reagent lactones were regenerated from acids by spraying with a solution p-toluenesulfonic acid in ether.

H. Separation of Diterpene and Steroidal Alkaloids

The alkaloids of this group possess mostly large complex molecules which cannot be transferred to the gaseous phase and thus the use of GC is very limited (see Table 7.23). For Solanum glycoalkaloids the solution lies in hydrolysis and in GC analysis of obtained aglycones. Bushway et al. (472) used a short packed column to determine solanidine in milk and compared GC and TLC methods. Aglycones of glycoalkaloids from Solanum tuber samples were separated on both packed and capillary columns (473) and retention indices of solanidine, tomatidine, and others were measured on a capillary column (474). In all these cases oven temperatures between 265 and 300°C were employed. The only steroidal alkaloid which can easily be determined by GC is conessine (see Figure 7.29). Duez et al. (475) compared GC and TLC (see Figure 7.30) for the determination of conessine and found GC suitable for cases in which a small number of samples is to be analyzed, whereas TLC will be advantageous in a large-scale analyses.

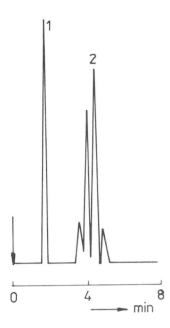

FIGURE 7.29 GC analysis of <u>Holarrhena floribunda</u> stem bark extract. Column 1 m × 2 mm packed with 3% SE-30 on Chromosorb W AW DMCS 100-120 mesh, helium 30 ml min^{-1}, 230 °C isothermal, FID. 1, codeine (internal std.); 2, conessine. (After Ref. 475.)

Diterpene alkaloids have been analyzed exclusively by liquid chromatography (see Table 7.23). Reversed phase LC using octadecyl-modified silica gel was applied to the separation of alkaloids from both <u>Aconitum</u> (476,477) and <u>Delphinium</u> (478) species. <u>Aconitum</u> bases were eluted with buffered organic modifier (see Figure 7.31), while for <u>Delphinium</u> alkaloids ion pairing was used. Cathedulines, which are sesquiterpenes with a macrolide bridge, were separated by means of gel permeation chromatography (479). In order to isolate three bases on a preparative scale, the use of recycling was necessary. The alkaloids were chromatographed on a polyvinylalcohol packing, whereas polystyrene packing failed to separate the mixture.

As stated above, liquid chromatography has also proved to be the best choice for the separation of glycoalkaloids. Bushway et al. (480) determined solanine and chaconine in potato tubers, peels, and sprouts. Three types of columns were tested: µBondapak C18, µBondapak NH$_2$, and "Carbohydrate" column (see Section II.B in

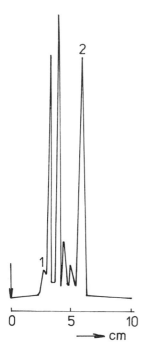

FIGURE 7.30 TLC scanning profile of <u>Holarrhena floribunda</u> stem bark extract. Plate 20 × 20 cm, silica gel 60F 254, mobile phase ethylacetate-hexane-diethylamine 75:24:6, absorption wavelength 500 nm. 1, codeine (internal std.); 2, conessine. (After Ref. 475.)

Chapter 5). Various mixtures of tetrahydrofuran, water, and acetonitrile served as the mobile phase in all experiments. The separation of a standard mixture is shown in Figure 7.32, where the two methods of detection are also compared. The analysis of extract from potato blossoms performed under identical conditions is shown in Figure 7.33. According to Bushway et al. (480), the best results were obtained with a Carbohydrate column. Later, Bushway et al. (472) used an amino column and almost unchanged conditions (mobile phase, flow rate) for the determination of glycoalkaloids in tater meal (472) and carbohydrate column (of another manufacturer) (481) for the comparison of extraction and clean-up procedures suitable for the analysis of potatoe tubers and products. Another method for the clean-up of potatoe extracts using ion pair separation on a disposable C18 cartridge was followed by RP-LC

Applications

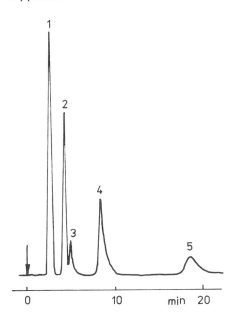

FIGURE 7.31 RP-LC analysis of <u>Aconitum crassicaule</u> alkaloids. Column 150 × 4.1 mm packed with Resolve C18 5 µm, mobile phase 1.8 mM ammonium carbonate-tetrahydrofuran 1:2, 1 ml min^{-1}, UV detection at 254 nm. 1, yunaconitine; 2, crassicauline A; 3, crassicautine; 4, foresaconitine; 5, crassicausine. (After Ref. 476.)

(483). A mobile phase based on acetonitrile-water was employed for the separation of alkaloids of <u>Solanum ptycanthum</u> (482), but the separation was carried out on silicic acid column. Based on selective complexing between tomatidine and sterols, Cziky and Hansson (484) suggested tomatidine bonded to silica gel as the stationary phase for affinity chromatography of sterols.

The group of steroidal alkaloids from <u>Veratrum</u> species has attracted considerable attention. Holan et al. (485) separated the mixture of ester alkaloids of <u>Veratrum sabadilla</u>, denoted as veratrine, by three techniques: on an analytical scale by NP-LC and RP-LC, and on a preparative scale by countercurrent droplet chromatography. The separation of <u>Veratrum</u> alkaloids by RP-LC can also be found in two other papers (486,487), where a µBondapak C18 column was eluted with aqueous acetonitrile (486) or with buffered methanol (487). Nevertheless, the applications of straight phase chromatography were also described (485,487,488). In two cases (485,488) the silica gel column was eluted with solvents usual

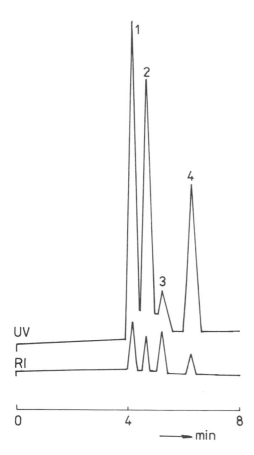

FIGURE 7.32 RP-LC separation of a mixture of glycoalkaloids. Column 300 × 4 mm packed with μBondapak NH_2 10 μm, mobile phase tetrahydrofuran-phosphate buffer (3.4 g KH_2PO_4/liter^{-1}-acetonitrile 50:25:25, 1 ml min^{-1}, UV detection at 208 nm and RI detection. 1, β-chaconine (2.5 μg); 2, α-chaconine (1.9 μg); 3, tomatine (3.1 μg); 4, α-solanine (2.3 μg). (After Ref. 480.)

Applications

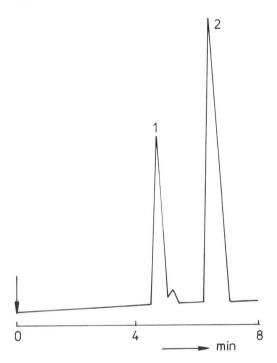

FIGURE 7.33 RP-LC analysis of glycoalkaloids from potato blossoms. UV detection at 208 nm, other conditions as given in Figure 7.31. 1, α-chaconine; 2, α-solanine. (After Ref. 480.)

to this technique: light petroleum with a small quantity of ethanol (485) or chloroform with the addition of methanol (488). In the third study the use of a mobile phase such as that used for RP-LC was reported (487).

Thin-layer chromatography proved its suitability for the separation of diterpene and steroidal alkaloids, especially on the preparative scale. Diterpene alkaloids of Delphinium species were separated by preparative centrifugal chromatography (478,489,490). Glycoalkaloids were separated by means of TLC for preparative (482) and identification (472) purposes. The determination of conessine by TLC followed by densitometry (475) was mentioned above. The TLC method, like GC, has proved its suitability for the determination of conessine in plant material. Similarly, Veratrum alkaloids were determined in plant material extract using TLC, either on silica gel or on alumina, and this technique was compared with column liquid chromatography (488).

REFERENCES

1. H. E. Harvey and R. M. Chell, Austr. J. Pharm. Sci., 10: 115 (1981).
2. R. K. Jhangiani and A. C. Bello, J. Assoc. Off. Anal. Chem., 68:523 (1985).
3. P. Majlat, P. Helboe, and A. K. Kristensen, Int. J. Pharm., 9:245 (1981).
4. C. Y. Ko, F. C. Marziani, and C. A. Janicki, J. Pharm. Sci., 69:1081 (1980).
5. V. B. Stuber and K. H. Müller, Pharm. Acta Helv., 57:181 (1981).
6. V. Das Gupta, J. Pharm. Sci., 69:110 (1980).
7. K. Wessely and K. Zech, High Performance Liquid Chromatography in Pharmaceutical Analyses, Hewlett-Packard GmbH, Böblingen, 1979, p. 31.
8. R. W. Beaver, J. E. Bunch, and L. A. Jones, J. Chem. Ed., 60:1000 (1983).
9. T. D. Wilson, J. Liq. Chromatogr., 9:2309 (1986).
10. M. Gazdag, G. Szepesi, and K. Csomor, J. Chromatogr., 243:315 (1982).
11. N. Muhammad and J. A. Bodnar, J. Liq. Chromatogr., 3:113 (1980).
12. P. Pietta, E. Manera, and P. Ceva, J. Chromatogr., 367:228 (1986).
13. B. F. Engelke and P. G. Vincent, J. Assoc. Off. Anal. Chem., 65:651 (1982).
14. G. Forni and G. Massarani, J. Chromatogr., 131:444 (1977).
15. A. W. Archer, J. Chromatogr., 351:595 (1986).
16. A. Raisi, E. Alipour, and S. Manouchelvi, Chromatographia, 21:711 (1986).
17. B. Pekić, Z. Lepojević, B. Slavica, and S. M. Petrović, Chromatographia, 21:227 (1986).
18. M. Ferrari and L. Verotta, J. Chromatogr., 437:328 (1988).
19. W. Zeng, W. Liang, and G. Tu, J. Chromatogr., 408:426 (1987).
20. K. Aramaki, T. Hanai, and H. F. Walton, Anal. Chem., 52: 1963 (1980).
21. Y. Nobuhara, S. Hirano, K. Namba, and M. Hashimoto, J. Chromatogr., 190:251 (1980).
22. N. R. Ayyangar and S. R. Bhide, J. Chromatogr., 436:455 (1988).
23. B. C. Pettitt, Jr., and C. E. Damon, J. Chromatogr., 242: 189 (1982).
24. M. Hutin, A. Cavé, and J. P. Foucher, J. Chromatogr., 268:125 (1983).

Applications

25. G. Holzer, L. H. Zalkow, and C. F. Asibal, J. Chromatogr., 400:317 (1987).
26. C. Bugatti, M. L. Colombo, and F. Tomè. J. Chromatogr., 393:312 (1987).
27. A. L. Christiansen, K. E. Rasmussen, and F. Tønnesen, J. Chromatogr., 210:163 (1981).
28. S. Borner and R. Brenneisen, J. Chromatogr., 408:402 (1987).
29. R. van der Heijden, P. J. Lamping, P. P. Out, R. Wijnsma, and R. Verpoorte, J. Chromatogr., 396:287 (1987).
30. R. van der Heijden, A. Hermans-Lokkerbol, R. Verpoorte, and A. Baerheim-Svendsen, J. Chromatogr., 396:410 (1987).
31. M. Montagu, P. Levillain, J. C. Chenieux, and M. Rideau, J. Chromatogr., 409:426 (1987).
32. T. Arvidsson, K.-G. Wahlund, and N. Daoud, J. Chromatogr., 317:213 (1984).
33. N. Daoud, T. Arvidsson, and K.-G. Wahlund, J. Chromatogr., 385:311 (1987).
34. N. Daoud, Acta Pharm. Suec., 24:39 (1987).
35. L. J. Cline Love, S. Zibas, J. Noroski, and M. Arunyanart, J. Pharm. Biomed. Appl., 3:511 (1985).
36. D. Westerlund, Chromatographia, 24:155 (1987).
37. H. Yoshida, I. Morita, G. Tamai, T. Masujima, T. Tsuru, N. Nakai, and H. Imai, Chromatographia, 19:466 (1984).
38. G. A. Kyerematen, L. H. Taylor, J. D. de Bethizy, and E. S. Vesell, J. Chromatogr., 419:191 (1987).
39. P. Jacob, III, B. A. Elias-Baker, R. T. Jones, and N. L. Benowitz, J. Chromatogr., 417:277 (1987).
40. G. Tamai, H. Yoshida, H. Imai, T. Takashina, K. Kotoo, T. Fuwa, Y. Tsuchioka, H. Matsuura, and G. Kajiyama, Chromatographia, 20:671 (1985).
41. G. W. Mihaly, K. M. Hyman, and R. A. Smallwood, J. Chromatogr., 415:177 (1987).
42. R. D. Todd, S. M. Muldoon, and R. L. Watson, J. Chromatogr., 232:101 (1982).
43. J. E. Wallace, S. C. Harris, and M. W. Peek, Anal. Chem., 52:1328 (1980).
44. J.-O. Svensson, A. Rane, J. Säwe, and F. Sjöqvist, J. Chromatogr., 230:427 (1982).
45. K. Isaksson and P. T. Kissinger, J. Liq. Chromatogr., 10:2213 (1987).
46. E. Bailey, P. B. Farmer, J. A. Peal, S. A. Hotchkiss, and J. Caldwell, J. Chromatogr., 416:81 (1987).
47. G. F. Kapke and R. B. Franklin, J. Liq. Chromatogr., 10:451 (1987).
48. S. Stavchansky, A. Combs, and M. Delgado, Anal. Lett., 19:639 (1986).

49. J. J. Lauff, J. Chromatogr., 417:99 (1987).
50. H.-U. Schultz, I. Trotnow, E. Kraas, and H. Hollandt, Chromatographia, 22:411 (1986).
51. J. S. Kennedy, B. W. Leduc, J. M. Scavone, J. S. Harmatz, R. I. Shader, and D. J. Greenblatt, J. Chromatogr., 422:274 (1987).
52. R. Chiou, R. J. Stubbs, and W. F. Bayne, J. Chromatogr., 422:281 (1987).
53. J. Tencheva, I. Yamboliev, and Z. Zhivkova, J. Chromatogr., 421:396 (1987).
54. D. E. M. M. Vendrig, J. J. M. Holthius, V. Erdélyi-Toth, and A. Hulshoff, J. Chromatogr., 414:91 (1987).
55. P. O. Edlund, J. Chromatogr., 226:107 (1981).
56. H. Humbert, J. Denouel, J. P. Chervet, D. Lavende, and J. R. Kiechel, J. Chromatogr., 417:319 (1987).
57. P. E. Nelson, S. L. Nolan, and K. R. Bedford, J. Chromatogr., 234:407 (1982).
58. C. Kim, Chromatographia, 22:303 (1986).
59. I. N. Papadoyannis and B. Cady, Anal. Lett., 19:1065 (1986).
60. J. Shah and W. D. Mason, Anal. Lett., 20:1493 (1987).
61. J. Shah and W. D. Mason, Anal. Lett., 20:881 (1987).
62. I. R. Tebbett, Chromatographia, 23:377 (1987).
63. C. Lora-Tamayo, T. Tena, and G. Tena, J. Chromatogr., 422:267 (1987).
64. S. Y. Wang, S. Y. Tham, and M. K. Poon, J. Chromatogr., 381:331 (1986).
65. M. C. Dutt, D. S.-T. Lo, D. L. K. Ng, and S.-O. Woo, J. Chromatogr., 267:117 (1983).
66. M. B. Kester, C. L. Saccar, and H. C. Mansmann, Jr., J. Chromatogr., 416:91 (1987).
67. M. B. Kester, C. L. Saccar, and H. C. Mansmann, Jr., J. Liq. Chromatogr., 10:957 (1987).
68. I. Ojanperä and E. Vuori, J. Liq. Chromatogr., 10:3595 (1987).
69. R. D. Barlow, P. A. Thompson, and R. B. Stone, J. Chromatogr., 419:375 (1987).
70. A. S. Christophersen, A. Biseth, B. Skuterud, and G. Gadeholt, J. Chromatogr., 422:117 (1987).
71. S. A. Hotchkiss and J. Caldwell, J. Chromatogr., 423:179 (1987).
72. Y. Tsuruta, K. Kohashi, and Y. Ohkura, J. Chromatogr., 224:105 (1981).
73. G. S. Clarke and M. L. Robinson, Anal. Proc., 22:137 (1985).
74. K. R. Lung and C. H. Lochmuller, J. Liq. Chromatogr., 9:2995 (1986).
75. J. N. Little, J. Liq. Chromatogr., 9:3033 (1986).

76. W. J. Hurst and R. A. Martin, in Advances in Laboratory Automatization-Robotics 1985 (J. R. Strimaitis and G. L. Hawk, eds.), Zymark Corporation, Hopkinton, MA, p. 117.
77. D. J. Myers, N. Szuminsky, and M. J. Levitt, in Advances in Laboratory Automatization-Robotics 1985 (J. R. Strimaitis and G. L. Hawk, eds.), Zymark Corporation, Hopkinton, MA, p. 71.
78. T. A. Gough and P. B. Baker, J. Chromatogr. Sci., 20:289 (1982).
79. A. Baerheim-Svendsen and R. Verpoorte, Chromatography of Alkaloids. Part A. Thin-Layer Chromatography. Journal of Chromatography Library Vol. 23A. Elsevier, Amsterdam 1983.
80. R. Verpoorte and A. Baerheim-Svendsen, Chromatography of Alkaloids. Part B. Gas-Liquid Chromatography and High-Performance Liquid Chromatography. Journal of Chromatography Library, Vol. 23B. Elsevier, Amsterdam 1984.
81. M. J. Knox, W. D. Clark, and S. O. Link, J. Chromatogr., 265:357 (1983).
82. R. E. Ardrey and A. C. Moffat, J. Chromatogr., 220:195 (1981).
83. S. Alm, S. Jonson, H. Karlsson, and E. H. Sundholm, J. Chromatogr., 254:179 (1983).
84. I. Barni Comparini, F. Centini, and A. Pariali, J. Chromatogr., 279:609 (1983).
85. W. A. König and K. Ernst, J. Chromatogr., 280:135 (1983).
86. I. W. Wainer, T. D. Doyle, F. S. Fry, Jr., and Z. Hamidzadeh, J. Chromatogr., 355:149 (1986).
87. O. Weller, J. Schulze, and W. A. König, J. Chromatogr., 403:263 (1987).
88. I. W. Wainer, T. D. Doyle, Z. Hamidzadeh, and M. Aldridge, J. Chromatogr., 268:107 (1983).
89. J. Gal, J. Chromatogr., 307:220 (1984).
90. C. Pettersson, A. Karlsson, and C. Gioeli, J. Chromatogr., 407:217 (1987).
91. J. R. Lloyd, M. L. Cotter, D. Ohori, and A. R. Oyler, Anal. Chem., 59:2533 (1987).
92. M. De Smet and D. L. Massart, J. Chromatogr., 410:77 (1987).
93. R. Gimet and A. Filloux, J. Chromatogr., 177:333 (1979).
94. M. Prokšová, K. Michalak, Z. Menyhardtová, and B. Proksa, Farm. Obzor, 54:257 (1985).
95. S. U. Alvi and F. Castro, J. Liq. Chromatogr., 9:2269 (1986).
96. R. C. George and J. J. Contario, J. Liq. Chromatogr., 11:475 (1988).

97. J. Veals, H. Kim, C. Korduba, D. Curtis, E. Durante, and C. Lin, J. Liq. Chromatogr., 11:417 (1988).
98. M. Nieder and H. Jaeger, J. Chromatogr., 424:73 (1988).
99. E. Brendel, I. Meinecke, E.-M. Henne, M. Zschunke, and C. de Mey, J. Chromatogr., 426:406 (1988).
100. K. Sagara, T. Oshima, and T. Misaki, Chem. Pharm. Bull., 31:2359 (1983).
101. H. Gulyás, G. Kemény, I. Hollósi, and J. Pucsok, J. Chromatogr., 291:471 (1984).
102. E. Lacey and R. L. Brady, J. Chromatogr., 315:233 (1984).
103. M. Lhermitte, J. L. Bernier, D. Mathieu, M. Mathieu-Nolf, F. Erb, and P. Roussel, J. Chromatogr., 342:416 (1985).
104. Y. Cai and L. He, Zhongcaoyao, 18:56 (1987); CA, 107: 141169b (1987).
105. R. F. Severson, K. L. McDuffie, R. F. Arrendale, G. R. Gwynn, J. F. Chaplin, and A. W. Johnson, J. Chromatogr., 211:111 (1981).
106. A. W. Garrison, J. D. Pope, and F. R. Allen, in Identification & Analysis of Organic Pollutants in Water (L. H. Keith, ed.), Ann Arbor Science, Ann Arbor, 1981, p. 517.
107. R. A. Andersen, T. R. Hamilton-Kemp, J. H. Loughrin, Ch. G. Hughes, D. F. Hildebrand, and T. G. Sutson, J. Agric. Food Chem., 36:295 (1988).
108. M. W. Ogden, J. HRC&CC, 11:428 (1988).
109. W. J. Chamberlain, W. S. Schlotzhauer, and O. T. Chortyk, J. Agric. Food Chem., 36:48 (1988).
110. D. Hill, W. Brande, D. Phillips, and O. Pomerleau, Anal. Lett., 16:355 (1983).
111. D. Blacke, Ch. Thevenon, M. Ciavatti, and S. Renaud, Anal. Biochem., 143:316 (1984).
112. R. A. Davis, J. Chromatogr. Sci., 24:134 (1986).
113. J. Godin, F. Girard, and G. Hellier, J. Chromatogr., 343:426 (1985).
114. J. A. Thompson, M.-S. Ho, and D. R. Petersen, J. Chromatogr., 231:53 (1982).
115. C. G. Norburg, J. Chromatogr., 414:449 (1987).
116. G. B. Neurath and F. G. Pein, J. Chromatogr., 415:400 (1987).
117. D. W. Armstrong, L. A. Spino, S. M. Han, J. I. Seeman, and H. V. Secor, J. Chromatogr., 411:490 (1987).
118. B. J. L. Sudan, C. Brouillard, C. Strehler, H. Strub, J. Sterboul, and J. Sainte-Laudy, J. Chromatogr., 288:415 (1984).
119. K. C. Cundy and P. A. Crooks, J. Chromatogr., 306:291 (1984).
120. K. C. Cundy, C. S. Godin, and P. A. Crooks, Drug Metab. Disp., 12:755 (1984).

121. K. C. Cundy, M. Sato, and P. A. Crooks, Drug. Metab. Disp., 13:175 (1985).
122. W. F. Pool and P. A. Crooks, Drug Metab. Disp., 13:578 (1985).
123. M. Sato and P. A. Crooks, Drug Metab. Disp., 13:348 (1985).
124. S. Mousa, G. R. van Loon, A. A. Houdi, and P. A. Crooks, J. Chromatogr., 347:405 (1986).
125. M. Horstmann, J. Chromatogr., 344:391 (1985).
126. A. T. Shulgin, J.-P. Jacob, III, N. L. Benowitz, and D. Lau, J. Chromatogr., 423:365 (1987).
127. J. A. Thompson, K. J. Norris, and D. R. Petersen, J. Chromatogr., 341:349 (1985).
128. M. Verzele, G. Redant, S. Qureshi, and P. Sandra, J. Chromatogr., 199:105 (1980).
129. D. E. Games, N. J. Alcock, J. van der Greef, L. M. Nyssen, H. Maarse, and M. C. ten Noever de Brauw, J. Chromatogr., 294:269 (1984).
130. M. Rathnawathie and K. A. Buckle, J. Chromatogr., 264:316 (1983).
131. K. M. Weaver, R. G. Luker, and M. E. Neale, J. Chromatogr., 301:228 (1984).
132. V. K. Attuquayefio and K. A. Buckle, J. Agric. Food Chem., 35:777 (1987).
133. R. J. Nash, W. S. Goldstein, S. V. Evans, and L. E. Fellows, J. Chromatogr., 366:431 (1986).
134. C. J. Briggs and K. J. Simons, J. Chromatogr., 257:132 (1983).
135. M. Ylinen, T. Naaranlahti, S. Lapinjoki, A. Huhtikangas, M.-L. Salonen, L. K. Simda, and M. Lounasmaa, Planta Med., 1986:85.
136. T. Hartmann, L. Witte, F. Oprach, and G. Toppel, Planta Med., 1986:390.
137. L. Witte, K. Mueller, and H. A. Arfmann, Planta Med., 1987: 192; CA, 107:112623q (1987).
138. R. Verpoorte, J. M. Verzeijl, and A. Baerheim-Svendsen, J. Chromatogr., 283:401 (1984).
139. R. W. Ross and C. A. Lau-Cam, J. Chromatogr., 370:403 (1986).
140. U. R. Cieri, J. Assoc. Off. Anal. Chem., 68:1042 (1985).
141. P. Leroy, A. Moreau, A. Nicolas, and B. Hosotte, J. HRC&CC, 6:514 (1983).
142. S. Paphassarang, J. Raynaud, R. P. Godeau, and A. M. Binsard, J. Chromatogr., 319:412 (1985).
143. B. Pekić, B. Slavica, Ž. Lepojević, and M. Gorunovic, Pharmazie, 40:423 (1985).
144. P. Duez, S. Chamart, M. Hanocq, L. Molle, M. Vanhaelen, and R. Vanhaelen-Fastré, J. Chromatogr., 329:415 (1985).

145. T. Zhang, J. Chromatogr., 315:287 (1984).
146. T.-Y. Zhang, D.-G. Cai, and Y. Ito, J. Chromatogr., 435:159 (1988).
147. C. Lora-Tamayo, M. A. Rams, and J. M. R. Chacon, J. Chromatogr., 374:73 (1986).
148. A. H. Lewin, S. R. Parker, and F. I. Carroll, J. Chromatogr., 193:371 (1980).
149. C. C. Clark, J. Assoc. Off. Anal. Chem., 61:683 (1978).
150. P. Jacob, III, B. A. Elias-Baker, R. T. Jones, and N. L. Benowitz, J. Chromatogr., 306:173 (1984).
151. S. Goenechea, G. Rücker, M. Neugebauer, and U. Zerell, Fresenius Z. Anal. Chem., 323:326 (1986).
152. J. M. Moore, D. A. Cooper, I. S. Lurie, T. C. Kram, S. Carr, C. Harper, and J. Yeh, J. Chromatogr., 410:297 (1987).
153. R. S. Schwartz and K. O. David, Anal. Chem., 57:1362 (1985).
154. C. M. Selavka, I. S. Krull, and I. S. Lurie, J. Chromatogr. Sci., 23:499 (1985).
155. I. S. Lurie, J. M. Moore, D. A. Cooper and T. C. Kram, J. Chromatogr., 405:273 (1987).
156. F. T. Noggle, Jr. and C. R. Clark, J. Assoc. Off. Anal. Chem., 66:151 (1983).
157. J. B. Crowther and J. D. Henion, Anal. Chem., 57:2711 (1985).
158. K. K. Kaistha and R. Tadrus, J. Chromatogr., 267:109 (1983).
159. J. K. Brown, R. H. Schingler, M. G. Chaubal, and M. H. Malone, J. Chromatogr., 87:211 (1973).
160. B. De Spiegeleer, W. van den Bossche, P. De Moerloose, and D. Massart, Chromatographia, 23:407 (1987).
161. H. Kanamori, I. Sakamoto, and M. Mizuta, Chem. Pharm. Bull., 34:1826 (1986).
162. M. Montagu, P. Levillain, J. C. Chenieux, and M. Rideau, J. Chromatogr., 331:437 (1985).
163. M. Montagu, P. Levillain, J. C. Chenieux, and M. Rideau, J. Chromatogr., 351:144 (1986).
164. D. V. McCalley, Chromatographia, 17:264 (1983).
165. D. V. McCalley, J. Chromatogr., 357:221 (1986).
166. E. Smith, J. Chromatogr., 299:233 (1984).
167. M. A. Johnston, W. J. Smith, J. M. Kennedy, A. R. Lea, and D. M. Hailey, J. Chromatogr., 189:241 (1980).
168. M. A. Johnston, A. R. Lea, and D. M. Hailey, J. Chromatogr., 318:362 (1985).
169. L. P. Valenti, J. Assoc. Off. Anal. Chem., 68:782 (1985).
170. C. Cavazzutti, L. Gagliardi, A. Amato, E. Gattavecchia, and D. Tonelli, J. Chromatogr., 257:166 (1983).

Applications

171. D. V. McCalley, J. Chromatogr., 260:184 (1983).
172. R. Verpoorte, T. Mulder-Krieger, M. J. Verzijl, and A. Baerheim-Svendsen, J. Chromatogr., 261:172 (1983).
173. C.-T. Chung and E. J. Staba, J. Chromatogr., 295:276 (1984).
174. A. T. Keene, L. A. Anderson, and J. D. Phillipson, J. Chromatogr., 260:123 (1983).
175. A. Hobson-Frohock and W. T. E. Edwards, J. Chromatogr., 249:369 (1982).
176. R. J. Robins, J. Payne, and M. J. C. Rhodes, Planta Med., 1986:220.
177. F. A. Mellon, J. R. Chapman, and J. A. E. Pratt, J. Chromatogr., 394:209 (1987).
178. L. N. Ace and B. Chaudhuri, Anal. Lett., 20:1483 (1987).
179. J. Vasiliades and J. M. Finkel, J. Chromatogr., 278:117 (1983).
180. M. Edstein, J. Stace, and F. Shann, J. Chromatogr., 278:445 (1983).
181. M. Arunyanart and L. J. Cline-Love, J. Chromatogr., 342:293 (1985).
182. G. Malikin, S. Lam, and A. Karmen, Chromatographia, 18:253 (1984).
183. C. Rosini, P. Altemura, D. Pini, C. Bertucci, G. Zullino, and P. Salvadori, J. Chromatogr., 348:79 (1985).
184. H. W. Sturrman, J. Köhler, and G. Schomburg, Chromatographia, 25:265 (1988).
185. M. W. Duncan, G. A. Smythe, and P. S. Clezy, Biomed. Mass Spectrom., 12:106 (1985).
186. M. W. Duncan, G. A. Smythe, M. V. Nicholson, P. S. Clezy, J. Chromatogr., 336:199 (1984).
187. R. L. St Claire, III, G. A. S. Ansari, and C. W. Abell, Anal. Chem., 54:186 (1982).
188. J. L. Cashaw, C. A. Geraghty, B. McLaughlin, and V. E. Davis, Anal. Biochem., 162:274 (1987).
189. S. R. Gautam, A. Nahum, J. Baechler, and D. W. A. Bourne, J. Chromatogr., 182:482 (1980).
190. P. J. Davis, J. Chromatogr., 193:170 (1980).
191. K. Sagara, Y. Ito, M. Ojima, T. Oshima, K. Suto, T. Misaki, and H. Irokawa, Chem. Pharm. Bull., 33:5369 (1985).
192. M. Hutin, A. Oztekin, A. Cave, and J. P. Foucher, J. Chromatogr., 265:139 (1983).
193. J.-P. Fells, P. Lechat, R. Rispe, and W. Cautreels, J. Chromatogr., 308:273 (1984).
194. J. Slavík and L. Slavíková, Collect. Czech Chem. Commun., 49:704 (1984).
195. J. Slavík, L. Slavíková, and L. Dolejš, Collect. Czech. Chem. Commun., 49:1318 (1984).

196. K. Tagahara, J. Koyama, T. Okatani, and Y. Suzuta, Chem. Pharm. Bull., 34:5166 (1986).
197. A. Baerheim-Svendsen, A. M. Van Kempen-Verleun, and R. Verpoorte, J. Chromatogr., 291:389 (1984).
198. T. Dzido, L. Jusiak, and E. Soczewinski, Anal. Chem., 31: 135 (1986).
199. T. Dzido, J. Chromatogr., 439:257 (1988).
200. Y. Hashimoto, M. I. Okada, A. Kato, K. Iwasa, K. Saiki, and N. Takao, Anal. Lett., 18:563 (1985).
201. Y. Hashimoto, M. I. Okada, U. Shome, and A. Kato, Anal. Lett., 19:2253 (1986).
202. K. M. Jensen, J. Chromatogr., 274:381 (1983).
203. M. Johansson, S. Eksborg, and A. Arbin, J. Chromatogr., 275:355 (1983).
204. N. P. Sahu and S. B. Mahato, J. Chromatogr., 238:525 (1982).
205. S. J. Bannister, J. Stevens, D. Musson, and L. A. Sternson, J. Chromatogr., 176:381 (1979).
206. J. E. Parkin, J. Chromatogr., 225:240 (1981).
207. M. J. Avram and C. A. Shanks, J. Chromatogr., 306:404 (1984).
208. K. Masumoto, Y. Tashiro, K. Matsumoto, A. Yoshida, M. Hirayama, and S. Hayashi, J. Chromatogr., 381:323 (1986).
209. P. O. Edlund, J. Chromatogr., 279:615 (1983).
210. G. J. Sewel, C. J. Soper, and R. T. Partiff, Anal. Chim. Acta, 163:237 (1984).
211. E. J. Cone, W. D. Darwin, and W. F. Buchwald, J. Chromatogr., 275:307 (1983).
212. K. W. Lewis, J. Chromatogr. Sci., 21:521 (1983).
213. J. A. Owen and D. S. Sitar, J. Chromatogr., 276:202 (1983).
214. A. W. Jones, Y. Blom, U. Bondesson, and E. Änggård, J. Chromatogr., 309:73 (1984).
215. R. H. Drost, R. D. Van Ooijen, T. Ionescu, and R. A. A. Maes, J. Chromatogr., 310:193 (1984).
216. R. A. R. Tasker and K. Nakatsu, Analyst (London), 111: 563 (1986).
217. J.-L. Janicot, M. Caude, and R. Rosset, Analusis, 14:441 (1986).
218. G. B. Cox and R. W. Stout, J. Chromatogr., 384:315 (1987).
219. M. G. Lee, J. Chromatogr., 312:473 (1984).
220. R. K. Jhangiani and A. C. Bello, J. Assoc. Off. Anal. Chem., 68:523 (1985).
221. E. J. Jarvi, J. C. Stolzenbach, and R. E. Larson, J. Chromatogr., 377:261 (1986).
222. J. Visser, G. Grasmeijer, and F. Moolenaar, J. Chromatogr., 274:372 (1983).

Applications

223. J.-O. Svensson, J. Chromatogr., 375:174 (1986).
224. Y. Kumagai, T. Ishida, and S. Toki, J. Chromatogr., 421:155 (1987).
225. S. M. Stubbs, R. Chiou, and W. F. Bayne, J. Chromatogr., 377:447 (1986).
226. P. E. Nelson, J. Chromatogr., 298:59 (1984).
227. F. Tagliaro, R. Dorizzi, M. Plescia, M. Pradella, S. Ferrari, and V. Lo Cascio, Fresenius Z. Anal. Chem., 317:678 (1984).
228. F. Tagliaro, A. Frigerio, R. Dorizzi, G. Lubli, and M. Marigo, J. Chromatogr., 330:323 (1985).
229. J. Zoer, P. Virgili, and J. A. Henry, J. Chromatogr., 382:189 (1986).
230. K. R. Bedford and P. C. White, J. Chromatogr., 347:398 (1985).
231. H. N. Al-Kaysi and M. S. Salem, Anal. Lett., 19:915 (1986).
232. F.-F. Wu and R. H. Dobberstein, J. Chromatogr., 140:65 (1977).
233. H. Neumann and H.-P. Meyer, J. Chromatogr., 391:442 (1987).
234. M. Prosek, M. Steblaj, A. Medja, and M. Pukl, Int. Lab., 1986 (June):74.
235. D. Furmanec, J. Chromatogr., 89:76 (1974).
236. A. R. Sperling, J. Chromatogr., 294:297 (1984).
237. H. Neumann, J. Chromatogr., 315:404 (1984).
238. I. Jane, J. Chromatogr., 111:227 (1975).
239. C. J. C. M. Laurent, H. A. H. Billiet, and L. de Galan, J. Chromatogr., 285:161 (1984).
240. H. A. H. Billiet, R. Wolters, L. de Galan, and M. Huizer, J. Chromatogr., 368:351 (1986).
241. S. Galewsky and C. L. Nessler, Chromatographia, 18:87 (1984).
242. W. R. Sisco, C. T. Rittenhouse, W. M. Maggio, Chromatographia, 20:289 (1985).
243. Unpublished results of authors.
244. N. R. Ayyangar and S. R. Bhide, J. Chromatogr., 366:435 (1986).
245. V. K. Srivastava and M. L. Maheshwari, J. Assoc. Off. Anal. Chem., 68:801 (1985).
246. L. W. Doner and A.-F. Hsu, J. Chromatogr., 253:120 (1982).
247. I. Lurie, J. Assoc. Off. Anal. Chem., 60:1035 (1977).
248. K. R. Khanna and S. Shukla, Planta Med., 1986:157.
249. E. Stahl and H. Jahn, Pharm. Acta Helv., 60:248 (1985); CA, 103:220894e (1985).
250. T. Veress, Magy. Kem. Foly., 92:54 (1986).
251. R. Dybowski and T. A. Gough, J. Chromatogr. Sci., 22:465 (1984).

252. N. K. Nair, V. Navaratnam, and V. Rajanda, J. Chromatogr., 366:363 (1986).
253. J. M. Moore, J. Chromatogr., 281:355 (1983).
254. J. M. Moore, A. C. Allen, and D. A. Cooper, Anal. Chem., 56:642 (1984).
255. H. Neumann and M. Gloger, Chromatographia, 16:261 (1982).
256. A. C. Allen, D. A. Cooper, J. M. Moore, M. Gloger, and H. Neumann, Anal. Chem., 56:2940 (1984).
257. J. M. Moore, A. C. Allen, and D. A. Cooper, Anal. Chem., 58:1003 (1986).
258. I. S. Lurie and A. C. Allen, J. Chromatogr., 317:427 (1984).
259. F. Munari and K. Grob, J. HRC&CC, 11:172 (1988).
260. P. C. White, I. Jane, A. Scott, and B. E. Connett, J. Chromatogr., 265:293 (1983).
261. I. S. Lurie and S. M. Carr, J. Liq. Chromatogr., 9:2485 (1986).
262. I. S. Lurie and K. McGuiness, J. Liq. Chromatogr., 10:2189 (1987).
263. H. J. G. M. Derks, K. van Twillert, and G. Zomer, Anal. Chim. Acta, 170:13 (1985).
264. H. J. G. M. Derks, K. van Twillert, D. P. K. H. Perboom-de Fauw, G. Zomer, and J. G. Loeber, J. Chromatogr., 370:173 (1986).
265. R. Gill, B. Law, C. Brown, and A. C. Moffat, Analyst, 110:1059 (1985).
266. L. L. Plotczyk and P. Larson, J. Chromatogr., 257:211 (1983).
267. R. J. Flanagan, G. C. A. Storey, R. K. Bhamra, and I. Jane, J. Chromatogr., 247:15 (1982).
268. B. Law, R. Gill, and A. C. Moffat, J. Chromatogr., 301:165 (1984).
269. I. Jane, A. McKinnon, and R. J. Flanagan, J. Chromatogr., 323:191 (1985).
270. D. W. Hill and K. J. Langner, J. Liq. Chromatogr., 10:377 (1987).
271. G. Musch, M. De Smet, and D. L. Massart, J. Chromatogr., 348:97 (1985).
272. G. Musch and D. L. Massart, J. Chromatogr., 370:1 (1986).
273. I. S. Lurie, S. M. Sottolano, and S. Blasof, J. Forensic Sci., 27:519 (1982).
274. J. K. Baker, R. E. Skelton, and Ch. Y. Ma, J. Chromatogr., 168:417 (1979).
275. B. B. Wheals, J. Chromatogr., 187:65 (1980).
276. A. F. Fell, B. J. Clark, and H. P. Scott, J. Chromatogr., 316:423 (1984).
277. C. M. Selavka and I. S. Krull, J. Liq. Chromatogr., 10:345 (1987).

278. E. G. Sundholm, J. Chromatogr., 265:285 (1983).
279. H. Moll and J. T. Clerc, Pharm. Acta Helv., 62:210 (1987); CA, 107:168162f (1987).
280. A. H. Stead, R. Gill, T. Wright, J. P. Gibbs, and A. C. Moffat, Analyst, 107:1106 (1982).
281. G. Musumarra, G. Scarlata, G. Romano, S. Clementi, and S. Wold, J. Chromatogr. Sci., 22:538 (1984).
282. G. Musumarra, G. Scarlata, G. Cirma, G. Romano, S. Palazzo, S. Clementi, and G. Giulietti, J. Chromatogr., 295:31 (1984).
283. G. Musumarra, G. Scarlata, G. Cirma, G. Romano, S. Palazzo, S. Clementi, and G. Giulietti, J. Chromatogr., 350:151 (1985).
284. T. Daldrup, F. Susanto, and P. Michalke, Fresenius Z. Anal. Chem., 308:413 (1981).
285. K. R. Brain and B. B. Thapa, J. Chromatogr., 258:183 (1983).
286. B. K. Chowdhury, S. K. Hirani, and D. Ngur, J. Chromatogr., 390:439 (1987).
287. B. K. Chowdhury, A. Mustapha, and P. Bhattacharyya, J. Chromatogr., 329:178 (1985).
288. B. K. Chowdhury, S. K. Hirani, and P. Bhattacharyya, J. Chromatogr., 369:258 (1986).
289. P. Bhattacharyya and S. S. Jash, J. Chromatogr., 298:200 (1984).
290. B. K. Chowdhurry, P. P. Rai, and P. Bhattacharyya, Chromatographia, 23:205 (1987).
291. R. V. Gerard and D. B. MacLean, Phytochemistry, 25:1143 (1986).
292. P. Lepom and K.-D. Robowsky, J. Chromatogr., 322:261 (1985).
293. C. Bicchi, A. D'Amato, and E. Cappelletti, J. Chromatogr., 349:23 (1985).
294. U. Candrian, J. Lüthy, P. Schmid, Ch. Schlatter, and E. Gallasz, J. Agric. Food Chem., 32:935 (1984).
295. T. Hartmann and G. Toppel, Phytochemistry, 26:1639 (1987).
296. H. J. Huizing, F. de Boer, H. Hendriks, W. Balraadjsing, and A. P. Bruins, Biomed. Environ. Mass Spectr., 13:293 (1986).
297. H. Hendriks, H. J. Huizing, and A. P. Bruins, J. Chromatogr., 428:352 (1988).
298. L. A. C. Pieters and A. J. Vlientinck, J. Liq. Chromatogr., 9:745 (1986).
299. H. Niwa, H. Ishiwata, and K. Yamada, J. Chromatogr., 257:146 (1983).
300. C. W. Qualls, Jr. and H. J. Segall, J. Chromatogr., 150:202 (1978).
301. B. Kedzierski and D. R. Buhler, Anal. Biochem., 152:59 (1986).

302. H. J. Segall, J. Liq. Chromatogr., 7 (Suppl. 2):377 (1984).
303. S. C. Jain and R. Sharma, Chem. Pharm. Bull., 35:3487 (1987).
304. D. Cheng and E. Röder, Planta Med., 1986:484.
305. C. R. Priddis, J. Chromatogr., 261:95 (1983).
306. G. M. Hatfield, D. J. Yang, P. W. Fergusson, and W. J. Keller, J. Agric. Food Chem., 33:909 (1985).
307. F. Tosun, M. Tanker, T. Özden, and A. Tosun, Planta Med., 1986:242.
308. T. Fujii, H. Kogen, S. Yoshifuji, and M. Ohba, Chem. Pharm. Bull., 33:1946 (1985).
309. A. Evidente, I. Iasiello, and G. Randazzo, J. Chromatogr., 281:362 (1983).
310. H. A. Claessens, M. Van Thiel, P. Westra, and A. M. Soeterbock, J. Chromatogr., 275:345 (1983).
311. M. F. Vergnes and J. Alary, Talanta, 33:997 (1986).
312. I. Stewart, J. Agric. Food Chem., 33:1163 (1985).
313. F. W. Karasek and H. Y. Tong, J. Chromatogr., 332:169 (1985).
314. F. T. Delbeke and M. Debackere, J. Chromatogr., 278:418 (1983).
315. F. T. Delbeke and M. Debackere, J. Chromatogr., 325:304 (1985).
316. B. Buszewski, K. Sebekova, P. Bozek, and D. Berek, J. Chromatogr., 367:171 (1986).
317. T. B. Vree, L. Riemens, and P. M. Koopman-Kimenai, J. Chromatogr., 428:311 (1988).
318. D. A. Dezaro, D. Dvorn, C. Horn, and R. A. Hartwick, Chromatographia, 20:87 (1985).
319. K. K. Verma, S. K. Sanghi, A. Jain, and D. Gupta, J. Pharm. Sci., 76:551 (1987).
320. J. C. Gfeller, R. Haas, J. M. Troendlé, and F. Erni, J. Chromatogr., 294:247 (1984).
321. S. A. Ashoor, G. J. Seperick, W. C. Monte, and J. Welty, J. Assoc. Off. Anal. Chem., 66:606 (1983).
322. Y.-B. Yang, F. Hevajans, and M. Verzele, Chromatographia, 20:735 (1985).
323. M. Dulitzky, E. de la Teja, and H. F. Lewis, J. Chromatogr., 317:403 (1984).
324. T. A. Tyler, J. Assoc. Off. Anal. Chem., 67:745 (1984).
325. F. Belliardo, A. Martelli, and M. G. Valle, Z. Lebensmitt. Unter. Forsch., 180:398 (1985).
326. H. Terada and Y. Sakabe, J. Chromatogr., 291:453 (1984).
327. W. J. Hurst, K. P. Snyder, and R. A. Martin, Jr., J. Chromatogr., 318:408 (1985).
328. Y.-N. Kim and P. R. Brown, J. Liq. Chromatogr., 10:2411 (1987).

Applications

329. A. Wahlländer, E. Renner, and G. Karlaganis, J. Chromatogr., 338:369 (1985).
330. B. Stavric, R. Klassen, and S. G. Gilbert, J. Chromatogr., 310:107 (1984).
331. R. Hartley, J. R. Cookman, and I. J. Smith, J. Chromatogr., 306:191 (1984).
332. R. Hartley, I. J. Smith, and J. R. Cookman, J. Chromatogr., 342:105 (1985).
333. M. J. Arnaud, Drug Metab. Disp., 13:471 (1985).
334. N. R. Scott, J. Chakraborty, and V. Marks, J. Chromatogr., 375:321 (1986).
335. S. H. Y. Wong, N. Marzouk, O. Aziz, ans S. Sheeran, J. Liq. Chromatogr., 10:491 (1987).
336. N. Grgurinovich, J. Chromatogr., 380:431 (1986).
337. B. K. Tang, T. Zubovits, and W. Kalow, J. Chromatogr., 375:170 (1986).
338. B. R. Dorrbecker, S. H. Merick, and P. A. Kramer, J. Chromatogr., 336:293 (1984).
339. M. B. Regazzi, R. Rondanelli, M. Calvi, E. Bre, and E. Restelli, Il Farmaco (Ed. Pr.), 40:259 (1985).
340. K. D. R. Setchell, M. B. Welsh, M. J. Klooster, F. W. Balistreri, and C. K. Lim, J. Chromatogr., 385:267 (1987).
341. E. Naline, B. Flouvat, C. Advenier, and M. Pays, J. Chromatogr., 419:177 (1987).
342. M. V. St-Pierre, A. Tesoro, M. Spino, and S. M. MacLeod, J. Liq. Chromatogr., 7:1593 (1984).
343. M. B. Kester, C. L. Saccar, M. L. Rocci, Jr., and H. C. Mansmann, Jr., J. Chromatogr., 380:99 (1986).
344. L. A. Ferron, J.-P. Weber, J. Hebert, and P. M. Bedard, Clin. Chem., 31:1415 (1985).
345. Y. H. Park, C. Goshorn, and O. Hinsvark, J. Chromatogr., 343:359 (1985).
346. G. Schumann, I. Isberner, and M. Oellerich, Fresenius Z. Anal. Chem., 317:677 (1984).
347. J. L. Bock, S. Lam, and A. Karmen, J. Chromatogr., 308:354 (1984).
348. Th. Dülffer, E. Hägele, and U. Herrmann, Fresenius Z. Anal. Chem., 324:327 (1986).
349. D. C. Turnell and J. D. H. Cooper, J. Chromatogr., 395:613 (1987).
350. J.-D. Huang, J. Pharm. Sci., 76:525 (1987).
351. G. E. Miller, L. L. Radulovic, and R. H. Dewit, Drug Metab. Disp., 12:154 (1984).
352. K. Sugiyama, M. Saito, T. Hondo, and M. Senda, J. Chromatogr., 332:107 (1985).
353. A. J. Berry, D. E. Games, and J. R. Perkins, J. Chromatogr., 363:147 (1986).

354. S. Yuen and C. A. Lau-Cam, J. Chromatogr., 329:107 (1985).
355. A.-M. Sjöberg and J. Rajama, J. Chromatogr., 295:291 (1984).
356. M. M. Mangino and R. A. Scanlan, J. Agric. Food Chem., 33:699 (1985).
357. M. Kneczke, J. Chromatogr., 198:529 (1980).
358. L. K. Unni and S. M. Somani, Drug Metab. Disp., 14:183 (1986).
359. S. M. Somani and A. Khalique, Drug Metab. Disp., 15:627 (1987).
360. S. M. Somani and A. Khalique, J. Anal. Toxicol., 9:71 (1985).
361. R. R. Brodie, L. F. Chasseaud, and A. D. Robbins, J. Chromatogr., 415:423 (1987).
362. R. Whelpton, J. Chromatogr., 272:216 (1983).
363. R. Whelpton and T. Moore, J. Chromatogr., 341:361 (1985).
364. K. Isaksson and P. T. Kissinger, J. Chromatogr., 419:165 (1987).
365. T. R. Bosin and K. F. Faull, J. Chromatogr., 428:229 (1988).
366. M. C. Pietrogrande, P. A. Borea, G. Lodi, and C. Bighi, Chromatographia, 23:713 (1987).
367. A. K. Bashir, M. S. F. Ross, and T. D. Turner, Fitoterapia, 57:190 (1986).
368. J. C. Cavin and E. Rodriguez, J. Chromatogr., 447:432 (1988).
369. D. J. Tweedie and M. D. Burke, Drug Metab. Disp., 15:74 (1987).
370. C. A. Erdelmeier, A. D. Kinghorn, and N. R. Rarnworth, J. Chromatogr., 389:345 (1987).
371. A. L. Christiansen and K. E. Rasmussen, J. Chromatogr., 270:293 (1983).
372. M. Wurst, M. Semerdžieva, and J. Vokoun, J. Chromatogr., 286:229 (1984).
373. R. Kysilka, M. Wurst, V. Pacáková, K. Štulík, and L. Haškovec, J. Chromatogr., 320:414 (1985).
374. M. Semerdžieva, M. Wurst, T. Koza, and J. Gartz, Planta Med., 1986:83.
375. M. Perkal, G. L. Blackman, A. L. Ottrey, and L. K. Turner, J. Chromatogr., 196:180 (1980).
376. R. Vanhaelen-Fastré and M. Vanhaelen, J. Chromatogr., 312:467 (1984).
377. J. Gartz, Pharmazie, 40:134 (1985).
378. A. Akbari, A. D. Jernigan, P. B. Bush, and N. H. Booth, J. Chromatogr., 361:400 (1986).
379. J. A. Owen, S. L. Nakatsu, M. Condra, D. H. Surridge, J. Fenemore, and A. Morales, J. Chromatogr., 342:333 (1985).

Applications

380. B. Diquet, L. Doare, and G. Gaudel, J. Chromatogr., 311: 449 (1984).
381. M. R. Goldberg, L. Speier, and D. Robertson, J. Liq. Chromatogr., 7:1003 (1984).
382. U. R. Cleri, J. Assoc. Off. Anal. Chem., 68:542 (1985).
383. U. R. Cieri, J. Assoc. Off. Anal. Chem., 66:867 (1983).
384. J. R. Lang, I. L. Honigberg, and J. T. Stewart, J. Chromatogr., 252:288 (1983).
385. J. R. Lang, J. T. Stewart, and I. L. Honigberg, J. Chromatogr., 264:144 (1983).
386. J. Wang and M. Bonakdar, J. Chromatogr., 382:343 (1986).
387. L. Gagliardi, G. Cavazzutti, A. Amato, and V. Zagarese, Anal. Lett., 17:423 (1984).
388. P. Duez, S. Chamart, M. Vanhaelen-Fastré, M. Hanocq, and L. Molle, J. Chromatogr., 356:334 (1986).
389. S. E. Manoilov, V. P. Komov, N. V. Kirillova, M. A. Strelkova, Yu. A. Samolin, L. S. Chernoborodova, and L. A. Nicolaeva, J. Chromatogr., 440:53 (1988).
390. R. L. Munier, P. Le Xuan, A. M. Drapier, and S. Meunier, Bull. Soc. Chim. Fr., 1982:250.
391. M. Weissenberg, A. Levy, I. Schaeffler, and E. C. Levy, J. Chromatogr., 452:485 (1988).
392. J.-P. Renaudin, J. Chromatogr., 291:165 (1984).
393. K. Hirata, A. Yamanaka, N. Kurano, K. Miyamoto, and Y. Miura, Agric. Biol. Chem., 51:1311 (1987).
394. W. Kohl, B. Witte, and G. Höfle, Planta Med., 47:177 (1983).
395. P. Morris, Planta Med., 1986:121.
396. T. Naaranlahti, N. Nordström, A. Huhtikangas, and M. Lounasmaa, J. Chromatogr., 410:488 (1987).
397. M. Verzele, L. de Taeye, J. van Dyck, G. de Decker, and C. de Pauw, J. Chromatogr., 214:95 (1981).
398. D. Drapeau, H. W. Blanch, and C. R. Wilke, J. Chromatogr., 390:297 (1987).
399. Y. Miura, K. Hirata, and N. Kurano, Agric. Biol. Chem., 51:611 (1987).
400. M. De Smet, S. J. P. Van Belle, G. A. Storme, and D. L. Massart, J. Chromatogr., 345:309 (1985).
401. M. De Smet, S. J. P. Van Belle, V. Seneca, G. A. Storme, and D. L. Massart, J. Chromatogr., 416:375 (1987).
402. D. E. M. M. Vendrig, J. Teeuwsen, and J. J. M. Holthuis, J. Chromatogr., 424:83 (1988).
403. J. A. Houghton, P. M. Torrance, and P. J. Houghton, Anal. Biochem., 134:450 (1983).
404. M. E. Spearman, R. M. Goodwin, and D. Kan, Drug Metab. Disp., 15:640 (1987).
405. V. S. Sethi and K. N. Thimmaiah, Cancer Res., 45:5382 (1985).

406. K. N. Thimmaiah and V. S. Sethi, Cancer Res., 45:5386 (1985).
407. P. Perera, T. A. van Beek, and R. Verpoorte, J. Chromatogr., 285:214 (1984).
408. P. Perera, G. Samuelsson, T. A. van Beek, and R. Verpoorte, Planta Med., 47:148 (1983).
409. P. Perera, F. Sandberg, T. A. van Beek, and R. Verpoorte, Planta Med., 49:28 (1983).
410. T. A. van Beek, R. Verpoorte, and A. Baerheim-Svendsen, Planta Med., 47:83 (1983).
411. T. A. van Beek, R. Verpoorte, and A. Baerheim-Svendsen, J. Nat. Prod., 48:400 (1985).
412. T. A. van Beek and R. Verpoorte, Fitoterapia, 56:304 (1985).
413. T. A. van Beek, R. Verpoorte, and A. Baerheim-Svendsen, J. Chromatogr., 298:289 (1984).
414. Y. Michotte and D. L. Massart, J. Chromatogr., 344:367 (1985).
415. W. Hammes and R. Weyhenmeyer, J. Chromatogr., 413:264 (1987).
416. B. Zsadon. L. Décsei, M. Szilasi, F. Tüdös, and J. Szejtli, J. Chromatogr., 270:127 (1983).
417. G. Szepesi, M. Gazdag, and R. Iváncsics, J. Chromatogr., 241:153 (1982).
418. G. Szepesi, M. Gazdag, and R. Iváncsics, J. Chromatogr., 244:33 (1982).
419. G. Szepesi and M. Gazdag, J. Chromatogr., 204:341 (1981).
420. G. Szepesi and M. Gazdag, J. Chromatogr., 205:57 (1981).
421. A. Amato, G. Cavazzutti, L. Gagliardi, M. Profili, V. Zagarese, F. Chimenti, D. Tonelli, and E. Gattavecchia, J. Chromatogr., 270:387 (1983).
422. R. M. Smyth, Analyst (London), 111:851 (1986).
423. P. K. Lala, Fitoterapia, 56:284 (1985).
424. M. Bogusz, J. Wijsbeek, J. P. Franke, and R. A. De Zeeuw, J. HRC&CC, 6:40 (1983).
425. G. M. Iskander, J. Strombom, and A. M. Satti, J. Liq. Chromatogr., 5:1481 (1982).
426. J. Hayakawa, N. Noda, S. Yamada, and K. Ucho, Yakugaku Zasshi, 104:57 (1984); CA 100:180156v (1984).
427. R. Dennis, J. Pharm. Pharmacol., 36:332 (1984).
428. A. A. M. Wahbi, M. A. Abounassif, and E. R. A. Gad-Kariem, Arch. Pharm. Chem., Sci. Ed., 15:87 (1987); CA, 107: 192389f (1987).
429. L. Alliot, G. Bryant, and P. S. Guth, J. Chromatogr., 232:440 (1982).
430. H. Li and J. Zhou, Yaowu Fenxi Zazhi, 2:326 (1982); CA, 98:166957r (1983).

Applications

431. H. Li and J. Zhou, Yaowu Fenxi Zazhi, 2:330 (1982); CA, 98:204487x (1983).
432. L. T. Hunter and R. E. Creekmur, Jr., J. Assoc. Off. Anal. Chem., 67:542 (1984).
433. J. J. L. Hoogenboom and C. R. Rammell, J. Assoc. Off. Anal. Chem., 68:1131 (1985).
434. M. MIshima, Y. Tanimoto, K. Oguri, and H. Yoshimura, Drug Metab. Disp., 13:716 (1985).
435. A. Hermans-Lokkerbol and R. Verpoorte, Planta Med., 1986: 299.
436. M. Hong, Yaowu Fenxi Zazhi, 3:79 (1983); CA, 99:76952b (1983).
437. M. Hong, S. Gao, and B. Wang, Yaowu Fenxi Zazhi, 3:155 (1983); CA, 99:146174a (1983).
438. Y. Wang, R. Yu, Q. Yang, T. Zhao, and S. Dong, Nanjing Yaoxueyuan Xuebao, 16:44 (1985); CA, 104:95574t (1986).
439. D. Gaudy and A. Peuch, J. Pharm. Belg., 42:251 (1987); CA, 108:62540h (1988).
440. L. Szepesy, I. Fehèr, G. Szepesi, and M. Gazdag, J. Chromatogr., 149:271 (1978).
441. T. Irie, G. Idzu, Y. Hashimoto, M. Ishibashi, and H. Miyazaki, Yakugaku Zasshi, 106:900 (1986).
442. H. Magg and K. Ballschmiter, J. Chromatogr., 331:245 (1985).
443. R. Gill and J. A. Key, J. Chromatogr., 346:423 (1985).
444. M. Wurst, M. Flieger, and Z. Reháček, J. Chromatogr., 174:401 (1979).
445. H. Pötter, M. Hülm, and Schumann, J. Chromatogr., 319:440 (1985).
446. P. J. Twitchett, S. M. Fletcher, A. T. Sullivan, and A. C. Moffat, J. Chromatogr., 150:73 (1978).
447. J. C. Young, Z.-J. Chen, and R. R. Marquardt, J. Agric. Food Chem., 31:413 (1983).
448. G. M. Ware, A. S. Carman, O. J. Francis, and S. S. Kuan, J. Assoc. Off. Anal. Chem., 69:697 (1986).
449. J. P. Chervet and D. Plas, J. Chromatogr., 295:282 (1984).
450. P. Spiegl and H. Viernstein, J. Chromatogr., 294:452 (1984).
451. L. Zecca, L. Bonnini, and S. R. Bareggi, J. Chromatogr., 272:401 (1983).
452. M. Žorz, A. Marušič, R. Smerkolj, and M. Prošek, J. HRC&CC, 6:306 (1983).
453. H. S. Nichols, W. H. Anderson, and D. T. Stafford, J. HRC&CC, 6:101 (1983).
454. J. DeRuyter, F. T. Noggle, Jr., and C. R. Clark, J. Liq. Chromatogr., 10:3481 (1987).
455. M. Wurst, M. Flieger, and Z. Řeháček, J. Chromatogr., 150:477 (1978).

456. C. Eckers, D. E. Games, D. N. B. Mallen, and B. P. Swann, Biomed. Mass Spectr., 9:162 (1982).
457. G. Szepesi, M. Gazdag, R. Iváncsics, K. Mihályfi, and P. Kovács, Pharmazie, 38:94 (1983).
458. A. S. Sidhu, J. M. Kennedy, and S. Deeble, J. Chromatogr., 391:233 (1987).
459. D. L. Dunn and R. E. Thompson, J. Chromatogr., 264:264 (1983).
460. D. L. Dunn, B. S. Scott, and E. D. Dorsey, J. Pharm. Sci., 70:446 (1981).
461. H. Bundgaard and S. H. Hansen, Int. J. Pharm., 10:281 (1982).
462. H. Bundgaard, E. Falch, C. Larsen, and T. J. Mikkelson, J. Pharm. Sci., 75:36 (1986).
463. H. Bundgaard, E. Falch, C. Larsen, G. L. Mosher, and T. J. Mikkelson, J. Pharm. Sci., 75:775 (1986).
464. A. Noordam, K. Waliszewski, C. Ollieman, L. Maat, and H. C. Beyerman, J. Chromatogr., 153:271 (1978).
465. J. J. O'Donnell, R. Sandman, and M. V. Drake, J. Pharm. Sci., 69:1096 (1980).
466. A. Noordam, L. Maat, and H. C. Beyerman, J. Pharm. Sci., 70:96 (1981).
467. H. Porst and L. Kny, Pharmazie, 40:23 (1985).
468. J. M. Kennedy and P. E. McNamara, J. Chromatogr., 212:331 (1981).
469. J. L. Van Ackeren, R. M. Venable, and I. W. Wainer, J. Assoc. Off. Anal. Chem., 67:924 (1984).
470. S. Yoshika, Y. Aso, T. Shibazaki, and M. Uchiyama, Chem. Pharm. Bull., 34:4280 (1986).
471. C. Durif, M. Ribes, G. Kister, and A. Peuch, Pharm. Acta Helv., 61:135 (1986).
472. R. J. Bushway, D. F. McGann, and A. A. Bushway, J. Agric. Food Chem., 32:548 (1984).
473. W. M. J. van Gelder, J. Chromatogr., 331:285 (1985).
474. W. M. J. van Gelder, H. H. Jonker, H. J. Huizing, and J. J. C. Scheffer, J. Chromatogr., 442:133 (1988).
475. P. Duez, S. Chamart, M. Vanhaelen, R. Vanhaelen-Fastré, M. Hanocq, and L. Molle, J. Chromatogr., 351:140 (1986).
476. P. Kulanthaivel and S. W. Pelletier, J. Chromatogr., 402:366 (1987).
477. H. Bando, K. Wada, M. Watanabe, T. Mori, and T. Amiya, Chem. Pharm. Bull., 33:4717 (1985).
478. W. Majak, R. E. McDiarmid, and M. H. Benn, J. Agric. Food Chem., 35:800 (1987).
479. I. Kubo, M. Kim, and G. de Boer, J. Chromatogr., 402:354 (1987).

Applications

480. R. J. Bushway, E. S. Barden, A. W. Bushway, and A. A. Bushway, J. Chromatogr., 178:533 (1979).
481. R. J. Bushway, J. L. Bureau, and J. King, J. Agric. Food Chem., 34:277 (1986).
482. A. C. Eldridge and M. E. Hockridge, J. Agric. Food Chem., 31:1218 (1983).
483. A. S. Carman, Jr., S. S. Kuan, G. M. Ware, O. J. Francis, Jr., and G. P. Kirschenheuter, J. Agric. Food Chem., 34: 279 (1986).
484. I. Csiky and L. Hansson, J. Liq. Chromatogr., 9:875 (1986).
485. G. Holan, W. M. P. Johnson, and K. Rihs, J. Chromatogr., 288:479 (1984).
486. C. A. Browne, F. R. Sim, I. D. Rae, and R. F. Keeler, J. Chromatogr., 336:211 (1984).
487. J. K. Reed, J. Gerrie, and K. L. Reed, J. Chromatogr., 356:450 (1986).
488. M. Sarsunová, V. Varinská, and K. Schmidt, Farm. Obz., 55:495 (1986).
489. H. K. Desai, B. S. Joshi, A. M. Panu, and S. W. Pelletier, J. Chromatogr., 322:223 (1985).
490. H. K. Desai, E. R. Trumbull, and S. W. Pelletier, J. Chromatogr., 366:439 (1986).

8
Conclusion

Here we will evaluate the status quo in the chromatographic analysis of alkaloids from the point of view of techniques and instrumentation used.

As has been already stated, liquid column chromatography holds the key position in alkaloid determination. It can be estimated that about 60-70% of all separations are presently performed using this technique. Of the four modes of liquid column chromatography which have been treated in this book (adsorption, reversed phase partition, ion exchange, and ion pair), only three are commonly used. If we attempt to express the relative abundance of single modes, then about 60% will belong to reversed phase partition, 25% to adsorption, and 15% to ion pair chromatography. Moreover, it seems that the dominant role of RP-LC is increasing— probably due to the flexibility of RP-LC systems, which is dependent on type and concentration of organic modifier along with the selection of concentration and pH of buffer. The use of buffers is especially advantageous in the chromatography of alkaloids, where the operation in acid, transient, or alkaline regions is possible.
It is probably also the reproducibility of RP-LC separation system that makes them attractive, although some objections must be raised concerning batch-to-batch reproducibility of RP packings. Looking at a detail of RP-LC experimentation, a movement toward the acidic region can be noticed because there, usually, a higher column efficiency can be achieved. Most problematic in this operation is the transient region, where the use of a buffer with a large buffering capacity is essential.

The separations on classical adsorbents are not so frequent in today's practice, but some of their inherent advantages are noteworthy. Silica gel is one of the cheapest packings, which is im-

Conclusion

portant in mass applications, with efficiency usually slightly better than that of materials obtained by its chemical derivatization. The variability of separation systems based on silica is remarkable ranging from adsorption chromatography of alkaloids as free bases through ion pair adsorption with the acidified mobile phase to ion exchange of both straight and reversed phase types. In this way different requirements set up by the type of sample and its pretreatment may be fulfilled. When used with methanolic mobile phases the same types of samples as in reversed phase chromatography can be analyzed. Difficulties in water content adjustement in the mobile and adsorbent phases are not so detrimental in the case of alkaloids owing to the mobile phases used. The importance of defining pH and salt content in the silica gel phase prior to the separations is now well recognized. The properties of alumina in high-performance chromatography, which are promising in many aspects, require re-evaluation.

The flexibility of ion pair chromatography is attributable to the possibility of effectively controlling the separation with small amounts of pairing reagent. The method retains all the advantages of reversed phase separations. By choosing the type of pairing reagent and its concentration, the retention of solutes can be changed in a relatively predictable manner. The role of other ionic components of the mobile phase should be kept in mind. The further elucidation of the mechanism involved is required.

In terms of the instrumentation, the columns should be mentioned first. For the most part, their dimensions vary within narrow limits—the length from 15 to 25 cm, the inner diameter from 4 to 5 mm. Sometimes short columns have been used in high-speed chromatography, but the application of microbored columns has been very rare. Practical separations often deal with plant extracts or biological samples, and these types of sample represent a great risk of column deterioration. Naturally, the risk increases with decreasing column diamater.

The detectors used in column liquid chromatography have undergone rapid development in recent years. Besides the diode array detector, which is becoming popular for alkaloid detection, voltammetric detectors have proven their suitability. Their high sensitivity, their compatibility with mobile phases used in RP-LC, and a certain selectivity obtained by either voltage selection or mobile phase composition adjustment have rendered voltammetric detectors the second most popular kind used.

A characteristic feature of thin layer chromatography (TLC) is its application to the identification of plant extracts and pharmaceuticals of plant origin. This is because the possibilities of TLC from the standpoint of qualitative analysis, e.g., several methods of detection successively used, color reaction in situ, two-dimensional development, are unsurpassed, and moreover the analysis is substantially cheaper. However, the good TLC scanners for quantitative analysis are expensive enough and give the linear response in a narrow range only, usually in one or two orders of solute concentration. Quantitative TLC should therefore be limited to samples where the concentrations of components of interest do not vary too much. This may be the case for alkaloids in plant extracts and pharmaceuticals.

Gas chromatography has its own limitations based on the volatility and thermal instability of analyzed alkaloids. However, it should be noted that this can often be overcome by sample derivatization or another operation, sometimes changing the entire solute structure, but the utility of such an operation is in question. Based on the author's assessment, it is better to analyze the compound as such if possible. There are sufficient opportunities to apply gas chromatography to the analysis of alkaloids where no or straight and rapid sample derivatization is possible. Many compounds of interest from a biochemical or forensic point of view, e.g., atropine, caffeine, cocaine, codeine, nicotine, etc., can be directly analyzed using GC. Owing to developments in capillary GC (splitless injection, immobilized stationary phases), fairly complex samples may be quantitatively analyzed with high sensitivity and accuracy, and without the necessity of removing the endogenous compounds too carefully. With the exception of very simple samples, such as some pharmaceuticals, capillary GC applications will predominate in the years to come.

Among other chromatographic methods, supercritical fluid chromatography has a chance of coming to wider use for alkaloid determination. Despite the fact that this method is not yet in common use, it may be applied first of all to the analysis of compounds of high molecular weight, e.g., glycoalkaloids, or to the determination of alkaloids that are difficult to detect by LC, e.g, some tropane alkaloids. The advantage of coupling supercritical fluid chromatography with mass spectrometry is evident and so, wherever possible from the standpoint of investment expenses, this combination should be used for the identification and quantitation of alkaloids.

Index

Absorption spectra:
  of alkaloids, 51-57
  of pairing reagents, 218
  of solvents, 172
  of sorbents for TLC, 286
Accacia prominens, 2
4-Acetoxy-3,6-dimethoxy-5-[2-(N-methylacetamido)-ethylphenanthrene], 428-430
Acetylaconitine, 555
17-O-Acetylajmaline, 523
5-Acetylamino-6-amino-3-methyluracil, 476
Acetylcodeine:
  by GC, 427
  by LC, 157, 158, 431-434, 586
  by TLC, 435
Acetylcorynoline, 294, 393
14-Acetyldelcosine, 300, 560
N-Acetyl-N-desethylaconitine, 556
Acetylechimidine, 452
7-Acetylechinatine, 451
O-Acetylharmol, 495
N-Acetylhydronornarceine, 430
Acetylisocorynoline, 393
Acetyllycopsamine, 296

3-Acetylmorphine:
  by GC, 427, 428, 585
  by LC, 433, 586
  in heroin, 427, 428, 433, 585
6-Acetylmorphine:
  in biological fluids, 401, 412, 434, 586
  derivatization to fluorescent product, 240, 241, 585
  by GC, 106, 401, 405, 427, 428, 584, 585
  in heroin, 420, 427, 428, 431-435, 584, 585
  by LC, 157, 201, 202, 412, 413, 420, 431-434
  in tissues, 405
  by TLC, 435
6-Acetylnalorphine, 401
17-O-Acetylnorajmaline, 523
6-Acetylnorcodeine, 428
O-Acetylnorharmol, 495
N-Acetylnorlaudanosine, 430
6-Acetylnormorphine, 427, 428
N-Acetylnornarcotine, 430
O-Acetylseneciphylline, 116, 296, 450
Acetylthebaol, 428, 430, 432
Aconitine, 22
  by LC, 555, 557

641

[Aconitine]
  in pharmaceuticals, 300, 560
  in plants, 557
  by TLC, 300, 560
Aconitum, 613
  Aconitum crassicaule, 555, 615
  A. falconeri, 556
  A. japonicum, 556
  A. napellus, 22
  A. yesoense, 22
Acylation:
  of alkaloids, 102
  of cocaine, 111
  of ecgonine, 111
  of morphine and codeine, 115
  of pilocarpine, 124
Adhatoda vasica, 453, 589
Adlumine, 398
Adsorption:
  on alumina, 154-159
  chromatography, 136-159
  coefficient, 136
  multisite on silica gel, 144
  on silica gel, 137-154
Agroclavine, 545, 546
Ajmalicine, 18
  by GC, 121
  by LC, 505, 507-509, 511, 603
  in plants, 121, 505, 507-509, 522-524, 603
  by TLC, 298, 522-524, 603
Ajmaline, 18
  by GC, 121, 601
  by LC, 505, 507, 510, 602, 603
  in pharmaceuticals, 507, 602, 603
  in plants, 121, 398, 505, 507, 510, 522, 523, 603
  by TLC, 398, 522, 523, 603
Akuammigine, 509
Alanginarckine, 461, 591
Alcuronium, 395, 575
Alfalfa, 121
Aljesaconitine A, 556

Alkaloids:
  boiling points, 42-48
  dissociation properties, 25-41
    in amphiprotic solvents, 37-41, 176-180
  electrochemical properties, 57-65, 244, 245
  melting points, 42-48
  physicochemical properties, 41-49
  pK_a values, 28, 29
  solubilities, 42-48
  spectral data UV-vis, 49-57
Alkylation:
  of alkaloids, 101
  of Cinchona alkaloids, 113
  of theobromine and theophylline, 119
Allococaine:
  by GC, 366, 566
  by LC, 370, 567, 569
Allocryptopine:
  by GC, 386, 574
  by LC, 392, 393
  in plants, 293, 386, 392, 393, 397, 574
  by TLC, 293, 294, 397
Alloecgonine, 366, 566
Allopseudococaine:
  by GC, 366, 566
  by LC, 370, 567, 569
Allopseudoecgonine, 366, 566
Alstonine, 17, 398, 603
Alumina:
  characterization, 154
  zero point of charge, 155
Amanita citrina, 321, 499
Amaryllidaceae alkaloids, 448-461, 592
  by GC, 117
  UV-vis absorption, 55
Amaryllis belladonna, 14
Ambelline, 117
6-Amino-5-(N-formylmethylamino)-1,3-dimethyluracil, 474, 488
6-Amino-5-(N-formylmethylamino)-1-methyluracil, 487

Index

6-Amino-5-(N-formylmethyl-amino)-3-methyluracil, 474
Ammodendrine, 461
Amurine, 390
Amurinine, 390
Anabasine:
 by GC, 105, 108, 354
 by TLC, 289
 in tobacco, 108, 354
Anagyrine, 117, 461
Analysis:
 qualitative GC, 125-127
 quantitative GC, 127-130
  external standard, 127
  internal standard, 127, 246
  standard addition, 127, 128
 quantitative LC, 245-247
  internal standard, 246
 quantitative TLC, 282-287
Anatabine, 105, 108, 354
Anchusa officinalis, 451
7-Angelyl/tiglyl dihydroechinatine, 451, 452
7-Angelyl/tiglyl echinatine, 451
β-Angelyl/tiglyl echinatine, 451
β-Angelyl/tiglyl trachelanthamine, 451
Angustifoline, 118, 452, 453
Anhalamine, 9, 104
Anhalonium lewinii, 2, 9
Anisodine, 370
Anisodus tangulicus, 370, 566
Anonamine, 458, 592
Apoatropine, 189, 365, 369
Apocodeine, 224
Apomorphine, 147, 224
Aporphine alkaloids, 113, 574
Apovincamine:
 enantiomers separation, 517, 518
 by LC, 145, 228, 520
Apovincaminic acid, 120
Apovincaminic acid ethyl ester (see Vinpocetine)
Apovincaminic acid methyl ester, 502

Apparicine:
 by CCC, 521
 by LC, 516, 517, 604
 in plants, 514, 517, 521, 604
 by TLC, 299
Areca catechu, 4
Arecaidine, 4, 235
Arecoline, 4, 235
Aricine, 17, 522
Aspidospermidine, 517
Aspidospermidine-type alkaloids, 605
Atropa belladonna, 6, 109, 289, 365, 368, 369, 566
Atropamine, 7
Atropine, 6
 dissociation constant, 28, 39
 by GC, 109, 365, 566
 by LC, 189, 368, 369, 423, 549, 566
 in pharmaceuticals, 289, 365, 368, 369, 549, 566
 in plants, 365, 369, 566
 thermal decomposition, 109
 by TLC, 289, 290

Baeocystin:
 by LC, 496, 498, 499, 601
 in mushrooms, 496, 498, 499, 500, 601
 by TLC, 500, 601
Balfourodinium, 375, 383, 568
Baphia racemosa, 448
Barley malt, 598
Belladonnine, 7
Benzoylaconine, 556
Benzoylecgonine, 7
 in biological fluids, 328, 367, 567
 electrochemical detection, 567
 by GC, 110, 111, 367, 567
 by LC, 371, 372, 567
 in street drugs, 371, 372
Benzoyltropeine, 7
Benzoyloxyurea, 476
Berberine, 10

[Berberine]
  by CCC, 531
  detection:
    fluorescence, 238
    UV-vis, 235
  by GC, 113
  by LC, 226, 389, 392, 574
  in pharmaceuticals, 226, 292, 294
  in plants, 226, 292, 389, 392, 574
  by TLC, 292, 294, 574
  in urine, 113

<u>Berberis</u>, 9, 292
Bicuculline, 294
Bisbenzylisoquinoline alkaloids, 386-398, 575
Boldine, 239, 393
Bromocryptine, 546
Bromovasicinone, 453
Browniine, 561
Brucine, 19
  by CCC, 531, 606
  by GC, 121, 605
  by LC, 393, 528
  in plants, 121, 393, 528, 605
  in pharmaceuticals, 121, 528, 532, 605
  by TLC, 532, 606
Brucine-N-oxide, 532
Bufotenine, 15, 120, 499
6-Butanoylmorphine, 405, 579

Caffeine (1,3,7-trimethylxanthine), 15, 462-489
  in biological fluids, 119, 201, 289, 297, 330, 331, 350, 381, 394, 592, 595-597
  in coffee, 201, 205
  in foodstuff, 341, 592, 594, 595, 598
  by GC, 106, 118, 119, 355, 452, 453, 527, 592
  in heroin, 428, 435
  by LC, 149, 157, 158, 190,

[Caffeine]
  193, 200, 201, 205, 348-350, 372, 381, 392, 394, 406, 412, 420, 425, 433, 585, 586, 592-597
  metabolites, 595, 596
  by paper chromatography, 598
  in pharmaceuticals, 312, 349, 406, 425, 593, 594, 598
  in plants, 119, 392, 592, 594
  by SFC, 154, 372, 567, 597
  by TLC, 278, 289, 297, 353, 435, 598

Calycotomine, 461
Canadine, 10, 293, 391
Candicine, 3
Canthin-2,6-dione, 499
Canthinone-type alkaloids, 601
Capacity ratio, 73, 83
  increments for organic carbon atoms, 152, 163
  relation to R_M value, 272
Capaurine, 391
Capsaicin, 362, 363
<u>Capsicum</u>, 362, 364, 565
Carbazole, 448, 454, 459
Carbazole alkaloids, 589
  by GC, 115, 448
  by LC, 454
  by TLC, 282, 296, 459
β-Carbolines:
  by LC, 493, 599, 601
  by TLC, 500, 599
<u>Castanospermum australe</u>, 448
<u>Catha edulis</u>, 557
Catharanthine, 508-512, 524
<u>Catharanthus roseus</u>, 298, 507-512, 603
Cathedulines E3, E4, and E5, 557, 613
C. bungeana, 294, 317, 393, 574
Cephaeline, 239, 295, 395
<u>Cephalis ipecacuanha</u>, 395
Cevadine, 558, 559
Chaconine:
  by LC, 557, 558, 613, 616, 617

Index 645

[Chaconine]
  in potatoes, 193, 557, 558,
    613, 616, 617
Chamoclavine, 545, 546
Chasmanthera dependens, 292,
  397, 574
Chelerythrine, 391, 393
Chelidonine:
  by LC, 391-393, 397
  by TLC, 293, 294, 297
Chelidonine-type alkaloids, 572,
  574
Chelidonium majus, 320, 391,
  392, 397, 574
7-(2-Chloroethyl)theophylline,
  464
8-Chlorotheophylline, 465, 468,
  472, 480, 483, 485, 486
Choisya ternata, 375, 383, 568
Chromophores, 50, 54-55, 234
Cinchona, 8, 195, 374, 378,
  379, 571
Cinchona alkaloids, 8, 374-375,
  568-572
  derivatization, 113
  detection:
    fluorescence, 279
    UV-vis, 54, 235
  diastereomers separation, 112
  by GC, 111-113
  by LC, 147, 148, 169, 184,
    187-189, 191
  by TLC, 290-292
Cinchona ledgeriana, 379-381,
  384, 572
Cinchonamine, 380
Cinchonidine, 8
  by GC, 111-113, 374, 568
  by LC, 148, 149, 187, 188,
    191, 375, 377-381, 569
  in pharmaceuticals, 374, 377
  in plants, 378-381, 571, 572
  by TLC, 291
Cinchonine, 8
  by GC, 111-113, 374, 568
  by LC, 148, 187, 188, 191, 375-
    381, 423, 569, 571, 572

[Cinchonine]
  in pharmaceuticals, 374, 377
  in plants, 378-381, 571, 572
  by TLC, 291
Cinchophylline, 380
Cinnamoylcocaine, 367, 371, 372
Citrus sinensis, 462, 469, 592
Claviceps purpurea, 20, 21, 545,
  546
Clavine alkaloids, 121, 545, 610
Cocaine, 7
  in biological fluids, 111, 328,
    366, 367, 566, 567
  electrochemical detection, 63,
    567, 588
  by GC, 106, 110, 111, 366,
    367, 566
  impurities in illicit, 567
  by LC, 156, 190, 370, 371,
    567, 569
  in pharmaceuticals, 290, 366,
    566, 567
  in plants, 111, 567
  by SFC, 154, 372, 567
  in street drugs, 111, 367, 371,
    372, 435, 443, 566, 567,
    588
  by TLC, 290, 373, 435, 443,
    567, 588
Coca leaves, 367
Cocoa:
  by GC, 119
  by LC, 469, 470, 595
  by paper chromatography,
    489, 598
Codeine, 11, 399-426
  anodic oxidation, 63
  in biological fluids, 115, 336,
    350, 430, 576, 577, 585
  in blood, 115, 334, 335
  on column acetylation, 577
  decomposition be GC, 577
  dissociation constant, 28, 39
  fluorescence detection, 238
  by GC, 106, 114, 115, 427,
    428, 430, 555, 576, 577,
    585, 613

[Codeine]
  in heroin, 427, 428, 431, 432, 435, 585
  metabolites, 577, 579
  by LC, 151, 157, 179, 185, 186, 190, 192, 201, 220, 222, 223, 348, 350, 430-433, 549, 563, 573, 579-586
  in opium, 114, 151, 223, 294, 432, 580, 584
  in pharmaceuticals, 295, 312, 313, 348, 549, 563, 580, 581
  in plants, 582, 583
  by SFC, 154, 372, 567, 580
  by TLC, 294, 295, 435, 562, 580
Codeinone, 420
Coffee:
  by GC, 118, 462, 592
  by LC, 205, 467, 469, 594
  by SFC, 487, 597
Cola spp., 469, 594
Colchiceine, 351
Colchicine, 3
  in biological fluids, 352, 564
  degradation products of, 564
  by LC, 351, 352, 392, 549
  in pharmaceuticals, 288, 353, 549, 564
  in plants, 316, 352, 353, 392, 564
  by TLC, 288, 564
Colchicoside, 316, 352
Colchicum, 316
Colchicum autumnale, 3, 352, 353, 564
Columbamine:
  by LC, 389
  by TLC, 293, 397, 532
Column equilibration:
  in adsorption chromatography, 142, 143, 149
  in ion pair chromatography, 213, 218, 226

[Column equilibration]
  in partition chromatography, 171, 175, 183
Compressibility factor, 70
Conessine:
  by GC, 555, 612, 613
  by TLC, 300, 562, 612, 614, 617
Coniine, 5
Conium maculatum, 4
Coptis chinensis, 292
Coptisine, 226, 294, 389
Coromandaline, 296
Coronaridine, 511, 517
Corybulbine, 391
Corydaline:
  by LC, 225, 389, 391
  by TLC, 294
Corydalis, 9, 294, 389
Corydine, 293
Corynantheidine, 298, 522
Corynanthe yohimbe, 16, 17
Corynantheine, 17, 522
Corynanthidine (see α-Yohimbine)
Corynanthine, 503, 522
d-Corynoline, 294, 317, 393
Cotinine:
  in biological fluids, 109, 204, 355-357, 359-362, 564, 565
  by GC, 105, 108, 109, 355-357
  by LC, 204, 358-362, 565
  in tissues, 108, 357, 362, 564
  in tobacco, 358
Cotinine-N-oxide, 361, 362
Countercurrent chromatography (CCC), 69
  of Strychnos alkaloids, 531, 606
  of Tabernaemontana alkaloids, 521, 604
  of tropane alkaloids, 370, 566
  of veratrine, 559, 615
C. pelletierana, 17
Crassicauline, 556, 615
Crassicausine, 556, 615
Crassicautine, 556, 615

Criglaucine, 117
Crinum glaucum, 117
Cryptopine, 573
  by LC, 190, 419, 421, 422, 424, 432, 584
  in pharmaceuticals, 421
  in opium, 422, 424, 432, 584
C. succirubra, 380, 572
C. succirubra x C. ledgeriana, 380
Cunninghamella fungus, 577
Curassavine, 296
Cuscohygrine, 365
Cusparine, 7
C. yanhusuo, 224
Cytisine, 14
  detection UV-vis, 235
  by GC, 118, 452
  in plants, 118, 452, 461
  by TLC, 461

Datura, 224, 566
D. barbeyi Huth., 560
Dead volume, 70, 84,
  in LC, 73, 160, 181
Dehydrocorydaline, 389
Dehydroemerine, 390
cis-Dehydroepivincamine, 145
Dehydroglaucine, 389, 390, 573, 575
Dehydrolupanine, 117
3,4-Dehydroreserpine, 505
Dehydrovincamine, 145
Delphatine, 561
Delphinine, 555
Delphinium, 613, 617
Delsoline, 300, 560
Demissidine, 554, 555
Dendrobium chrysanthemum, 12
11-Deoxojervine, 559
Deoxyaconitine, 300, 555, 560
Derivatization (see also Acylation, Alkylation, and Trimethylsilylation), 100-103, 286, 287, 295

N-Desacetylcolchicine, 351
Desacetylvinblastine, 513, 514, 515
Desacetylvinblastine-N-oxide, 515
  succinyl derivative of, 515
Desacetylvincristine, 514, 515
  3-deoxyderivative, 516
Deserpidine, 61, 62, 522
O-Desmethylquinidine, 383
Detectors:
  in GC, 97-99
    electron capture, 98
    flame ionization, 97
    nitrogen-phosphorus, 97, 127
  in LC, 229-245
    absorbance, 230-236
    electrochemical, 189, 242-245
    fluorescence, 237-241
    refractive index, 236-237
Diacetylnorcodeine, 428, 429
3,6-Diacetylnormorphine, 427, 428
Diamorphine, 114, 201, 221
3,6-Dibutanoylmorphine, 405
Dictamnine, 8, 235, 374
Dictamnus albus, 8, 567
Didehydroheroin, 428
Didehydronarcotine, 428
Dihydrocapsaicin, 362-364
Dihydrochelerythrine, 144
Dihydrocinchonidine (see Hydrocinchonidine)
Dihydrocinchonine (see Hydrocinchonine)
Dihydrocodeine, 414
Dihydrocorynantheine, 522
10,11-Dihydrodiolquinidine, 382
10,11-Dihydrodiolquinidine glucuronide, 382
10,11-Dihydrodiolquinidine oxide, 382
Dihydroepicinchonidine, 291
Dihydroepicinchonine, 291

Dihydroepiquinidine, 291
Dihydroepiquinine, 291
Dihydroergocornine:
 in biological fluids, 541
 by GC, 534, 607
 by LC, 179, 537-541, 607, 609
 in pharmaceuticals, 534, 537, 539-541, 547, 607, 609
 by TLC, 547, 607
Dihydroergocristine:
 fluorescence detection, 239
 by GC, 534, 607
 by LC, 239, 504, 505, 537-542, 602, 607, 609
 in pharmaceuticals, 504, 505, 534, 537, 539-541, 547, 602, 607, 609
 in plasma, 542
 by TLC, 547, 607
Dihydroergocryptine:
 by GC, 534, 607
 by LC, 537-541, 607, 609
 by TLC, 547, 607
Dihydroergoptine, 539
Dihydroergosine, 543
Dihydroergotamine:
 in biological fluids, 333, 337, 541, 542
 detection, 239, 245
 by LC, 193, 239, 537, 541, 542
Dihydroergotoxine:
 in biological fluids, 534, 543, 607, 610
 degradation products, 610
 by GC, 534, 607
 by LC, 543, 607, 609, 610
 in pharmaceuticals, 609, 610
 by TLC, 607
3,7-Dihydro-1-ethyl-3-(2-hydroxypropyl)-1$\underline{H}$-purine-2,6-dione, 475
6,7-Dihydro-7-hydroxy-1-hydroxymethyl-5$\underline{H}$-pyrrolizidine, 456

Dihydrolysergic acid, 540, 541
Dihydromorphine:
 in biological fluids, 409
 derivatization, 240
 by LC, 409, 411, 413
Dihydromorphinone, 409, 414
Dihydronudaurine, 390
Dihydroquinidine (see Hydroquinidine)
Dihydroquinine (see Hydroquinine)
Dihydroquinoline alkaloids, 568
Dihydrosanquinarine, 398
Dihydroxytetrahydroisoquinolines, 573
3,4-Dimethoxy-β-phenylethylamine, 346
$\underline{N},\underline{N}$'-Dimethylnicotine, 204, 359
1,3-Dimethyluric acid:
 in biological fluids, 120, 472, 474, 479-482, 488, 597
 by GC, 120
 by LC, 472, 474, 479-482, 484, 485, 597
 by TLC, 488
1,7-Dimethyluric acid, 474, 488
1,9-Dimethyluric acid, 479
3,7-Dimethyluric acid, 472, 485, 487
1,7-Dimethylxanthine (see Paraxanthine)
1,3-Dimethylxanthine (see Theophylline)
3,7-Dimethylxanthine (see Theobromine)
D. innoxia, 369, 373, 566
2,3'-Dipyridyl, 354, 355
Dissociation constants, 26
 of alkaloids, 28-29, 32, 37-39
 of buffers, 33-35, 37
Distribution coefficient, 72
Diterpene alkaloids, 554-562, 612-617
 by TLC, 300
 UV-vis absorption, 55
D. metel, 7

Index

D. nuttallium, 557, 561
Dopamine, 386, 387
Doronine, 460
Dregamine, 299
Drugs, 436-447, 573, 587, 589
 antiarrythmic, 146
 antidepressant tricyclic, 152
 by GC, 106, 107, 111, 126,
  127, 436-438, 587, 589
 by LC, 420, 434, 438-443,
  479, 549, 563, 587-589
 by TLC, 443-447, 488, 588,
  589
D. staphisagria, 23, 300, 560
D. stramonium, 369, 373, 566
Dynamic ion exchange, 153, 181,
 209
Dyphilline (Dihydroxypropyl-
 theophylline), 335, 465,
 479, 484

Eburnamonide, 517
Eburnane alkaloids (see Vinca
 alkaloids)
Ecgonine:
 in biological fluids, 367, 567
 detection, 567
 by GC, 110, 111, 366, 367,
  371, 566, 567
Echimidine, 296, 452
Echinatine, 451
Efficiency improvement:
 in adsorption chromatography,
  146, 147, 149
 in ion exchange chromatography, 198, 201-203
 in ion pair chromatography,
  222, 224, 226
 in partition chromatography,
  163, 174, 182-184
Ellipticine, 238, 459, 589
Eluotropic series:
 in adsorption chromatography,
  138, 139
 in ion exchange chromatog-

[Eluotropic series]
  raphy, 197, 202
 in ion pair chromatography,
  217
 in partition chromatography,
  159, 170
Elymoclavine, 545, 546
Emetine, 12
 in biological fluids, 395, 575,
  576
 fluorescence detection, 221,
  239
 by LC, 179, 180, 193, 221,
  395, 575, 576
 by TLC, 295
 TLC of epimers, 461, 591
Emilia sonchifolia, 460
Enantiomer separation:
 of eburnane and other alkaloids, 227, 229, 605
 of ephedrines, 184, 189, 194,
  342, 346
 of nicotine and analogs, 358,
  564
 packings for, 174, 194, 572
 of tetrahydroberberine alkaloids, 391, 574
 by TLC, 268
Enprophylline, 475
Ephedra, 349, 364
E. vulgaris, 2
Ephedrine, 3
 in biological fluids, 336, 351
 detection, 239, 244
 by GC, 104, 106, 107, 342
 GC of diastereoisomers, 342,
  346
 in heroin, 435
 by LC, 104, 106, 107, 149,
  193, 348-351, 372, 423,
  425, 563, 573
 LC of diastereoisomers, 184,
  189, 194, 342, 346, 347
 in pharmaceuticals, 288, 313,
  348, 349, 425, 563
 by TLC, 278, 288, 289, 353,

[Ephedrine]
   435
Epiajmalicine, 508
Epiamurinine, 390
Epicinchonidine, 291
Epicinchonine, 291
Epicorynoline, 393
Epigalathamine, 458
19-Epiiboxygaine, 299, 516, 604
Epilupinine, 118, 452
3-Epiquinamine, 380
Epiquinidine:
   by LC, 376, 378, 569
   by TLC, 291
Epiquinine:
   by LC, 376, 378, 569
   by TLC, 291
Epivincamine, 145, 228, 520
   LC of optical isomers, 229, 518
Epivindolinine, 509
Epivoacangine, 516
19-Epivoacristine, 299, 604
Equation:
   Clausius-Clapeyron, 74
   Hendersson-Hasselbach, 27, 31
   Kubelka-Munk, 283, 285
   rate in GC and LC, 78
   van't Hoff, 172
Ergine, 545
Ergocornine, 21
   by LC, 535-538, 608
   in fermentation product, 538
Ergocorninine, 21
   by LC, 227, 535-537
   in fermentation product, 537
Ergocristine, 21
   in biological fluids, 539
   detection, 239
   by LC, 145, 239, 535-539
   in plants, 537
Ergocristinine, 227, 535-538
Ergocryptine, 21
   by LC, 535-538, 543, 544, 546, 547, 608
   in wheat, 538

Ergocryptinine, 21
   by LC, 227, 535-538
   in wheat, 538
Ergometrine (see Ergonovine)
Ergometrinine (Ergonovinine, Ergobasinine), 535, 537, 538
Ergonovine (Ergometrine, Ergobasine), 20
   by GC, 121
   by LC, 222, 529, 535, 537, 538
   in pharmaceuticals, 222
   in wheat, 538
Ergosine, 21
   by LC, 535, 536, 543, 544
Ergosinine, 21, 535, 536
Ergostine, 535
Ergostinine, 535
Ergot (see also Claviceps purpurea), 537, 538, 606, 607, 610
Ergot alkaloids, 534-547, 598, 606-610
   absorption UV-vis, 50, 52, 55
   in biological fluids, 332
   2-bromo derivatives, 537
   detection, 235, 281
   dihydrogen derivatives, 156
   by GC, 121, 123
   by LC, 190, 192, 193, 227, 518
Ergotamine, 21
   in biological fluids, 539
   detection, 239
   isomerization, 608
   by LC, 145, 239, 535-539
   in plants, 537
   in wheat, 538
Ergotaminine, 21
   in biological fluids, 539
   detection, 239
   isomerization, 608
   by LC, 227, 239, 535-539
   in plants, 537
   in wheat, 538

Index

Ergothioneine, 22, 611
E. rotundifolium, 451
Erythroxylon coca, 7
Eserine (see Physostigmine)
Eseroline, 491-493, 599
Ethylbenzatropine, 369
Ethylmorphine:
   in biological fluids, 401, 410
   by GC, 401
   by LC, 407, 410, 563
   in pharmaceuticals, 563
7-Ethyltheophylline, 471
Etofylline, 485
Eupatorium cannabinum, 451
Eupaverine, 295
Europine, 460
Evoxine, 375
Ezochasmanine, 561

γ-Fagarine, 374
Falconerine, 556
Falconerine acetate, 556
Festuclavine, 545, 546
Foresaconitine, 556, 615
N-Formyldeacetylcolchicine, 289
Forsythia suspensa, 292
Fumaria, 317, 398, 574
Furoquinoline alkaloids, 567, 568
Fusarium solani, 573

Galanthamine:
   in biological fluids, 332, 457, 458, 592
   metabolites, 592
   by LC, 457, 458, 592
Galanthaminone, 458
Galipea officinalis, 7
Garryine, 300, 560
Geissoschizine, 299
Genista anatolica, 461, 591
Glaucine, 9
   in biological fluids, 390, 574
   detection, 239
   by LC, 225, 239, 389, 390,

[Glaucine]
   393, 573, 575
   in plants, 225, 294, 317, 389, 390, 393, 573, 574
   by TLC, 294
Glaucium flavum, 317, 389, 573
Glycoalkaloids (see Steroid alkaloids)
Glycosmis pentaphylla, 448, 454
Glycozolidine, 448, 454, 459
Glycozoline, 448, 454, 459
Gramine, 15, 120, 461
   nitrosation products, 490, 598
Gramodendrine, 453, 461
Grewia villosa, 494, 599
G. squamigerum, 292, 293, 396, 574

Haemanthamine, 14
Haemanthus puniceus, 14
Half-wave potential, 59
Halostachine, 346
Haplophyllum alkaloids:
   catalytic waves, 64
Harmaline:
   detection, 239
   by GC, 121
   by LC, 239, 494
   by TLC, 298
Harmalol, 239, 494
Harman, 16
   in biological samples, 490, 495, 599
   by GC, 121, 490, 599
   by LC, 493-495, 599
   in plants, 494
   by TLC, 298, 500, 599
Harmine, 16
   detection, 239
   by GC, 121
   by LC, 239, 494, 495
   metabolites, 601
   by TLC, 298
Harmol, 239, 494, 495
   metabolites, 601
Heliconiini butterflies, 494, 601

Heliotridine, 460
Heliotropium curassavicum, 296
H. ellipticum, 460
Heptazolidine, 448, 454, 459
Heroin, 427-435, 573, 582, 584, 587
  anodic oxidation, 63
  in biological fluids, 410, 578, 585
  in blood, 150
  by GC, 106, 111, 114, 115, 578, 584, 585
  by LC, 150, 156-158, 190, 191, 201, 221, 410, 420, 581, 585, 586
  by TLC, 294, 295, 443, 585
Heterophyloidine, 561
Higenamine, 387
H. indicum, 13
Hirsutine, 502, 526, 605
Histamine, 239, 338
H. niger, 369, 566
Hoerhammericine, 509
Hokbusin A, 556
Holarrhena floribunda, 300, 555, 562, 613, 614
Homatropine, 368
Homocapsaicin, 362
Hordenine, 598
  by GC, 104, 120
  by LC, 346
  by TLC, 288, 289
Hydrastine, 398
Hydrastinine:
  electrochemical reduction, 61
Hydrocinchonidine:
  by LC, 148, 187, 188, 375-379, 569, 571
  in plants, 378, 379, 571
  by TLC, 291
Hydrocinchonine:
  by LC, 187, 188, 375-379, 381, 569
  in plants, 378, 379, 381
  by TLC, 291
Hydroquinidine:
  in biological fluids, 290, 381

[Hydroquinidine]
  by LC, 148, 187, 188, 375-381, 569, 571
  in pharmaceuticals, 377
  in plants, 378-381, 571
  by TLC, 290, 291
Hydroquinine:
  in biological fluids, 383
  by CCC, 531
  by LC, 148, 187, 188, 375-381, 383, 433, 569-571, 586
  in pharmaceuticals, 377
  in plants, 378-381, 571
  by TLC, 291
Hydroxycolchicine, 289
3-Hydroxycotinine, 357, 361
8'-Hydroxydihydroergotamine, 541
9-Hydroxyellipticine, 459
β-Hydroxyethyltheophylline, 472, 473, 475-477, 479, 481-486, 597
6-Hydroxyharman, 494
Hydroxylupanine, 118, 452, 453
Hydroxy-7-methoxyharman, 495
8-Hydroxymethyltheophylline, 465
3-Hydroxyquinidine, 383
Hydroxysenkirkine, 458, 592
13-Hydroxysparteine:
  by GC, 118, 452
  in plants, 118, 452, 461
  by TLC, 462
4-Hydroxystrychnine, 533
20-Hydroxytabersonine, 509
Hygrine, 12, 365
Hyoscine, 224, 566
Hyoscyamine, 6
  by CCC, 370
  by GC, 365
  by LC, 224, 369, 566
  by TLC, 289, 372, 566
Hyoscyamus, 289
Hypoxanthine, 200

Ibogaine, 509

Ibogamine, 299
Imidazole alkaloids, 548-553, 610-612
　by GC, 124
　by TLC, 299, 300
Indicine N-oxide, 13
Indirect detection:
　in ion exchange chromatography, 200, 205, 206
　in ion pair chromatography, 228, 229
Indole alkaloids (see also Ergot alkaloids, Rauwolfia alkaloids, Strychnos alkaloids, Tabernaemontana alkaloids, Vinca alkaloids, and Yohimbine alkaloids), 490-526, 598-610
　absorption UV-vis, 55
　by GC, 120-123
　by LC, 149, 156, 194
　by TLC, 297-299
Indolizidine alkaloids:
　polyhydroxy, 448, 589
Inocybe aeruginascens, 497, 500, 501, 601
Integerrimine:
　by GC, 116, 449, 450
　by LC, 455
　by SFC, 458, 592
　by TLC, 296
Integerrimine N-oxide, 451, 455
Ion exchange:
　on alumina, 155-158
　on chemically bonded packing, 181, 182
　chromatography, 194-206
　packing for, 199
　on silica gel, 149-153
Ion exclusion, 180, 181, 196, 202
Ionic strength, 35
　in adsorption chromatography, 150, 151, 156, 392, 406, 414
　in ion exchange chromatography, 197, 201, 202

[Ionic strength]
　in ion-pair chromatography, 211, 217, 375
　in partition chromatography, 178, 180, 181, 348, 360, 406, 511
Ipecac (see Cephaelis ipecacuanha)
Iphigena indica, 353
Isoajmalicine, 121, 298
Isoajmaline, 121, 509
Isoapocodeine, 224
3-Isobutyl-1-methylanthine, 464, 482
Isochamoclavine, 546
Isocinchophyllanine, 379
Isocodeine, 399
Isocolchicine, 351
Isocorynantheidine, 298
Isocorynoline, 294
Isoemetine, 461, 591
α-Isolupanine, 117, 118, 452, 453
Isolysergic acid, 543, 544
Isolysergic acid diethylamide, 543
Isomethuenine, 516, 604
Isopilocarpic acid, 549-553
　by LC, 551-553, 611
　by TLC, 553, 612
Isopilocarpine, 22, 548-553
　detection, 235
　by GC, 548, 611
　by LC, 200, 490, 549-553, 611
　by TLC, 553, 612
8-Isopropylscopolamine, 370
Isoptelefonium, 383
Isoquinoline alkaloids (see also Opium alkaloids and Phthalidisoquinoline alkaloids), 386-398, 572-589
　absorption UV-vis, 50, 54, 231
　by GC, 113-115
　by TLC, 292-295
3-Isoreserpine, 505
Isorhoeadine, 398
Isorubijervine, 562
Isosetoclavine, 545
β-Isovalerylechinatine, 451

Isovallesiachotamine, 509
Isovincanole, 145, 228

Jacobine, 116, 296, 450
  acetate adduct, 451
Jacoline, 450
Jaconine, 450
Jacozine, 116, 296, 450
  acetate adduct, 451
  water adduct, 450
Jatrorrhiza palmata, 292, 397, 574
Jatrorrhizine, 226, 293, 397
Jervine, 559, 562
Jesaconitine, 557

Koenidine, 448, 454, 459
Koenimbine, 448, 454, 459
Kokusaginine, 383

L. arbustus subsp. calcaratus, 117, 453, 461, 591
L. argentus, 117
Lasiocarpine, 460
Lasiocarpine N-oxide, 460
Laudanosine, 388, 422, 572
L. australianum, 449
L. clavatum var. boronicum, 449
L. deuterodensum, 449
Leurosidine, 512, 603
Leurosine, 511, 512, 603
  N-formyl derivative, 516
L. fastigiatum, 449
L. niger, 13
Lobelia inflata, 5
Lobeline, 5, 61, 349
Lochnerine, 511
Lonchocarpus costariciensis, 449
LSD (see Lysergic acid diethylamide)
Lumicolchicine, 289, 351
Lupanine:
  by GC, 117, 118, 452, 453

[Lupanine]
  by TLC, 461
Lupine alkaloids (Quinolizidine alkaloids), 448-461, 590, 591
  absorption UV-vis, 55
  catalytic waves, 64
  by GC, 116-118
Lupinine, 13
Lupinus angustifolius, 117, 452, 591
Lusitanine, 461
Lycoctonine, 561
Lycopersicum esculentum, 23
Lycopodium, 589
Lycopsamine, 296
Lycopsamine N-oxide, 296
Lycorine, 14, 117, 457, 592
Lysergamide, 543, 544
Lysergene, 545,
Lysergic acid, 425, 543-545
Lysergic acid diethylamide (Lysergide, LSD):
  in biological fluids, 544, 610
  detection, 588
  by GC, 535, 610
  by LC, 543, 544, 610
Lysergic acid methylpropylamide:
  by GC, 535, 610
  by LC, 543, 544
Lysergide (see Lysergic acid diethylamide)
Lysergides, 610

Macleaya cordata, 385, 393, 574
Magnolia, 9
Mammillaria microcarpa, 346
Mass spectrometry:
  coupling with GC, 99
  for acylated mescaline, 105
  for Amaryllidaceae alkaloids, 117
  for berberine, 113
  for lupine alkaloids, 116, 117
  for nicotine, 354, 357
  for purine bases, 120

Index

[Mass spectrometry]
  for quinine and quinidine, 112
  for quinoline alkaloids, 374
  for Senecio alkaloids, 116, 296
  for tropane alkaloids, 366, 367
  coupling with LC, 230
    for Cinchona alkaloids, 381, 572
    for ephedrines, 342, 348
    for pepper and capsicum oleoresins, 362, 565, 568
Mecambrine, 390
Mesaconitine:
  by LC, 555, 556
  by TLC, 300, 560
Mescaline, 2
  acylation, 105
  by GC, 104, 105
  by LC, 156, 346
Metanephrine, 289
Metanicotine, 105
3-Methoxycanthin-2,6-dione, 499
10-Methoxycanthin-6-one, 499
10-Methoxycinchonamine, 380
9-Methoxyellipticine, 459
4-Methoxyephedrine, 346
6-Methoxyharman, 494
Methoxytryptamine, 510
12-Methoxyvoaphylline, 299, 516, 604
Methuenine, 299
Methylatropine, 368
8-O-Methyl-14-benzoylaconine, 555
3-Methylcarbazole, 448, 454, 459
Methylcodeine, 295
N-Methylcotinine, 204, 359
Methylecgonine, 6
N-Methylephedrine, 349
6-Methylergoline-8-carboxylic acid, 539, 540
Methylergometrine, 539
N-Methylflindersine, 144
N-Methylhistamine, 338

β-N-Methylisocorypalminium hydroxide, 293
Methyllycaconitine, 557, 561
N-Methylmescaline, 104
Methylnicotine:
  detection, 565
  by LC, 204, 359-361
N-Methylnornicotine, 359, 360
Methylreserpate, 505
Methylscopolamine, 368
N-Methyltyramine, 346
Methyluric acid, 485
1-Methyluric acid:
  by GC, 120
  by LC, 474, 479-482, 484, 597
  by TLC, 488
3-Methyluric acid, 479, 487
7-Methyluric acid, 474, 485, 487
1-Methylxanthine:
  by LC, 474, 476, 479
  by TLC, 488
3-Methylxanthine:
  by GC, 120
  by LC, 472, 474, 479-482, 484, 485, 487, 597
7-Methylxanthine:
  in biological fluids, 474, 487
  by LC, 470, 471, 474, 479, 487
Methysergide, 543, 544
Metocurine, 396, 575
Micellar chromatography, 215, 226, 572
Minovincinine, 509
Mitragyna stipulosa, 502, 526, 605
Morphinans, 574, 584
Morphine, 11, 399-426, 572, 573, 579-583
  absorption spectra, 57
  anodic oxidation, 62, 63
  in biological fluids, 115, 294, 295, 329, 336, 399-403, 407-414, 416, 430, 434, 577, 578, 585
  in blood, 115, 334, 335, 402, 403, 412-414

[Morphine]
  decomposition, 576
  derivatization, 239, 579
  detection, 238, 244, 579, 580, 588
  by GC, 114, 115, 427-430, 576-580, 584, 585
  glucuronides, 227, 335, 410
  in heroin, 427, 428, 431, 432, 435, 584, 585
  hydrolysis, 585
  by LC, 151, 157, 158, 185, 186, 190-192, 201, 202, 219-223, 227, 348, 430-434, 584-586
  metabolites, 577, 579
  in opium, 114, 151, 156-158, 223, 294, 405, 418, 420, 422-426, 432, 580, 584, 585
  in pharmaceuticals, 221, 295, 312, 348, 406, 407
  in plants, 419, 421, 581-583
  in tissues, 403
  by TLC, 294, 295, 416, 435
Morphine alkaloids (see also Opium alkaloids), 399-416, 573
Morus nigra, 449
Mukonal, 448, 459
Multiflorine, 118, 452
Murrayazoline, 448, 454, 459
M. tetrancistra, 346
Myosmine, 105, 108, 354

Nalorphine:
  in biological fluids, 336, 400, 401, 412
  derivatization, 240
  by GC, 400, 401, 578
  by LC, 408, 411-413
Narceine, 11, 572
  anodic oxidation, 63
  by LC, 190, 192, 223, 419, 422, 425, 426, 455, 580-582

[Narceine]
  in opium, 223, 422, 425, 580
Narcissus, 14
α-Narcotine (Noscapine), 11, 572
  absorption UV-vis, 50, 52
  anodic oxidation, 63
  in biological fluids, 394, 574
  detection, 235, 244
  dissociation constant, 29, 38, 39
  by GC, 417, 418, 427
  in heroin, 427, 433-435
  hydrolysis, 574
  by LC, 151, 157, 158, 185, 186, 190, 191, 201, 220, 222, 223, 227, 394, 418-426, 432-434, 574, 582-586
  in opium, 151, 157, 223, 294, 417, 418, 420, 422-426, 432, 584
  in pharmaceuticals, 421
  in plants, 421, 582, 583
  by TLC, 294, 295, 435
Narcotinic acid (Noscapinic acid), 227, 394, 574
Neopetasitenine, 457
Neophytadiene, 108, 354
Neosenkirkine, 458, 592
Neostigmine, 224, 599
Nicotiana, 3
Nicotine, 3, 564-566
  in biological fluids, 108, 204, 289, 297, 328, 336, 355, 356, 359-361, 564, 565
  derivatization, 109, 565
  dissociation constant, 29, 39
  enantiomers, 564
  by GC, 105, 107-109, 354-357, 564
  by LC, 203, 204, 358-362, 558, 565
  metabolites, 564
  in tissues, 108, 357, 362, 564
  by TLC, 289, 297
  in tobacco, 105, 107, 108, 316, 354, 358, 564, 565
  smoke, 107, 564

Index

Nicotine N-oxide (Nicotine N'-oxide):
  by LC, 204, 359-362
  in tissues, 362, 565
  in urine, 204, 359, 360, 362
Nicotyrine, 105, 289
N-Nitrosonornicotine, 355
Noragroclavine, 546
Norchamoclavine, 546
Norcodeine:
  in biological fluids, 399, 401, 414
  by GC, 399, 401, 414, 427
  in heroin, 427
Norephedrine:
  enantiomers, 342, 346, 347
  by GC, 104, 106, 107
  by LC, 342, 348-350, 423
  by TLC, 289, 352
Norepinephrine, 386, 387
Norethylmorphine, 407
Norharman, 495, 599
Norhydrocapsaicin, 362-364
Norhyoscyamine, 370
Norisosetoclavine, 546
Normorphine:
  in biological fluids, 399, 401, 409, 410
  derivatization, 240
  by GC, 399, 401, 427
  in heroin, 427
  by LC, 407, 409, 410, 413
Nornicotine:
  by GC, 105, 108, 354, 355
  by LC, 204, 359
  by TLC, 289
Norpseudoephedrine (Nor-ψ-ephedrine), 104, 289, 350
Norsanguinarine, 398
Nortropine, 109
Noscapine (see α-Narcotine)
Noscapinic acid (see Narcotinic acid)
Novacine, 121
Nudicauline, 557
Number of effective plates, 76, 85
Number of theoretical plates, 75, 83

Ochrosia elliptica, 322, 459
Opium, 580, 581
  by GC, 114, 405, 417, 418
  by LC, 150, 151, 156, 188, 222, 223, 420, 422-425, 432, 584
  sample preparation, 318
  by TLC, 294, 295
Opium alkaloids, 399-426, 572, 573, 579-583
  absorption UV-vis, 54, 231
  anodic oxidation, 63
  detection, 235, 282
  by GC, 114, 115
  by LC, 150-152, 156, 157, 184, 186, 188, 190-192, 201, 219, 220, 223, 226
  in plants, 318, 319
  by TLC, 294, 295
Otonecine, 457
Otosenine, 457, 458, 592
Oxindole alkaloids, 605
17-Oxolupanine, 118, 452
Oxonitine, 556
Oxyphylline (7-(2-Hydroxyethyl)theophylline), 465

Packings for LC (see also Silica gel, Alumina):
  chemically bonded, 159, 164-168
  ion exchangers, 198-200
  swelling, 198, 200
  polymeric, 159, 170, 193, 194
  properties, 162, 164-169, 174, 175, 184
  residual silanols, 163, 169, 182, 183, 216
  stability, 169, 175, 200, 218, 226
  variability of, 221, 224
Palliclavine, 546

Palmatine:
  detection, 235
  by LC, 226, 389
  in plants, 226, 294, 389, 397
  by TLC, 293, 294, 397
Panaeolina foenisecii, 321, 499
Panaeolus rickenii, 496
Papaver bracteatum, 295, 415, 416, 580
Papaverine, 9, 399-426, 572
  in biological fluids, 113, 388, 573
  in blood, 190
  dissociation constant, 31
  dissociation curve, 30
  electrochemical reduction, 61, 62
  by GC, 106, 113, 115, 580
  in heroin, 427, 434, 435
  by LC, 185, 186, 190-192, 201, 220, 222, 223, 227, 348, 388, 389, 393, 394, 432-434, 549, 573, 575, 580, 582-586
  in opium, 157, 158, 223, 294, 417, 418, 420, 422-426, 432, 580, 584
  in plants, 393, 421, 582, 583
  by TLC, 290, 294, 295, 435
Papaverrubine A, 319, 398
Paraxanthine (1,7-Dimethylxanthine):
  in biological fluids, 330, 331, 337, 471-475, 477, 479, 483, 484, 486, 488
  by LC, 471-475, 477, 479, 482-486, 596
  by TLC, 488
Paullinia cupana, 469, 594
P. bohemica, 497, 500
P. callosa, 501, 601
P. cubensis, 321, 499
Peak asymmetry factor, 76
Peak potential, 58
Peganum harmala, 16
Pelletierine, 5

Penniclavine, 546
Pepper:
  by GC, 109, 110, 358, 565
  by LC, 362, 363, 565, 568
  preparation, 316
Perakine, 524
Perivine:
  by LC, 516, 517, 604
  in plants, 299, 321, 322, 516, 517, 604
  by TLC, 299
Periwinkle (see Catharanthus roseus)
Perloline, 454, 589
Petasitenine, 457
Petasites albus, 116
Peyote alkaloids, 104
pH effects:
  in adsorption chromatography, 143, 150, 151, 155, 156, 406, 414
  on dissociation of alkaloids, 27, 30, 40, 41
  in ion exchange chromatography, 197, 200-202, 359
  in ion pair chromatography, 216, 222, 225, 392, 506
  in partition chromatography, 176-179, 193, 348, 360, 406, 422, 433, 467, 517
  in TLC, 273
Phalaris, 120
Phenanthrenisoquinoline alkaloids, 572, 573, 576-582
Phenanthridine alkaloids, 320
Phenylephrine, 348, 415
Phenylethylamine derivatives (see also Peyote alkaloids), 342, 346-353
  by GC, 104, 105
  by TLC, 288, 289
Pholcodine, 336, 401
Phthalidoisoquinoline alkaloids, 317, 386-398
P. hybridus, 116, 296
Physostigmine, 598, 599

Index

[Physostigmine]
  in biological fluids, 330, 492, 493, 503
  decomposition, 599
  detection, 245, 599
  $\underline{N},\underline{N}$-dimethyl derivative, 491, 492
  by LC, 226, 392, 490-493, 503, 598-600, 611
  metabolites, 491, 493, 599
  in pharmaceuticals, 490, 611
  in plants, 392
Pilocarpic acid, 549-553
  esters and diesters of, 550, 611
  by LC, 549, 551-553, 611
  by TLC, 553, 612
Pilocarpine, 22, 548-553
  in biological fluids, 226
  degradation, 548, 551-553, 611
  detection, 235, 237, 612
  by GC, 124, 611
  by LC, 200, 226, 490, 611
  in pharmaceuticals, 300, 548, 549, 551, 552, 599, 611
  by TLC, 299, 300, 611, 612
Pilocarpus, 21
Piper nigrum, 4
Piperanine, 363
Piperettin, 568
Piperidine alkaloids (see Pyridine and piperidine alkaloids)
Piperine, 4
  by GC, 109, 110, 358, 565
  isomers, 193, 565
  by LC, 362, 363, 565, 568
  in pepper and pepper extract, 109, 110, 193, 289, 316, 358, 363, 565, 568
  by TLC, 289
Piperolein A, Piperolein B, 363, 568
Piperylin, 568
Piptadenia peregrina, 15

P. jaborandi, 21, 610
P. japonicus, 457
Plate height, 76
  reduced, 76
Platydesminium, 375, 383, 568
Platyphylline, 116, 296, 450
Pleiocorpamine, 509
Poppy straw, 419, 421, 426, 581, 583
Potato (see Solanum)
P. pilosum, 390, 574
Preparation:
  of biological samples, 315, 323-341
  of pharmaceutical formulations, 311-313
  of plant materials, 313-322
  of sample, 309-341
    automated, 338-341
P. rhoeas, 319
Procaine, 157, 586
$\underline{N}$-$\underline{n}$-Propylajmaline, 147, 238
$\underline{N}$-$\underline{n}$-Propylnorapomorphine, 224
Protoberberine alkaloids:
  by TLC, 574
Protocevine (see Veracevine)
Protopine, 10, 574
  by LC, 392, 393
  in plants, 293, 392, 393, 397, 398
  by TLC, 293, 294, 397, 398
Proxyphylline, 474
P. semilanceata, 298, 320, 321, 496-501, 588, 601
Pseudocinchona africana, 17
Pseudococaine:
  by GC, 366, 566
  by LC, 370, 567, 569
Pseudoecgonine, 366, 566
Pseudoephedrine (ψ-ephedrine), 3
  in biological fluids, 350, 563, 564
  detection, 564
  enantiomers, 342, 346, 347
  by GC, 104
  by LC, 349, 350, 372, 563

[Pseudoephedrine]
  in pharmaceuticals, 349, 563
  by TLC, 289, 564
Pseudomorphine, 221, 240, 406
Pseudotropine (ψ-tropine), 109
Psilocin:
  detection, 238
  by LC, 150, 202, 203, 496-499, 601
  in plants, 320, 321, 496-500, 601
  by TLC, 500, 501, 601
Psilocybe, 150, 202
Psilocybin:
  detection, 238
  by LC, 150, 202, 203, 496-499, 601
  in plants, 320, 321, 496-501, 601
  by TLC, 500, 501, 601
P. somniferum, 9-11, 319, 420, 424, 572, 573, 580, 582
P. subaeruginosa, 498
Psychotria ipecacuanha, 12
Punica granatum, 4
Purine bases, 462-489, 592-598
  absorption UV-vis, 55
  in biological fluids, 330
  by GC, 118-120
  by LC, 227
  by SFC, 610
  by TLC, 297
Pyridine and piperidine alkaloids (see also Tobacco alkaloids and Tropane alkaloids), 354-364, 565
  by GC, 105-111, 448, 589
  by TLC, 289, 290
Pyridocarbazole alkaloids, 322, 589
Pyridostigmine, 224, 599
Pyroclavine R, 545
Pyrrolidine alkaloids:
  polyhydroxy, 448, 589
Pyrrolizidine alkaloids (see Senecio alkaloids)

Quebrachamine, 194, 517
Quebrachamine-type alkaloids, 517, 605
Quinamine, 380
Quinidine, 8
  absorption UV-vis, 56
  in biological fluids, 290, 324, 328, 329, 381-385, 572
  fluorescence, 147, 239
  by GC, 111-113, 374, 568
  by LC, 147-149, 156, 180, 187, 188, 191, 239, 324, 375-383, 549, 569-572
  in pharmaceuticals, 374, 376, 377, 549, 570
  in plants, 378-381, 571, 572
  by TLC, 290, 291, 384, 385, 572
Quinidine N-oxide, 383
Quinidinone, 148, 379, 383
Quinine, 8
  in biological fluids, 324, 329, 381-383, 572
  by CCC, 531
  detection, 239
  by GC, 111-113, 374, 568
  in heroin, 434, 435
  by LC, 148, 149, 179, 180, 187, 188, 191, 239, 324, 375-383, 421, 423, 433, 434, 528, 529, 549, 569-572, 586
  in pharmaceuticals, 324, 329, 381-383, 572
  in plants, 378-381, 571, 572
  by TLC, 291, 435
Quininone, 376, 569
Quinitoxine, 376, 569
Quinoline alkaloids (see also Cinchona alkaloids), 54, 374, 375, 567-572
  absorption UV-vis, 50, 54
  by GC, 111-113
  by TLC, 290-292
Quinolizidine alkaloids (see Lupine alkaloids)

Ratio gas/liquid in GLC, 89
Rauwolfia, 18, 298
Rauwolfia alkaloids, 502-526, 598, 601-603
  absorption UV-vis, 55
  by GC, 120, 121
  by LC, 201
  in plants, 120, 121
  by TLC, 297
Rauwolfia canescens, 18
R. vomitoria, 121, 507, 523, 603
Rauwolscine (see α-Yohimbine)
Reagent:
  Dragendorff's, 280
  Ehrlich's, 281
  iodoplatinate, 280
Remija, 8, 571
Rescinnamine:
  in biological fluids, 506
  by GC, 601
  by LC, 222, 505-507, 603
  in plants, 398, 505, 507, 522, 523, 603
  by TLC, 398, 522, 523, 603
Reserpiline, 522, 523
Reserpine, 18
  in biological fluids, 503, 506, 521
  derivatization, 239, 241, 602
  detection, 61, 62, 245, 602
  by GC, 601
  by LC, 185, 201, 202, 423, 503-509, 521, 601-603
  in pharmaceuticals, 504, 505, 602
  in plants, 505, 507, 522, 523, 602, 603
  by TLC, 522, 523, 603
Reserpinine (Raubasinine), 505, 522, 523
Resolution, 71
Retention data:
  in GC, 117, 122, 123
  in LC, 145, 157, 177, 179, 180, 225
  in TLC, 289, 291-293, 298, 299

Retention index, 126
Retention ratio (see Selectivity ratio)
Retention time, 70
  adjusted, 70
  relative, 125
Retention volume, 70, 84
  adjusted, 70, 84
Retronecine, 13
Retrorsine:
  analog, 450
  by GC, 116, 449
  by LC, 455, 591
  in plants, 116, 296, 449, 455, 460, 591
  by SFC, 458, 592
  by TLC, 296, 460
Retrorsine N-oxide, 455
R_F values, 271
  of alkaloids from Glaucium squamigerum, 293
  of Cinchona alkaloids, 291
  of colchicine alkaloids, 289
  of Ephedra alkaloids, 289
  of indole alkaloids, 298
  of Tabernaemontana alkaloids, 299
  of tropane alkaloids, 290
Rhoeadine, 319, 398
Rhoeagine, 398
Ribalinium, 383
Ricinine, 4, 235
Ricinus communis, 4
Roemerine, 390
R. serpentina, 18, 505, 522, 523, 602, 603
Rubijervine, 562
Rubremetine, 395, 575
Rubreserine, 599
  by LC, 226, 491-493
R. vomitoria, 121, 507, 523, 603

Salsola richteri, 9
Salsoline, 244

Salsolinol:
  detection, 244, 573
  in food, 386, 387, 573
  by GC, 386, 573
  by LC, 387, 573
Sandwichine, 121
Sanguinaria canadensis, 393, 574
Sanguinarine:
  by LC, 391-393
  by TLC, 294, 397
S. anonymus, 320, 458, 590, 592
S. brevicaule, 555
Scopine, 6, 7
Scopolamine:
  by CCC, 370
  decomposition, 109
  detection, 244
  by GC, 109, 365, 566
  by LC, 368, 369
  in plants, 109, 289, 365, 369, 370, 372, 566
  by TLC, 289, 290, 372
Scopoline, 109
Scoulerine, 293
Selectivity:
  in adsorption chromatography, 152
  in ion exchange chromatography, 197
  in ion pair chromatography, 217, 221, 224
  in partition chromatography, 163, 172, 174-176, 194
  values, 170
  ratio, 74
Senecio, 13, 451
Senecio alkaloids (Pyrrolizidine alkaloids), 448-461, 590
  absorption UV-vis, 55
  detection, 281, 282
  diastereomers separation, 296
  by GC, 116
  $\underline{N}$-oxides, 191
  by TLC, 296

Senecio alpinus, 450, 590
Senecionine:
  by GC, 116, 449, 450
  isomers of, 450
  by LC, 455, 591
  in plants, 116, 296, 449, 450, 455, 460, 591
  by SFC, 458, 592
  by TLC, 296, 460
Senecionine $\underline{N}$-oxide:
  by GC, 451
  by LC, 455, 456
  by TLC, 460, 590
Seneciphylline:
  by GC, 116, 450
  by LC, 455, 591
  by TLC, 296
Seneciphylline $\underline{N}$-oxide, 451, 455
Senecivernine, 116, 293
Senkirkine:
  by LC, 457
  by SFC, 458, 592
  by TLC, 460
Serpentine:
  by LC, 505, 508, 510, 512, 603
  in plants, 508, 510, 523, 524, 603
  by TLC, 522-524, 603
Setoclavine, 545, 546
Silanization:
  of GC columns, 90, 116, 124
  of GC supports, 92
Silica gel:
  characterization, 137
  dissolution at higher pH, 153, 169
  ion exchange, 143, 149, 150
  layers in TLC, 267
  properties, 140-141
  silanol groups, 137, 142, 143
Simaba multiflora, 298, 499, 601
S. inaequidens, 116, 296, 449, 460, 590
Skimmianine, 374, 375, 383
S. leptophyes, 555

Index

S. nux-vomica, 19, 121, 528, 532, 605, 606
Solamargine, 558, 561
Solanidine, 23
  by GC, 124, 554, 555, 612
  in milk, 124, 300, 554, 561, 612
  by TLC, 300, 561, 612
  in tuber samples of Solanum, 555, 612
α-Solanine:
  by LC, 193, 557, 558, 613, 616, 617
  in potatoes, 193, 558, 613
  by TLC, 561
Solanthrene, 534, 554
Solanum, 23, 124, 554, 557, 558, 612-614, 617
Solasodiene, 554, 555
Solasodine, 124, 554, 555
Solasonine, 558
Solvent:
  demixing, 272-274
  isohydric, 149
  physical data, 138, 139
  polarity, 170, 172
  selectivity, 172, 217
  solubility parameter of, 161, 172, 277
  strength, 159, 170-172, 193, 202, 217
    parameter, 136
  UV cutoff, 138, 139, 172, 235
S. oplocense, 555
Sparteine, 14
  catalytic waves, 64
  dissociation constant, 29, 39
  by GC, 117, 118, 452
Spartioidine, 455
S. spegazzinii, 555
Spermostrychnine, 20, 121
S. piersiana, 529, 606
S. pierriana, 529, 606
S. psilosperma, 19
S. ptycanthum, 558, 561, 615
Staphidine, 300, 560

Staphisine, 300, 560
Stationary phase (see also Packings for LC):
  in GC, containing KOH, 92, 104
  presaturation, 183, 273
  in RP-TLC, 268
  in TLC, 267-269
Stemmadenine, 299
Sternbergia lutea, 457, 592
Steroid alkaloids (Glycoalkaloids, Solanum alkaloids), 554-562, 612-617
  absorption UV-vis, 55
  detection, 234, 237, 282
  by GC, 124, 125
  by LC, 152, 191, 193
  by TLC, 300
Streptomycetes, 577
Strychnine, 19, 605
  in biological material, 121, 222, 529, 530, 605, 606
  by CCC, 531, 606
  derivatives, 527, 605
  by GC, 121, 527, 605
  in heroin, 157, 158, 420, 432
  by LC, 157, 158, 190, 222, 420, 423, 432, 528-530, 605
  metabolites, 531, 533, 606
  in pharmaceuticals, 121, 528, 531, 532, 605
  in plants, 121, 528, 529
  by TLC, 278, 297, 353, 531-533, 606
Strychnine N-oxide, 531-533
Strychnos alkaloids, 527-533, 598, 605, 606
  absorption, UV-vis, 55
  by GC, 121
Strychnos icaja, 121
Strychnospermine, 20, 121
Stylophorum diphyllum, 294, 396, 574
Stylopine, 293, 391
S. sucrense, 555

Supercritical fluid chromatography, 68, 154
  coupling with MS, 230, 372, 487, 546, 567, 572, 580, 597
  with FID or UV detection, 372, 458, 487, 546, 590, 592, 597
Support for stationary phase in GC, 90, 91
S. vernei, 555
S. vulgaris, 455, 460, 590, 591
Swainsona, canescens, 449
Symphytine, 296
Symphytine N-oxide, 296
Symphytum officinalis, 116, 191, 296
Synephrine, 289

Tabernaemontana, 321, 525, 604
Tabernaemontana alkaloids, 502-526, 604
  by CCC, 521
  by LC, 226
  sample preparation, 321, 322
  by TLC, 297, 299, 525
Tabernaemontana chippii, 525
Tabernaemontanine, 299, 321, 322, 517
Tabersonine, 299, 509
Tacamine, 299
T. dichotoma, 297, 516, 525, 604
T. divariatica, 322, 517, 521, 525, 604
Tea:
  by GC, 462, 467-469, 594
  by LC, 205
T. elegans, 521, 525
Temperature:
  in adsorption chromatography, 149, 379
  in ion exchange chromatography, 198, 201
  in ion pair chromatography, 217, 218

[Temperature]
  in partition chromatography, 171-174, 511
Tetrahydroalstonine:
  by GC, 121
  by LC, 508, 509
  by TLC, 298
Tetrahydroberberine, 113, 202
  alkaloids, 388
  enantiomers, 574
Tetrahydrocolumbamine, 225, 389
Tetrahydrocoptisine, 294
Tetrahydrocorysamine, 294, 391
Tetrahydrojatrorrhizine, 391
  benzyl derivative, 391
  13-methyl derivative, 391
Tetrahydropalmatine:
  by LC, 225, 389, 391
  by TLC, 294
Tetrahydropapaveroline:
  detection, 244, 573
  by LC, 387, 388, 573
Tetrahydroprotoberberine, 391
Tetrahydrorhombifoline, 118, 452
Tetraphyllicine, 121, 522
Thalictricavine, 391
Thebaine:
  anodic oxidation, 63
  by GC, 405, 417, 418, 580
  in heroin, 435
  by LC, 151, 157, 183, 186, 190, 192, 201, 220, 223, 415, 419-426, 432, 580, 582-584
  in opium, 151, 157, 223, 294, 405, 417, 418, 420, 422-426, 432, 580, 582, 584
  in plants, 295, 415, 416, 421, 580, 583
  by TLC, 294, 295, 416, 435, 580
Theobromine (3,7-Dimethylxanthine), 15
  in biological fluids, 330, 471-475, 479, 483, 487, 595, 596

Index

[Theobromine]
  in coca, 341, 489, 595, 598
  in foodstuffs, 467, 470
  by GC, 119
  by LC, 200, 202, 465-475, 483, 487, 595
  methylation, 119
  by paper chromatography, 489, 598
  by SFC, 487, 597
Theophylline (1,3-Dimethylxanthine), 15
  in biological fluids, 119, 120, 201, 330, 331, 335, 337, 350, 464, 471-477, 479-488, 593, 595-597
    with automated procedure, 341
  detection, 244
  in foodstuffs, 467, 470, 595
  by GC, 119, 464, 593
  by LC, 200-202, 348-350, 465-467, 470-477, 479-487, 570, 595-597
  metabolites, 595-597
  methylation, 119
  in pharmaceuticals, 349, 594, 598
  by SFC, 487, 597
Thermopsine, 117
Thin-layer chromatography, 266-308
  centrifugal preparative, 276, 398, 560, 561
  ion pair, 273, 274, 295
  overpressured, 276-278
    of canthinones, 499
    of doping agents, 278, 297, 353, 564, 598
    of strychnine, 606
  technique, 267-287
Tobacco:
  by GC, 105, 107, 108, 354, 564
  by LC, 358, 565
  preparation, 316
Tobacco alkaloids, 354-362

[Tobacco alkaloids]
  in biological fluids, 108
  by GC, 105-109
  by TLC, 289
Tomatidine, 23
  as bonded phase for sterols, 615
  by GC, 554, 555, 612
Tomatine, 557, 616
T. orientalis, 322, 521, 525
T. psorocarpa, 525
Triacetylnormorphine, 428, 429
Trimethylallantoin, 488
Trimethylcolchicinic acid, 351
Trimethylsilylation:
  of alkaloids, 100
  of Amaryllidaceae alkaloids, 117
  of Cinchona alkaloids, 113
  of ergometrine, 121
  of isoquinoline alkaloids, 113
  of morphine alkaloids, 114, 115
  of Rauwolfia alkaloids, 121
  of Senecio alkaloids, 116
  of tropane alkaloids, 109
  of Vinca alkaloids, 120
Trimethyluric acid (1,3,7-Trimethyluric acid), 471, 474, 488
1,3,7-Trimethylxanthine (see Caffeine)
Tropane alkaloids, 365-373, 566, 567
  absorption UV-vis, 54
  detection, 234, 236
  by GC, 109-111
  ion pairs with picric acid, 147
  by LC, 147
  by TLC, 289, 290
Tropic acid, 368, 369
Tropine, 5
  dissociation constant, 29, 39
  by GC, 109, 566
Truxilline, 367, 371, 567
Tryptamine:
  by GC, 120

[Tryptamine]
  by LC, 499, 508, 530
Tryptamine-type alkaloids, 601
Tryptophan, 479
Tubocurarine:
  in biological fluids, 336, 396, 575
  by LC, 202, 395, 396, 575
  by TLC, 290, 295
Tubotaiwine, 517
T. undulata, 525
Tyramine, 346

Uncaria, 120, 297
Uncarine, 298
Uric acid methylderivatives, 480
Usaramine, 455

Vallesiachotamine, 509
Vasicine, 453, 589
  analogs, 453, 454, 589
  stability in solution, 589
Vasicinone, 453, 589
V. californicum, 559
Veatchine, 300, 560
Velocity of mobile phase:
  linear, 70
  reduced, 77
Veracevine, 24, 48, 559
Veramine, 562
Veratramine, 559
Veratridine, 558, 559
Veratrine, 558, 559, 615
Veratroylzygadenine, 562
Veratrum, 559, 562, 615, 617
Vinblastine:
  in biological fluids, 332, 513, 514, 603
  degradation, 604
  derivatives, 515
  isolation, 603
  by LC, 508, 510-515, 603, 604

[Vinblastine]
  in plants, 508, 510-512, 603
Vinca alkaloids (Eburnane alkaloids), 502-526, 604-605
  dimeric, 603-604
  enantiomers, 227
  by GC, 120
  by LC, 144, 145, 190, 227, 517-519
  in pharmaceuticals, 312
Vinca minor, 519, 604
Vincadifformine, enantiomers, 194, 517
Vincadine, 517
Vincamenine, 145, 228, 520
Vincamine:
  in biological fluids, 502, 521, 604, 605
  enantiomers, 194, 229, 517, 605
  by GC, 502, 604
  by LC, 145, 227, 229, 517-521, 605
  in pharmaceuticals, 520, 521, 605
  in plants, 519
  by TLC, 298, 299
Vincaminic acid ethyl ester, 145
Vincamone, 145, 228, 520
Vincanole, 145, 228, 520
Vincine, 298, 519
Vincristine:
  in biological materials, 332, 512, 514, 603
  degradation, 604
  isolation, 605
  by LC, 510-515, 603-605
  in plants, 510, 511
Vindesine, 512, 513
Vindoline, 508, 510-512, 524
Vindolinine, 509, 511
Vinpocetine (apovincaminic acid ethyl ester):
  enantiomers, 518
  by GC, 120, 502, 604
  by LC, 145, 228, 520

Index

[Vinpocetine]
  in pharmaceuticals, 312, 520
  in plasma, 120, 332, 520, 604
Voaphylline, 517, 521
Vobasine:
  by LC, 516, 517, 604
  by TLC, 299
Vomicine, 121, 533
Vomilenine, 524
V. rosea (see Catharanthus roseus)
V. sabadilla, 615
V. viride, 24

Wheat, 538, 608

Xanthine:
  derivatives, 465, 477, 479, 480
  by ion exchange chromatography, 200, 202
Xanthine bases (see Purine bases)

Xanthocercis zambesiaca, 449

Yohimbene (see 3-Epi-α-yohimbene)
Yohimbic acid, 503
Yohimbine, 16
  in biological fluids, 503, 504, 601
  catalytic waves, 64
  dissociation constant, 29, 39
  by GC, 601
  by LC, 179, 185, 190, 423, 503-505, 509, 601
  in pharmaceuticals, 505
  in plants, 522
  by TLC, 522
α-Yohimbine (Rauwolscine), 522
δ-Yohimbine (see Ajmalicine)
Yohimbine alkaloids (see also Indole alkaloids), 55, 502-526, 598, 601, 603
Yunaconitine, 556, 615